Biomass to Renewable Energy Processes

Biomass to Renewable Energy Processes

Edited by Jay Cheng

CRC Press
Taylor & Francis Group
Boca Raton   London   New York

CRC Press is an imprint of the
Taylor & Francis Group, an **informa** business

CRC Press
Taylor & Francis Group
6000 Broken Sound Parkway NW, Suite 300
Boca Raton, FL 33487-2742

Library of Congress Cataloging-in-Publication Data

Biomass to renewable energy processes / editor, Jay Cheng.
    p. cm.
  Includes bibliographical references and index.
  ISBN 978-1-4200-9517-3 (alk. paper)
  1. Biomass energy. 2. Renewable energy sources. I. Cheng, Jay. II. Title.

TP339.B5767 2009
662'.88--dc22

2009022591

Visit the Taylor & Francis Web site at
http://www.taylorandfrancis.com

and the CRC Press Web site at
http://www.crcpress.com

Contents

Preface..vii
Editor...ix
Contributors..xi

1. **Introduction** 1
   Jay J. Cheng

2. **Biomass Chemistry** 7
   Deepak R. Keshwani

3. **Biomass Resources** 41
   Ziyu Wang and Deepak R. Keshwani

4. **Biomass Logistics** 71
   Matthew W. Veal

5. **Kinetics and Microbiology of Biological Processes** 135
   Jay J. Cheng

6. **Anaerobic Digestion for Biogas Production** 151
   Jay J. Cheng

7. **Biological Process for Ethanol Production** 209
   Jay J. Cheng

8. **Biological Process for Butanol Production** 271
   Jian-Hang Zhu and Fangxiao Yang

9. **Chemical Conversion Process for Biodiesel Production** 337
   Dong-Zhi Wei, Fangxiao Yang, and Erzheng Su

10. **Thermochemical Conversion of Biomass to Power
    and Fuels**...437
    *Hasan Jameel, Deepak R. Keshwani, Seth F. Carter,
    and Trevor H. Treasure*

Index 491

Preface

A reliable and sustainable energy supply has been a major cause of concern for the global community, especially in the last decade, for the following reasons: the continuously increasing consumption of fossil fuels, our current major energy source, to support economic growth has had a significant impact on the global climate change; the price of fossil fuels has experienced huge fluctuation; and there is concern about the availability of fossil fuels in the near future. To respond to this energy supply crisis, a lot of effort has been made to explore renewable energy production technologies around the world, including hydroelectric, geothermal, wind, solar, and biomass.

Biomass energy products are generated from agricultural crops and residues, herbaceous and woody materials, and organic wastes. These materials can either be directly combusted for energy production or processed into energy products such as bioethanol, biodiesel, and biogas, which are then used as transportation fuels or for the production of electricity and heat. Although biomass energy production is a fast-growing area because of its renewable nature and the abundance of materials, its market share in total energy production is still very small (less than 5%). Currently, commercial biomass energy products mainly include bioethanol extracted from sugarcane and corn, biodiesel extracted from plants and waste oils, biogas produced from organic waste materials, and electricity and heat generated from the direct combustion of wood chips. The industry is very vulnerable to the fluctuation of feedstock prices. New technologies need to be developed to convert abundant biomass, such as lignocellulosic materials, into energy products in a cost-effective and environmentally friendly manner in order to tremendously increase the market share of biomass energy. The exploration of novel feedstocks is also important for the healthy development of the biomass energy industry.

This book describes fundamental principles and practical applications of biomass energy production processes for engineering and science students, and for professionals who are interested in this area and are well acquainted with the basic biological, chemical, and engineering principles. It explains the principal theories of biological processes, biomass materials and logistics, and technologies for bioenergy products, such as biogas, ethanol, butanol, biodiesel, and synthetic gases. Authors from several engineering disciplines have contributed to this book. I would like to thank them all for their hard work. I would also like to thank other colleagues who helped us in one way or another to make this book a reality.

Jay J. Cheng
Raleigh, North Carolina

Editor

Jay J. Cheng is a professor of bioprocessing and environmental engineering in the biological and agricultural engineering department at North Carolina State University, Raleigh. He joined the North Carolina State University faculty as an assistant professor in 1997. Prior to this, he served as a lecturer and the fermentation division director in the food engineering department at Jiangxi Polytechnic University (now renamed Nanchang University, China) and as a research associate at the University of Cincinnati, Ohio. He received his BS in chemical engineering from the Jiangxi Institute of Technology in China, his MS in biological engineering from St. Cyril & Methodius University in Macedonia, and his PhD in environmental engineering from the University of Cincinnati.

Professor Cheng has received several major awards including the Top Honors Award for Visual Aids in Teaching from Jiangxi Polytechnic University in 1988, first place (in the PhD category) in the Water Environment Federation Student Paper Competition in 1995, and a poster presentation award (first place) at the Water Environment Federation's 73rd Annual Meeting in 2000. He became a Fulbright scholar in 2005.

Dr. Cheng has taught courses in the areas of fermentation, environmental engineering, and bioenergy during his professional career. His research interests include brewing, anaerobic digestion of waste materials, nutrient recovery from wastewater with aquatic plants, and biomass conversion to biofuel. He has published over 30 articles in refereed scientific journals and presented more than 60 papers at national and international technical conferences on these topics. He has been invited to present over 25 keynote speeches and seminars in China, Europe, and the North and South Americas. He has served as an associate editor for the *Journal of Environmental Engineering* and on more than 15 international and national professional committees.

Contributors

Seth F. Carter
Wood & Paper Science Department
North Carolina State University
Raleigh, North Carolina

Jay J. Cheng
Biological and Agricultural
   Engineering Department
North Carolina State University
Raleigh, North Carolina

Hasan Jameel
Wood & Paper Science Department
North Carolina State University
Raleigh, North Carolina

Deepak R. Keshwani
Biological Systems Engineering
   Department
University of Nebraska-Lincoln
Lincoln, Nebraska

Erzheng Su
Chemical Engineering Department
East China University of Science
   and Technology
Shanghai, China

Trevor H. Treasure
Wood & Paper Science Department
North Carolina State University
Raleigh, North Carolina

Matthew W. Veal
Biological and Agricultural
   Engineering Department
North Carolina State University
Raleigh, North Carolina

Ziyu Wang
Biological and Agricultural
   Engineering Department
North Carolina State University
Raleigh, North Carolina

Dong-Zhi Wei
Qingdao Institute of Bioenergy and
   Bioprocess Technology
Chinese Academy of Sciences
Qingdao, China

Fangxiao Yang
Resodyn Corporation
Butte, Montana

Jian-Hang Zhu
Qingdao Institute of Bioenergy and
   Bioprocess Technology
Chinese Academy of Sciences
Qingdao, China

1

Introduction

Jay J. Cheng

CONTENTS

1.1 Objectives of This Book..1
1.2 Renewable Energy versus Fossil Fuel Energy 2
1.3 Life Cycle Assessment..4
1.4 Problems..5
References..6

1.1 Objectives of This Book

This book is written as a textbook for a graduate course in renewable energy production for both graduate and senior undergraduate students in the areas of agricultural, biological, chemical, and environmental engineering as well as crop, food, plant, and wood sciences. It is also intended to be a reference book for professional engineers and scientists who are interested in the processes of converting biomass into renewable energy sources.

This book introduces fundamental principles and practical applications of biomass-to-renewable energy processes, including biological, chemical, and thermochemical processes. Chemical properties of a variety of biomass are presented in this book. Resources of biomass that can be utilized for renewable energy production are also presented in this book, including their production and basic characteristics. Logistics of biomass handling such as harvesting, transportation, and storage are also included in this book. Biological processes include anaerobic digestion of waste materials for biogas and hydrogen production, and bioethanol and biobutanol production from sugars, starch, and cellulose. Pretreatment technologies, enzymatic reactions, fermentation, and microbiological metabolisms and pathways are presented and discussed in this book. The chemical process of biodiesel production from plant oils, animal fats, and waste oils and fats is described and discussed in this book. Thermal processes include combustion, gasification, and pyrolysis of woody biomass and of agricultural residues. Engineering principles of biomass combustion, gasification, and pyrolysis, and potential end-products are discussed in this book.

1.2 Renewable Energy versus Fossil Fuel Energy

Energy consumption has increased steadily over the last century as world population has grown and more countries have become industrialized since 1900. Fossil fuel, especially crude oil, is currently the predominant energy source around the world. Table 1.1 shows the top five crude oil–consuming countries in 2005 (Energy Information Administration, 2006). However, the reserves of fossil fuel are limited and will be depleted in the near future at its current consumption rate. Although there are debates about the exact year of peak oil production, it is generally believed that it would occur before 2025 after which significant reduction of oil production should be expected. Moreover, burning fossil fuels causes environmental concerns such as greenhouse gas (GHG) emission, which is generally believed as the major reason for global climate change. Because the world economy depends on oil, the consequences of inadequate oil availability could be severe. Therefore, there is a great interest in exploring alternative energy sources. The negative environmental impact of burning fossil fuels reminds us that the alternative energy source should be sustainable and environment-friendly. Energy production from biomass such as crop, herbaceous, woody, and waste materials has a great advantage over fossil fuels. The former is renewable annually or in several years, while the latter needs thousands or millions of years for reproduction. The energy production from biomass releases CO_2, which is believed as a major GHG to cause global climate change, but the CO_2 is utilized for biosynthesis during the growth of biomass. Thus, using biomass for

TABLE 1.1

Top Crude Oil–Consuming Countries in the World in 2005

Country	Crude Oil Consumption (Million Barrels/Day)
United States	20.3
China	6.2
Japan	6.1
India	2.2
South Korea	2.2

energy production can have a balanced CO_2 production and consumption or little net CO_2 release, compared to a huge discharge of CO_2 from burning fossil fuels. Currently, energy production from biomass is only a small portion of the total energy production. The sources of the 2005 energy consumption in the United States include 85% fossil fuel, 8.2% nuclear energy, 3.3% biomass, 2.7% hydroelectric power, 0.34% geothermal energy, 0.18% wind power, and 0.07% solar energy (Energy Information Administration, 2006). The resources of biomass materials such as crops, grasses, wood, agricultural residues, and organic wastes are quite abundant, so there is a great potential to substantially increase energy production from biomass materials.

Ethanol production from dedicated crops or agricultural residues is one form of renewable energy that addresses the critical need for sustainable fuels. There is already a well-developed market for ethanol in the United States; over 6.5 billion gal of ethanol were produced in 2007 according to the Renewable Fuels Association. Ethanol is primarily used as an additive to gasoline to improve emissions and boost octane. Biogas is generally produced from organic waste materials such as sewage sludge, agricultural wastes, industrial wastes, and municipal solid wastes. Proper treatment of these waste materials is necessary to protect our environment from pollution. At the same time, biogas production from the treatment facilities provides a renewable energy source. A number of commercial biogas production plants have been in operation utilizing sewage sludge, animal manure, and municipal solid wastes, and biogas is used for either electricity generation or direct combustion for heat production. Biodiesel is produced from vegetable oils and animal fats, and used as an alternative fuel of petroleum diesel for buses and trucks. The major benefits of producing and using bioenergy generated from biomass are as follows:

1. **Energy Independence.** Crude oil or natural gas reserves are located in limited regions in the world. Many countries have to rely on importing oils from limited oil-producing countries. For example, the U.S. Energy Information Administration reported that the United States imported 56% of its overall demand of petroleum in 2003, two-thirds of which were from the politically unstable Persian Gulf. The reliance makes the United States vulnerable to supply disruptions and nonmarket-related price instability, thereby jeopardizing the nation's energy and economic security according to the U.S. Congress's record that a dependence on foreign oil of greater than 50% is a peril to the country (NREL, 2000). On the other hand, biomass is almost everywhere. Renewable energy production from biomass could protect the oil-deficit countries from depending on foreign oil, and generate local jobs.

2. **Air Quality.** Oxygenated fuels such as ethanol typically promote more complete combustion as compared to fossil fuels. More

complete combustion translates into fewer emissions, particularly carbon monoxide (CO) emissions (National Science and Technology Council, 1997).

3. **Water Quality.** Ethanol is replacing methyl tertiary butyl ether (MTBE) as the preferred octane enhancer in the United States because of concerns over MTBE's persistence in the environment and possible negative effects to the water quality and human health. Ethanol breaks down quickly and spills of ethanol do not pose a critical threat to the environment.

4. **GHG Emissions.** The combustion of fossil fuels results in a net increase in the emission of GHGs (primarily carbon dioxide) into the atmosphere. According to the U.S. Energy Information Administration, the U.S. transportation sector is responsible for approximately one-third (1/3) of all carbon dioxide emissions. The impact of GHGs on global climate change is of increasing concern around the world. The use of biogas, bioethanol, and biodiesel as energy sources can significantly reduce the net GHG emissions. For example, use of E85 (85% ethanol and 15% gasoline) can reduce the net emissions of GHGs by as much as 25% on a fuel-cycle basis as compared to gasoline (Wang, 1999).

1.3 Life Cycle Assessment

The purpose of the life cycle assessment (LCA) of renewable energy production is to evaluate the environmental impacts of a product or process. LCA normally involves energy balance and GHG emissions. Energy balance determines the net energy production of a renewable energy product or process, while GHG emissions indicate its impact on climate changes. The first step to conduct an LCA is to define a boundary for the assessment, for example, from corn cultivation to ethanol combustion for corn-based fuel ethanol production process. All the unit operations inside the boundary need to be analyzed for energy input and output as well as GHG emissions. Again using corn-based fuel ethanol production process as an example, the unit operations generally include corn cultivation, corn harvest, transportation of corn to ethanol plant, conversion of corn to fuel ethanol, ethanol fuel distribution, and combustion of fuel ethanol to provide energy. Corn cultivation involves mainly land preparation, seeding, fertilization, irrigation, and weed control. Corn harvest includes corn kernel collection and residue handling. The conversion of corn to fuel ethanol consists of corn grinding, saccharification, fermentation, distillation, dehydration, and by-product processing. Similar procedures can be applied to the LCA of lignocellulose-based ethanol and

TABLE 1.2

Impact of Common GHGs on Global
Climate Change

Greenhouse Gas	Global Warming Potential or CO_2 Equivalent
CO_2	1.0
CH_4	21.0
N_2O	310.0

Source: U.S. Environmental Protection Agency/U.S. Greenhouse Gas Inventory Program/Office of Atmospheric Programs, *Greenhouse Gases and Global Warming Potential Values,* 2002.

biodiesel production processes. For biogas production from organic waste materials, LCA usually needs to include the cost of the waste disposal to prevent environmental pollution as an offset to the biogas production process.

The major GHG emission in the energy production processes is CO_2. Other GHGs related to the unit operations of renewable energy production include methane (CH_4) and nitrous oxide (N_2O). The impact of the GHGs on global climate change can be expressed with global warming potential (GWP) or CO_2 equivalent. Their typical values are listed in Table 1.2.

In general, biogas production from waste materials needs little energy input or has a high net energy production. It also has a high impact on GHG emission reduction. A big challenge for biogas as a renewable energy source is its low energy density, which makes it economically unfavorable to transport from one location to another. Both corn- and lignocellulose-based ethanol productions are generally believed as net energy-producing processes. However, corn-based ethanol production has limited reduction in GHG emission, mainly because the corn residues are not utilized in the process, but lignocellulose-based ethanol production has significant reduction in GHG emission. Biodiesel usually has a much better net energy production than bioethanol, but it has limited impact on GHG emission because only a portion of the oil plant, that is, seeds, is utilized in biodiesel production.

1.4 Problems

1. How were the fossil fuels (petroleum, natural gas, and coal) formed? How long did it take for the formation? Discuss the alternative energy sources that are currently used in the world and their advantages and disadvantages.

2. Anaerobic lagoons are widely used for municipal and agricultural wastewater treatment. However, there are increasing concerns about GHG emissions from the lagoons. Thus, there have been efforts in covering the anaerobic lagoons to collect biogas that is used for energy generation or simply flared. Explain why the actions would help reduce the GHG emission and the effect of the GHG.

3. Switchgrass is considered as a promising biomass crop for renewable energy production, especially bioethanol as transportation fuel. Conduct a LCA of switchgrass-to-bioethanol process.

References

Energy Information Administration, U.S. Department of Energy. 2006. http://www.eia.doe.gov/

National Renewable Energy Laboratory (NREL). 2000. *Biofuels for Sustainable Transportation*. DOE/GO-102000-0812.

National Science and Technology Council. 1997. Committee on Environment and Natural Resources, Executive Office of the President of the United States. Interagency assessment of oxygenated fuels. A science report. http://www.ostp.gov/galleries/NSTC%20Reports/Interagency%20Assessment%20of%20Oxygenated%20Fuels%201997.pdf

U.S. Environmental Protection Agency/U.S. Greenhouse Gas Inventory Program/Office of Atmospheric Programs. 2002. *Greenhouse Gases and Global Warming Potential Values*. http://epa.gov/climatechange/emissions/downloads09/InventoryUSGhG1990-2007.pdf

Wang, M. Q. 1999. Fuel-cycle greenhouse gas emissions impacts of alternative transportation fuels and advanced vehicle technologies. *Transportation Research Record*, 1664, 9–17.

2

Biomass Chemistry

Deepak R. Keshwani

CONTENTS

2.1   Introduction 7
2.2   Review of Organic Chemistry 8
    2.2.1   Alkanes 8
    2.2.2   Alkenes 9
    2.2.3   Alkynes 10
    2.2.4   Aromatic Compounds 11
    2.2.5   Alcohols 12
    2.2.6   Compounds Containing Carbonyl Groups 12
    2.2.7   Ethers 13
2.3   Review of Carbohydrate Chemistry 15
    2.3.1   Monosaccharides 15
    2.3.2   Disaccharides and Glycosidic Bonds 17
2.4   Starch 20
2.5   Cellulose 23
2.6   Hemicellulose 26
2.7   Lignin 29
2.8   Pectin 31
2.9   Vegetable Oil 32
2.10 Extractives 35
2.11 Problems 38
References 39

2.1 Introduction

Biomass-to-renewable energy processes involve the synthesis or break-down of organic compounds that constitute different types of biomass. These organic compounds range in their complexity and contain numerous functional groups that influence the ultimate structure and chemistry of biomass. Familiarity with these organic compounds and their functional

7

groups is important in understanding and developing biological and chemical processes that convert biomass into renewable energy and value-added products. This chapter summarizes relevant concepts from organic and carbohydrate chemistry followed by a discussion of the structure and chemistry of different types of biomass.

2.2 Review of Organic Chemistry

2.2.1 Alkanes

Alkanes are hydrocarbons that are exclusively made up of single bonds between carbon and hydrogen. They are called saturated hydrocarbons because of the absence of double or triple bonds. Saturated acyclic hydrocarbons have the general empirical formula C_nH_{2n+2}. The simplest alkanes are methane ($n=1$), ethane ($n=2$), and propane ($n=3$), whose structures are shown in Figure 2.1. When $n \geq 4$, more than one compound is possible for the same empirical formula. These compounds are called structural isomers and differ in the arrangement of the carbon atoms (straight chain or branched).

When $n=4$, there are two distinct compounds with the empirical formula of C_4H_{10}: normal butane (*n*-butane) and isobutane (Figure 2.2). Similarly,

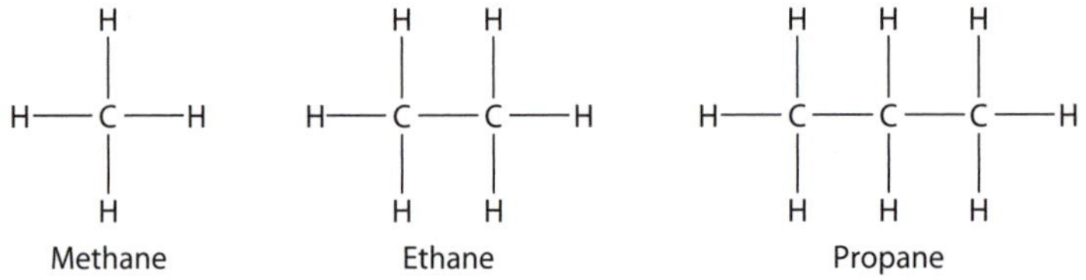

FIGURE 2.1
Structures of simple alkanes.

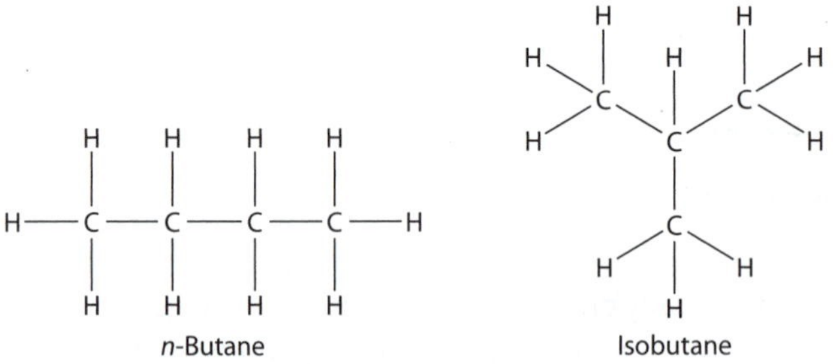

FIGURE 2.2
Structural isomers of butane (C_4H_{10}).

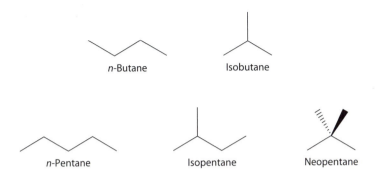

FIGURE 2.3
Simplified representations of the structural isomers of butane (C_4H_{10}) and pentane (C_5H_{12}).

when $n=5$ (pentane), there are three structural isomers with the empirical formula of C_5H_{12}. The number of structural isomers rapidly increases as the number of carbon atoms increases. For example, C_7H_{16} has nine structural isomers and $C_{10}H_{22}$ has 75 structural isomers.

Since including every carbon and hydrogen atom in structures can be cumbersome, representations of organic compounds are simplified by using chains of carbon atoms with hydrogen atoms assumed implicit. Figure 2.3 shows the simplified representations of the structures of n-butane and isobutane, along with the three structural isomers of pentane (n-pentane, isopentane, and neopentane). Note the use of wedged bonds in the case of isopentane. These bonds are often used in place of plain lines to provide a three-dimensional perspective of substituent groups attached to a central atom.

2.2.2 Alkenes

Alkenes are hydrocarbons that contain at least one double bond between adjacent carbon atoms. The simplest acyclic alkenes contain only one such double bond and have the general formula empirical C_nH_{2n}. The simplest examples of such alkenes are ethene ($n=2$) and propene ($n=3$), shown in Figure 2.4.

Structural isomers of alkenes are obtained by changing the position of the double bonds or by changing the way the carbon atoms are joined to each other. The presence of double bonds can lead to another type of isomerism: geometric isomerism (*cis–trans* isomerism). However, the presence of a double bond is not a guarantee for such geometric isomerism. In order for such isomerism to be exhibited, each carbon atom involved in the double bond must be attached to different functional groups. For example, Figure 2.5 shows the structural isomers with the

$H_2C=CH_2$ $H_2C=CH-CH_3$

Ethene   Propene

FIGURE 2.4
Structures of simple alkenes.

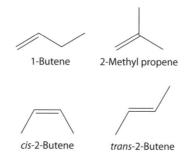

FIGURE 2.5
Structural and geometric isomers of butene (C_4H_8).

empirical formula C_4H_8: 1-butene, 2-butene, and 2-methyl propene. However, 2-butene exhibits geometric isomerism as well. The *cis* isomer of 2-butene has the methyl (CH_3) substituent groups oriented on the same side of the double bond and the *trans* isomer has the methyl groups on opposite sides of the double bond.

2.2.3 Alkynes

Alkynes are hydrocarbons that contain at least one triple bond between adjacent carbon atoms. The simplest alkynes contain only one such triple bond and have the general formula C_nH_{2n-2}. The simplest examples of such alkynes are ethyne and propyne for which no structural isomers are possible. As shown in Figure 2.6, there are two structural isomers possible for butyne and pentyne based on the position of the triple bond. However, alkynes do

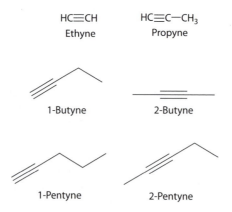

FIGURE 2.6
Structures of ethyne (C_2H_2), propyne (C_3H_4), and isomers of butyne (C_4H_6) and pentyne (C_5H_8).

not exhibit geometric isomerism because of the linear nature of the carbon–carbon triple bond.

2.2.4 Aromatic Compounds

Alkanes, alkenes, and alkynes are examples of aliphatic compounds, which are characterized by the fact that electron pairs in the molecules either belong to a single atom or are shared between a pair of atoms. Compounds in which some electron pairs cannot be assigned to a specific atom or pair of atoms are termed aromatic compounds. These compounds typically contain a conjugated ring of alternating single and double bonds.

The most commonly discussed aromatic compound is benzene, which has a molecular formula of C_6H_6. There are two structures for benzene since it is impossible to specify the exact location of the double bonds (Figure 2.7). Neither structure is an adequate representation since the carbon–carbon bonds are neither single nor double. The two structures are collectively termed resonance structures. Typically, benzene is represented by a six-carbon ring with a circle inside to indicate the equal distribution of the double bonds between all the carbon atoms.

Aromatic organic compounds can be heterocyclic (one or more atoms in the aromatic ring is not carbon), polycyclic (two or more aromatic rings fused together), or substituted (functional group attached to one or more carbon atoms of the ring in place of a hydrogen atom). See Figure 2.8 for examples of each type of organic aromatic compounds.

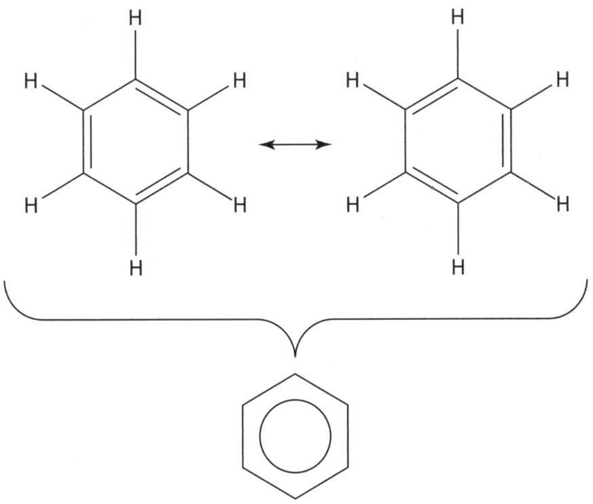

FIGURE 2.7
Resonance structures of benzene.

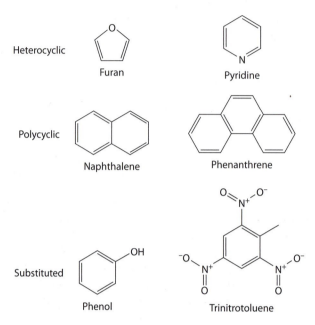

FIGURE 2.8
Examples of heterocyclic, polycyclic, and substituted aromatic compounds.

2.2.5 Alcohols

An alcohol is an organic compound in which a hydroxyl (OH) group is attached to at least one carbon atom. A primary alcohol is one in which the hydroxyl-bearing carbon atom is connected to only one other carbon atom. A secondary alcohol has two carbon atoms connected to the hydroxyl-bearing carbon atom and a tertiary alcohol has three carbon atoms connected to the hydroxyl-bearing carbon atom. The structures of some alcohols are shown in Figure 2.9.

Methanol (CH_3OH) and ethanol (C_2H_5OH) are the simplest examples of primary alcohols. As is the case with alkanes, when $n \geq 3$, structural isomers are possible based on the position of the OH group. For example, n-propanol is a primary alcohol and isopropanol is a secondary alcohol.

2.2.6 Compounds Containing Carbonyl Groups

A carbonyl group is a functional group that contains a carbon atom double-bonded to an oxygen atom (C=O). Organic compounds that contain carbonyl groups include aldehydes, ketones, carboxylic acids, and esters (Figure 2.10).

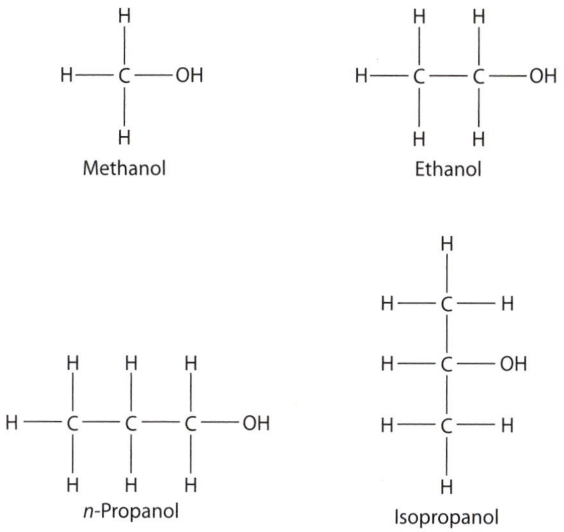

FIGURE 2.9
Structures of primary and secondary alcohols.

Aldehydes are organic compounds that contain a terminal carbonyl group in which there is at least one hydrogen atom connected to the carbonyl carbon. The general formula for aldehydes is R–(CO)–H, where R is a substituent group, which can be a hydrogen atom, aliphatic or aromatic. Ketones have two carbon atoms attached to the carbonyl carbon and have the general formula R_1–(CO)–R_2, where R_1 and R_2 can be aliphatic or aromatic groups. Carboxylic acids contain a carbonyl group bonded to a hydroxyl group and have the general formula R–COOH, where R can be an aliphatic or aromatic group. Esters contain a carbonyl group bonded to an oxygen atom and have the general formula (R_1–COO–R_2) where R_1 can be a hydrogen atom, aliphatic or aromatic group and R_2 can be an aliphatic or aromatic group.

2.2.7 Ethers

Ethers are organic compounds that contain an oxygen atom connected to two carbon atoms via single bonds. They have the general formula R_1–O–R_2, where R_1 and R_2 can be aliphatic or aromatic groups. Symmetric (simple) ethers have identical R groups attached to the oxygen atom and asymmetric (mixed) ethers have different R. Structures of some ethers are shown in Figure 2.11.

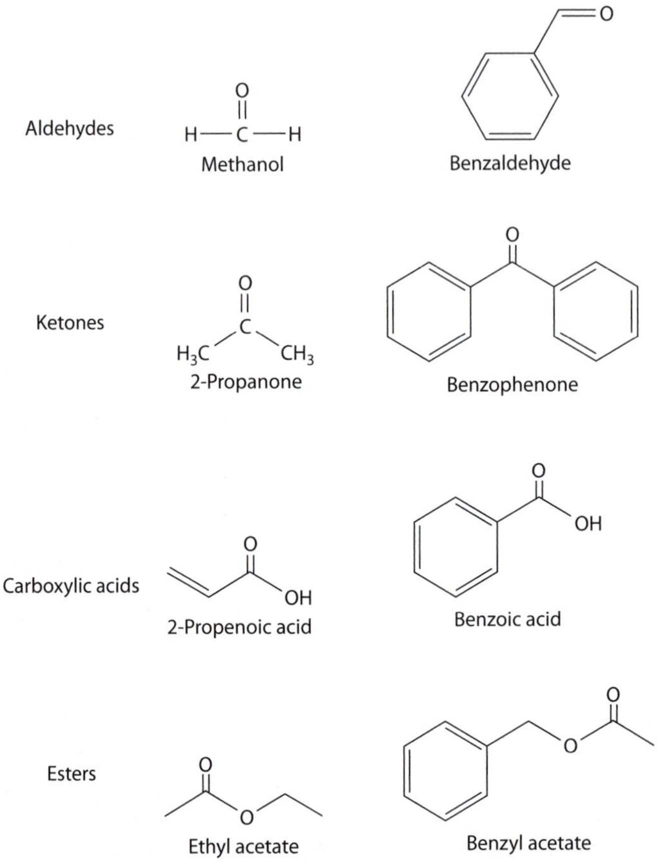

| Aldehydes | Methanol | Benzaldehyde |

| Ketones | 2-Propanone | Benzophenone |

| Carboxylic acids | 2-Propenoic acid | Benzoic acid |

| Esters | Ethyl acetate | Benzyl acetate |

FIGURE 2.10
Examples of organic compounds containing carbonyl groups.

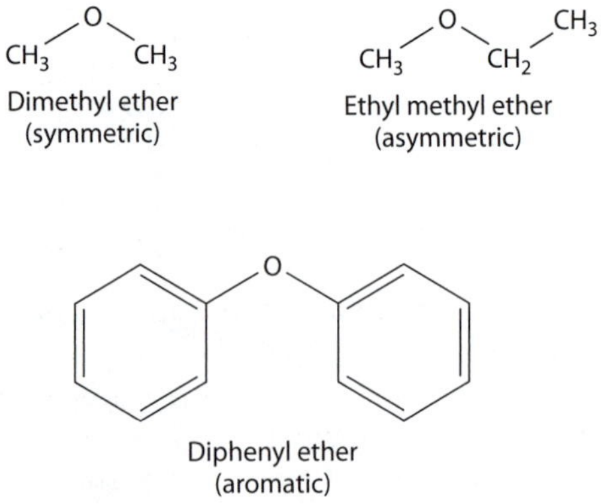

Dimethyl ether
(symmetric)

Ethyl methyl ether
(asymmetric)

Diphenyl ether
(aromatic)

FIGURE 2.11
Examples of ethers.

2.3 Review of Carbohydrate Chemistry

2.3.1 Monosaccharides

Carbohydrates are organic compounds that contain carbonyl functional groups with numerous hydroxyl groups attached to the non-carbonyl carbon atoms. Monosaccharides are the simplest carbohydrates and typically have the chemical formula $C_n(H_2O)_n$. They can be classified based on the number of carbon atoms. Trioses have three carbon atoms, tetroses have four carbon atoms, pentoses have five carbon atoms, hexoses have six carbon atoms, and so forth (Figure 2.12).

Monosaccharides can also be classified based on the nature of the carbonyl group. Aldoses are monosaccharides that contain an aldehyde functional group and ketoses are monosaccharides that contain a ketone functional group. For example, glucose is an aldohexose and fructose is a ketohexose (Figure 2.12).

A third method of classification is based on the stereochemistry of the chiral carbon farthest away from the aldehyde or ketone functional group. A chiral carbon is one that is bonded to four different substituent groups. A molecule that exhibits chirality cannot be superimposed on its mirror image. Fisher projections are used to show the structural configuration about the chiral carbon atom. Fisher projections are simple representations of organic molecules where all bonds are depicted as horizontal or vertical lines. In the case of aldoses, the carbon atoms in the main chain are numbered starting from the terminal aldehyde group. For ketoses, the numbering starts at

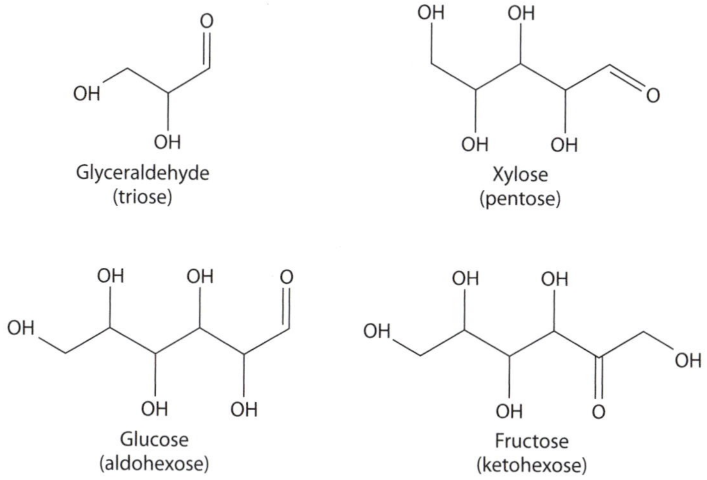

FIGURE 2.12
Examples of monosaccharides depicted as open-chain structures.

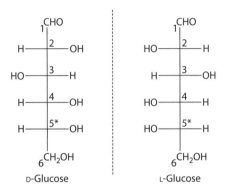

FIGURE 2.13
Fisher projections of D-glucose and L-glucose.

the terminal carbon atom closest to the ketone functional group. Figure 2.13 shows the Fisher projections of the two stereoisomers of glucose. The chiral carbon atom farthest away from the aldehyde is C5 (marked with*). When the hydroxyl group is to the right of the chiral carbon, the isomer is called D-glucose. When the hydroxyl group is to the left of the chiral carbon, the isomer is called L-glucose. D-sugars are more common in natural systems.

So far, the structures of monosaccharides have been depicted in a chain form. However, in solution, monosaccharides form ring structures that are in equilibrium with the open-chain forms. In general, aldehydes and ketones react with alcohols to form hemiacetals and hemiketals. Since monosaccharides contain several hydroxyl groups and one aldehyde or ketone group within the same molecule, they form cyclic hemiacetals and hemiketals. Most monosaccharides form five-member rings called furanoses or six-member rings called pyranoses. Smaller (<4) and larger (>6) hemiacetals and hemiketals are thermodynamically unstable.

The ring structure of monosaccharides is favored in solution and the mechanism for ring formation is similar for most sugars. For example, the D-glucopyranose ring is obtained when the oxygen of the hydroxyl group attached to the C5 carbon in the chain form links with the carbonyl carbon (Figure 2.14).

In the process, the hydrogen atom from the OH group of the C5 carbon is transferred to the carbonyl oxygen to create a new OH group. This rearrangement can result in two possible pyranoses: α-D-glucopyranose and β-D-glucopyranose. The difference between the two ring structures is in the orientation of the newly formed OH group on the C1 carbon. α-D-glucopyranose has the new OH group projected below the plane of the ring and β-D-glucopyranose has the new OH group projected above the plane of the ring structure. Isomers, such as these, differ only in their configuration about the carbonyl carbon atom and are called anomers. At equilibrium, a solution of D-glucose contains 64% of the β anomer and 36% of the α anomer.

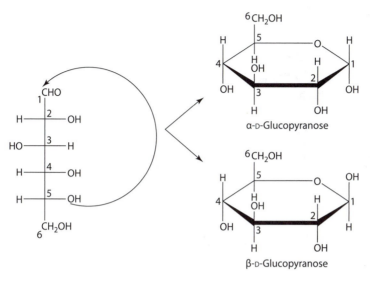

FIGURE 2.14
Hemiacetal formation in D-glucose resulting in α and β anomers of D-glucopyranose.

The ring structures of monosaccharides in Figure 2.14 are drawn as convenient planar representations called Hayworth projections. However, a more sterically accurate representation of β-D-glucopyranose and α-D-glucopyranose is shown in Figure 2.15, where the structures are shown in chair conformations. α-D-glucopyranose is less stable because of the diaxial interaction between the OH group of the C1 carbon and H atom of the C3 carbon, which are both projected below the plane of the ring. Monosaccharides can also form boat conformations, which are energetically less stable. The boat conformation of β-D-glucopyranose is shown in Figure 2.15.

2.3.2 Disaccharides and Glycosidic Bonds

Disaccharides consist of two simple sugars linked via a glycosidic bond. Examples of common disaccharides include maltose (made up of two glucose units), sucrose (made up of one glucose and one fructose unit), lactose (made up of one glucose and one galactose unit), and cellobiose (made up of two glucose units). The reaction involves the formation of a linkage between the anomeric carbon (C1 carbon, adjacent to the oxygen of the sugar ring) of a monosaccharide and the oxygen atom from the OH group of another monosaccharide. The resulting linkage is called a glycosidic bond. The reaction is accompanied by the loss of one water molecule for every glycosidic bond and is therefore called a dehydration or condensation reaction.

Glycosidic bonds between monosaccharides are classified based on the stereochemistry of the OH group on the anomeric carbon and the specific

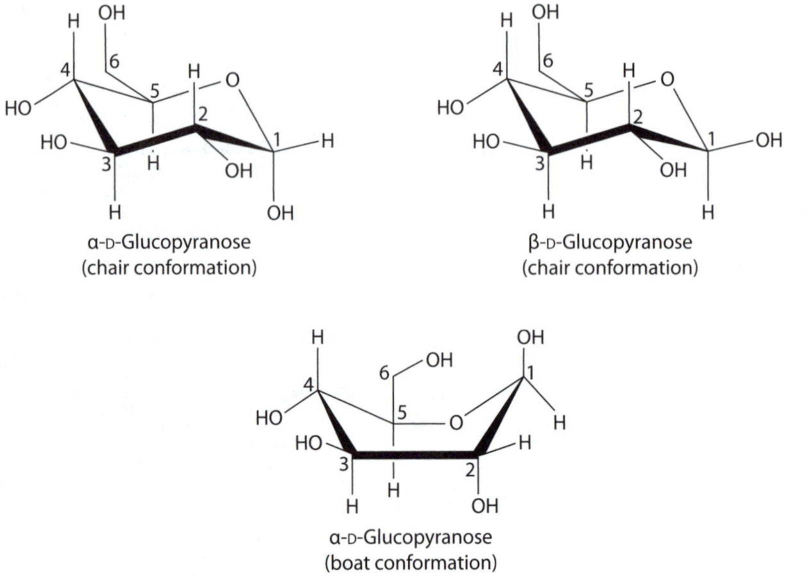

FIGURE 2.15
Chair and boat conformations for D-glucopyranose.

carbon atoms involved in the bond formation. For example, maltose consists of two α-D-glucopyranose molecules linked via an alpha bond between the C1 carbon of one molecule and the oxygen attached to the C4 carbon of the other molecule. Hence, the resulting linkage is a α-1,4 glycosidic bond (Figure 2.16).

The bond is called an alpha bond as it involves a hydroxyl group oriented in the alpha position. Cellobiose consists of two β-D-glucopyranose molecules linked via a beta bond between the C1 carbon of one molecule and the oxygen attached to the C4 carbon of the other molecule. Hence, the resulting linkage is a β-1,4 glycosidic bond (Figure 2.16). Other examples of glycosidic linkages include α-1,2 between glucose and fructose to form sucrose, α-1,1 between two glucose units to form trehalose, β-1,6 between two glucose units to form gentiobiose and α-1,6 between two glucose units to form isomaltose.

Disaccharides are part of a larger group called oligosaccharides, which consist of 2–10 monosaccharide units connected via glycosidic bonds. An example of an oligosaccharide is raffinose (Figure 2.17), which is a trisaccharide composed of galactose, fructose, and glucose units.

Polysaccharides are relatively larger and more complex carbohydrates. They are polymers made up of numerous monosaccharides (up to several 1000 units) joined by a single type or several types of glycosidic bonds. Polysaccharides can be homopolymers like starch and cellulose that are made

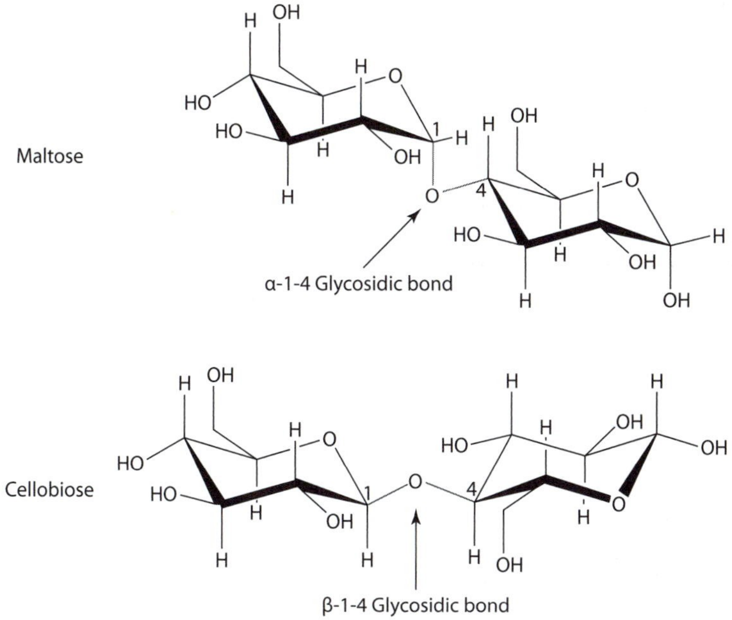

FIGURE 2.16
Glycosidic bonds between glucose molecules resulting in maltose and cellobiose.

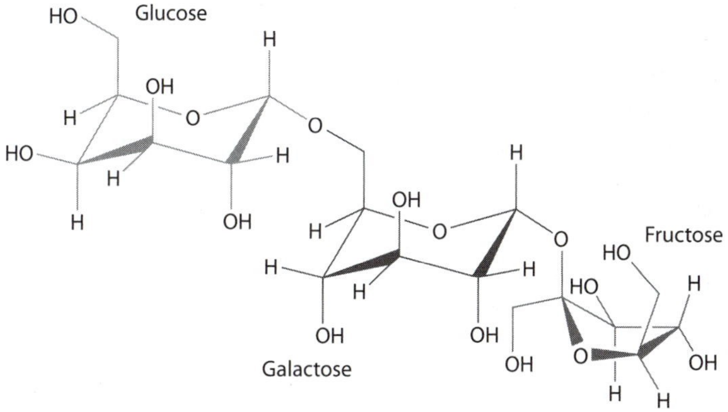

FIGURE 2.17
Structure of raffinose, a trisaccharide consisting of glucose, galactose, and fructose.

up of a single type of monosaccharide or heteropolymers such as hemicelluloses that are made up of more than one type of monosaccharide. The structure and function of these polysaccharides is influenced in large part by the nature of the glycosidic linkages discussed in this section.

2.4 Starch

Starches are the most abundant form of food reserve in plants. It is the main polysaccharide reserve produced by all higher plants. The word starch has its origins from the Anglo-Saxon *stercan* (or *sterchen*), which literally means to stiffen. This origin is quite appropriate since starch has historically been used as a glue and a sizing agent in early paper production. The major sources for starch production vary geographically and include rice, wheat, potatoes, and corn. In addition to being an integral part of the human diet, applications of starches include use as a thickening agent in its basic form, production of adhesives, paper, and textiles and as a feedstock for the chemical and bio-ethanol industry.

Starch is typically stored as granules in roots, tubers, fruits, and seeds. While the granular nature is a universal occurrence, starches can occasionally be produced in amorphous forms, as is the case in certain lichens. While starch granules primarily consist of carbohydrates, minor constituents such as proteins, lipids, phosphate groups, and other plant cell-wall materials are also present. The granular nature of starch is often used to distinguish between different sources of starch by the examination of properties such as granule size and shape, gelatinization temperature, and degree of swelling in different solvents.

In chemical terms, starch is a polysaccharide with a chemical formula of $(C_6H_{10}O_5)_n$ that is made up of individual α-D-glucopyranose units that are linked via glycosidic bonds. While all starches ultimately hydrolyze into α-D-glucose, there are two distinct polymers present: amylose and amylopectin. Together, these two polymers represent 97%–99% of the dry weight of starch. Depending on the source, starches typically contain 20%–25% amylose and 75%–80% amylopectin. However, depending on specific commercial applications, breeding and genetic engineering have resulted in plants that produce starches that have high amylose content or are completely amylose free. Table 2.1 shows the typical amylose and amylopectin contents in starch from different sources.

The amylose fraction of starch is a homopolymer of D-glucopyranose units linked via α-1,4 glycosidic bonds and typically has a molecular weight ranging from 10^5 to 10^6. Depending on the source, the number of D-glucopyranose units ranges from a few hundred to several thousand per chain. Two D-glucopyranose units linked via the α-1,4 glycosidic bonds make up maltose, the disaccharide repeat unit of amylose chains (shown in Figure 2.18).

The chain length and consequently the molecular weight of amylose also vary depending on the maturity of the starch. For example, potato starch can exhibit 40%–50% variability in chain length depending on the season. Occasional occurrences of α-1,6 linkages to phosphate groups have been observed. However, these groups have minimal effect on the properties and behavior of amylose.

TABLE 2.1

Amylose and Amylopectin Content (% of Total Starch) in Different Sources of Starch

Source	Amylose	Amylopectin
Corn (yellow dent)	26–31	69–74
Corn (waxy)	<1	>99
Corn (high amylose)	55–80	20–45
Potato	20–27	73–80
Sweet potato	20	80
Tapioca	17	83
Wheat	27–31	69–73

Sources: Data from Thomas, D.J. and Atwell W.A., *Starches*, Eagan Press, St. Paul, MN, 1999; Whistler, R.L. and Smart, C.L., *Polysaccharide Chemistry*, Academic Press, New York, 1953.

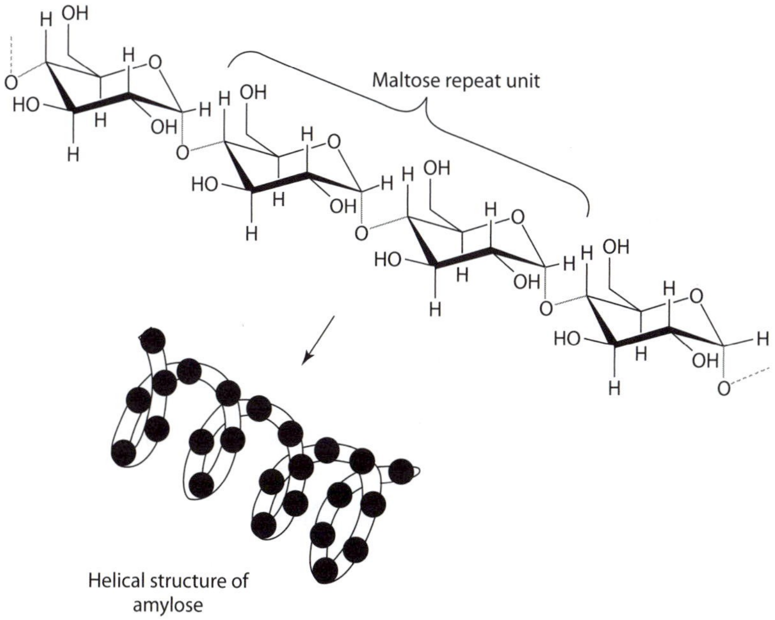

FIGURE 2.18
Amylose structure showing a chain of α-1,4-linked D-glucopyranose units resulting in a helix.

As a result of the angle of the α-1,4 glycosidic bonds, amylose takes on a helical shape (Figure 2.18) and is semicrystalline in nature. There are typically six glucose residues per turn of the helix. The interior of the helical structure is hydrophobic and allows the formation of clathrate complexes with free fatty acids. The complexation phenomenon of the helical structure of amylose is the basis for the iodine test for starches. The formation of inclusion complexes between iodine and amylose results in the characteristic blue color. The formation of complexes by amylose is affected by factors such as temperature, pH, and contact time with the complexing agent. While amylose typically exists as a random coil, complexation results in a regular structure that contains six D-glucose units per turn of the helix. In solution, amylose exhibits retrogradation, which is the formation of hydrogen bonds between individual chains. The result is the formation of insoluble particles with high molecular weights.

The amylopectin fraction of starch is a branched polymer with a much larger molecular weight (10^7–10^9) than that of amylose. In fact, amylopectin is one of the largest naturally occurring molecules. Amylopectin is made up of D-glucopyranose units linked via α-1,4 glycosidic bonds with nonrandom α-1,6-linked branching (shown in Figure 2.19).

While the branching is nonrandom, it has been observed to occur at least once every 20–30 D-glucopyranose units. About 5% of the total glycosidic

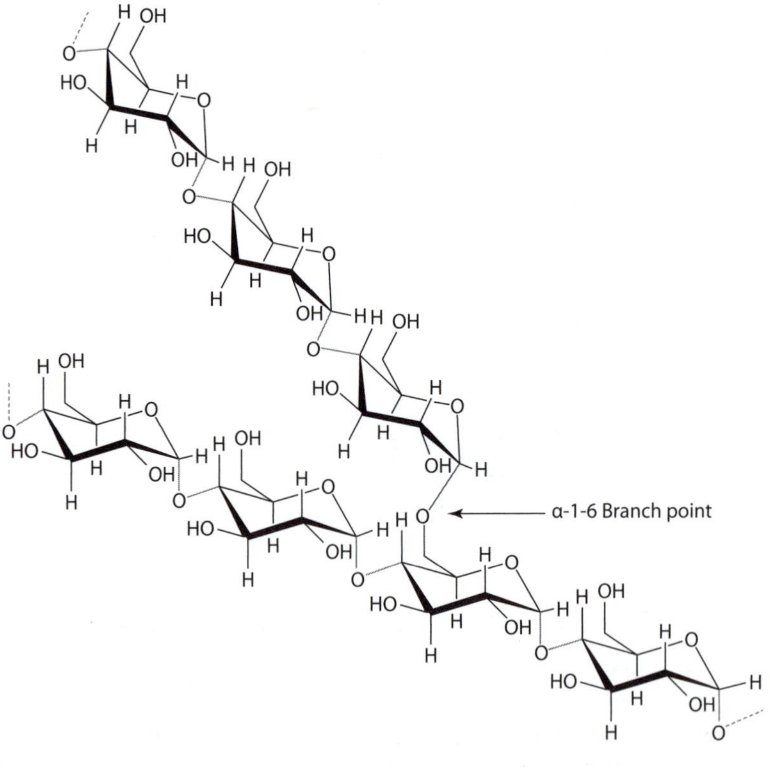

FIGURE 2.19
Structure of amylopectin showing a chain of α-1,4-linked D-glucopyranose units with α-1,6 branching.

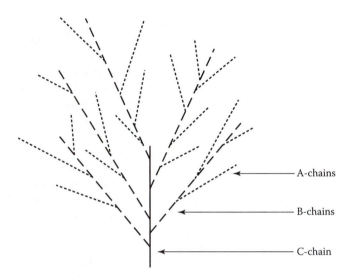

FIGURE 2.20
Treelike depiction of amylopectin showing the different types of polymer chains.

linkages are involved in the formation of branches. The branching in amylopectin causes it to be relatively noncrystalline in comparison to amylose. The chain length of branches is typically 17–26 D-glucopyranose units.

Three types of chains exist in amylopectin: A-chains that are outer unbranched chains attached to the rest of molecule via α-1,6 linkages, B-chains that are the inner branched chains to which A-chains are attached, and the C-chain to which all the B-chains are attached. The C-chain contains the only reducing end group in a molecule of amylopectin. The treelike model is a common depiction of the structure of amylopectin (shown in Figure 2.20).

However, the nature of the branching and the relative number of different types of chains is still discussed in research literature. Unlike amylose, amylopectin does not tend to form clathrate complexes with lipids or iodine. However, limited amount of binding between iodine and amylopectin results in a reddish-purple coloration and is thought to occur possibly via surface complexation.

2.5 Cellulose

Cellulose is the most abundant naturally occurring organic material. It is the primary structural element in plant cell walls and constitutes about a third of all plant matter. The term cellulose was appropriately applied when the

French chemist Anselme Payen identified it as the fundamental constituent of fibrous tissue in plant cell walls. Although, the use of cellulose as a raw material for chemicals has only been exploited since the twentieth century, it has been used in the form of cotton, linen, and pulp since ancient times. Cellulose is the major component in paper, cardboard, and textiles made from plant fibers. Other applications of cellulose include its use as a building insulation material and production of a wide range of commercial products such as cellophane, rayon, carboxymethyl cellulose, and nitrocellulose. The purest form of natural cellulose is cotton fiber, which is about 98% pure cellulose. Other fiber sources include flax, hemp, and jute. Woody biomass (40%–50% cellulose), agricultural residues (30%–45% cellulose), and grasses (25%–50% cellulose) offer a large resource pool for value-added applications of cellulose including its use as a potential long-term bioethanol feedstock.

In chemical terms, cellulose is a linear polysaccharide consisting of β-D-glucopyranose units that are linked via glycosidic bonds and has a chemical formula of $(C_6H_{10}O_5)_n$. However, unlike starch, the glycosidic linkage in the case of cellulose is β-1,4. Depending on the source, each cellulose chain can contain up to 15,000 D-glucopyranose units. The repeat unit is cellobiose: two D-glucopyranose units linked via a β-1,4 glycosidic bond. Figure 2.21 shows the structure of cellulose.

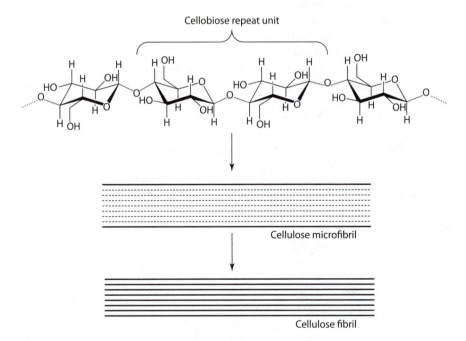

FIGURE 2.21
Structure of cellulose showing β-1,4-linked D-glucopyranose units forming linear chains that make up the microfibril which in turn constitutes the fibril.

The angle of the β-1,4 glycosidic bond results in linear cellulose chains that enable strong intra and intermolecular hydrogen bonding. This creates highly tensile microfibrils that are aggregates of cellulose chains containing both crystalline and amorphous regions. Groups of microfibrils constitute cellulose fibrils, which are part of wood fibers. Properties such as the degree of crystallinity, the degree of polymerization, and the microfibril width vary based on the source, age, and pretreatment of the fiber.

Several distinct forms of cellulose are known to exist based on the interaction between individual cellulose chains. Cellulose I is the native form in which individual chains are organized parallel to each other. Two types of intramolecular hydrogen bonding occur within the same chain. The first type is between the endocyclic oxygen (oxygen atom in the ring) and the hydrogen atom in the OH group of the C3 carbon (of a neighboring glucose). The second type is between the oxygen atom in the OH group of the C6 carbon and the hydrogen atom in the OH group of the C2 carbon (of a neighboring glucose). Between cellulose chains, there is a single intermolecular hydrogen bond between the hydrogen atom in the OH group of the C6 carbon and the oxygen atom in the OH group of the C3 carbon atom. Figure 2.22 shows the hydrogen bonding patterns in cellulose chains. These hydrogen

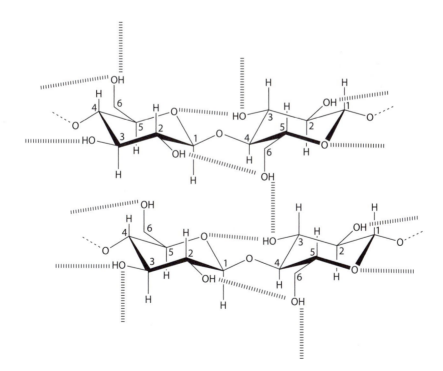

FIGURE 2.22
Intra- and intermolecular hydrogen bonding in cellulose that contributes to its crystallinity.

bonds enable the formation of microfibrils from individual cellulose chains resulting in a high degree of crystallinity.

Naturally occurring cellulose typically exhibits the cellulose I structure. Cellulose II is the nonnative form (mercerized cellulose) where individual chains are organized antiparallel to each other. While intramolecular hydrogen bonding in cellulose II is the same as in cellulose I, two additional intermolecular hydrogen bonds between adjacent cellulose chains exist making cellulose II thermodynamically more stable. Cellulose II can be obtained from cellulose I by regeneration with water or alcohol or mercerization with sodium hydroxide. Other forms of cellulose (III and IV) can be also produced from chemical treatments of cellulose I and II.

2.6 Hemicellulose

Hemicellulose is a word used to collectively describe all major polysaccharides present in plant cell walls except cellulose and pectin. Historically, hemicelluloses were thought to be intermediates in the biosynthetic pathways of cellulose. However, it is now known that hemicelluloses have distinct biosynthetic pathways. Hemicelluloses are known to form covalent bonds with functional groups in lignin and interact with cellulose via hydrogen bonds. The amount of hemicelluloses present is about 25%–30% in softwoods and hardwoods, 20%–25% in agricultural residues, and 15%–30% in grasses. Current applications include furfural production and the use of β-glucans derived from hemicelluloses as dietary fibers. Improvements in purification and bioconversion of hemicelluloses will enable their use in the production of other value-added products such as ethanol, xylitol, and butanediol.

Unlike cellulose, which is a linear homopolymer of glucose units, hemicelluloses are branched heteropolymers with monomer units that include pentoses (arabinose and xylose), hexoses (glucose, galactose, mannose, rhamnose, and fucose), and uronic acids (galacturonic, glucuronic, and methylglucuronic). Apart from being heteropolymers, other features that distinguish hemicelluloses from cellulose include a low degree of polymerization (~200), amorphous structure, and low thermal stability.

Nearly all hemicelluloses are either xylans or glucomannans. Xylans are heteropolysaccharides that consist of a homopolymeric backbone of β-1,4-linked D-xylopyranose units with random side chains of residues such as that arabinose, glucuronic acid, ferulic acid, and acetic acid. Glucomannans are heteropolysaccharides that consist of a copolymeric backbone of β-1,4-linked D-glucopyranose and D-mannopyranose units and may contain random side chains of galactose residues.

In hardwoods and grasses, the major hemicelluloses are xylans and the minor hemicelluloses are glucomannans. Hardwood xylans (Figure 2.23)

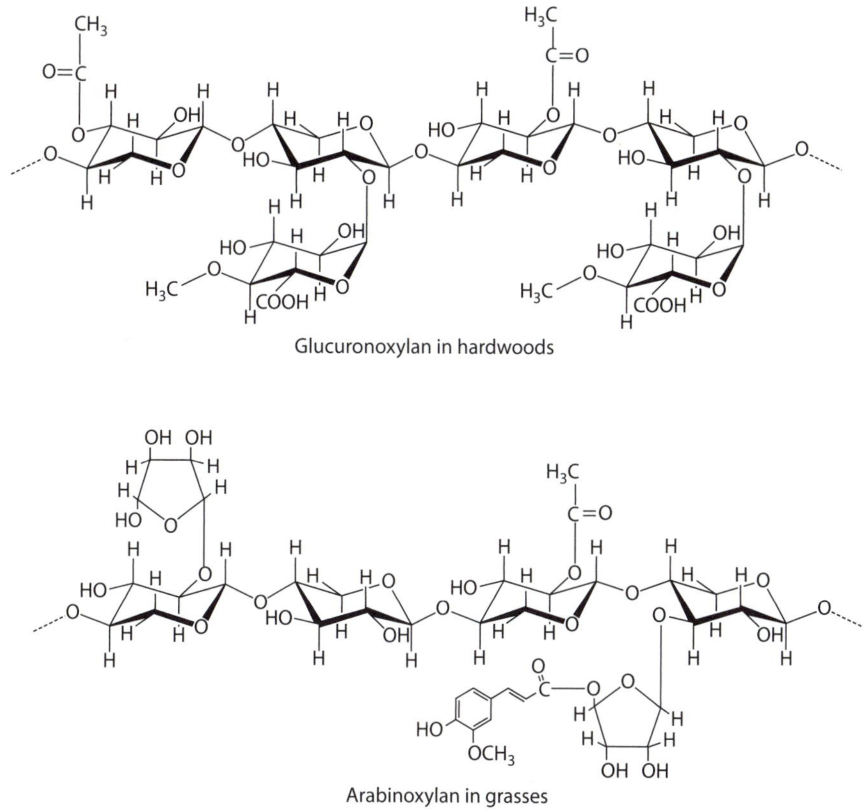

Glucuronoxylan in hardwoods

Arabinoxylan in grasses

FIGURE 2.23
General structures of xylans (major hemicellulose) of hardwoods and grasses.

are referred to as glucuronoxylans due to the occurrence of methylated α-D-glucuronic acid residues attached at the C2 position in 10% of the xylopyranose units of the main chain.

Hardwood xylans also contain a large number of acetyl groups attached at the C2 and C3 position of xylopyranose units of the main chain. Xylans in grasses are referred to as arabinoxylans due to the occurrence of a large number of α-L-arabinofuranose residues attached at the C2 and C3 position of xylopyranose units of the main chain. Xylans in grasses (Figure 2.23) also contain a large number of acetyl groups attached at the C2 and C3 position of xylopyranose units of the main chain and phenolic groups attached to arabinose residues. Glucomannans in hardwoods and grasses (shown in Figure 2.24) have a glucose to mannose ratio of 1:2 and rarely contain any substituent groups.

In softwoods, the major hemicelluloses are galactoglucomannans and the minor hemicelluloses are arabinoglucuronoxylans. Galactoglucomannans

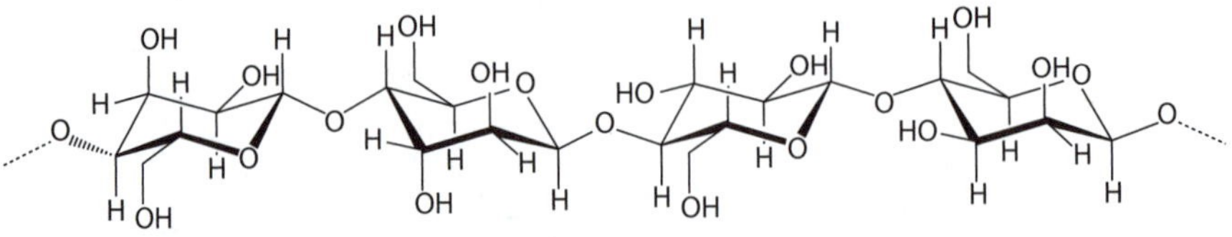

FIGURE 2.24
General structure of glucomannan (minor hemicellulose) in hardwoods and grasses.

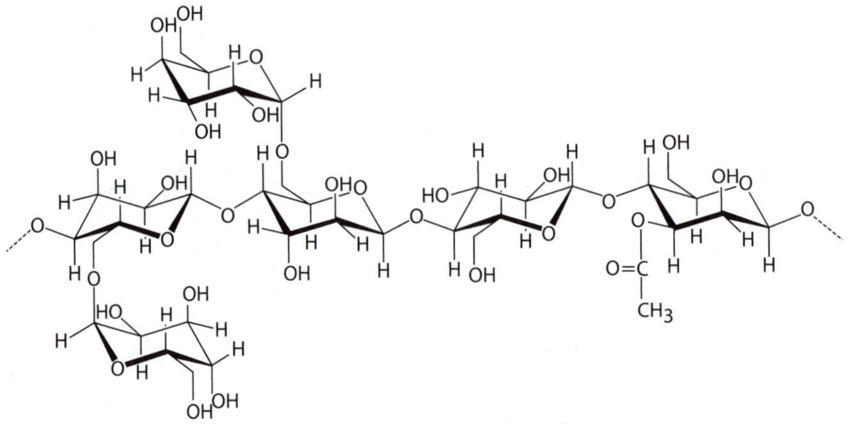

Galactoglucomannan in softwoods

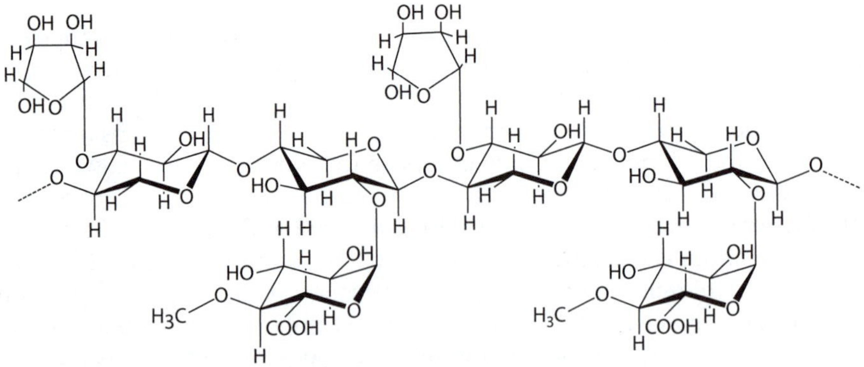

Arabinoglucuronoxylan in softwoods

FIGURE 2.25
General structure of galactoglucomannan (major hemicellulose) and arabinoglucuronoxylan (minor hemicellulose) in softwoods.

in softwoods (Figure 2.25) typically have a glucose to mannose ratio of 1:3 with large amounts of α-D-galactopyranose residues attached at the C6 position of either glucopyranose or mannopyranose units of the main chain and some acetyl groups attached at the C2 or C3 position of mannopyranose units of the main chain. Arabinoglucuronoxylans in softwoods (Figure 2.25) contain methylated α-D-glucuronic acid residues and α-L-arabinofuranose residues typically attached at the C2 and C3 positions of xylose molecules, respectively.

2.7 Lignin

Lignin is the second most abundant organic polymer after cellulose, and the name is derived from the Latin word for wood, *Lignum*. It is a characteristic component of the cell walls of higher plants where it plays a vital role in vascular transport and confers mechanical strength. The amount of lignin ranges from 15% to 25% for most grasses and hardwoods and up to 40% for softwoods. In addition to being biologically significant, lignin also has economic significance in several industries. A large amount of lignin is separated from wood in the paper industry and is typically burnt as a fuel to provide energy in paper mills. Modified lignin removed from pulping processes have been used as dispersing agents in oil-well and cement applications, adhesive agents for pelletization of animal feed, tanning agents in the leather industry, and raw material for vanillin production.

In chemical terms, lignin is defined as a highly cross-linked aromatic polymer of phenylpropane units with a molecular weight in excess of 10,000 units. There is considerable debate and research on the overall structure of lignin and several models have been proposed. The biosynthesis of phenylpropane units, which are the fundamental building blocks of lignin starts with the conversion of glucose into aromatic amino acids via the shikimic acid pathway. The cinnamic acid pathway then converts these amino acids into three distinct phenylpropane units: *p*-coumaryl alcohol, coniferyl alcohol, and sinapyl alcohol. These phenylpropane compounds are called monolignols as they are the fundamental units of lignin. The structures of these monolignols are shown in Figure 2.26.

Monolignols differ in the number of methoxy (OCH_3) groups attached to the phenolic unit. Sinapyl alcohol has two methoxy groups, coniferyl alcohol has one methoxy group, and *p*-coumaryl alcohol has none. Softwood lignins are primarily made up of coniferyl alcohol units (>95%). Hardwood lignins are made up of both coniferyl alcohol and sinapyl alcohol units in ratios that vary among different species. Lignins in grasses are typically made up of all three monolignol units.

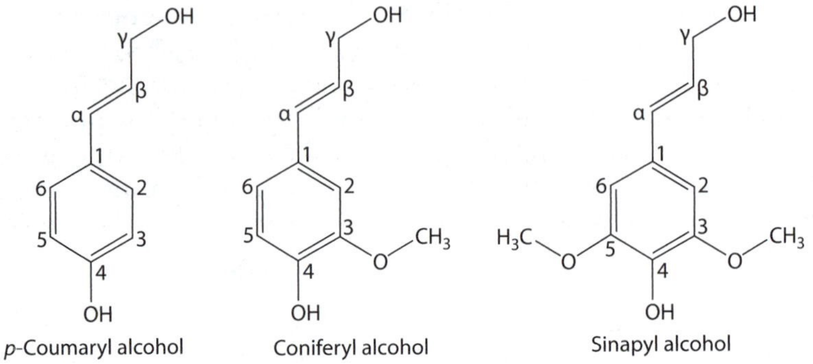

FIGURE 2.26
Structures of monolignols, the fundamental building blocks of lignin.

The polymeric structure of lignin is formed by the enzyme-catalyzed dehydrogenation of individual monolignols. The process entails the formation of resonance stabilized phenoxy radicals via electron transfer mechanisms. Figure 2.27 shows five resonance structures for the phenoxy radical from coniferyl alcohol. Each of these radicals can couple with other radicals to form dimers, trimers, and oligomers of lignin. Repeated polymerization of these oligomers results in the formation of the macromolecular structure of lignin.

In Figure 2.26, the carbon atoms on the alkyl units of the monolignols are labeled α, β, and γ and the carbon atoms on the phenol units of the monolignols are numbered 1–6. This labeling of the carbon atoms helps to categorize the numerous types of linkages that can be formed between two monolignol units. For example, a β–O–4 linkage implies that a bond is formed between the β carbon of the alkyl unit and the oxygen atom attached to the C4 carbon of the phenolic unit. Linkages can be ether bonds or carbon–carbon bonds. However, more than two-thirds of linkages in most plants and trees are ether bonds. Examples of some ether and carbon–carbon bonds in lignin are shown in Figure 2.28. The most common linkage is β–O–4, and it is this bond that is targeted during delignification in pulping processes.

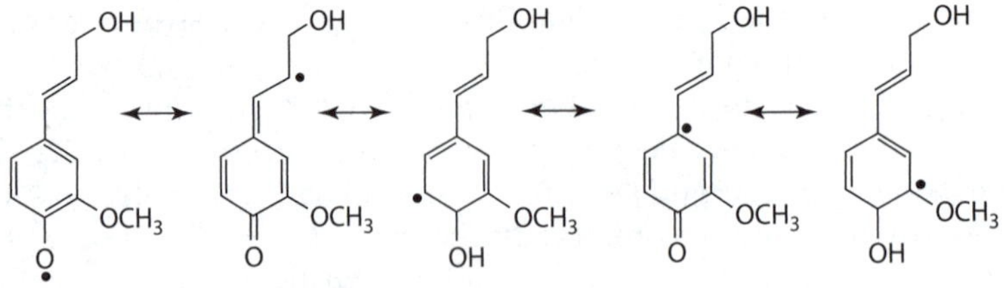

FIGURE 2.27
Resonance structures of the phenoxy radical from coniferyl alcohol.

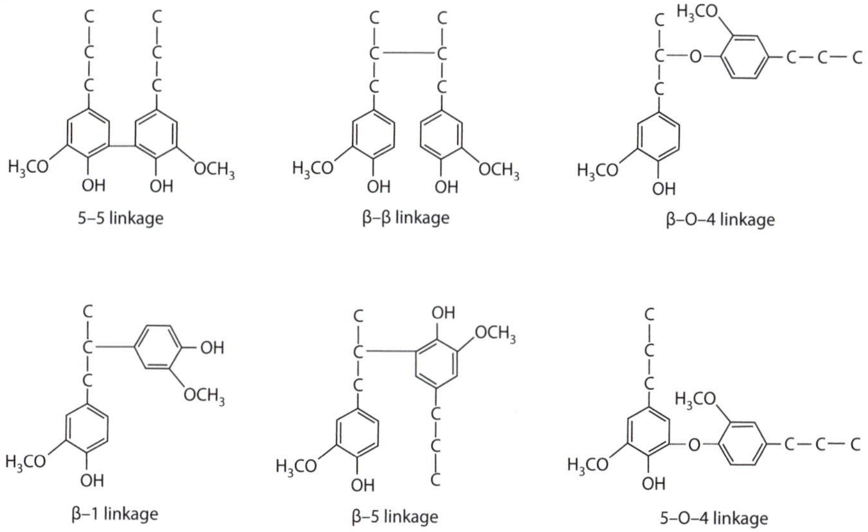

FIGURE 2.28
Examples of carbon–carbon and ether linkages typically found in lignin.

2.8 Pectin

Pectin is a word used to collectively describe a family of polysaccharides found in both the cell wall and intercellular layers of plant tissues. The name is derived from the Greek word *Pektikos*, which signifies coagulation. The amount of pectic polysaccharides in the cell walls is less than 5% for softwoods and hardwoods, 2%–10% for grasses, and up to 35% for flowering and fruit-bearing plants. The biological functions of pectin include maintenance of cell-wall structure, cell-to-cell adhesion and signaling, and development of fruits. On a commercial scale, pectin is a by-product of the citrus and cider industries and is primarily used as a gelling, thickening, and stabilizing agent for food and cosmetic applications.

Pectins are a family of heteropolymeric and homopolymeric polysaccharides. These polysaccharides are dominated by galacturonic acid units, which account for approximately 70% of all monomeric units in pectin. The dominant polysaccharide in the pectin family is homogalacturonan (Figure 2.29), which accounts for approximately 65% of all pectins. Homogalacturonan is a homopolymer of approximately 100 α-1,4-linked galacturonic acid units.

Rhamnogalacturonans, xylogalacturonans, and apiogalacturonans are other pectins present in smaller amounts and are structurally more complicated with several cross-linked side chains of polysaccharides made up of galactose, arabinose, xylose, and fucose. For example, rhamnogalacturonan

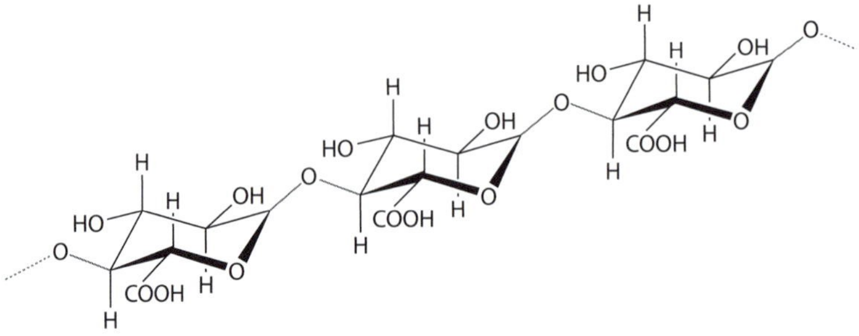

FIGURE 2.29
General structure of homogalacturonan, the major pectin polysaccharide.

has a backbone of α-1,4-linked galacturonic acid units with L-rhamnose present at intervals of about eight units. Approximately half of the rhamnose units are linked to arabinan, arabinogalactan, and galactan side chains.

2.9 Vegetable Oil

Vegetable oils (and fats) are lipid materials found in plants. While oils and fats can be obtained from animal sources, about 71% of the total edible oil (and fat) production is from plant sources. Vegetable oils and fats are mainly used for human consumption. They add texture and flavor to other food ingredients. Vegetable oils with high flash points (canola, sunflower, peanut, etc.) are heated to cook other foods. Other applications include animal feed, medicinal and cosmetic products, industrial lubricants, and biodiesel production. While some vegetable oils such as virgin olive oil are suitable for consumption after extraction, most oils need to undergo processing to remove impurities.

Vegetable oils (and fats) primarily consist of fatty acids triglycerides (typically 96%–98%). Glycerol (1,2,3-propanetriol) is a sugar alcohol that has three hydroxyl groups, as shown in Figure 2.30.

Triglycerides are formed when all the three hydroxyl groups of glycerol form ester bonds with three fatty acid molecules. The esterification process is a condensation reaction that releases water molecules in the process. Most triglycerides are defined as complex or mixed since two or three different fatty acids are involved in the esterification. Monoacid triglycerides are relatively rare. Diglycerides are formed when two hydroxyl groups are esterified and monoglycerides are formed when only one hydroxyl group is esterified. The structures of tri-, di-, and monoglycerides are shown in Figure 2.30.

Fatty acids are aliphatic compounds with a carboxylic acid functional group. Most natural fatty acids contain 4–28 carbon atoms (most often an

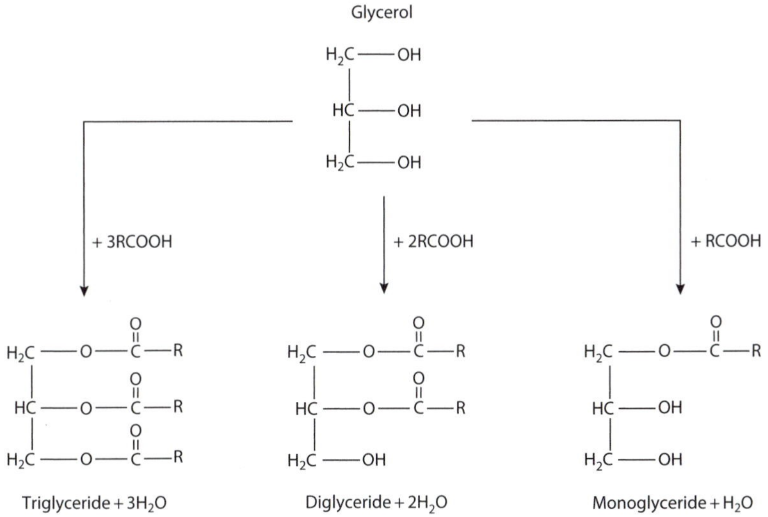

Glycerol

$$H_2C - OH$$
$$HC - OH$$
$$H_2C - OH$$

+ 3RCOOH

+ 2RCOOH

+ RCOOH

Triglyceride + 3H$_2$O

Diglyceride + 2H$_2$O

Monoglyceride + H$_2$O

FIGURE 2.30
Formation of tri-, di-, and monoglcyerides via esterification between glycerol and fatty acids.

even number). Based on chain length, they are classified as short chain (4–8 carbon atoms), medium chain (8–14 carbon atoms), and long chain (more than 14 carbon atoms). Fatty acids are also classified based on the presence or absence of double bonds. Saturated fatty acids contain no double bonds, monounsaturated fatty acids contain one double bond, and polyunsaturated fatty acids contain two or more double bonds. Some major fatty acids (by composition) typically seen in vegetable oils are linoleic acid (18 carbon atoms with 2 double bonds), linolenic acid (18 carbon atoms with 3 double bonds), oleic acid (18 carbon atoms with 1 double bond), palmitic acid (16 carbon atoms with no double bonds), and stearic acid (18 carbon atoms with no double bonds). Figure 2.31 shows the structures of these fatty acids, and Table 2.2 shows their amount in different oilseeds.

The presence of double bonds in the aliphatic chains of fatty acids results in *cis–trans* isomerism discussed previously. Figure 2.32 show the *cis* and *trans* isomers of oleic acid. The *cis*-isomer of oleic acid has a pronounced kink in the structure. The presence or absence of these kinks in fatty acids affects the fluidity of oils and fats. The kinks created by *cis*-isomers negatively affect the packing of fatty acid chains and result in triglycerides that are fluid at normal temperatures. Triglycerides containing fatty acids in the *trans* form are thermodynamically more stable and melt at higher temperatures.

As mentioned previously, triglycerides account for 96%–98% of the total composition of most vegetable oils. Two percent to 4% of vegetable oils consist of a wide range of other compounds that are classified as major non-triglycerides and minor non-triglycerides. Major non-triglycerides include

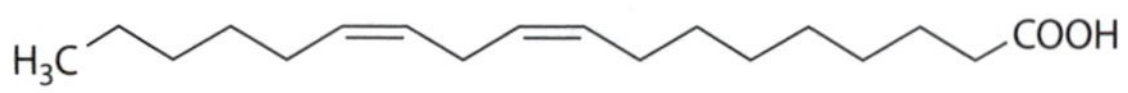

Linoleic acid

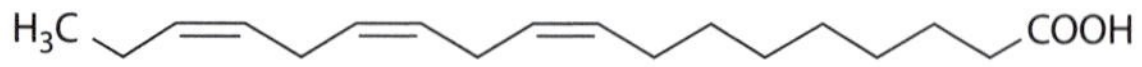

Linolnic acid

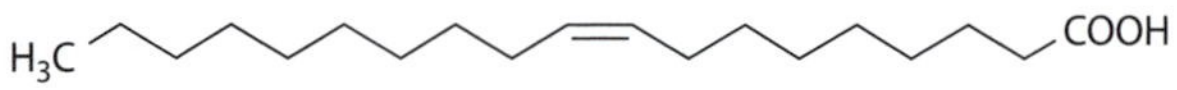

Oleic acid

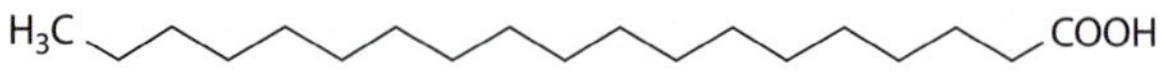

Palmitic acid

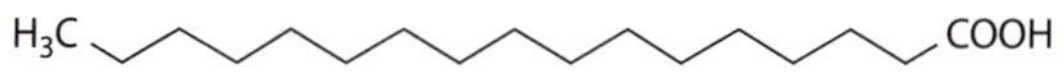

Stearic acid

FIGURE 2.31
Examples of fatty acids typically seen in vegetable oils.

TABLE 2.2

Content of Major Fatty Acids (% of Crude Oil) in Different Oilseeds

Source	Linoleic	Linolenic	Oleic	Palmitic	Stearic
Soybean	50–57	5–10	18–29	10–13	3–6
Canola	11–23	5–13	52–67	1–6	1–3
Cottonseed	56–62	<1	14–22	18–27	2–4
Sunflower	48–74	<1	13–40	5–8	2–7
Peanut	14–43	<1	36–67	8–14	2–5
Palm	6–12	<1	36–44	40–48	3–7

Sources: Data from Firestone, D., *Physical and Chemical Characteristics of Oils, Fats, and Waxes*, 2nd edn., AOCS Press, Champaign, IL, 2006; Salunkhe, D.K. et al., *World Oilseeds: Chemistry, Technology, and Utilization*, Van Nostrand Reinhold, New York, 1992.

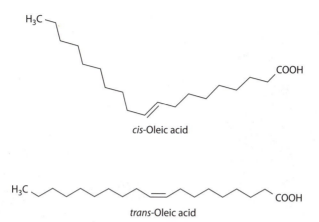

FIGURE 2.32
cis–trans isomerism in oleic acid.

free fatty acids, diglycerides, monoglycerides, and phospholipids. Minor non-triglycerides are present in very small amounts (parts per million) and include trace metals, tocopherols, sterols, and pigments such as carotene and chlorophyll.

Free fatty acids are those that are released from triglycerides, which then become either mono- or diglycerides. This process can occur via the action of lipase enzymes during storage of oilseeds, extraction of oil from oilseeds, and during storage of crude vegetable oil. Most crude vegetable oils typically contain less than 2% free fatty acids and less than 1% mono- and diglycerides. An exception is palm oil, which can contain up to 4% free fatty acids, up to 2% monoglycerides, and up to 7% diglycerides.

Phospholipids (also called gums or phosphatides) have a structure similar to triglycerides. In place of one fatty acid chain, there is a phosphate group that is often attached to organic molecules such as choline, inositol, and serine. Figure 2.33 shows the general structure of phospholipids and organic molecules that may be attached to the phosphate group. Most crude vegetable oils typically contain 1%–4% phospholipids, which are removed during oil processing for human consumption and fuel production.

2.10 Extractives

The term extractives is used to describe a wide range of compounds present in hardwoods, softwoods, and grasses that can be extracted using polar and nonpolar solvents. The amount and composition of extractives vary by plant type, geographical location, and season. Extractives are associated with the

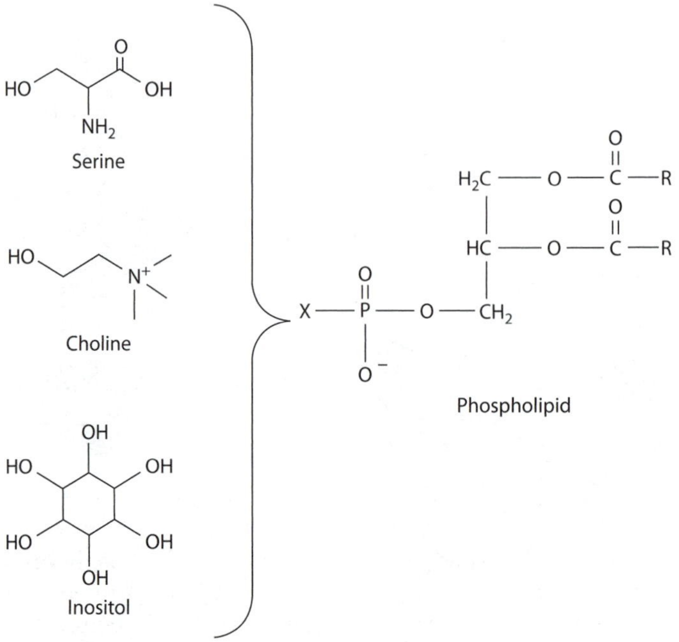

FIGURE 2.33
General structure of phospholipids with examples of organic molecules that may be attached.

fragrance and color of lignocellulosic biomass and enhance their durability by increasing resistance to decay and insects. While wood extractives can cause operational and quality problems in the pulp and paper industry, they also yield useful products such as turpentine and tall oil.

Most woody biomass, grasses, and agricultural residues contain 2%–10% extractives by weight. The amount of extractives can be much higher in tropical regions and in specific parts of woody biomass such as bark or heartwood. Extractives are much lower in concentration than other cell-wall polymers, but the variability among species provides a basis for characterization and chemotaxonomy. The three main groups of extractives are terpenes and terpenoids, fats and waxes, and phenolic (and related) compounds. Small amounts of straight-chain alkanes, free fatty acids, and alcohols can also be present.

Terpenes and their derivatives are a large group of compounds that are characterized by the presence of two or more isoprene (2-methyl butadiene) units in their structures. They have a general formula of $(C_{10}H_{16})_n$ and are categorized as monoterpenes ($n=1$), sesquiterpenes ($n=1.5$), diterpenes ($n=4$), and polyterpenes ($n>4$). Terpenoids are derivatives of terpenes containing hydroxyl, carbonyl, carboxylic acid, and ester functional groups. Figure 2.34 shows the structures of some terpenes and terpenoids present in extractives.

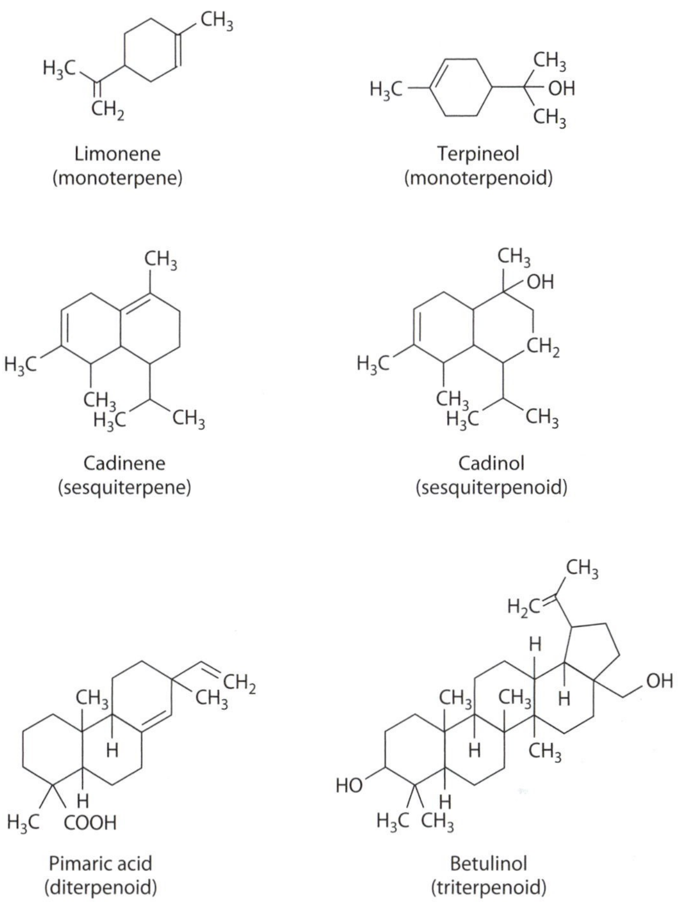

FIGURE 2.34
Examples of terpenes and terpenoids found in extractives.

The fats in extractives are mostly dominated by triglycerides with smaller amounts of mono- and diglycerides. In general, most of the triglycerides in extractives contain fatty acids with chain lengths of 16–22 carbon atoms. They can be saturated or contain 1–3 double bonds. Waxes are esters of fatty acids with much larger alcohols. These alcohols can be straight-chain alcohols (16–30+ carbon atoms), sterols, or terpene alcohols.

Phenolic (and related) extractives are a heterogeneous group of compounds that are characterized by the presence of phenolic units in their structures. They include simple phenols, lignans (dimers of phenylpropane units), flavenoids, stilbenes, and tropolones. Figure 2.35 shows the structures of some phenolic extractives.

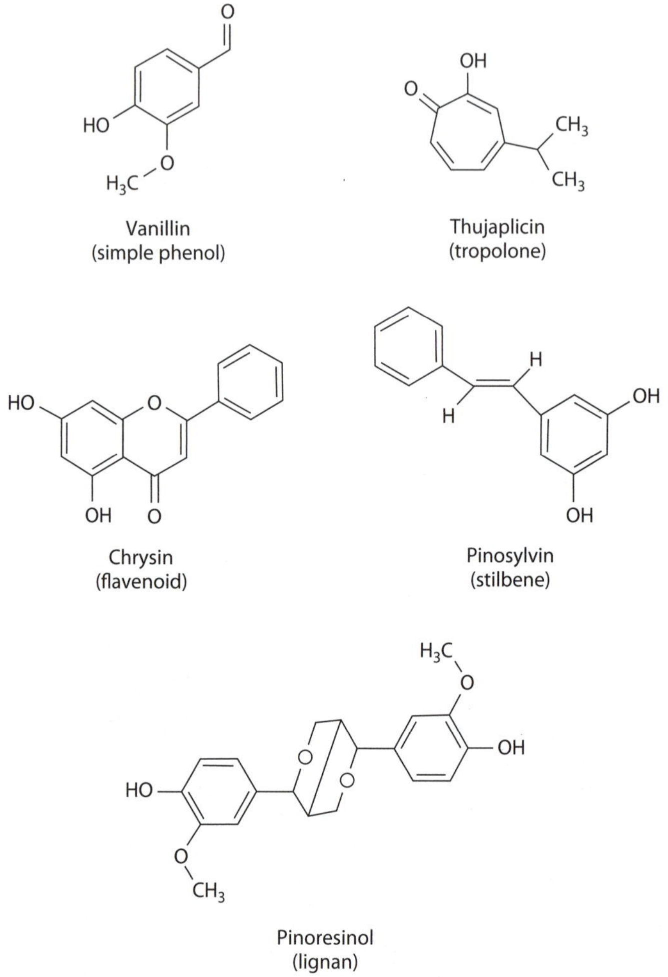

Vanillin
(simple phenol)

Thujaplicin
(tropolone)

Chrysin
(flavenoid)

Pinosylvin
(stilbene)

Pinoresinol
(lignan)

FIGURE 2.35
Examples of phenolic (and related) compounds found in extractives.

2.11 Problems

1. Draw all possible structural isomers for hexane, an alkane with an empirical formula of C_6H_{14}.

2. Explain *cis–trans* isomerism. What are the requirements for such isomerism to occur?

3. Differentiate between aliphatic and aromatic organic compounds and explain the phenomena of resonance structures in aromatic compounds.

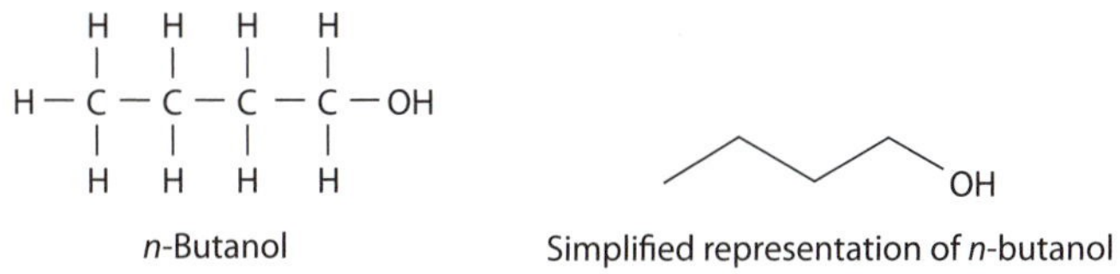

FIGURE 2.36
Structure of *n*-butanol.

4. The structure for *n*-butanol (a primary alcohol) is shown in Figure 2.36. Rearrange the structure to produce isomers that are secondary and tertiary alcohols.

5. Draw the α and β anomers of the pyranose ring formed from the open-chain structure of fructose shown in Figure 2.12.

6. What are the differences between amylose and amylopectin?

7. Explain why cellulose exhibits more crystallinity than amylose or amylopectin.

8. What are the differences between hemicellulose and cellulose?

9. Distinguish between the major and minor hemicelluloses of hardwoods, grasses, and softwoods.

10. Define resonance in the context of aromatic compounds and explain the role of resonance structures in the biosynthesis of lignin.

11. The cloud point is an indicator of biodiesel performance at low temperatures. It is defined as the temperature at which crystals start to form in the fuel. As the temperature is further lowered, the fuel starts to solidify. Explain why biodiesel produced using palm oil has a much higher cloud point than biodiesel produced from canola or soybean.

References

Adler, E. 1977. Lignin chemistry—past, present, and future. *Wood Science and Technology*, 11(3): 169–218.

Boxer, R. J. 1997. *Essentials of Organic Chemistry*. Dubuque, IA: WCB.

Carey, F. A. 2006. *Organic Chemistry*. 6th edn. Dubuque, IA: McGraw-Hill.

ChemDraw Ultra 11. 2008. Cambridge, MA: Cambridge Soft Corporation.

Clarke, A. J. 1997. *Biodegradation of Cellulose: Enzymology and Biotechnology*. Lancaster, PA: Technomic Publishing Company.

Fan, L. T., M. M. Gharpuray, and Y. Lee. 1987. *Cellulose Hydrolysis*. Berlin: Springer-Verlag.

Fengel, D. and G. Wegener. 1984. *Wood: Chemistry, Ultrastructure, Reactions*. Berlin: W. de Gruyter.

Firestone, D. 2006. *Physical and Chemical Characteristics of Oils, Fats, and Waxes*. 2nd edn. Champaign, IL: AOCS Press.

Fishman, M. L. and J. J. Jen. (eds.) 1986. *Chemistry and Function of Pectins*. Washington, DC: American Chemical Society.

Galliard, T. (ed.) 1987. *Starch-Properties and Potential*. Chichester, U.K.: Published for the Society of Chemical Industry by Wiley.

Gupta, M. K. 2008. *Practical Guide for Vegetable Oil Processing*. Urbana, IL: AOCS Press.

Hon, D. N. (ed.) 1996. *Chemical Modification of Lignocellulosic Materials*. New York: Marcel Dekker.

Kettunen, P. O. 2006. *Wood Structure and Properties*. Uetikon-Zuerich/Enfield, NH: Trans Tech Publications Ltd.

Mohnen, D. 2008. Pectin structure and biosynthesis. *Current Opinion in Plant Biology*, 11(3): 266–277.

Murphy, D. J. (ed.) 2005. *Plant Lipids: Biology, Utilisation and Manipulation*. Oxford, U.K.: Blackwell Publishing.

National Renewable Energy Laboratory. 2004. *Biomass Feedstock Composition and Properties Database*, http://www.nrel.gov/biomass/energy_analysis.html.

O'Leary, M. H. 1976. *Contemporary Organic Chemistry: Molecules, Mechanisms, and Metabolism*. New York: McGraw-Hill.

Pearl, I. A. 1967. *The Chemistry of Lignin*. New York: Marcel Dekker.

Roberts, J. C. 1996. *The Chemistry of Paper*. Cambridge, U.K.: Royal Society of Chemistry.

Saha, B. C. 2003. Hemicellulose bioconversion. *Journal of Industrial Microbiology & Biotechnology*, 30(5): 279–291.

Salunkhe, D. K., J. K. Chavan, R. N. Adsule, and S. S. Kadam. 1992. *World Oilseeds: Chemistry, Technology, and Utilization*. New York: Van Nostrand Reinhold.

Schubert, W. J. 1965. *Lignin Biochemistry*. New York: Academic Press.

Sjeostreom, E. 1993. *Wood Chemistry: Fundamentals and Applications*. 2nd edn. San Diego, CA: Academic Press.

Stick, R. V. 2001. *Carbohydrates: The Sweet Molecules of Life*. San Diego, CA: Academic Press.

Thomas, D. J. and W. A. Atwell. 1999. *Starches*. St. Paul, MN: Eagan Press.

Whistler, R. L. and C. L. Smart. 1953. *Polysaccharide Chemistry*. New York: Academic Press.

Zaikov, G. E. (ed.) 2005. *Chemistry of Polysaccharides*. Lieden: VSP International Science.

3

Biomass Resources

Ziyu Wang and Deepak R. Keshwani

CONTENTS

3.1 Sugar Crops 42
    3.1.1 Sugarcane 42
    3.1.2 Sugar Beet 43
    3.1.3 Sweet Sorghum..44
3.2 Starch Crops 45
    3.2.1 Corn 45
    3.2.2 Wheat..46
    3.2.3 Potato 47
    3.2.4 Sweet Potato..48
3.3 Agricultural Residues 48
    3.3.1 Corn Stover 49
    3.3.2 Wheat Straw..50
    3.3.3 Rice Straw...50
3.4 Herbaceous Biomass...51
    3.4.1 Switchgrass 51
    3.4.2 *Miscanthus* 53
    3.4.3 Coastal Bermuda Grass 55
3.5 Woody Biomass..57
3.6 Oilseeds..60
    3.6.1 Soybean 60
    3.6.2 Rapeseed (Canola) 61
    3.6.3 Sunflower 62
    3.6.4 Oil Palm..63
    3.6.5 Waste Edible Oil..63
3.7 Problems...64
References...64

3.1 Sugar Crops

3.1.1 Sugarcane

Sugarcane is classified in the genus *Saccharum*, tribe Andropogoneae, and family Poaceae. Brazil is the largest producer of sugarcane (514 million tons in 2007). A large portion of Brazil's sugarcane crop is used to produce ethanol for gasoline–ethanol (gasohol) fuel blends. The current yield of sugarcane in Brazil is 74 tons ha^{-1}. The second largest producer of sugarcane is India (356 million tons in 2007) where it is primarily used to produce jaggery, refined sugar, and alcoholic beverages. China is the third leading producer of sugarcane with a yield of 106 million tons in 2007. Other major sugarcane-producing countries are Thailand, Pakistan, Mexico, Colombia, Australia, the United States, and the Philippines. The annual production of sugarcane in the United States is around 28 million tons, most of which is grown in Florida, Hawaii, Louisiana, and Texas (Food and Agricultural Organization of the United Nations: FAO, 2008).

Sugarcane occurs as stalks bunched in stools of 5–50 stalks or evenly scattered. Sugarcane can be grown in regular rotation with other crops. Cultivation requires a tropical or subtropical climate devoid of a killing freeze (below 28°F), with a minimum of 60 cm (24 in.) of annual rainfall (Clarke and Edye, 1996). Stem cuttings have become the most common method of propagation for sugarcane. Time to maturity can range from 8 to 24 months. The temperature should be sufficiently high for a period of at least 8 months to enable rapid growth, after which a period of slow growth occurs. Sugar content increases during this period of slow growth and ripeness is defined based on when the sugar content reaches its maximum value (Martin et al., 2006). After the crop has been harvested, sugarcane is permitted to regrow from the old root stock once or several times in order to obtain several harvests from the original planting. This repeated cropping procedure is known as ratooning and may last for 6 or 7 years. The vigor of the sugarcane and local conditions determine the number of ratoons. For example, two to four ratoons are typical in the United States, but four to five ratoons are typical in Australia (Martin et al., 2006). At the end of ratooning, the roots of the crop are ploughed out and the field can be replanted with sugarcane or another crop right away, or after a period of fallow (Cheesman, 2004).

Sugarcane is highly efficient in converting solar energy to organic matter. However, strong sunlight and plentiful water is needed to meet the considerable growth potential. While the growing period requires well-distributed rainfall (or irrigation), the preharvest ripening period requires moderately dry conditions. However, abundant sunlight is needed throughout the entire season. In fact, it has been shown that the number of sunshine hours positively correlates with the yield of sugar from sugarcane (Garside et al., 1997).

For germination, the optimum temperature is 20°C–32°C and cane sets fail to sprout below 10°C. A diurnal maximum temperature of 35°C and a minimum temperature of 18°C with an average temperature of around 26°C are appropriate for tillering. Temperatures outside this range along with cloudy and short days negatively impact tillering. During the grand growth phase, a warm climate with sufficient irrigation and absence of high wind velocity is ideal. However, cool and dry weather is favorable for ripening. Ideal weather conditions for ripening are daytime temperatures of 23°C–30°C, nighttime temperatures of 7°C–14°C, and a low relative humidity of around 50%–55% (Verma, 2004).

The chemical composition of sugarcane is normally analyzed based on its different parts. For stalks, water (71%) is the most abundant component. Other components of stalks (on a wet basis) include 15% total sugars (14% sucrose), 12% lignocellulose, and 0.5% ash. The primary component in leaves and tops is also water (74%). Tops and leaves contain a small amount of total sugars (0.2%) and no sucrose. However, they contain more lignocellulosic material than stalks. Ash accounts for around 2% of the total mass of tops and leaves, with 2% comprising other minor components (Rivacoba and Morin, 2002).

3.1.2 Sugar Beet

Sugar beet, a biennial root vegetable, contains a high concentration of sucrose in its root (Table 3.1). Although sugar beet can be grown in a wide variety of climates, it is most commonly grown in regions with temperate climates

TABLE 3.1

Sucrose and Lignocellulose Content (% Dry Basis) of Sugarcane, Sugar Beet, and Sweet Sorghum

Biomass	Sucrose	Lignocellulose
Sugarcane (stalks)	49	42
Sugar beet	68	20
Sweet sorghum	45	44

Sources: Data from Rivacoba, R.S. and Morin, R.B., Sugarcane and sustainability in Cuba, in Funes, F. et al. (eds.), *Sustainable Agriculture and Resistance: Transforming Food Production in Cuba*, Food First Books, Oakland, CA, 2002; Cheesman, O., *Environmental Impacts of Sugar Production: The Cultivation and Processing of Sugarcane and Sugar Beet*, CAB International, Cambridge, MA, 2004; Claassen, P.A.M. et al., *The 2nd World Conference on Biomass for Energy, Industry and Climate Protection*, Rome, Italy, 2004, p. 1522.

(Draycott, 2006). The average sugar (sucrose) content of sugar beet is around 17% of the total raw biomass, and the percentage can vary depending on the specific variety and growing condition. In addition to sucrose, sugar beet also contains 75% water, 5% lignocellulosic material, and 3% glucose and fructose (Cheesman, 2004). As of 2007, the world's three largest sugar beet producers were France (32.3 million tons), United States (31.9 million tons), and Russia (29 million tons). In 2008, approximately one million acres of U.S. cropland was used for sugar beet production (FAO, 2008).

3.1.3 Sweet Sorghum

Sweet sorghum is an annual plant that has a wide range of uses that include production of sugar, syrup, fuel, and roofing applications. Sweet sorghum is able to thrive under drier and warmer conditions than many other crops, yet provide biomass yields. The primary sweet sorghum production region is Asia, which accounts for 33% of total world production. North America is the second largest producer at 23% of total world production (Kim and Dale, 2004). The juice from sweet sorghum (Figure 3.1) is composed of sucrose (56%), glucose (30%), and fructose (14%), and its solid components consist of cellulose (15%–25%), hemicellulose (35%–50%), and lignin (20%–30%) (Phowchinda et al., 1997). The ash component is reported to be around 3% (Claassen et al., 2004).

FIGURE 3.1
Sweet sorghum juice. (Courtesy of E. Nicole Hill, North Carolina State University, Raleigh, NC.)

3.2 Starch Crops

3.2.1 Corn

Corn (*Zea mays* L.), or maize, is the third leading cereal crop in the world, following wheat and rice. Corn is a tall (1.5–3.5 m) annual plant belonging to the family Gramineae and tribe *Maydeae* (Farnham et al., 2003). The worldwide corn production has tremendously increased since 1955 due to an increase in the land area allocated for corn production and increases in overall yields per unit area of land (Farnham et al., 2003). In 2007, global annual corn production was around 800 million tons with a harvested area of 150 million hectares and an average yield of 5 tons ha^{-1}. The United States accounted for over 40% of the total global corn harvest in 2007. Other top corn producers include Argentina, Brazil, Canada, China, France, India, Indonesia, and Italy (FAO, 2008).

Corn is typically planted in the spring when warm temperatures are conducive for growth. The growth of corn plants is reduced when air and soil temperatures are low. In order to overcome slow root growth and the potential for reduced nutrient uptake under adverse conditions, starter fertilizers are utilized to provide an accessible nutrient source for corn growth (Zublena and Anderson, 1994). In addition to temperature, soil moisture is another key factor that determines the adaptability of corn to an area. The growth of corn requires plentiful moisture because of its long growing season and yields can vary based on rainfall patterns (Sopher et al., 1973). The third key environmental factor is sunlight, which is partially responsible for optimum yields (Farnham et al., 2003). Many of the corn varieties grown in the United States and Canada are hybrids. This reflects the fact that high yields of corn depend on not only environmental and field factors but also genetic characteristics (Wagger, 1985).

Corn can be grown through a number of different physical arrangements or methods including strip cropping, intercropping, continuous corn system, and corn–soybean rotation. Historically, Native Americans planted corn in hills (or mounds) using several approaches that included planting just corn, or planting corn along with beans that provided nitrogen and squashes that prevented the growth of weeds and excessive evaporation. While the hill technique is still being used in some Native American reservations, corn is now typically planted in rows that enable cultivation while the plant is young.

The predominant component of corn (Figure 3.2) is starch, which accounts for 64%–78% of the total mass on a dry basis. Other carbohydrates include cellulose fiber (8.3%–11.9%), pentosans (5.9%–6.6%), sucrose (0.5%–3.3%), and monomeric sugars such as glucose, fructose, and raffinose in minor amounts (Watson, 1982; Allen, 1979). Protein content of the corn kernel can range from 8% to 18% (Bressani and Mertz, 1958). Higher protein contents can lead to

FIGURE 3.2
Corn grain. (Courtesy of F. John Hay, University of Nebraska-Lincoln, Lincoln, NE.)

a decrease in the corn yield by as much as 30% as well as a decrease in the overall starch content and kernel size (Dudley, 1974). Corn protein is composed of albumin (7%), globulin (10%), prolamin or zein (39%), and glutelin (35%) (Salunkhe et al., 1992). The fat content of corn varies from 1% to 6% (Quckenbush et al., 1961). The fat content can affect other components as well. For example, a higher fat content is associated with reduced starch content but superior protein quality (Weber and Alexander, 1975; Salunkhe et al., 1992). The lipids (fats) in corn are mainly composed of triglycerides (80%) and phospholipids (8%), with other minor components that include diglycerides (1%–3%), sterols (1%–6%), sterylesters (1%–3%), glycolipids (2%–5%), free fatty acids (<1%), and small amounts of waxes, pigments, and odor-active compounds (Salunkhe et al., 1992). The mineral content in corn ranges from 1% to 4% and includes major dietary minerals such as phosphorus (0.27%), potassium (0.38%), magnesium (0.17%), sulfur (0.14%), and calcium (0.03%) (Watson, 1982). The vitamin content of corn is lacking because most of the water-soluble vitamins are bound to cell components (Salunkhe et al., 1992). The typical composition of dent corn is shown in Table 3.2. However, composition does vary depending on the type of seed used during planting (Hopkins, 1974).

3.2.2 Wheat

Wheat (Figure 3.3) is another important cereal crop and can be grown in a wide range of climates. Asia and Europe are the two primary regions for wheat production, followed by North America. China is the single largest producer of wheat in the world with an annual production of 110 million tons in 2007. The second largest producer is India (75 million tons). Other

TABLE 3.2

Typical Composition (% Dry Basis) of Corn, Potato, Sweet Potato, and Wheat

Biomass	Starch	Protein	Fat	Ash
Corn (dent)	72.4	9.6	4.7	1.43
Potato	74.7	9.8	0.44	5.2
Sweet potato	60–70	5.0	1.0	3.0
Wheat	77.1	12.6	4.8	3.7

Sources: Data from Earle, F.R. et al., *Cereal Chem.*, 23(5), 504, 1946; USDA, USDA National Nutrient Database, 2008. Available from http://www.nal.usda.gov/fnic/foodcomp/search/; Woolfe, J.A., *Sweet Potato: An Untapped Food Resource*, Cambridge University Press, New York, 1992.

FIGURE 3.3
Wheat. (Courtesy of Chengci Chen, Montana State University, Bozeman, MT.)

major wheat producers include the United States (54 million tons), Russia (49 million tons), and France (33 million tons) (FAO, 2008). Wheat is mainly composed of carbohydrates (mostly starch), which accounts for around 77.1% (dry basis) of total biomass. Other components include protein (12.6%), fat (4.8%), and ash (3.7%) (USDA, 2008).

3.2.3 Potato

Potato, the fourth largest global food resource, is a starchy, tuberous peren-nial crop. The growth of potato is widely distributed across 140 countries, most of which have tropical and subtropical climates (Beukema and van der Zaag, 1990). Global potato production in 2007 was 322 million tons, with

FIGURE 3.4
Purple sweet potato. (Courtesy of E. Nicole Hill, North Carolina State University, Raleigh, NC.)

China being the dominant contributor at 72 million tons (FAO, 2008). Other major producers include Russia (37 million tons), India (26 million tons), and the United States (18 million tons). The dominant component in potato is water (about 79%). The carbohydrate content is around 18%, of which 86% is starch. The amounts of protein, fiber, fat, and ash are reported to be 2.02%, 2.2%, 0.09%, and 1.08%, respectively (USDA, 2008).

3.2.4 Sweet Potato

Sweet potato (Figure 3.4), native to the tropical parts of South America, is another starch-based perennial crop that has attracted attention as a food source since late 1980s. Global production of sweet potato reached around 126.3 million tons in 2007, 80% of which came from China. In the United States, North Carolina is the leading state in sweet potato production. In 2007, North Carolina accounted for nearly 40% of total sweet potato production in the United States (FAO, 2008). Approximately 80%–90% of the dry weight of sweet potato is carbohydrates (60%–70% starch and 10%–20% simple sugars). The protein content is about 5%, lipid content is about 1%, and ash content is about 3% (Woolfe, 1992).

3.3 Agricultural Residues

Agricultural residues refer to either crop residues or processing residues. The former are disposed parts of crops after harvest, while the latter are leftovers from the processing of the harvested portions of agricultural crops for uses such as food, fiber, and feed. Crop residues like corn stover (Figure 3.5), wheat straw (Figure 3.6), and rice straw are abundant, cheap, and readily available for energy production and other value-added applications. The United States alone generates around 500 million tons of agricultural residues annually (Milbrandt, 2005).

3.3.1 Corn Stover

Corn stovers that remain in a field after corn has been harvested include stalks (50%), leaves (22%), cobs (15%), and husks (13%). Corn stover is usually

FIGURE 3.5
Corn stover. (Courtesy of F. John Hay, University of Nebraska-Lincoln, Lincoln, NE.)

FIGURE 3.6
Wheat straw bales. (Courtesy of Chengci Chen, Montana State University, Bozeman, MT.)

produced at a rate of 1 kg per kg of corn grain on a dry basis (Kim and Dale, 2004). This production rate translates to about 2000 kg of corn stover per bushel of corn (Heaton et al., 2008). The United States generates about 80–100 million dry tons of corn stover per year (Kadam and McMillan, 2003). The three major components are cellulose (35%–40%), hemicellulose (17%–35%), and lignin (7%–18%) (Table 3.3).

3.3.2 Wheat Straw

Like corn stover, the three major components of wheat straw are cellulose, hemicellulose, and lignin whose dry weight percentages range from 33% to 50%, 24% to 36%, and 9% to 17%, respectively (Table 3.3). Additionally, wheat straw contains about 20% non–cell wall materials (pectin, protein, etc.) and 1%–2% ash (Schmidt and Bjerre, 1997). Wheat straw is produced at a rate of 1–3 tons per acre under intensive farming conditions, with a dry weight ratio of grain to residue of 1:2.7 (Reitz, 1976). It is estimated that the United States generates approximately 82 million tons of wheat straw annually (Kadam and McMillan, 2003).

3.3.3 Rice Straw

Rice straw is among the most abundant lignocellulosic waste materials in the world and is the major residue from rice production. Rice straw includes culms (stems), leaf blades, leaf sheaths, and the remains of the panicle after threshing (Juliano, 1985). Rice straw can contain 36%–47% cellulose, 19%–25% hemicellulose, and 10%–24% lignin (Table 3.3). These values are comparable to those of other cereal crop residues (Garrote et al., 1999). However, rice straw has higher silica content than most other cereal straw. The annual global production of rice straw is about 731 million tons, with a global distribution as follows: Africa (20.9 million tons), Asia (667.6 million tons), Europe (3.9 million tons), North America (37.2 million tons), and Oceania (1.7 million tons).

TABLE 3.3

Typical Composition (% Dry Basis) of Corn Stover, Wheat Straw, and Rice Straw

Biomass	Cellulose	Hemicellulose	Lignin
Corn stover	35–40	17–35	7–18
Wheat straw	33–50	24–36	9–17
Rice straw	36–47	19–25	10–24

Source: Data from Garrote, G. et al., *Holz als Roh- und Werkstoff,* 57(3), 191, 1999.

3.4 Herbaceous Biomass

Herbaceous biomass for bioenergy applications typically refers to annual and perennial plant species that can be cultivated to produce feedstock for solid, liquid, and gaseous forms of energy. The richness in the diversity of herbaceous biomass makes them an attractive feedstock for energy applications regardless of geography. Of particular interest are perennial grasses that can produce high biomass yield for several years. While such grasses have been used for centuries for forage on farms, they are now increasingly viewed as viable long-term feedstocks for energy production. Some characteristics of these types of biomass that contribute to this viability include high yield potential, high carbon content in the biomass, and overall positive environmental impacts.

In 1978, the U.S. Department of Energy (DOE) established a program to develop bioenergy feedstocks and evaluated 35 herbaceous crops, including 18 perennial grasses between 1985 and 1989 on marginal crop lands. The study indicated that switchgrass was the perennial grass with the most potential to be an energy feedstock (Cherney et al., 1990). In fact, switchgrass was identified by the U.S. DOE as a model herbaceous energy crop in 1991 (McLaughlin, 1992). Similar efforts were made in Europe to evaluate the potential for different types of herbaceous biomass and research related to perennial grasses is primarily focused on giant reed, *Miscanthus*, reed canary grass, and switchgrass (Lewandowski et al., 2003).

While there are many concerted research efforts that emphasize specific types of perennial grasses such as switchgrass and *Miscanthus*, the importance of other region-specific perennial grasses should not be minimized. For example, coastal Bermuda grass is still among the best perennial pasture and hay grasses for much of the southeast United States and is grown by many hog farmers in the region as part of nutrient management plans to remove nitrogen and phosphorous from swine wastes (Sun and Cheng, 2005). Consequently, familiarity with the agricultural practices could ease the establishment of this perennial grass as an energy feedstock for the region. Sections 3.4.1 through 3.4.3 discuss the characteristics of switchgrass, *Miscanthus*, and coastal Bermuda grass as representative examples of perennial herbaceous energy crops.

3.4.1 Switchgrass

Switchgrass (*Panicum virgatum* L.) (Figure 3.7) is a warm-season perennial grass native to North America and Central America and is also known to occur in Africa and South America (Stubbendieck et al., 1997). It can grow to a height of more than 3 m and have a root depth of more than 3.5 m (Weaver, 1968). The environmental benefits associated with switchgrass include the potential for significant carbon sequestration, nutrient recovery from runoff,

FIGURE 3.7
Switchgrass. (Courtesy of Jiele Xu, North Carolina State University, Raleigh, NC.)

soil remediation, and provision of habitats for grassland birds (Keshwani and Cheng, 2009). Lewandowski et al. (2003) and Vogel et al. (2002) noted that conventional equipment for seeding, crop management, and harvesting can be used to integrate switchgrass into existing farming operations.

There are two general ecotypes of switchgrass: lowland ecotypes (such as *Kanlow* and *Alamo*) that are vigorous, tall, thick-stemmed, and adapted to wetter conditions, and upland ecotypes (such as *Cave-in-rock* and *Trailblazer*) that are shorter, thinner-stemmed, and adapted to drier conditions (Gunter et al., 1996). Switchgrass cultivars are adapted to specific ecological regions and plant hardiness zones in which they or their parental germplasms have evolved and are also photoperiod sensitive (Benedict, 1940). Switchgrass has considerable cold tolerance and stands have been maintained even in Canada and in Southern England (El Bassam, 1998). In general, switchgrass does well on a wide variety of soil types. It is drought tolerant and also tolerant to wet areas. It can grow on soils ranging from sand to clay loam and can tolerate soils with pH values ranging from 4.9 to 7.6 (Wolf and Fiske, 1995). However, Jung et al. (1988) state that establishment and growth is better if soils are limed to neutral pH.

Switchgrass is established by seed (planted 1–2 cm deep) and recommended seeding rates are 200–400 pure live seeds per square meter (Vogel, 1987). Temperature gradient studies with different switchgrass cultivars indicated that optimal germination temperature is between 27°C and 30°C (Dierberger, 1991). The seedbed has to be free of weeds since competition with weeds is a major reason for stand failure of switchgrass (Moser and Vogel, 2003). While phosphorous and/or potassium are often added to benefit seedling growth, nitrogen application is not common since it will stimulate excessive weed growth. However, post-establishment, economically viable yields will require annual nitrogen application of 50–100 kg ha^{-1} (Wright, 1994).

Fike et al. (2006) stated that harvest management strategies depend on specific cultivars and noted that cultivars such as *Cave-in-Rock* and *Shelter* had higher yields with two-cut harvesting in comparison to lowland cultivars such *Alamo* and *Kanlow*. However, other studies have indicated that multiple-cut harvesting of both upland and lowland cultivars of switchgrass provide lower long-term yields (Beaty and Powell, 1976; Koshi et al., 1982). Parrish and Fike (2005) concluded that while properly timed two-cut harvesting of some upland varieties may yield more biomass, a one-cut harvest is appropriate for most varieties and noted that if two harvests are taken, more nitrogen must be applied to compensate for removal during the mid-season harvest. Limited regrowth (McMurphy et al., 1975) and stand susceptibility to defoliation (Belesky and Fedders, 1995) of switchgrass also point to the appropriateness of single-cut harvesting.

A wide range of biomass (dry matter) yields have been reported in literature for switchgrass from field trials in different parts of the United States. Examples of published yields include 6 tons per acre in the midwestern United States (Hopkins et al., 1995), 10 tons per acre in Texas (Sanderson et al., 1996), and 15 tons per acre in Oklahoma (Thomason et al., 2004). Although these high yields are site-specific for test plots, Parrish and Fike (2005) stated that with well-adapted cultivars from breeding and biotechnology research, annual yields of over 6 tons per acre are feasible for lands that receive annual rainfall of at least 70 cm with nitrogen applications of 50 kg N ha^{-1} per year.

Typical cellulose, hemicellulose, and lignin content in different varieties of switchgrass are shown in Table 3.4. In general, cellulose ranges from 28% to 37%, hemicellulose ranges from 25% to 30%, and lignin ranges from 15% to 20%. For combustion applications, a low amount of ash is desirable. McLaughlin et al. (1996) reported that the ash content for switchgrass ranges from 2.8% to 7.6% and is generally lower than that of coal.

3.4.2 *Miscanthus*

Miscanthus (*Miscanthus* spp.) is a genus that collectively refers to about 14–17 species whose genetic origins are in southeast Asia. The genus is related to the sugarcane family and is found throughout a wide range of tropical and subtropical climates from southeast Asia to the Pacific Islands (Greef and Deuter, 1993). *Miscanthus* was originally introduced into Europe as an ornamental garden grass and is now the focus of several research efforts to establish it as a bioenergy crop.

TABLE 3.4

Composition (% Dry Basis) of Different Switchgrass Varieties from
NREL's Biomass Feedstock Composition and Properties Database

Switchgrass Variety	Cellulose	Hemicellulose	Lignin
Alamo—whole plant	33	26	17
Alamo—leaves	28	24	15
Alamo—stems	36	27	17
Blackwell—whole plant	34	26	18
Cave-in-rock—whole plant	33	26	18
Cave-in-rock—leaves	30	24	16
Cave-in-rock—stems	36	27	18
Kanlow—leaves	32	25	17
Kanlow—stems	37	26	18
Trailblazer	32	26	18

There are two general types of rhizomes: *Miscanthus sacchariflorus* has a broad creeping thick-stemmed rhizome and *Miscanthus sinensis* rhizome is tuft forming and has thinner stems (Lewandowski et al., 2003). *Miscanthus* × *giganteus*, a hybrid of these two rhizomes, is widely used in Europe for biomass productivity trials as part of the *Miscanthus* productivity network (Heaton et al., 2004). *Miscanthus* can reach a height of over 2 m during the establishment year and up to 4 m in subsequent years in Europe and as a high as 7–10 m in China, with root penetration of over 1 m into the soil (El Bassam, 1998).

Unlike switchgrass, establishment by seed may be problematic since some *Miscanthus* species do not produce viable seeds. El Bassam (1998) describes alternatives that include plant division, stem cuttings, rhizome cuttings, and micropropagation via tissue culture. In fact, the sterile hybrid *Miscanthus* × *giganteus* is typically propagated via rhizome cuttings or by tissue culture. A detailed description of these propagation methods is provided by Christian and Haase (2001). The optimum planting density is one to two plants per square meter, with a recommended planting depth of 5–7 cm for rhizome cuttings (El Bassam, 1998). Just like switchgrass, weed control in the establishment and initial years is important for *Miscanthus*.

Although nutrient requirements for *Miscanthus* are low, the addition of nitrogen, potassium, and phosphorous may be needed to preserve high biomass yields. Rutherford and Heath (1992) reported the amounts of different nutrients needed to maintain a proper soil balance for *Miscanthus*: 50 kg ha^{-1} of nitrogen, 45 kg ha^{-1} of potassium (K_2O), 21 kg ha^{-1} of phosphorous (P_2O_5), 25 kg ha^{-1} of sulfur, 13 kg ha^{-1} of magnesium, and 25 kg ha^{-1} of calcium. Lewandowski et al. (2003) noted that at locations where sufficient nitrogen mineralization from soil organic matter has occurred, fertilization can be avoided or limited to 50–70 kg ha^{-1} on an annual basis. Hotz et al. (1996) concluded that soil suitable for growing maize is also likely to be suitable for *Miscanthus* and noted the importance of soil characteristics such as high water-holding capacity, high organic matter content, and good air movement.

Miscanthus should be harvested only once a year since multiple cutting would overexploit the rhizomes and result in the death of the stands (Lewandowski et al., 2003). The time of harvest depends on local conditions. El Bassam and Huisman (2001) state that in northern Europe, harvesting should be carried out in early spring when moisture content of the biomass is low and in southern Europe, a late autumn harvest is preferable to avoid biomass losses due to adverse winter conditions. Clifton-Brown et al. (2001) summarized yield data from several studies and concluded that in northern Europe, annual yields (dry matter) ranged from 6 to 10 tons per acre and in southern Europe, where irrigation was required, annual yields ranged from 10 to 15 tons per acre.

Typical amounts of cellulose, hemicellulose, and lignin present in *Miscanthus* are shown in Table 3.5. In addition to the higher cellulose content, *Miscanthus* also differs from switchgrass in that the ash content is reported to be no more than 4% (Lewandowski et al., 2003), which would indicate that it may be more suitable for combustion applications.

3.4.3 Coastal Bermuda Grass

Bermuda grass (*Cynodon* spp.) is a warm-season perennial grass that was introduced into the United States in the early 1700s (Mueller et al., 1993). Although *Cynodon* species have a wide geographic distribution, they occur in greatest abundance in tropical and subtemperate environments. Currently, Bermuda grass is grown on 10–15 million acres in the southern United States for forage and hay production (Boateng et al., 2007). Some species grow to a height of only 15–20 cm while others grow as tall as 1 m or more (Taliaferro et al., 2004) (Figure 3.8).

Coastal Bermuda grass (Figure 3.8) is a hybrid between a common "cotton patch" variety found near Tifton, GA and a variety introduced from South Africa. It was released in 1943 by the USDA-Agricultural Research

TABLE 3.5

Composition (% Dry Basis) of *Miscanthus* from Published Studies

Reference	Cellulose	Hemicellulose	Lignin
de Vrije et al. (2002)	38	24	25
Papatheofanous et al. (1996)	43	27	22
Sorensen et al. (2008)	40	18	25

FIGURE 3.8
Coastal Bermuda grass hay. (Courtesy of Jay Cheng, North Carolina State University, Raleigh, NC.)

Service and the Georgia Agricultural Extension Station and was specifically designed for high productivity across the southern United States with up to twice the yield of naturalized common strains (Burton, 1948). Bermuda grass cultivars such as coastal are relatively easy to establish, produce high biomass yields per unit area, and have high tolerance to biotic and abiotic stresses (Taliaferro et al., 2004). Additionally, it has been noted that nutrient removal by hybrid Bermuda grasses exceeds that for most other grasses (Burns et al., 1990). Hence, coastal Bermuda grass is widely grown by hog farmers in the southeast United States as part of nutrient management plans to remove nitrogen and phosphorous from swine wastes (Sun and Cheng, 2005).

Coastal Bermuda grass will grow well in a variety of soil types including sands, loams, silts, and clays as long as internal soil drainage is good (Mueller et al., 1993). Since coastal Bermuda grass is a hybrid variety and does not produce fertile seeds, it is commonly established from sprigs, which include tillers, rhizomes, stolons, and root portions, with a recommended planting rate of 20–60 bushels of sprigs per acre (planted 3–5 cm deep) in rows spaced 0.9 m apart (Taliaferro et al., 2004). Each bushel typically contains approximately 1000 sprigs (Mueller et al., 1993). As is the case with switchgrass and *Miscanthus*, weed control during establishment is critical.

While fertilization with nitrogen, phosphorous, and potassium are required, nitrogen has the greatest influence on biomass yield and even influences the amount of other nutrients to sustain biomass productivity at specific nitrogen levels (Taliaferro et al., 2004). For Bermuda grass grown in the upper Coastal Plains of North Carolina, dry matter yields were 0.5 tons per acre when no nitrogen was applied (Mueller et al., 1993). The report noted that yields increased to 2.6, 5.0, and 6.7 tons per acre as the amount of nitrogen applied was increased to 100, 200, and 400 lb per acre. Based on data from Prine and Burton (1956), a clipping interval of six to eight weeks maximized annual dry matter yields for coastal Bermuda grass grown in Tifton, GA.

In addition to the above-mentioned yields, 6–10 tons per acre has also been cited as expected annual dry matter yields for coastal Bermuda grass in a positive soil environment (Holtzapple et al., 1994; Newman, 2008). The cellulose and hemicellulose content in coastal Bermuda grass is reported to be 26% cellulose and 31% hemicellulose, respectively (Holtzapple et al., 1994). Lignin content in Bermuda grass is reported to be approximately 20% and ash content is reported to range from 4% to 7% (Sun and Cheng, 2005; Wang, 2008).

3.5 Woody Biomass

Woody biomass (Examples in Figures 3.9 and 3.10) is divided into two categories: softwoods and hardwoods. Softwoods are gymnosperms with needlelike leaves, and are commonly referred to as evergreens; pine, and spruce are examples of softwoods. Softwoods are the dominant source of lignocellulosic materials in the northern hemisphere. In contrast to softwoods, hardwoods are angiosperms with broad leaves and are also referred to as flowering plants. Typical hardwood examples are poplar, willow, and oak (Bowyer et al., 2007; Brown, 2003). While 92% of the world's softwood forests are in North America, Russia, and Europe, hardwoods are more widely distributed and only a third of the world's hardwood forests are in the aforementioned regions (Bowyer et al., 2007).

While most species of pine exhibit good growth in acidic soils, some are also known to tolerate calcareous soils that are relatively alkaline in nature. While soil with good drainage properties (like sandy soil) is preferred, there are exceptions such as Lodgepole pine, which can grow on poorly drained wet soil. There are several species of pine that can tolerate extreme climates. For example, Siberian Dwarf pine is well-adapted to frigid environments and Turkish pine is well adapted to arid environments. On the other hand, some species are susceptible to adverse climates. For example, Loblolly pine, the most economically important tree species in the southeast United States does not tolerate extremely cold conditions (Aubrey et al., 2007).

Among the different types of woody biomass, poplar (*Populus* spp.) is of particular interest since it is being used as a short rotation coppice crop. The capacity for fast growth has resulted in the cultivation of poplar for direct combustion, gasification, and production of briquettes. For example, in the

FIGURE 3.9
Pine trees. (Courtesy of Jay Cheng, North Carolina State University, Raleigh, NC.)

FIGURE 3.10
Wood chips. (Courtesy of Jay Cheng, North Carolina State University, Raleigh, NC.)

United Kingdom, poplar is typically grown for 2–5 years and then harvested (annual yield of 5–6 dry tons per acre) for direct combustion (Aylott et al., 2008; El Bassam, 1998). Hybrid poplar yields in North America are reported to be around 5 dry tons per acre (Brown, 2003). Most species of poplar exist in regions with temperate climates and typically grow to heights ranging from 15 to 35 m. Examples of specific species of poplar prevalent in different parts of the world are *Populus alba* (southern and central Europe), *Populus tremula* (Europe and Asia), and *Populus tremuloides* (North America) (El Bassam, 1998).

In general, poplar grows well on soil types ranging from sandy to loamy-clay as long as soil pH is near neutral. While irrigation is not required, it is reported to be beneficial for poplar grown in regions that receive less than 60 cm of annual precipitation (Johansson et al., 1992). Poplar species intended for use as short rotation crops are typically propagated via plantlets or cuttings that are planted vertically into the soil in double rows with 125 cm spacing between each set of double rows and 75 cm spacing within each double row (El Bassam, 1998). Weed control during the establishment and second years are important as weeds can reduce dry matter yield by as much as 20% (Parfitt et al., 1992).

On an elemental basis, wood primarily consists of carbon, hydrogen, and oxygen. Carbon (49%) is the major element on a dry weight basis, followed by oxygen (44%) and hydrogen (6%). While ash typically accounts for less than 1% of the total dry matter, it can be as high as 3.0%–3.5% in some tropical woody species (Fengel and Wegner, 1984). In general, hardwoods have lower ash content than softwoods, but contain more acetyl groups (Brown, 2003). As indicated in Table 3.6, both softwoods and hardwoods contain approximately 40%–50% cellulose (on a dry basis). Hemicellulose content in softwoods is around 11%–20% with 5%–10% glucomannan and 15%–20% in hardwoods with no glucomannan. The lignin content in softwoods is higher than that in hardwoods. It should be noted that these polymeric components are not uniformly organized and distributed within the wood cell wall, and their concentrations vary from one morphological region to another.

TABLE 3.6

Typical Composition (% Dry Basis) of Hardwoods and Softwoods

Biomass	Cellulose	Hemicellulose	Lignin
Hardwoods	40–50	15–20	20–25
Softwoods	40–50	11–20	27–30

Source: Data from Saka, S., Structure and chemical composition of wood as a natural composite material, in Shiraishi, N. et al. (eds.), *Recent Research on Wood and Wood-based Materials*, Elsevier Applied Science, New York, 1993.

3.6 Oilseeds

3.6.1 Soybean

Soybean, a species of legume native to East Asia, is an annual plant that has been used as a food source for centuries and has even seen use for medicinal purposes in China. The dominance of soybean as a global oilseed resource can be attributed to its advantageous agronomic characteristics, high oil content for food and fuel purposes, and highly nutritive protein content. The top five producers of soybean are the United States, Brazil, Argentina, China, and India with respective annual production rates (in 2007) of 70.7, 58.2, 45.5, 15.6, and 9.4 million tons in 2007 (FAO, 2008). Soybean production can be maximized when grown in moist fertile loams at an optimum temperature of around 25°C (Martin et al., 2006). A day length ranging from 12 to 14 h is considered to be optimum for blooming and a pH range of 6–6.5 is desirable for most soybean cultivars (El Bassam, 1998).

The soybean seed is made up of three structural components: cotyledon (90% dry basis), hull (7%–8%), and hypocotyls (2.3%). All three structural components contain proteins, oils, carbohydrates, and ash. As a percent of the total weight of the seed on a dry basis, proteins account for 40%, carbohydrates account for 34%, oils account for 21%, and ash accounts for 5% (Table 3.7). The carbohydrates present include both soluble carbohydrates such as sucrose, stachyose, and raffinose and insoluble carbohydrates such as cellulose, hemicellulose, and pectin. Other minor components of the seed include phospholipids, sterols, and minerals. The exact composition of the seed is influenced by both genetic and environmental factors (Salunkhe et al., 1992). The amount of a particular component can be influenced by other components. For example, Krober and Cartter (1962) noted a negative correlation between the protein content and the nonprotein components in soybean seeds.

Soybean oil primarily consists of neutral lipids, of which triacylglycerols (are the primary component accounting for 95%–97% of the oil.

TABLE 3.7

Composition (% Dry Basis) of Different Oil Seeds

Oilseed	Oil	Protein	Carbohydrates	Ash
Soybean	21	40	34	5
Rapeseed (Canola)	38–50	36	14	4–6
Sunflower	44–51	17–19	15–20	>1

Sources: Data from Wang, T., Soybean oil, in Gunstone, F.D. (ed.), *Vegetable Oils in Food Technology: Composition, Properties, and Uses,* Blackwell Publishing, Boca Raton, FL, 2002; Salunkhe, D.K. et al., *World Oilseeds: Chemistry, Technology, and Utilization.* Van Nostrand Reinhold, New York, 1992.

Phospholipids (1%–3%) and free fatty acids (<1%) are minor lipid components. Some unsaponifiable matter such as phytosterols and tocopherols are also present in small amounts. In addition, there are trace amount of metals such as iron (1–3 ppm) and copper (0.03–0.05 ppm) (Pryde, 1980). The fatty acid profile of soybean oil is dominated by unsaturated fatty acids that include linoleic acid, oleic acid, and linolenic acid (Table 3.8).

3.6.2 Rapeseed (Canola)

Rapeseed, a bright yellow flowering member of the family Brassicaceae, is grown primarily for the production of animal feed and vegetable oil. Rapeseed is prevalent in Europe given its preference for a cool and moist climate. Rapeseed includes a variety of plant species. Canola (Figure 3.11) refers to a specific type of rapeseed species characterized by low erucic acid and low glucosinolate contents. Rapeseed has become the third leading source of vegetable oil in the world during the last two decades. The worldwide production of rapeseed (including canola) reached to 49.5 million tons in 2007. China is the largest producer of rapeseed with a production of 10.4 million tons, followed by Canada as the second largest producer (8.9 million tons), and India (7.1 million tons). Another two major producers are Germany and France (FAO, 2008).

As with soybean, the chemical composition of the seeds in rapeseed species is also influenced by genetic and environmental factors. In general, rapeseeds contain twice as much oil (38%–50%, dry basis) and slightly less protein (36%). Albumins and globulins are the two major proteins in rapeseeds with globulins accounting for about 70% of the total protein content. However, the protein content varies considerably between species and cultivars. The soluble carbohydrates in rapeseed include sucrose (7.4%), stachyose (2.5%), and raffinose (0.33%), and the insoluble carbohydrates include cellulose (4%–5%), pectins (4%–5%), hemicellulose (3%), and starch (1%) (Salunkhe et al., 1992).

TABLE 3.8

Fatty Acid Profiles in Different Oils

Source	Linoleic Acid	Oleic Acid	Linolenic Acid	Palmitic Acid
Soybean	53.2	23.4	7.8	11
Rapeseed	21	61	11	7
Oil palm	10–11	39–40	Very low	44–45
Waste edible oil	53.29	27.75	N/A	11.73

Sources: Data from Orthoefer, F.T., Vegetable oils, in Hui, Y.H. (ed.), *Bailey's Industrial Oil and Fat Products*, 5th edn., Wiley, New York, 1996; Lin, S.W., Palm oil, in Gunstone, F.D. (ed.), *Vegetable Oils in Food Technology: Composition, Properties, and Uses*, Blackwell Publishing, Boca Raton, FL, 2002; Shah, V. et al., *Biotechnol. Prog.*, 23(2), 512, 2007.

FIGURE 3.11
Canola. (Courtesy of Chengci Chen, Montana State University, Bozeman, MT.)

The constituents of rapeseed oil include triacylglycerols (91.8%–99.0%), phospholipids (up to 4.3%), free fatty acids (0.5%–1.8%), unsaponifiables (0.5%–1.2%), tocopherols (700–1000 ppm), chlorophylls (5–55 ppm), and sulfur (5–35 ppm) (Mag, 1990). In general, most natural rapeseed varieties contain about 50% erucic acid as a percentage of the total oil. Because of the toxicity of erucic acid to humans, rapeseed varieties with low erucic acid levels have been developed through genetic methods. Canola is a prominent example of such a variety. These modified types of rapeseeds are also characterized by higher levels of oleic acid, linoleic acid, and linolenic acid than those of the natural cultivars. Canola oil contains 61% oleic acid, 21% linoleic acid, 11% α-linolenic acid, and 7% saturated fatty acids (CanolaInfo, 2005). Tocopherols and sterols are the two main unsaponifiable components present in small amounts in Canola oil. Pigments such as carotenoids (around 2–3 ppm), primarily β-carotene are also found in canola oil (Salunkhe, 1992).

3.6.3 Sunflower

Sunflower (*Helianthus*) is an annual plant native to the Americas. Following soybean, palm, and canola, sunflower is the fourth largest oil source globally. The annual production of sunflower in 2007 is around 27 million tons (FAO, 2008). Apart from being prevalent in Russia and most of Europe, other top sunflower-producing countries include Argentina, China, India, South Africa, Turkey, and the United States (Gupta, 2002). However, sunflower is considered to be geographically limited due to specific climatic and soil requirements such as long day length providing sufficient sunlight and soil that is fertile, moist, and well drained with good structure (Gupta, 2002).

Sunflower seeds contain 44%–51% oil (dry basis), 17%–19% protein, 15%–20% fiber residues, and less than 1% ash. A number of micronutrients such as tocopherols (~650 ppm), sterols (~0.3%), phospholipids (~0.8%), carotenoids (~1.1 ppm), sterol-esters, waxes, chlorophyll, and trace metals exist in sunflower seeds as well. Major fatty acids present in sunflower oil are palmitic, stearic, oleic, and linoleic acids, which account for 60%–70% of the total fatty acids. Minor fatty acids include palmitoleic, linolenic, arachidic, behenic, and lignoceric acids.

3.6.4 Oil Palm

Oil palm, a tropical plant native to South Africa, is now grown in Africa, Asia, and Central and South America as an important source of edible oil. Two types of oil can be extracted from the plant: palm oil from the mesocarp and palm kernel oil from the kernel inside the nut. Palm oil is semisolid at room temperature and is composed primarily of neutral lipids with a small amount of phospholipids and glycolipids. Carotenoids and tocopherols are the major unsaponifiable materials present in palm oil. Some sterols, waxes, and hydrocarbons are also present (Salunkhe, 1992). Unlike the fatty acid profiles of other vegetable oils discussed thus far, palm oil contains equal amounts of saturated and unsaturated fatty acids. The major fatty acids are oleic acid (39%–40%, dry basis), palmitic acid (44%–45%), linoleic acid (10%–11%), and linolenic acid, which is present in small amounts (Lin, 2002). Malaysia is the leading producer of palm oil in the world, followed by Indonesia and Nigeria. Other palm oil producers include Colombia, Thailand, Papua New Guinea, Ivory Coast, and Ecuador (Gunstone, 2002).

3.6.5 Waste Edible Oil

Waste edible oil, just as the name implies, is an oil-rich substance leftover from oil that has been used in cooking foods and is unsuitable for its original purpose. This waste oil can be obtained from fast food restaurants, industrial deep fryers in food-processing plants, snack food factories, and even household kitchens. Waste edible oil contains substantially more free fatty acids and water and less triacylglycerols than fresh edible oils obtained from oilseeds. A typical fatty acid profile for waste edible oil from restaurants includes linoleic acid (53%), oleic acid (28%), and palmitic acid (11.73%) (Shah et al., 2007). Currently, more than 15 million tons of waste edible oil is generated annually in the world with the United States accounting for 10 million tons. Europe, China, Canada, Japan, and Malaysia also generate a large amount of waste edible oil every year (Gui et al., 2008).

3.7 Problems

1. Assuming an ethanol yield of 0.51 g per g of carbohydrate (all types), compare the ethanol production potential (per dry ton) from corn, potato, sweet potato, sweet sorghum, and wheat. When appropriate, include all major carbohydrates from the entire crop. (Note: clearly state all assumptions and cite sources related to biomass composition.)

2. What are the major differences between hardwoods and softwoods? Discuss the relevance (if any) of these differences for bioenergy production.

3. Assuming an ethanol yield of 0.51 g per g of carbohydrate for both cellulose and hemicellulose, calculate the land requirements to produce 500,000 gal of ethanol from the following biomass: switchgrass, *Miscanthus*, coastal Bermuda grass, and poplar. (Note: clearly state all assumptions and cite sources for data related to yields and composition.)

4. Explain why overapplication of nitrogen during the establishment year could result in poor stand quality or even stand failure for switchgrass.

5. What are some advantages and disadvantages associated with the current reliance on corn as the major feedstock for ethanol production in the United States? Comment on how other carbohydrate-rich types of biomass might alleviate some concerns associated with corn.

6. Differentiate between canola and natural rapeseed varieties.

7. Based on information from Chapters 2 and 3, comment on why palm oil could be a more suitable raw material for biodiesel production in South Asia rather than in northern Europe.

References

Allen, R. D. 1979. Ingredient analysis table. *Feedstuffs* 51: 31.

Aubrey, D. P., M. D. Coleman, and D. R. Coyle. 2007. Ice damage in Loblolly pine: understanding the factors that influence susceptibility. *Forest Science* 53(5): 580.

Aylott, M. J., E. Casella, I. Tubby, N. R. Street, P. Smith, and G. Taylor. 2008. Yield and spatial supply of bioenergy poplar and willow short-rotation coppice in the U.K. *New Phytologist* 178(2): 358–370.

Beaty, E. R. and J. D. Powell. 1976. Response of switchgrass (*Panicum Virgatum* L.) to clipping frequency. *Journal of Range Management* 29(2): 132–135.

Belesky, D. P. and J. M. Fedders. 1995. Warm-season grass productivity and growth-rate as influenced by canopy management. *Agronomy Journal* 87(1): 42–48.

Benedict, H. M. 1940. Effect of day length and temperature on the flowering and growth of four species of grasses. *Journal of Agricultural Research* 61: 661–671.

Beukema, H. P. and D. E. van der Zaag. 1990. *Introduction to Potato Production*. 2nd edn. Wageningen: Pudoc.

Boateng, A. A., W. F. Anderson, and J. G. Phillips. 2007. Bermuda grass for biofuels: Effect of two genotypes on pyrolysis product yield. *Energy and Fuels* 21(2): 1183–1187.

Bowyer, J. L., R. Shmulsky, and J. G. Haygreen. 2007. *Forest Products and Wood Science: An Introduction*. 5th edn. Ames, IA: Blackwell Publishing.

Bressani, R. and E. T. Mertz. 1958. Studies on corn proteins. 4. Protein and amino acid content of different corn varieties. *Cereal Chemistry* 35(3): 227–235.

Brown, R. C. 2003. *Biorenewable Resources: Engineering New Products from Agriculture*. 1st edn. Ames, IA: Iowa State Press.

Burns, J. C., L. D. King, and P. W. Westerman. 1990. Long-term swine lagoon effluent applications on coastal Bermuda grass. 1. Yield, quality, and element removal. *Journal of Environmental Quality* 19(4): 749–756.

Burton, G. W. 1948. Coastal Bermuda grass. Cir. 10. Georgia Agricultural Experimental Station, Athens.

CanolaInfo. 2005. *Comparison of Dietary Facts*. Manitoba, Canada: CanolaInfo.org. Available from http://www.canolainfo.org/media/pdfs/dietary-fat-chart.pdf

Cheesman, O. 2004. *Environmental Impacts of Sugar Production: the Cultivation and Processing of Sugarcane and Sugar Beet*. Cambridge, MA: CAB International.

Cherney, J. H., K. D. Johnson, J. J. Volenec, E. J. Kladivko, and D. K. Greene. 1990. *Evaluation of Potential Herbaceous Biomass Crops on Marginal Croplands: (1) Agronomic Potential*. Oak Ridge, TN: Oak Ridge National Laboratory.

Christian, D. G. and E. Haase. 2001. Agronomy of *Miscanthus*. In Jones, M. B. and M. Walsh (eds.), *Miscanthus for Energy and Fibre*. London: James & James Science Publishers Ltd.

Claassen, P. A. M., T. de. Vrije, and M. A. W. Budde. 2004. Biological hydrogen production from sweet sorghum by thermophilic bacteria. In *The 2nd World Conference on Biomass for Energy, Industry and Climate Protection*, Rome, Italy, 1522–1525.

Clarke, M. A. and L. A. Edye. 1996. Sugar beet and sugarcane as renewable resources. In Fuller, G. T. A., D. D. McKeon, and D. D. Bills (eds.), *Agricultural Materials as Renewable Resources: Nonfood and Industrial Applications*. Washington, DC: American Chemical Society.

Clifton-Brown, J. C., S. P. Long, and U. Jorgensen. 2001. *Miscanthus* Productivity. In Jones, M. B. and M. Walsh (eds.), *Miscanthus for Energy and Fibre*. London: James & James Science Publishers Ltd.

de Vrije, T., G. G. de Haas, G. B. Tan, E. R. P. Keijsers, and P. A. M. Claassen. 2002. Pretreatment of Miscanthus for hydrogen production by *Thermotoga elfii*. *International Journal of Hydrogen Energy* 27(11–12): 1381–1390.

Dierberger, B. S. 1991. *Switchgrass germination as influenced by temperature chilling, cultivar, and seed lot*. M.S. Thesis. University of Nebraska, Lincoln, NE.

Draycott, A. P. 2006. *Sugar Beet*. Ames, IA: Blackwell Publishing.

Dudley, J. W. 1974. *Seventy Generations of Selection for Oil and Protein in Maize*. Madison, WI: Crop Science Society of America.

Earle, F. R., J. J. Curtis, and J. E. Hubbard. 1946. Composition of the component parts of the corn kernel. *Cereal Chemistry* 23(5): 504–511.

El Bassam, N. 1998. *Energy Plant Species: Their Use and Impact on Environment and Development*. London: James & James Science Publishers Ltd.

El Bassam, N. and W. Huisman. 2001. Harvesting and storage of *Miscanthus*. In Jones, M. B. and M. Walsh (eds.), *Miscanthus for Energy and Fibre*. London: James & James Science Publishers Ltd.

FAO (Food and Agricultural Organization). 2008. Crops production. FAO Statistical Databases. Available from http://faostat.fao.org/

Farnham, D. E., G. O. Benson, and R. B. Pearce. 2003. Corn perspective and culture. In White, P. J. and L. A. Johnson (eds.), *Corn: Chemistry and Technology*, 2nd edn. St. Paul, MN: American Association of Cereal Chemists.

Fengel, D. and G. Wegener. 1984. *Wood: Chemistry, Ultrastructure, Reactions*. Berlin, NY: de Gruyter.

Fike, J. H., D. J. Parrish, D. D. Wolf, J. A. Balasko, J. T. Green, M. Rasnake, and J. H. Reynolds. 2006. Switchgrass production for the upper southeastern USA: Influence of cultivar and cutting frequency on biomass yields. *Biomass and Bioenergy* 30(3): 207–213.

Garrote, G., H. Dominguez, and J. C. Parajo. 1999. Hydrothermal processing of ligno-cellulosic materials. *Holz als Roh- und Werkstoff* 57(3): 191–202.

Garside, A. L., M. A. Smith, L. S. Chapman, A. P. Hurney, and R. C. Magarey. 1997. The yield plateau in the Australian sugar industry 1970–1990. In Keating, B. A. and J. R. Wilson (eds.), *Intensive Sugarcane Production: Meeting the Challenge Beyond 2000: Proceedings of the Sugar 2000 Symposium, Brisbane, Australia, August 20–23, 1996*, 103–124. Oxon, U.K.; New York: CAB International.

Greef, J. M. and M. Deuter. 1993. Syntaxonomy of *Miscanthus x Giganteus*. *Angewandte Botanik* 67(3–4): 87–90.

Gui, M. M., K. T. Lee, and S. Bhatia. 2008. Feasibility of edible oil vs. non-edible oil vs. waste edible oil as biodiesel feedstock. *Energy* 33(11): 1646.

Gunstone, F. D. 2002. Production and trade of vegetable oils. In Gunstone, F. D. (ed.), *Vegetable Oils in Food Technology: Composition, Properties, and Uses*. Boca Raton, FL: Blackwell Publishing.

Gunter, L. E., G. A. Tuskan, and S. D. Wullschleger. 1996. Diversity among populations of switchgrass based on RAPD markers. *Crop Science* 36(4): 1017–1022.

Gupta, M. K. 2002. Sunflower oil. In Gunstone, F. D. (ed.), *Vegetable Oils in Food Technology: Composition, Properties, and Uses*. Boca Raton, FL: Blackwell Publishing.

Heaton, E. A., R. B. Flavell, P. N. Mascia, S. R. Thomas, F. G. Dohleman, and S. P. Long. 2008. Herbaceous energy crop development: Recent progress and future prospects. *Current Opinion in Biotechnology* 19(3): 202–209.

Heaton, E., T. Voigt, and S. P. Long. 2004. A quantitative review comparing the yields of two candidate C-4 perennial biomass crops in relation to nitrogen, temperature and water. *Biomass and Bioenergy* 27(1): 21–30.

Holtzapple, M. T., E. P. Ripley, and M. Nikolaou. 1994. Saccharification, fermentation, and protein recovery from low-temperature afex-treated Coastal Bermuda grass. *Biotechnology and Bioengineering* 44(9): 1122–1131.

Hopkins, C. G. 1974. Improvement in the chemical composition of the corn kernel. In Dudley, J. W. (ed.), *Seventy Generations of Selection for Oil and Protein in Maize*. Madison, WI: Crop Science Society of America.

Hopkins, A. A., K. P. Vogel, K. J. Moore, K. D. Johnson, and I. T. Carlson. 1995. Genotype effects and genotype by environment interactions for traits of elite switchgrass populations. *Crop Science* 35(1): 125–132.

Hotz, A., W. Kuhn, and S. Jodl. 1996. Screening of different *Miscanthus* cultivars in respect of their productivity and usability as a raw material for energy and industry. In Chartier, P. et al. (eds.), *Biomass for Energy and the Environment: Proceedings of the 9th European Bioenergy Conference, Copenhagen, Denmark, June 24–27, 1996*. Oxford, U.K.; New York: Pergamon.

Johansson, H., S. Ledin, and L. S. Forsse. 1992. Practical energy forestry in Sweden: A commercial alternative for farmers. In Hall, D. O. et al. (eds.), *Biomass for Energy and Industry*, Bochum, Germany: Ponte Press.

Juliano, B. O. 1985. Rice hall and rice straw. In Juliano, B. O. (ed.) *Rice: Chemistry and Technology*, 2nd edn. St. Paul, MN: American Association of Cereal Chemists.

Jung, G. A., J. A. Shaffer, and W. L. Stout. 1988. Switchgrass and big bluestem responses to amendments on strongly acid soil. *Agronomy Journal* 80(4): 669–676.

Kadam, K. L. and J. D. McMillan. 2003. Availability of corn stover as a sustainable feedstock for bioethanol production. *Bioresource Technology* 88(1): 17–25.

Keshwani, D. R. and J. J. Cheng. 2009. Switchgrass for bioethanol and other value-added applications: A review. *Bioresource Technology* 100(4): 1515–1523.

Kim, S. and B. E. Dale. 2004. Global potential bioethanol production from wasted crops and crop residues. *Biomass and Bioenergy* 26(4): 361.

Koshi, P. T., J. Stubbendieck, H. V. Eck, and W. G. McCully. 1982. Switchgrasses— Forage yield, forage quality and water-use efficiency. *Journal of Range Management* 35(5): 623–627.

Krober, O. A. and J. L. Cartter. 1962. Quantitative interrelations of protein and non-protein constituents of soybeans. *Crop Science* 2:171–172.

Lewandowski, I., J. M. O. Scurlock, E. Lindvall, and M. Christou. 2003. The development and current status of perennial rhizomatous grasses as energy crops in the U.S. and Europe. *Biomass and Bioenergy* 25(4): 335–361.

Lin, S. W. 2002. Palm oil. In Gunstone, F. D. (ed.), *Vegetable Oils in Food Technology: Composition, Properties, and Uses*. Boca Raton, FL: Blackwell Publishing.

Mag, T. 1990. Further processing of canola and rapeseed oils. In Shahidi, F. (ed.), *Canola and Rapeseed: Production, Chemistry, Nutrition, and Processing Technology*. New York: Van Nostrand Reinhold.

Martin, J. H., R. P. Waldren, and D. L. Stamp. 2006. *Principles of Field Crop Production*, 4th edn. Upper Saddle River, NJ: Pearson Prentice Hall.

McLaughlin, S. B., 1992. New switchgrass biofuels research program for the Southeast. In *Proceedings of the Annual Automobile Technology Development Contractors' Coordination Meeting*, Dearborn, MI, November 2–5, 111–115.

McLaughlin, S. B., R. Samson, D. Bransby, and A. Wiselogel. 1996. Evaluating physical, chemical and energetic properties of perennial grasses as biofuels. In *Bioenergy '96: partnerships to develop and apply biomass technologies: Proceedings of the 7th National Bioenergy Conference, September 15–20, 1996*. Nashville, TN: Southeastern Regional Biomass Energy Program.

McMurphy, W. E., C. E. Denman, and B. B. Tucker. 1975. Fertilization of native grass and weeping lovegrass. *Agronomy Journal* 67(2): 233–236.

Milbrandt, A. 2005. A geographic perspective on the current biomass resource availability in the United States. NREL/TP-560–39181. Golden, CO: National Renewable Energy Laboratory.

Moser, L. E. and K. P. Vogel. 2003. Switchgrass, big bluestem and indianagrass. In Barnes, R. F., D. A. Miller, and C. J. Nelson (eds.), *Forages: An Introduction to Grassland Agriculture*, 6th edn. Ames, IA: Iowa State Press.

Mueller, J. P., J. T. Green Jr., D. S. Chamblee, J. C. Burns, J. E. Bailey, and R. L. Brandenburg. 1993. *Bermuda Grass Management in North Carolina*. Raleigh, NC: N.C. Cooperative Extension Service.

Newman, Y. C. 2008. *Bermuda Grass: A Quick Reference*. Document SS AGR 264. Gainsville, FL: Florida Cooperative Extension Service.

NREL Biomass feedstock composition and properties database. Available from http://www.nrel.gov/biomass/energy_analysis.html.

Orthoefer, F. T. 1996. Vegetable oils. In Hui, Y. H. (ed.), *Bailey's Industrial Oil and Fat Products*, 5th edn. New York: Wiley.

Papatheofanous, M. G., E. G. Koukios, G. Marton, and J. Dencs. 1996. Characterization of *Miscanthus sinensis* potential as an industrial and energy feedstock. In Chartier, P. et al. (eds.), *Biomass for energy and the environment: Proceedings of the 9th European Bioenergy Conference, Copenhagen, Denmark, June 24–27, 1996*. Oxford, U.K.; New York: Pergamon.

Parfitt, R. I., G. M. Arnold, and A. Foulkes. 1992. Weed control in new plantations of short-rotation willow and poplar coppice. *Aspects of Applied Biology* 29: 419–424.

Parrish, D. J. and J. H. Fike. 2005. The biology and agronomy of switchgrass for biofuels. *Critical Reviews in Plant Sciences* 24(5–6): 423–459.

Phowchinda, O., M. L. Delia-Dupuy, and P. Strehaiano. 1997. Alcoholic fermentation from sweet sorghum: Some operating problems. In *The 9th Annual Meeting of the Thai Society for Biotechnology and the 2nd JSPS-NRCT-DOST-LIPP-VCC Seminar on Biotechnology: An Essential Tool for Future Development*. Thailand.

Prine, G. M. and G. W. Burton. 1956. The effect of nitrogen rate and clipping frequency upon the yield, protein content and certain morphological characteristics of coastal Bermuda grass (*Cynodon dactylon*, (L) Pers.). *Agronomy Journal* 48: 296–301.

Pryde, E. H. 1980. Composition of soybean oil. In Erickson, D. R. et al. (eds.), *Handbook of Soy Oil Processing and Utilization*. Champaign, IL: AOCS Press.

Quckenbush, F. W., J. C. Firch, W. J. Raburn, M. McQuistan, E. N. Petzold, and T. E. Karg. 1961. Composition of corn-analysis of carotenoids in corn grain. *Journal of Agricultural and Food Chemistry* 9(2): 132–135.

Reitz, L. P. 1976. *Wheat in the United States*. Agricultural Information Bulletin 386. Washington, DC: USDA Agricultural Research Service.

Rivacoba, R. S., and R. B. Morin. 2002. Sugarcane and sustainability in Cuba. In Funes, F. et al. (eds.), *Sustainable Agriculture and Resistance: Transforming Food Production in Cuba*. Oakland, CA: Food First Books.

Rutherford, I. and M. C. Heath. 1992. *The Potential of Miscanthus as a Fuel Crop*. Energy Technology Support Unit (ETSU) report B1354, Harwell, U.K.

Saka, S. 1993. Structure and chemical composition of wood as a natural composite material. In Shiraishi, N., H. Kajita, and M. Norimoto (eds.), *Recent Research on Wood and Wood-based Materials*. New York: Elsevier Applied Science.

Salunkhe, D. K., J. K. Chavan, R. N. Adsule, and S. S. Kadam. 1992. *World Oilseeds: Chemistry, Technology, and Utilization*. New York: Van Nostrand Reinhold.

Sanderson, M. A., R. L. Reed, S. B. McLaughlin, S. D. Wullschleger, B. V. Conger, D. J. Parrish, D. D. Wolf, C. Taliaferro, A. A. Hopkins, W. R. Ocumpaugh, M. A. Hussey, J. C. Read, and C. R. Tischler. 1996. Switchgrass as a sustainable bioenergy crop. *Bioresource Technology* 56(1): 83–93.

Schmidt, A. S. and A. B. Bjerre. 1997. Pretreatment of agricultural crop residues for conversion to high-value products. In Campbell, G. M., C. Webb, and S. L. McKee (eds.), *Cereals: Novel Uses and Processes.* New York: Plenum Press.

Shah, V., M. Jurjevic, and D. Badia. 2007. Utilization of restaurant waste oil as a precursor for sophorolipid production. *Biotechnology Progress* 23(2): 512–515.

Sopher, C. D., R. J. McCracken, and D. D. Mason. 1973. Relationships between drought and corn yields on selected south Atlantic coastal plain soils. *Agronomy Journal* 65(3): 351–354.

Sorensen, A., P. J. Teller, T. Hilstrom, and B. K. Ahring. 2008. Hydrolysis of *Miscanthus* for bioethanol production using dilute acid presoaking combined with wet explosion pre-treatment and enzymatic treatment. *Bioresource Technology* 99(14): 6602–6607.

Stubbendieck, J. L., S. L. Hatch, and C. H. Butterfield. 1997. *North American Range Plants.* 5th edn. Lincoln, NE: University of Nebraska Press.

Sun, Y. and J. J. Cheng. 2005. Dilute acid pretreatment of rye straw and Bermuda grass for ethanol production. *Bioresource Technology* 96(14): 1599–1606.

Taliaferro, C. M., F. M. Bouquette Jr., and P. Mislevy. 2004. Bermuda grass and stargrass. In Moser, L. E., B. L. Burson, and L. E. Sollenberger (eds.). *Warm-Season (C₄) Grasses.* Madison, WI: American Society of Agronomy: Crop Science Society of America: Soil Science Society of America.

Thomason, W. E., W. R. Raun, G. V. Johnson, C. M. Taliaferro, K. W. Freeman, K. J. Wynn, and R. W. Mullen. 2004. Switchgrass response to harvest frequency and time and rate of applied nitrogen. *Journal of Plant Nutrition* 27(7): 1199–1226.

USDA. 2008. Nutrients in wheat and potato. USDA National Nutrient Database. Available from http://www.nal.usda.gov/fnic/foodcomp/search/.

Verma, R. S. 2004. *Sugarcane Production Technology in India.* Lucknow: International Book Distributing Co.

Vogel, K. P. 1987. Seeding rates for establishing big bluestem and switchgrass with preemergence atrazine applications. *Agronomy Journal* 79(3): 509–512.

Vogel, K. P., J. J. Brejda, D. T. Walters, and D. R. Buxton. 2002. Switchgrass biomass production in the Midwest USA: Harvest and nitrogen management. *Agronomy Journal* 94(3): 413–420.

Wagger, M. G. 1985. Corn growth and development. In Anderson, J. R. and M. G. Wagger (eds.), *Corn Production Systems in North Carolina.* Raleigh, NC: Agricultural Extension Service.

Wang, T. 2002. Soybean oil. In Gunstone, F. D. (ed.), *Vegetable Oils in Food Technology: Composition, Properties, and Uses.* Boca Raton, FL: Blackwell Publishing.

Wang, Z., 2008. Alkaline pretreatment of coastal Bermuda grass for bioethanol production. M.S. Thesis. North Carolina State University, Raleigh, NC.

Watson, S. A. 1982. Corn: Amazing maize: General properties. In Wolf, I. A. (ed.), *Handbook of Processing and Utilization in Agriculture.* Boca Raton, FL: CRC Press.

Weber, E. J. and D. E. Alexander. 1975. Breeding for lipid composition in corn. *Journal of the American Oil Chemists' Society* 52(9): 370–373.

Weaver, J. E. 1968. *Prairie Plants and Their Environment: A Fifty-Year Study in the Midwest.* Lincoln, NE: University of Nebraska Press.

Wolf, D. D. and D. A. Fiske. 1995. *Planting and Managing Switchgrass or Forage, Wildlife, and Conservation.* Virginia Polytechnic Institute and State University. Extension Publication 418–013.

Woolfe, J. A. 1992. *Sweet Potato: An Untapped Food Resource.* New York: Cambridge University Press.

Wright, L. L. 1994. Production technology status of woody and herbaceous crops. *Biomass and Bioenergy* 6(3): 191–209.

Zublena, J. P. and J. R. Anderson. 1994. *Starter Fertilizers for Corn Production.* Raleigh, NC: Cooperative Extension Service.

4

Biomass Logistics

Matthew W. Veal

CONTENTS

4.1 Machine System Performance Analysis 75
     4.1.1 Machine Capacity 75
     4.1.2 Machine Rate Analysis 78
         4.1.2.1 Fixed Costs 79
         4.1.2.2 Variable Costs 82
     4.1.3 Equipment Selection Criteria 88
4.2 Harvesting 92
     4.2.1 Forage, Grass, and Hay Harvest 93
     4.2.2 Agricultural Residue Recovery 102
     4.2.3 Forest Biomass Equipment 105
4.3 Transportation 108
     4.3.1 Trucking 111
     4.3.2 Rail 118
     4.3.3 Maritime Shipping.. 121
     4.3.4 GIS Analysis of Transportation Systems 122
4.4 Storage 124
4.5 Densification 129
4.6 Problems 130
References 133

The feasible conversion of biomass into an energy resource or value-added product is determined in part by the ability to develop efficient and economical processes that harvest, collect, and store biomass feedstocks. Biomass logistics is a general concept used to analyze and manipulate the flow of materials from the production site (i.e., an agricultural field or a forest) to the point of conversion or use (i.e., the biorefinery). A successful biomass logistics network will insure the timely delivery of materials, prevent excessive degradation of the feedstock, and process the crop into a form that increases the conversion efficiency. Logistics involves the integration of biomass collection systems, material storage facilities, transportation networks, and additional processing systems that attempt to either maintain or improve the physical

FIGURE 4.1
The movement of corn from an agricultural field starts with harvest by a grain combine (left), then the grain is transported by truck (middle) for storage in grain bins (right).

properties of the material. Figure 4.1 provides an example of a logistics system used to collect and transport grains from an agricultural field.

The type of crop, residue, or waste material utilized as a raw material in the conversion process along with the mechanized systems employed to harvest and transport this system have a significant impact on the cost and sustainability of the energy conversion system. Equipment selection, biomass collection and storage procedures, transportation network considerations, and postharvest handling practices are typically biomass parameters that must be considered prior to the arrival of biomass at a processing facility. Decisions related to these preconversion parameters have a significant impact on the biorefinery in terms of delivered feedstock cost, processing characteristics, final product cost, and overall system sustainability. Logistics describes all of the activity associated with biomass that occurs between cultivation and conversion, and its impact on the overall dynamics of any biomass-based energy system is significant.

Logistics is often an overlooked aspect in the discussion of bioenergy crops and processes. The methods employed to harvest, collect, store, and transport a crop from a field or forest to the biorefining destination have the potential to add significant costs to a system. The costs associated with the logistics network can more than double the delivery price of the biomass material, thereby altering the economics to the point that the crop can no longer be considered a viable bioenergy source. Switchgrass has been reported to have a production costs ranging between $33.03/ton to $36.09/ton. This cost may seem reasonable; however, production costs only include the costs associated with establishing the crop in the field and cultivating it over the growing season. In the case of switchgrass, an additional 60% cost could be added to the production cost based on choosing a given harvest, transport, and delivery system.

Biomass is generally a light, fluffy, or low-density material at the time of harvest. This is particularly true of forage, hay, grass, and plant residue materials that are considered among the most plentiful bioenergy crops. A logistics network for removing forage, hay, and grass material from a field to a centralized storage facility is shown in Figure 4.2. Notice that there are many

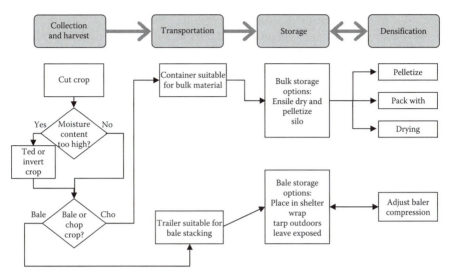

FIGURE 4.2
A flowchart illustrating the potential process that can be used to process forage, grass, and hay crops in the field.

decisions that impact the flow of material in the network. One aspect that remains somewhat hidden in this flowchart is how the decisions to involve different processes in the system will impact the cost of the operation. For example, baling the material is a relatively low output system but the equipment involved is low cost compared to a forage chopper. The forage chopper may be more expensive, but the output of the machine will be higher. When studying a logistics system, it is critical that the expenses associated with a specific operation can be justified based on the amount of material processed.

A general rule to keep in mind when considering a logistics system is that the expense associated with moving biomass from the field to the biorefinery will increase at each instance the biomass is handled or at each additional step added to a supply chain. Other factors that increase the cost of delivery includes the material's relative size, density, moisture content, the complexity of the equipment involved, and the specialization of the system. As the uniqueness of a system or machine increases, the expenses associated with the operation of the system will also increase. Biomass is inherently a cheap material that has a relatively low energy density. This low-value characteristic of biomass requires high volume processing of the material in order to maintain financial viability. Applying a "low value" descriptor to biomass should not lead to negative opinions, rather the abundance and relative ease of collecting biomass gives it an advantage over other energy resources. The collection of biomass does not require the application of expensive technology such as photovoltaic cells or involve the costly exploration equipment required to drill or mine natural resources like petroleum or coal. The systems used to collect, harvest, and transport this material must be designed to keep expenses at a minimum so that biomass can maintain its competitive advantage relative to other energy resources.

As an example of the type of cost inflation that can occur when biomass is collected, consider the case of a woody biomass harvest. If two forest stands were stocked with an identical amount of biomass by weight and one stand was made up of 500 trees with an average diameter of 20 cm and the other stand has 125 trees, 50 cm in diameter, then the stand with larger trees will be less expensive to harvest and the harvest will be completed in a more timely matter. Harvesting the smaller stand will required more trees to be handled; leading to more interaction between steps in the supply chain. Also, transportation efficiency will be reduced because stacking small trees creates lots of void space or air pockets on the truck's trailer. Another example of how economics impacts the use of biomass as an energy resource can be found in algae production for aqua oils. Algae have been cultivated and harvested for many years because they are a source of many building block compounds used in the pharmaceutical and nutraceutical industries. When these compounds are used to manufacture prescription drugs or dietary supplements, the value of the algal material is tens of thousands of dollars per ton and considerable costs can be used to cultivate the algae because the return on investment is significant. However, if the algal oils are used for biodiesel production, these cultivation costs are no longer justified because of the relatively low value of diesel fuel compared to a pharmaceutical product.

This chapter will discuss the steps that are encountered in a biomass supply chain and discuss important considerations that must be addressed when designing a logistics network. The movement of crop material from the field to the factory can negatively impact the practicality of using a particular biomass resource for the generation of energy or production of a value-added production. This material movement does little to add value to the biomass, however, it can become cost prohibitive if the logistic network is not optimized. Biomass begins its journey in the supply chain during the harvest and collection activities that occur in the field. Once the biomass is collected, it must be transported to a facility for processing into an energy source. At the factory, the material must be adequately stored to prevent degradation prior to use at the processing facility. Because biomass is a light, fluffy material, there are opportunities at each of these three major steps along the supply chain to densify the material to make subsequent operations more efficient. Not only is it important to analyze each of these supply chain steps, but the interactions between these entities can also impact the overall performance of the logistics network. Through better understanding of logistics networks, it is possible to create efficient logistics networks by maximizing material throughput, increasing system utilization, and keeping an eye on costs.

4.1 Machine System Performance Analysis

The objective of this chapter is to describe potential systems that can be included in a biomass supply chain network. In addition to describing how these systems operate and providing some information about the parameters required for successful operation, the criteria for evaluating systems will be discussed in detail. Because logistics plays an important role in the economics for the biomass producer, the conversion process, and the consumer, it is important to understand how a logistics system is analyzed. Logistics deals with the movement of goods and materials through a shipping network; therefore, the volume of material moved, the rate of this movement, and the cost associated with the movement are the important parameters to focus on rather than the design criteria used to conceive a specific machine. Logistics is interested in the performance and interactions of a system of machines not the design aspects.

Typically, machine systems, whether they involve collection/harvest activities or transportation, are evaluated on two key criteria: the volume of material moved per unit of time and cost per unit volume of material processed. For the purposes of a logistics system, discussion will be limited to the machine systems used to harvest, collect, and process the material in the field. Capacity and machine rate are the descriptors used to describe how the machine system performs relative to the evaluation criteria. Capacity refers to the quantity of work completed by the machine per unit of time. Machine rate refers to the costs associated with the operation of the equipment and includes both the overhead cost and the cost of any consumables spent during operation.

4.1.1 Machine Capacity

The capacity of agricultural and forestry equipment is usually described by the amount of area covered per unit of time or the volume of material processed per unit of time. Regardless of which descriptor is applied, these capacity measurements allow the machine's operating cost to be assigned to the volume of material processed in a given work period. Capacity helps determine the rate at which material will be processed in the field and also allows an equipment manager to determine how machine systems should interact. By balancing the capacities of various machine systems, it is possible to create a nearly seamless network of machines that move material from the field to the factory with few delays and in an economical manner.

Field capacity is a relatively simple and direct calculation that can be used to quickly assess how much material a system of machines can process in the field. The units on field capacity are typically hectares per hour. There are three factors used to determine the field capacity of the machine:

1. Speed—average rate of travel (km/h)
2. Width—distance across processing part of the machine (m)
3. Utilization or efficiency—ratio of time that a machine is involved in productive work versus the amount of time that a machine is scheduled to work

The formula for determining field capacity is

$$FC = \frac{sw}{10} * U \qquad (4.1)$$

where
  FC is the field capacity, ha/h
  s is the field speed, km/h
  w is the processing width, m
  U is the equipment utilization or field efficiency, %

When the utilization or efficiency is assumed to be 100%, the result of the field efficiency equation is a term known as theoretical field efficiency. The theoretical field efficiency can never be achieved in the field because delays, mechanical breakdowns, and servicing requirements of the equipment cannot be eliminated. However, the theoretical field capacity can be used to quickly assess potential bottlenecks in the harvest system or as a quick tool to size equipment for field operations.

Utilization refers to the ratio of time that a machine is involved in productive work versus the amount of time that a machine is scheduled to work. The key concept in the utilization definition is "productive work performed by the machine." The person operating the machine can be productive when the machine is not necessarily carrying out the intended job. For instance, a tractor's work function may be to shuttle biomass from a piece of harvesting equipment to a truck at the roadside. The tractor's production would most likely be measured in volume processed per unit of time. If the tractor is occasionally needed to pull trucks out of the mud when they depart for the biorefinery, then the tractor is no longer performing productive work related to its job task, but the operator is productive and would be compensated for their effort.

There are several factors that keep machines from achieving 100% utilization rates and utilization rates vary by machine system, implement size, and climatic conditions. Undoubtedly, there will be idle time for repair and maintenance of the machine as well as breaks needed by the operator during the course of their working day. By understanding and planning for potential delays in the system, it is possible to eliminate or reduce the impact of these disruptions in production. Among the most common reasons for equipment not being utilized to its fullest potential are

1. Unused capacity—overlapping passes by a mower or not using full cutting width of a forage chopper

2. Fill procedures—stopping to fill chemical application tanks on a machine

3. Unload procedures—stopping to unload a grain combine

4. Turning time—generally implements behind a tractor are disengaged when the machine is turning at the end of a row

5. Following manufacturer suggested adjustment/service/maintenance procedures

6. Operator rest stops

7. Incorrectly matched systems—one machine with access working capacity has to stop to wait on another machine

8. Equipment failure/breakdowns

9. Weather conditions

Actual or effective field capacity is the term given to the field efficiency actually seen in the field. Table 4.1 provides a summary of values that can be used to determine field efficiencies for a variety of equipment used for biomass harvest and collection. An important point to note is that the size of equipment is usually based on the capacity needed to complete a field operation in a given amount of time. However, there needs to be additional consideration for more random variables such as weather, breakdowns, and

TABLE 4.1

Typical Range for Field Operation Parameters for Biomass Harvest Equipment

Equipment Type	Operating Speed (km/h)	Field Efficiency Values (%)	Estimated Life (h)
Harvest Implements			
Baler, rectangular baler	5–10	65–80	2000
Baler, large rectangular	7–14	70–90	3000
Baler, round	5–20	50–75	1500
Forage harvester	3–10	55–80	2500
Mower	5–12	75–90	2000
Mower conditioner	5–9	75–90	2500
Rake	6–12	75–90	2500
Forage wagon	—	—	3000
Self-Propelled Harvest Equipment			
Cotton picker	3–6	65–80	3000
Forage harvester	2.5–12	55–85	4000
Grain combine	3–6.5	65–85	3000

additional requirements for the machine. Very few agricultural or forestry operations use equipment like tractors for a single task; they may be called on to assist with other operations in the same window of time.

Example 4.1:

A 60 kW tractor is used to operate a mower conditioner with a cutting width of 3.5 m. The mower is cutting a 100 ha switchgrass field with an average yield of 15 ton/ha.

Find the field capacity for this operation, the volume of material processed per hour, and the amount of time required to cut the field.

From Table 4.1, the mower conditioner operates at 8 km/h with a utilization rate of 80%. Now apply Equation 4.1.

$$FC = \frac{s\,w}{10} \times U = \frac{8\,\text{km/h} \times 3.5\,\text{m}}{10} \times 0.80 = 2.24\ \text{ha/h}$$

Material capacity = 2.24 ha/h × 15 ton/ha = 33.6 ton/h
Time required to mow the field = 100 ha/2.24 ha/h = 44.64 h

Both field and material capacity are important measures that aid in balancing machine systems to prevent bottlenecks in the supply chain. If a machine that has a material capacity of 40 ton/h interacts with a machine with a material capacity of 20 ton/h, the system will not achieve optimal efficiency. The utilization rate of the larger machine will drop and the resulting cost per processed ton of material will increase. To balance the system, it would be necessary to add a second smaller machine.

4.1.2 Machine Rate Analysis

Field capacity and utilization are important measures of the operating efficiency of a machine system; however, the ability to assign a cost to the production of biomass cannot be realized. Machine rate analysis estimates the average hourly cost associated with the ownership and operation of equipment over the productive life of the machine. Matthews (1942) first proposed the idea in the classic book *Cost Controls in the Logging Industry*. Machine rate analysis generates a dollars per hour cost for operating machine. The machine rate can then be combined with field efficiency values and biomass yields to calculate a cost per unit volume processed ($/ton) or a cost per unit area processed ($/ha). An important point about machine rate analysis is it only provides an average estimate, which suggests that the actual cash flow associated with the machine's operation may be higher or lower at a given time. There are variations in the values of components of the machine rate calculation that are associated with the age of the machine. An older machine would experience higher repair costs than a newer machine, but the depreciation of the newer machine would be higher.

The American Society of Biological and Agricultural Engineers (ASABE) has developed a set of standard equations that allows for uniform calculation of the costs associated with equipment ownership. Two standards are of particular importance, ASAE D497.5 Agricultural Machinery Management Data and ASAE EP496.3 Agricultural Machinery Management. Tables 4.1 and 4.2 contain information and coefficients that can be used to assist with the calculation of cost components associated with machine-operating expenses. The cost components associated with equipment expenses fall into two main categories: fixed and variable costs.

4.1.2.1 Fixed Costs

Fixed costs are expenses that do not vary with the use of the entity. Fixed costs are also known as ownership costs and represent expenditures that will occur whether or not a machine is used to complete a task. An example of a fixed cost is the payment required to a finance company will be owed whether the machine is at work in the field or has been parked for three months in the equipment shelter. Fixed costs are broken into three individual components for the sake of analysis: depreciation, interest, and taxes.

Depreciation refers to equipment's decline in value that occurs as the machine ages, its technology becomes obsolete, and weather wears out the machine. The salvage value and the economic life of the machine must be defined before a machine's depreciation can be calculated. Economic

TABLE 4.2

Coefficients Required for Ownership and Operating Cost Equations

Equipment Type	C_1	C_2	C_3	RMF_1	RMF_2
Harvest Implements					
Baler, rectangular baler	0.852	0.101	0	0.23	1.8
Baler, large rectangular	0.852	0.101	0	0.10	1.8
Baler, round	0.852	0.101	0	0.43	1.8
Forage harvester	0.791	0.091	0	0.15	1.6
Mower	0.756	0.067	0	0.47	1.85
Mower conditioner	0.756	0.067	0	0.17	1.8
Rake	0.791	0.091	0	0.17	1.4
Forage wagon	0.943	0.111	0	0.16	1.6
Self-Propelled Harvest Equipment					
Cotton picker	1.132	0.165	0.0079	0.11	1.8
Forage harvester	1.132	0.165	0.0079	0.03	2.0
Grain combine	1.132	0.165	0.0079	0.04	2.1
Tractor, <60 kW	0.981	0.093	0.0058	0.007	2.0
Tractor, 60–112 kW	0.942	0.100	0.0008	0.007	2.0
Tractor, >112 kW	0.976	0.119	0.0019	0.007	2.0

life refers to the length of time in years over which the hourly machine rate will be estimated. Economic life may not equal the machine's service life as the machine may be traded or sold before it is no longer serviceable. Salvage value refers to the machine's residual value at the end of its economic life. Salvage values can be estimated by studying used equipment sales and auctions. Salvage value after n years of use can be calculated as a percentage of the purchase price using Equation 4.2. The salvage value is a function of the age of the machine in years and the average hours of accumulated use.

$$SV_n = 100\left[C_1 - C_2(n^{0.5}) - C_3(h^{0.5})\right]^2 \tag{4.2}$$

where
SV$_n$ is the salvage value after n years, % of purchase price
C_1 C_2, C_3 are the coefficients from Table 4.1
n is the age of machine, years
h is the average annual use, h

Example 4.2:

Find the salvage value for the tractor from Example 4.1. Assume the tractor was purchased for $80,000, has 400h of annual use, and an economic life of 15 years.
  From Table 4.2, the C_1, C_2, and C_3 coefficients required for the salvage value equation are 0.941, 0.100, and 0.0008 for the tractor. Using Equation 4.2, it is possible to find the salvage value as a percentage of the original purchase price.

$$SV_{15} = 100\left[0.941 - 0.100(15^{0.5}) - 0.0008(400^{0.5})\right]^2 = 29\%$$

The salvage value of the machine is $80,000 × 0.29 = $23,216.

Even if the machine is owned to the point that it is no longer serviceable, there can still be a salvage value because the metal and working parts can be sold to a scrap yard for recycling. Typically, straight-line depreciation is applied to agricultural and forestry equipment. This method applies the depreciation value evenly across the machine's economic life. To find the depreciation value, the difference between the purchase price and salvage value is determined, then this value is divided by the economic life.
  Interest refers to the charge associated with the use of money to purchase or finance equipment. If the equipment is financed, the interest value is equivalent to the interest rate assigned to the loan used to buy the machinery. If the machine is purchased outright, then interest can be considered a lost opportunity cost. In other words, if the money used to buy the machine had been invested in some other entity, what would the return on that investment look like? The expected rate of return from the alternative investment strategy

would be the interest value for this scenario. The interest value should be adjusted for inflation because inflation leads to a situation where loans can be repaid using devalued currency. If a machine is purchased using a loan that charges 7.5% interest and inflation is anticipated to be 3% annually, the interest value used for machine rate calculations would be the difference in the two values, 4.5%.

Taxes, housing, and insurance make up the final fixed cost component. These three values combined are generally a fraction of interest and depreciation costs. Because of the capital investment required to purchase agricultural and forestry equipment, insurance is recommended to protect against a machine becoming damaged due to fire, weather, or vandalism. Annual insurance payments range between 0.25% and 0.5% of the machine's purchase price. Taxes will vary greatly by region and some regions provide exemptions from property taxes for this type of equipment. In a machine rate analysis, taxes can range from 0% to 1.0% of the machine's purchase price. Finally, housing takes into account the costs associated with providing shelter for a machine from the elements. Equipment housing options are highly variable and leaving a machine unsheltered, exposed to the elements, is a no cost housing option. Housing a machine should reduce repair and maintenance costs and is a recommended practice. Housing costs range from 0% to 0.75% of the machine's purchase price. As a general rule, tax, housing, and insurance costs are combined when they are estimated. Values between 1% and 2% of the equipment's purchase price are used to estimate the contribution of these three costs to the overall fixed cost estimate.

Total annual ownership costs can be approximated using Equation 4.3. This equation accounts for depreciation, interest, taxes, housing, and insurance. The equation yields the average annual ownership costs as a percentage of the machine's purchase price.

$$C_o = 100 \left[\frac{1-SV}{L} + \frac{1+SV}{2} i + TIH \right] \tag{4.3}$$

where
C_o represents the average annual ownership costs, % of purchase price
SV is the salvage value after L years, % of purchase price
L is the economic life, years
i is the interest rate, % expressed as decimal
TIH is the percent of purchase price required for taxes, insurance, and housing

Example 4.3:

Find the average annual ownership cost associated with the tractor from Examples 4.1 and 4.2. Again assume the tractor was purchased for $80,000, has 400 h of annual use, has an economic life of 15 years, and an interest rate of 7% can be assumed after adjustments for inflation.

Using Equation 4.3, it is possible to find the estimated annual ownership cost as a percentage of the original purchase price.

$$C_o = 100\left[\frac{1-0.29}{15} + \frac{1+0.29}{2}0.05 + 0.015\right] = 9.5\%$$

The annual cost of ownership for the tractor is $80,000 × 0.095 = $7566/year
If the tractor operates 400 h/year then the average hourly ownership cost is $7566/400 h = $18.91.

4.1.2.2 Variable Costs

Variable costs are expenses that change based on equipment usage. Variable costs are also known as operating costs, and these costs will be zero when the machine is not in use. An example of a variable cost is fuel and lubrication. A machine that is parked at the equipment storage facility will not burn fuel or require lubrication. Variable costs breakdown into four distinct categories: repair and maintenance, labor, fuel, and lubrication.

As the machine ages, it may be involved in accidents, weather will corrode parts, other parts will wear from repetitive use, and some systems will simply fail after reaching their service life. The costs associated with repairing and maintaining equipment vary widely based on the type of machine, the manufacturer, the operating climate, preventative maintenance activities, and parts availability in a region. Using service manuals, service records, and discussing with equipment dealers is perhaps the best method to determine the costs associated with repairing and maintaining a specific machine. However, there are methods that have been developed to estimate repair and maintenance costs. As a machine ages, the repair and maintenance costs will increase. These costs will continue to accumulate well beyond the original purchase price of the machine. Equipment owners are often presented with a dilemma as the machine ages; because the financial resources to buy a new machine may not exist therefore it is necessary to pay relatively high repair costs.

The rate at which repair and maintenance costs accumulate varies dramatically between machine and implement types. Typically, the more involved a piece of equipment becomes with the biomass material, the quicker it will accumulate repair costs. Also, repair and maintenance costs are calculated as a percentage of the purchase price of the machine; therefore, an expensive machine would generate a higher cash value available in a given year to complete repairs compared to a less expensive machine. Figure 4.3 illustrates both of these points as the accumulated repair and maintenance costs for a round baler, 80 kW tractor, and grain combine have been graphed. The round baler accumulated repair costs the quickest relative to its purchase price. The round baler is in constant contact with biomass during operation, which increases wear and tear on the machine, compared to the agricultural tractor, which does not have a direct interaction with the biomass. The main difference between the round baler and grain combine is the ability for the

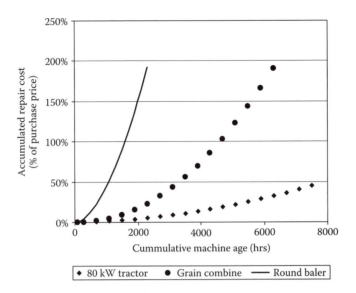

FIGURE 4.3

Comparison of accumulated repair and maintenance cost of select agricultural machines.

purchase price to offset the impact of the repair cost. A 5% accumulated repair and maintenance costs would correspond to $1250 for the round baler and approximately $11,000 for the grain combine. Obviously, the grain combine is more expensive to maintain but because it is more expensive, it will take longer for the accumulated repair and maintenance costs to surpass the original purchase price of the machine.

Equation 4.4 approximates a projection of the accumulated repair and maintenance costs. The equation requires an estimate of the accumulated hours on the machine. Typically, this value will be the product of the average annual use and the economic life of the machine. Unlike the other equations, this equation must be modified when used equipment purchases are considered. The accumulated repair costs must be calculated at the time of purchase, this value will then be subtracted from the projected accumulated cost. This modification is necessary because the new owner would not be responsible for repair and maintenance costs that were incurred prior to ownership.

$$C_{RM} = (RMF_1)P\left[\frac{h}{1000}\right]^{(RMF_2)} \qquad (4.4)$$

where

C_{RM} is the accumulated repair and maintenance costs, $

RMF_1, RMF_2 are the coefficients from Table 4.2

P is the equipment purchase price, $

h is the accumulated hours of use

Example 4.4:

Find the cumulative repair cost for the 60 kW tractor used in the previous examples, all of the previous assumptions remain intact. Also calculate the average hourly repair cost for the tractor at the end of this economic life.

Using Equation 4.4 and values from Table 4.2, it is possible to find the estimated cumulative repair and maintenance costs for the tractor following a 15 year economic life with an average of 400 h annual use.

$$C_{RM} = (0.007) \times \$80,000 \left[\frac{(15 \text{ year} \times 400 \text{ h/year})}{1000} \right]^{(2.0)} = \$20,160$$

The average hourly repair cost for this tractor at the end of its economic life is $\$20,160 \div (15 \text{ year} \times 400 \text{ h/year}) = \$20,160 \div 6000 \text{ h} = \$3.36/\text{h}$.

Fuel costs are associated with the operation of agricultural or forestry equipment in their work environment. The fuel consumption rate can easily be estimated using the rated engine horsepower for the machine being used to complete work. Using the average fuel consumption rate and the current market price for fuel, it is possible to calculate the cost of fuel on an hourly basis.

$$Q_{avg} = X_{FT} * P_{PTO} \tag{4.5}$$

where
Q_{avg} is the average fuel consumption, L/h
P_{PTO} is the rated engine power, kW
X_{FT} is the fuel type coefficient, 0.305 for gasoline fuel, 0.222 for diesel fuel

The method described above is a conservative measure as it assumes the engine's power is being fully utilized. ASABE Standard D497.5 provides a more detailed method for determining the fuel consumption, which accounts for the ratio of available horsepower to the horsepower required for the field operation.

Lubrication costs account for expenses associated with maintaining proper engine lubrication through oil changes. There are formulas (see ASAE D497.5) that attempt to estimate the volume of oil consumed by an engine as it operates. These formulas provide excessive detail for cost analysis; therefore, lubrication costs are assumed to be 15% of the fuel cost. This constant value can be modified for the engine's age (older engines tend to burn more oil) or based on operating parameters of the engine. Equipments requiring additional filters or consuming relatively high values of hydraulic fluid are examples of equipment requiring adjustments to the standard lubrication cost constant.

Example 4.5:

Find the fuel and lube cost for the 60 kW tractor used in the previous examples; all of the previous assumptions remain intact. Assume the price of diesel fuel is $0.92/L.

Using Equation 4.5, it is possible to find the estimated average hourly fuel consumption for the engine. The fuel price has been specified and lubrication will add an additional 15% to the final fuel cost.

$$Q_{avg} = 0.222 \times 60 = 13.32 \, L/h$$

The average hourly fuel cost is 13.32 L/h × $0.92/L = $12.25
The average hourly fuel and lubrication cost is $12.25 × 1.15 = $14.09.

The final variable cost to consider is labor. Labor costs represent the hourly rate paid to employees of the operation. These costs should include the wage, taxes, and any fringe benefits such as insurance or retirement compensation. Labor in agricultural production is somewhat seasonal as personnel may be hired for specific seasonal activities such as planting or harvesting. Because labor is attached to the machinery but people tend to work longer than the machine during a scheduled work shift. Personnel may be required to clean equipment, travel to the field, or service the machine prior to operation. To account for these additional activities, the labor rate is adjusted by 10%–20% depending on the extent of the employee's activity beyond actually operating the machine.

As both a final example and a way to summarize the machine rate analysis, Tables 4.3 and 4.4 have been constructed. The tables illustrate how the cost components should be arranged, and it clarifies which variables are inputs versus those that are calculations. All values on the left side of the table are variables that come from the previous tables, manufacturer data, or personnel experience. The tables also include the mower conditioner unit that was discussed in Example 4.1. The mower conditioner is only an implement; therefore, there are no fuel, lubrication, or labor costs associated with this machine. These operating costs are eliminated because the tractor expense calculation has accounted for these values. The purchase price of the mower conditioner is $27,500, the implement has a 10 year economic life, and the average annual use will be approximately 200 h/year.

The final value from Tables 4.3 and 4.4 indicates that the tractor and mower conditioner cost $50.16/h and $15.89/h to own and operate, respectively. The total system would have a total cost of $66.05/h. This cost may seem high, and they should be recalculated to reflect the actual amount of biomass or land area processed in that time. In Example 4.1, both the field capacity (2.24 ha/h) and the material capacity (33.6 ton/h) were calculated. If the machine rate is divided by a capacity value, it is possible to acquire either the cost per volume of processed material or the cost per unit of area covered by

TABLE 4.3

Total Ownership and Operating Costs of a 60 kW Tractor

Ownership Cost Variables			Ownership Costs
Purchase price (PP)	$80,000	Salvage value	Equation 4.2 29.5% of PP or $23,216
Economic life	15 years	Total annual ownership cost	Equation 4.3 9.5% of PP or $7566
Interest rate	5%	**Average hourly ownership cost**	**Equation 4.3 ÷ Avg. annual hours** **$18.91/h**
Annual taxes, insurance, housing fees as % of purchase price	1.5%		**Operating Costs**
Annual use	400 h	Hourly repair cost	Equation 4.4 ÷ total accumulated hours at end of economic life $3.36/h
C_1, C_2, C_3	0.941, 0.100, 0.0008	Hourly fuel and lubrication cost	Equation 4.5*fuel price *1.15 $14.09/h
Operating Cost Variables		Labor cost	Labor rate * 1.15 $13.80/h
RM_1, RM_2	0.007, 2.0	**Average hourly operating costs**	**$31.25/h**
Price of fuel	$0.92/L		
Tractor power rating	60 kW	**Total hourly cost for the 60 kW tractor**	**$50.16/h**
Labor rate	$12/h		

the machine. Carrying out this calculation shows this equipment setup can process material at a rate of $1.97/ton or cover the field at a rate of $29.49/ha.

As mentioned earlier, the replacement strategy for agricultural and forestry equipment is a difficult decision to make. However, there is one sound method that can be employed to help make this decision for an equipment manager. It requires two values to be plotted (Figure 4.4): the accumulated hourly rate of the machine and the annual hourly rate. The accumulated hourly rate is found by summing all costs associated with the ownership and operation of the equipment then dividing this summation by the total hours accumulated by the machine. The accumulated hourly rate will initially decline as the impact of depreciation and interest on the costs decrease. However, this decline will slow and eventually reverse itself once the repair costs for the machine become the primary expense.

The annual hourly cost to operate the equipment follows a trend similar as the accumulated hourly cost. Annual hourly cost refers to the cost to operate the equipment in a single year divided by the hours the machine is operated in that

TABLE 4.4

Total Ownership and Operating Costs of a 3.5 m Mower Conditioner

Ownership Cost Variables		Ownership Costs	
Purchase price (PP)	$27,500	Salvage value	29.6% of PP or $8142
Economic life	10 years	Total annual ownership cost	11.5% of PP or $3102
Interest rate	5%	**Average hourly ownership cost**	**$7.75/h**
Annual taxes, insurance, housing fees as % of purchase price	1.0%	**Operating Costs**	
Annual use	200 h	Hourly repair cost	$8.14/h
C_1, C_2, C_3	0.756, 0.067, 0.0	**Average hourly operating costs**	**$8.14/h**
Operating Cost Variables			
RM_1, RM_2	0.17, 1.8	**Total hourly cost for the 3.5 m mower conditioner**	**$15.89/h**

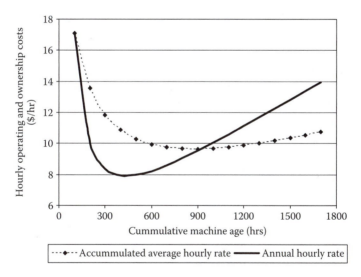

FIGURE 4.4
Accumulated average and annual hourly rates for a 3.5 m mower conditioner purchased for $27,500.

particular year. The most noticeable difference is the steep decline in annual hourly rate between years 1 and 2. This decline is due to the severe depreciation penalty that occurs when new equipment initially leaves the dealership. The first year depreciation value can be as high as 30% of the original purchase price. Careful inspection of Figure 4.4 shows the annual and accumulated hourly rates intersect. The number of operating hours corresponding to the intersection (approximately 900 h in the figure) is the optimal time to replace the equipment with a new version of a similar model. This intersection represents the point in a machine's life where the repair costs become the major operating expense and this condition will continue to deteriorate with time.

The logistics system will be required to move large quantities of biomass from the field or forest to a refinery and this activity will involve the use of many machines. Through field capacity and machine rate analysis, it is possible to determine the expenses associated with the movement and infield processing of biomass. The mower conditioner example that was just illustrated is straightforward, but the analysis is time-consuming given the fact that the mower conditioner would be one of four potential machines working in an agricultural field to take a standing switchgrass crop and process it into bales for transportation on the public road system. This analysis would need repeating for a hay rake, a baler, and a wagon used to haul bales from the field for the final costs associated with extracting the material from the field to be established.

4.1.3 Equipment Selection Criteria

Both the machine's capacity and the machine rate can be used to determine the optimal machine size required (Equation 4.6) to complete a task in a timely manner at the lowest cost. When harvesting biomass, one of the first considerations must be the timeliness of the operation. Determining the rate material must be removed from the field or forest will impact feedstock quality, scheduling of additional logistics components, and ultimately the operation of the biorefinery. Calculating the capacity required to complete a task in a specific amount of time requires an estimate of the number of days that the task should be completed. Also, the probability of having a working day during that period must be estimated. There are data sets available from a variety of sources that discuss the probability of having a working day in a specific region. These probabilities can range from less than 30% during months with high rainfall to nearly 70% during dry months. With agricultural and forestry operations, it is important to keep in mind that a significant rainfall on a single day can keep a machine from operating for several days as the soil or biomass dries.

$$FC_R = \frac{A}{BG(\text{pwd})} \tag{4.6}$$

where
  FC_R is the required field capacity, ha/h
  A is the area, ha

B is the number of days available to accomplish the task

G is the expected work shift each day, h/day

pwd is the probability of a working day

After determining the required field capacity, it may indicate that the current machine system is not adequate to complete the operation. This presents a manager with a difficult decision as either the number of days for the task should be extended or a larger machine should be purchased. A manager can calculate the cost incurred for each incremental increase in the machine's capacity ($/h-ha). This unit price function (Equation 4.7) requires careful study of equipment prices because the incremental cost to increase the width of the machine's processing mechanism. For instance, mowing equipment may cost an additional $5000 for each meter of cutting width added to the device. The calculations should be made with considerations given to the equipment sizes that are available from the manufacturer. The trend has been for agricultural equipment to get larger with time; however, there are restrictions both regulatory and technological that limit the width of the implement.

$$K_p = \frac{10P_w}{sU} \qquad (4.7)$$

where

K_p is the unit price function, $/ha-h

P_w is the price per unit width increase, $/m

s is the field speed, km/h

U is the utilization or field efficiency, %

Example 4.6:

In Example 4.1, calculations revolving around a 60 kW tractor operating in conjunction with a 3.5 m wide mower conditioner found 33.6 ton/h in a 100 ha field. A request has been forwarded asking for the cutting capacity should be increased to 45 ton/h so the biorefinery can be properly maintained. Determine the required field capacity and make an assessment regarding the cost associated with the increased capacity.

First, the annual field capacity can be calculated using Equation 4.6. Using local knowledge, it is assumed that the harvest is occurring during a fairly dry period of the year, so the probability of a working day is 75%. Also, from Example 4.1, this field yields 15 ton/ha.

First, establish the amount of time it would take to harvest the field if the material capacity for the machine was increased to 45 ton/h.

15 ton/ha × 100 ha = 1500 tons of switchgrass to harvest

1500 tons ÷ 45 ton/h = 33.33 h to harvest

The product of the variables B and G should equal to 33.33 h. In other words, scheduling three 11 h work shifts would achieve the necessary throughput as would scheduling four 8.33 h work shifts.

$$FC_R = \frac{100 \text{ ha}}{33.33 \text{ h } (0.75)} = 4 \text{ ha/h}$$

The field capacity required to meet the increase in production demand is 4 ha/h. Now the question becomes does this make sense to implement from an economic standpoint? After extensive discussion with equipment dealers, it is clear that increasing the width of a mower conditioner costs $5000/m. Using previous values established for Example 4.1, it is possible to estimate how an increase in equipment capacity will impact the bottom line (see Equation 4.7).

$$K_p = \frac{10(\$5000)}{8.0 \text{ km/h}(0.80)} = \frac{\$7813}{\text{ha-h}}$$

The K_p result indicates that it will cost $7813 to increase the production of the system by one unit of capacity. This value could then be distributed across the biomass harvest volume to determine its impact on a dollars per ton basis. Also, before this result is accepted, manufacturer's data should be checked to make certain that a larger mower conditioner is available for use. A final point to keep in mind is field capacity is a function of machine speed and utilization. If strategies can be developed to increase speed or efficiency of the operation, it will be possible to increase field capacity at zero additional cost. Also, if the field activity can be moved to a drier part of the year, the probability of working day value will increase and positively influence the field capacity at no additional cost.

As a final exercise in equipment selection, there is an equation that exists to determine the optimal field capacity based on the economic factors driving both the equipment selection decision and the marketing opportunities for the crop. Equation 4.8 is a valuable equation for determining the proper size of the equipment to complete a timely and economic harvest. An important aspect of the equation, the cost term on the far right of the square root, is a cost factor for timeliness of harvest. Many agricultural crops are subject to degradation if harvest activities are not completed in an efficient manner. This timeliness factor will cause the equation to recommend slightly larger and more costly machines; however, the benefit should be timely delivery of a higher quality and, therefore, more valuable crop. Figure 4.5 illustrates how the timeliness factor impacts the optimization result. If the optimization is calculated on purely economic considerations, the optimal capacity is 2.0 ha/h. The optimal capacity increases to 2.65 ha/h when economics as well as timeliness are considered in the optimization. With the larger machine, it is unlikely that weather delays and breakdowns will affect the performance of the system as radically as the smaller machine. The extra capacity

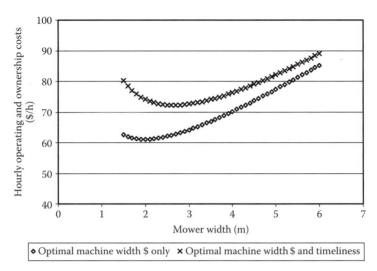

FIGURE 4.5
Optimization of mower conditioner cutting width using economic considerations as well as economic considerations coupled with a timeliness factor.

of the larger machine buffers the system's production against short-term disruptions.

$$M_{OC} = \sqrt{\frac{100A}{C_oK_p}\left[L + T_o + \frac{K_3AYV}{4G(pwd)}\right]} \tag{4.8}$$

where
M_{oc} is the optimal capacity, ha/h
C_o is the cost of ownership, %
K_p is the unit price function, $/ha-h
A is the area, ha
L is the labor rate, $/ha
T_o is the tractor operating cost, $/ha
K_3 is the coefficient (0.018 for hay harvest, 0.006 for grain harvest, 0.002 for other harvest)
Y is the yield per unit area, ton/ha
V is the value of yield, $/ton
G is the expected work shift each day, h/day
pwd is the probability of a working day

Example 4.7:

Solve Equation 4.8 for the example 60 kW tractor and 3.5 m mower conditioner used throughout this chapter.

The majority of these values have been calculated in previous examples. The only modifications that are necessary are the values for labor and tractor cost must be converted to dollars per hectare and the annual cost of ownership must be expressed as a percentage. Be sure to note in the example that the percentage is input into the equation, not the decimal form of the equation.

$$M_{OC} = \sqrt{\frac{100(100\,ha)}{(9)(\$7813)}\left[\frac{\dfrac{\$6.16}{ha} + \dfrac{\$22.39}{ha}}{+\dfrac{(0.018)(100\,ha)(15\,ton/ha)(15\,ton/ha)(\$25/ton)}{4(10\,h)(0.80)}}\right]}$$

$M_{OC} = 2.65\,ha/h$

This result indicates that a slight increase in capacity is needed to achieve the optimal field capacity; currently the system has a field capacity of 2.24 ha/h. The 0.41 ha/h increase could be achieved by increasing the utilization rate of the machine to 95% (unlikely), increasing the tractor's field speed to 9.5 km/h (maybe), or buying a mower conditioner with a 4.14 m cutting width (impossible, this dimension does not exist). The true solution is to identify strategies that affect cutting width, field speed, and efficiency. A probable solution would be to buy a 4 m mower conditioner, increase the average field speed to 8.5 km/h, and maintain a utilization rate of at least 78%.

As a concluding thought on equipment selection, it is important to know that implements should be paired with tractors that have a rated engine power similar to the implement's power requirement. A tractor that is too small for the implement will dramatically reduce system efficiency and cause excess strain on the tractor leading to higher repair costs. A tractor that is too large for an implement is generating excess cost due to a higher expenses associated with depreciation, interest, insurance, fuel, lubrication, and repair. ASABE Standard EP496.3 Agricultural Machinery Management provides useful formulas for determining the required horsepower for particular implements. However, this is a level of detail that is not necessary as nearly all equipment manufacturers publish power requirements for the implements they produce. This information is readily available on equipment manufacturer websites and at dealer locations. When an analysis of a tractor-implement pairing is conducted, the size of the implement cannot be arbitrarily changed to increase field capacity without insuring that the tractor used in the analysis is still capable of providing sufficient power to the newly considered, larger implement.

4.2 Harvesting

Thus far this chapter has focused on methods that can be used to study the performance of machines and processes that make up the logistics network.

In order to carry out a complete analysis of a logistics system, a description of the individual components of the biomass supply chain is required as well as knowledge of some details about the machinery and processes that assist with the movement of biomass from the field or forest to a biorefinery. The first step in the biomass supply chain is the collection of biomass from an agricultural field, forest, or industrial facility. The majority of agricultural and forest crops harvested as an energy resource enter the logistics supply chain at the point of harvest. Harvesting involves the mechanical gathering or collection of biomass material. Production agriculture for food, feed, and fiber has established numerous mechanized systems for the collection of agricultural crops. Harvesting operations related to forestry biomass also have well-established mechanized systems specifically designed for cutting and transporting woody biomass. Many of the conventional agricultural and forestry harvesting systems can be used for biomass harvest and collection activities with minor changes in operating protocols. There are distinct differences between the operation of agricultural equipment and forestry equipment; therefore, discussion of these systems will be separate.

4.2.1 Forage, Grass, and Hay Harvest

Many of the forage, grass, and hay crops begin their journey through the logistics supply chain when they are harvested in the field using a mower, a forage chopper, or sickle bar cutter. Because mechanized systems for the harvest of forage and hay crops have been well established, there are a number of harvest options. Figure 4.6 illustrates both the equipment and processing options available for hay, grass, and forage harvest and collection activities. Selection of one harvest process and implement depends on the operational costs, the machine's efficiency, and the form of the crop following the harvest operation. Special consideration should be given to the condition of the crop at the time of harvest when equipment is selected. Both grain and grass crops are susceptible to lodging (i.e., being blown over by wind) and becoming tangled with other crops in the field. Lodged and tangled crops will dramatically decrease the efficiency of equipment that operates by singulating the rows into individual units along the equipment's length, like some forage choppers. Another common problem is that weather conditions can seriously impede field drying operations. If the crop is baled wet, it will spoil and letting the crop dry excessively in the field will make the crop brittle and lead to harvest losses.

All forage, grass, and hay crops must be cut to begin the harvesting process. The majority of these cutting implements use rely on shear force or an impact force to severe the plant from its roots. A typical shear force cutting device is the sickle bar, which shears material when a reciprocating knife comes into contact with a fixed counter-shear or ledger plate. The key to all cutting mechanisms is they must generate enough shear to severe the biomass in the field. Rotary mowers are an example of impact force cutters as

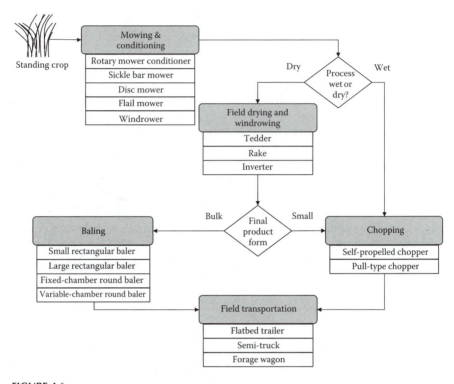

FIGURE 4.6
Processing operations and associated equipment of harvest, collection, and field transportation activities for forage, grass, and hay crops.

a rotating blade has sufficient velocity to slice through material when contact is made. It is critical that the mower is able to sustain a minimum knife velocity, which is a function of the crop's physical properties, to prevent the mower from stalling during field operations.

Cutter bar or sickle bar mowers (Figure 4.7) are the most common cutting implement that operate on the principle of shear force cutting. The operation of a cutter bar mower is very similar to a pair of scissors. A cutter bar mower is composed of two primary sections: knives and guards. The knife segment moves linearly, back and forth inside of the guards. The guards are stationary and provide protection for the knives and structural support for the cutter bar. Cutting occurs when the plant is pinned against a guard and the knife's blade severs the crop material. In very dense crops, it is not uncommon to add an additional cutter bar segment and remove the guards. Two cutter bars working back and forth in opposite directions will increase the capacity of the mower. The major advantage of a cutter bar mower over other mower types is that the cutting of the crop is not as aggressive as it may be with other mowers. Aggressively cutting the crop leads to damage and creates a high volume of clippings. Clippings refer to any material that

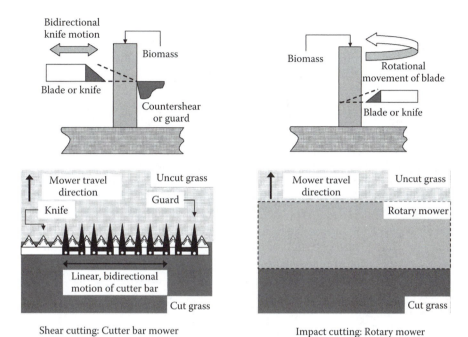

FIGURE 4.7
Examples of the shear and impact cutting mechanisms.

can pass through a 5 cm screen and are major components of harvest losses associated with mowing. Clippings are too small for mechanized equipment to handle, so they either fall to the ground or are swept away by wind. The disadvantage of cutter bars is they are capacity limited; therefore, they have been surpassed by rotary mowers and windrowers as the primary means to cut forage, grass, and hay on large commercial agricultural operations.

Rotary mowers use rapidly spinning disks to cut crop material based on the impact force operating theory. The majority of rotary mowers used for forage, grass, and hay cutting are composed of a series of disks that spin along the horizontal plane (Figure 4.7). Compared to a rotary mower using a single horizontal blade for cutting (i.e., a residential lawnmower), mowing implements comprising a series of disks reduce the incident rate of crop injury and damage. A flail mower is a type of rotary mower in which the axis of rotation is along the vertical plane. Rotary mower blades are allowed to pivot on a spinning disk. Centrifugal force holds the blades in cutting position when the mower is operating at its rated speed; however, in the event the mower encounters a solid object the blades can recoil and prevent damage to the expensive drive mechanisms used to synchronize the disks. Rotary mowers spin the blades at very high revolutions that allow faster tractor speeds during harvest and as a result have increased field capacity. The major disadvantages of rotary mowers are they require more horsepower than a cutter bar mower and there is a higher probability the mower can pick up and throw a solid object. Thrown objects pose a threat to the health and well-being of equipment operators and bystanders.

Conditioning is a process that is often carried out simultaneously with mowing operations. Conditioning involves cracking, crushing, or bruising plant material at the time of harvest in order to facilitate improved field drying. Conditioning has proven an important step in the production of forage and hay crops because it decreases field loss, improves moisture conditions, and improves the crop's recovery from adverse weather conditions. A mower conditioner cuts a crop then passes it through a series of rollers to complete the conditioning step (Figure 4.8).

Different roller designs exist to change the conditioning characteristic based on the type of crop harvested. A windrower is a large, self-propelled machine that cuts forage material, conditions it, and collect the material in a windrow. Because of increased size and dedicated function of a windrower, it can cut a wider swath than towed implement and they generally have greater material handling capacity. A swather is almost identical to a windrower in form and function, except that the conditioning equipment has been removed from the machine. Swathers are usually used to cut and windrow small grains to facilitate more efficient combine harvest operations.

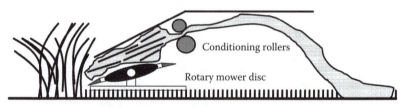

Mower conditioner

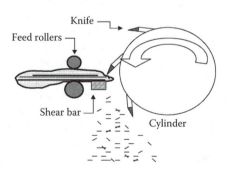

Forage chopping cylinder

FIGURE 4.8
Examples of a mower conditioner and forage chopping cylinder.

Following mowing operations, the grass crop will either be spread across the entire field or the mower's cutting mechanisms will place the cut material in long narrow rows or windrows. If the material is spread across the field in a thin layer, field drying will occur quickly but the volume of material requires concentration into windrows so balers and forage choppers can operate efficiently. If the crop is already in a windrow, the baler or chopper can operate efficiently, but the field drying will not be completed fast enough. If the material is windrowed by the mower, the top of the windrow will dry at an acceptable rate, but the material on the bottom, lying on the ground will maintain high moisture content. Baling wet straw will lead to bale spoilage. Also, once the crop is mowed, it is vulnerable to weather conditions, such as rain, which can introduce additional moisture to the plant or undo any field drying that has already taken place.

There are three hay tools available to address the in-field material distribution and moisture content concerns; rakes, inverters, and tedders. A rake is the primary tool used to move cut material into windrows and to turn windrows to promote increased drying. There are generally two types of rakes, parallel bar side-delivery, and finger wheels. A side delivery rake has a series of parallel bars with spring tines that sweep the material to the trailing end of the rake. A side delivery rake is angled relative to the forward movement of the tractor to allow the material to exit the rake and form a windrow. If the rake was not angled sufficiently or was positioned perpendicular to the tractor's travel direction, the forage crop simply accumulated under the rake and exits at the ends sporadically. A finger wheel rake uses large wheels with spring-loaded tines attached to the edges to move hay into a windrow. As the tractor moves forward, the wheels lightly touch the ground and spin. The turn motion moves the forage material and like the side delivery rake, it exits at the trailing end of the rake to form a windrow. Like the side delivery rake, a wheel rake is positioned at an angle relative to the tractors forward motion.

Regardless of the type of rake, the angle of the rake behind the tractor will determine the shape of the windrow. Windrows can be light and fluffy or tighter and packed. Determining the best shape depends on the subsequent operation, for instance light, fluffy windrows are good for drying hay quickly but are not ideal for baling. Rakes can be used to gather crop material spread across the field, combine smaller windrows created by the mower into larger windrows to improve baling or chopping efficiency or to rearrange windrows so additional material can be exposed to sunlight to improve field drying. Regardless of the objective of a raking operation, it is important to monitor the moisture content of the material. If the material is too dry, it will be brittle and break apart during raking operations. If the material has not been allowed adequate field drying time prior to raking, then the moisture content will be high. Raking at high moisture will create heavily packed windrows that will take a long time to dry, and if the windrow is not allowed to dry, any bales made from wet material are more susceptible to spoilage. A forage moisture content of 40% is recommended at the time of raking to facilitate drying without incurring excess harvest loss.

Tedders are implements towed behind a tractor that are used to fluff and spread material in a thin layer across the field. Tedding forage reduces field drying times; however, it adds additional production costs to the harvest system. Therefore, tedding is usually used if weather conditions have either prevented drying or increased the moisture content of the hay. For instance, rainfall on a windrow could lead to harvest loss or spoilage concerns. A tedder could be employed in this case to spread the windrow in a thin layer so that it will dry quickly when weather conditions improve. A windrow inverter is another towed implement that inverts a windrow so the damp bottom part of the windrow is exposed to sunlight to speed field drying. Windrow inverters do not significantly reposition the material in the field, and they are not as aggressive in the handling of the crop as some other tools.

Once biomass is windrowed, the final field operation for forage, grass, and hay crops alters the form of the material so it is better packaged for transportation. Baling and chopping are the two primary means to process windrowed material to prepare it for transportation and storage. Baling can be considered both a harvest process and a densification process. Baling provides a means to compress loose hay, forage, or biomass material into individual units that aid further movement of these material through the logistics network and densifies the material to improve handling efficiency. There are two shapes of bales, rectangular and round bales; however, the size of these bales can vary greatly based on the specifications of the equipment. Round bales can range in size from a 1.22 m wide by 1.22 m diameter to 1.52 m wide by 1.83 m diameter. Large rectangular bales can range from 0.91 m wide by 0.91 m tall by 1.83 m long bale to a 1.22 m wide by 1.22 m tall by 2.44 m long bale. The largest round and square bales may weigh in excess of 1 ton depending on type of crop and moisture properties at the time of harvest. It should be noted that there are small rectangular balers that produce bales weighing up to 70 kg. These small balers work well for small farmers and ranchers; however, due to material handling expenses they are not viable for biomass collection.

The primary difference between these two bail types is the round bale has some ability to prevent moisture from entering the bale. This allows round bales to be stored outdoors for longer periods of time before they must be moved. A "crust" forms around the outer edge of the round bale. This protective layer helps direct water around the outer perimeter of the bale and into the ground. Round bales are susceptible to moisture damage along the flat ends, but there are storage arrangements that reduce this risk. Rectangular bales are usually moved under shelter shortly after they are created to prevent moisture from spoiling the biomass. Wet bales are very dangerous to store, because the moisture promotes the growth of microorganisms. These microorganisms can metabolize the crop and cause considerable losses. But potentially more dangerous, the activity of these microorganisms can generate sufficient heat to cause the hay bales to spontaneously combust damaging crop, structure, and potentially risk human life.

While susceptibility to moisture is a significant concern by paying careful attention to storage practices, this issue can be resolved. Storage is actually the greatest advantage of square bales over round bales. Square bales can be stacked more efficiently, requiring less space, and have the ability to use straightforward material handling procedure. Round bales do not stack easily and leave lots of void space when stacked. As a result, it is impossible to maximize trucking capacity and additional storage volume is required by a biorefinery to store an equivalent weight of round bales compared to rectangular bales. Another difference in the two types of bales is the cost of the baler itself. A round baler is approximately 30% of the cost of a large rectangular baler. This is not necessarily a disadvantage for the rectangular baler, because throughput and transportation costs/opportunities can make up for this difference. Large square balers have a higher field capacity, and they have a significant advantage when long-distance transport via semitruck or rail is required.

The two primary types of round balers are fixed and variable chamber balers (Figure 4.9). Both of these balers require power from a tractor to operate the equipment and both are equipments with pickup mechanisms to feed material into the implement. While the resulting bale is similar in appearance for both round bales, there are differences in how the bale is built by each baler. An illustration providing detail about the operation of fixed- and variable-chamber is shown in Figure 4.9. A fixed chamber baler has a series of belts or rollers around the perimeter of the baling chamber. As the chamber fills, the material is tossed around until it begins to roll. The chamber continues to fill until the pressure inside the bale chamber increase to a predetermined value indicating a complete bale is formed. Because the material is initially tumbled around the baler prior to rolling, the density of these bales is variable. The inner portion of the bale has a lower bulk density than the outside.

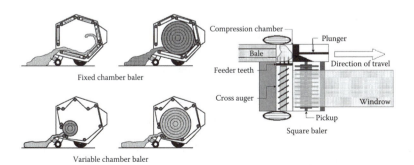

FIGURE 4.9
Biomass baling options and their operation.

The majority of round balers are variable-chamber balers. The series of moving belts starts the hay rolling as it is fed into the machine. However, unlike in the fixed-chamber baler, these belts also restrict the volume available for hay to fill inside the chamber. Springs, idler arms, and hydraulic pressure are mechanisms used to control the tension on the belts. As the chamber fills with hay, it creates pressure that reacts against the belt tension, eventually causing the belts to expand. The bale continues to build until either there is no room left for further belt expansion or the bale reaches a predetermined size specified by the operator. Because the expansion of belts is a function of the amount of hay in the chamber, the ejected bale has a more consistent density. Both types of round baler wrap the finished bale with twine, wire, or netting once the bale has achieved a predetermined size specified by the operator. This tying procedure is sufficient to keep the bale intact during transportation and storage.

Rectangular balers (Figure 4.9) operate much differently from round balers after the material is gathered and feed into the machine by a pickup mechanism. In a rectangular baler, the incoming material comes into contact with a cross auger that moves the biomass across the machine from the pickup side to the baling side of the implement. At this point, feeder teeth push material into the compression chamber. These feeder teeth insure an even distribution of biomass within the compression chamber. Next the plunger compresses the biomass, removing as many air pockets or voids as possible. As the plunger compresses the material, a knife cuts any material that has a presence in both the cross auger and the compression chamber space. This cutting activity provides dimensional consistency for the bales. Once the hay charge is fully compressed, fingers called hay dogs hold the compressed hay in place, thereby preventing expansion. The plunger retracts, a new hay charge is loaded, and the process repeats until the length of compressed hay is equal to the desired bale length. A tying mechanism then places twine or wire around the bale so that it can maintain its dimensions during storage and transportation. Figure 4.9 provides a detailed drawing of the mechanisms used in producing a rectangular biomass bale. Not shown is the precompression chamber built into some large rectangular bales that allows the baler to compress the crop at greater pressures.

The productivity of all balers is based on the amount of material in the windrow (kg/m) and the forward speed that can be achieved by the baler during the baling operation (km/h). The forward velocity of the baler will be equal to the forward field speed of the tractor. Typically, balers are rated based on the number of bales per hour. Because bales per hour is a customary unit for evaluating balers, the ownership and operating costs are often expressed in dollars per bale as well. All of these values can be converted to a weight per hour or dollar per hour value as long as the average bale weight can be determined. Bale weights are highly variable because they depend, in large part, on the moisture content of biomass in the windrow. Field moisture contents can vary significantly across a single field. There are sources that list typical bale weights for various forage, grass, and hay crops.

Chopping is another option for forage, grass, and hay material that has been cut and windrowed. A forage chopper, also known as a silage chopper, is special type of cutting device that was developed to permit the bulk handling of forage crops. Forage harvesters are either pull-type that require a tractor to provide operating power and forward movement or self-propelled. Forage choppers have different types of feeding heads that can be attached to the front of the machine. A windrow pickup head feeds material that has been cut and collected in a windrow into the chopper. Row crop heads and rotary heads have additional functionality because these feeding heads have the ability to cut standing crop and feed it directly into the chopper (Figure 4.10). A row crop head singulates and processes material in each individual row. The rotary head operates independent of row layout by cutting a continuous swath. A windrow pickup head is required for chopping dry material, because the other feeding heads process a standing crop that would still have significant moisture in the plant. Regardless of whether the machine is pull-type or self-propelled, a series of rollers feed material from the biomass pickup portion of the implement to the chopping mechanism. The feeding rate can be modified by changing the angular velocity of the feed rollers. As the feed rate changes, the length of the chopped biomass pieces will change.

The majority of forage choppers use a rotating cylinder equipped with a series of knives and a shear bar to cut material into small segments (Figure 4.8). The knives on the rotating cylinder can be continuous across

FIGURE 4.10
A pull-type forage harvester chopping sweet sorghum.

the length of the cylinder or they are in shorter segments and arranged in a checkerboard pattern. Knives can be removed to change the cutting length of the machine. The shear bar is a static knife that is responsible for as much cutting as the rotating knives. The majority of the cutting completed by the stationary knife occurs along the knife's vertical edge. Once the material is chopped, it is blown into a forage wagon or open-top truck for transport to a storage facility. Chopping forage is a simple process to implement for the harvest and collection of biomass. This system represents a straightforward harvest process that minimizes the required pieces of equipment and has the potential to process material much faster than baling operations. The major disadvantages are that the cutting and feeding of biomass in a forage chopper requires a significant amount of power, so much so that forage chopping is one of the largest consumers of energy of all agricultural operations. Also, storage of wet biomass is more complicated and potentially more expensive because wet biomass is more susceptible to rot and soluble sugars will breakdown quickly if they are not utilized.

4.2.2 Agricultural Residue Recovery

Agricultural residues are often discussed as a biomass resource that can be collected either at the time of the primary crop harvest or in a subsequent operation. The benefit of collecting agriculture residues at the time of harvest is that the harvesting of the higher value primary crop can be used to subsidize most of the biomass collection operation. The residues from grain crops are often mentioned as a biomass resource because they interact with a grain combine during harvest (Figure 4.11). Corn stover and wheat straw are among the most important crop residues that are listed as potential biofuels feedstocks.

Typically, during grain combine operation, a portion of the plant material is fed into the machine and a series of devices inside the combine work to separate the grain from the rest of the plant material. The grain remains on the combine, stored in a tank, while the plant material or material other than grain (MOG) exits the rear of the machine. The material exiting the rear of the combine can then be collected and processed using equipment developed for forage, grass, and hay field processing. Figure 4.12 illustrates the processing and equipment options available for the collection of agricultural residues.

Because the different plant components have varying values as a bioenergy feedstock, a new term, fractionization, is introduced. Fractionization involves the separation of plant material into different grouping that could be based on economic value, physical properties, bioenergy conversion potential, etc. Fractionization can be a beneficial process to incorporate in biomass recovery from both an economic and sustainability standpoint. Fractionization is important in the case of grain harvest because it allows high-value material such as the grain kernel to be recovered and sold to generate the maximum

FIGURE 4.11
Wheat straw processed during combine harvest is an agriculture residue that is usually collected for animal feed and there are options to use it as a bioenergy feedstock.

income. Another important component of fractionization is the potential to segregate biomass and allow the collection of only components that show potential as an energy feedstock. From an agronomic perspective, it may harm the productivity of an agricultural field to remove all of the biomass. Leaving some biomass in the field is important for soil erosion protection, nutrient management, and maintaining adequate organic matter levels in the soil. Fractionization is the process required to allow sustainable agricultural practices while agricultural residues are collected for bioenergy needs.

There has been a significant amount of research and development related to the collection of corn stover. Corn stover is the plant material that remains on a field's surface following combine operations to collect grain. The amount of corn stover can be estimated from a field using a term called the harvest index. A harvest index represents the ratio of grain yield to total biomass yield and a typical value for corn is around 50%. Based on a harvest index value of 50%, a corn plant produces an equal amount of grain and biomass on a dry matter basis. Generally, corn stover collection requires the use of a shredding attachment in front of the baler's pickup because the corn plants are not cut low to the ground. A corn head on a grain combine is primarily concerned with pulling the ear off the plant and corn heads are not operated as low to the ground as conventional mowing implements. This relatively high operating height for the corn head also reduces the likelihood of

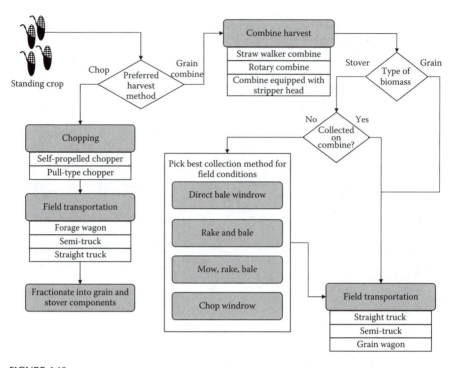

FIGURE 4.12

Processing operations and associated equipment of harvest, collection, and field transportation activities for grain crop residues.

feeding rocks or other solid debris into the machine. A result of this higher operation is that there is substantial stubble left in the field; shredders help break up this stubble to improve baling operations. Shredders also insure relatively consistent particle size to improve bale formation.

Stover collection systems can be categorized as one-pass or multi-pass systems. In a one-pass system, the grain combine would thresh and separate grain as well as fractionize and collection stover. Stover exiting the rear of the machine can be directly baled, cobs may be separated and captured in a segregated grain tank, or stover can be chopped and blown into a forage wagon. Single pass systems are under development, and their viability is not fully recognized to date. A multi-pass system collects stover material after the combine operation is complete. Efforts to quantify stover collection efficiency by hay and forage equipment found that baling only has a relatively low recovery rate, 41.1%, compared to raking and baling (57.2% stover recovery) and mow, rake, and bale operations (65.7% stover recovery) (Prewitt et al., 2007). This same study found that creating a windrow directly behind the combine has the potential to recover up to 74.1% of the available stover. Additional investigation by Sokhansanj and Turhollow (2004) found that round bale systems are $2.00/ton cheaper to produce, transport, and store than a large square bale. The higher capital cost and requirements for additional raking lead to the higher cost for the square bale.

The simultaneous harvest of grain and collection of biomass may not be the ideal operational situation for a grain combine. Combine designs have been optimized to manage biomass throughput in the machine so the machine can operate faster, with greater threshing efficiency, and minimize harvest losses. Adding operations to a grain combine or increasing the throughput of nongrain material can seriously impact the field capacity, grain loss, and threshing efficiency values established for a machine. Because the combine has one of the highest machine rates for an agricultural machine, any decrease in the volume of grain processed will be harmful to a farming operation, unless the value of the collect biomass exceeds any operating losses. The value of the biomass must cover both the expense incurred through biomass collection and any additional costs by inhibiting a grain combine from doing its primary work function, threshing grain.

4.2.3 Forest Biomass Equipment

Harvesting and preparing forest biomass requires specialized equipment due to the increased size of the material, the rugged conditions, and required throughput to maintain an economically viable operation. Potential sources for forest biomass include trees, logging residues such as tops and limps, and short rotation woody crops grown specifically for energy production. There are established systems for harvesting woody biomass from the forest; however, these systems were designed for processing matured trees with diameters in excess of 50 cm. Conventional forest harvesting equipment is divided into two major methodologies: tree-length systems and cut-to-length (CTL) systems (Figure 4.13). Tree-length systems keep the tree intact through the harvest and transportation, with minor infield processing carried out to remove limbs and tops to facilitate transportation. A CTL system divides or merchandises the tree into various products based on the diameter of the tree. Both of these systems have the potential to collect and harvest woody biomass; however, these systems are relatively expensive to operate and the ability to assemble a profitable biomass recovery system using conventional tree-harvesting equipment can be difficult. Figure 4.14 displays machinery that is commonly used to carry out both whole-tree and CTL processing operations.

In a tree-length system, trees are harvested using a feller-buncher, which cuts a tree using a large spinning disk saw and then lays the tree in a pile or bunch. Next, a skidder uses a large grapple to pick up these tree piles at one end and drag them to a central location at the logging site called the logging deck. Along the way, skidders may push the trees in reverse through a delimbing gate to break off the limbs and tops. If a delimbing gate is not employed, the limbs and tops will be removed either manually with a chainsaw or using a pull through delimber and loader. The loader will pick

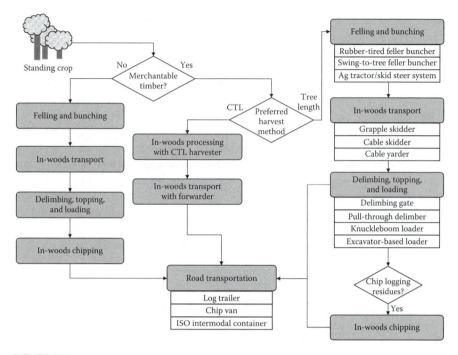

FIGURE 4.13
Processing operations and associated equipment of harvest, collection, and field transportation activities for forest biomass.

FIGURE 4.14
An excavator modified for CTL harvesting (left), a tracked, grapple skidder and rubber-tired feller buncher (middle), and a pull-through delimber equipped with a topping saw (right) are common pieces of equipment used to process woody biomass.

individual trees from the pile pulled to the deck by the skidder. The loader can sort the trees according to size, perform the delimbing and topping operations, and place the trees on the truck for shipping.

A CTL system uses fewer machines and provides an opportunity to divide a single tree into several products. The CTL system relies on two machines: a harvester and a forwarder. The harvester cuts the tree, removes the limbs, and cuts the tree into various length pieces. The decision on how to cut the tree into individual length is based on the highest value product that can be produced from the tree segment. After the tree is cut to length, the tree segments are left on the ground. Next, a forwarder uses a grapple to pick up the tree segments and place them in a bunk, which secures the material off the ground. The forwarder transports the woody material from the forest to the truck for on-road transport. The forwarder is also responsible for unloading its bunk and placing the trees onto the truck.

The advantage of a CTL system versus a tree-length system is the reduced number of machines, greater marketing opportunities, reduced environmental impact, and the system operates "cold." Unlike a tree-length system, the harvester and forwarder can operate at their own pace and the interaction between these machines is not as critical. The forwarder could potentially arrive on site after the harvester has finished its operation, and there would be no significant impact on productivity. The disadvantages of CTL systems are the equipment requires a significant capital investment, a high-skilled operator is needed to manage the complexity of the machine, and the throughput of the system is reduced. The higher machine rate and reduced productivity can be overcome by the increased profit that comes from the greater merchandising opportunities. When biomass is used for energy, these marketing opportunities are diminished meaning the potential for profitability of utilizing this system for harvest.

Regardless of the harvest system used for the extraction of woody biomass, chippers can be added to either system to provide additional biomass recovery options. Traditionally, in-woods chippers have been operated to produce wood chips for pulp and paper operations (Figure 4.15).

It would be possible to set up an in-woods chipping operation to process whole trees for biomass as well, assuming a competitive price structure for energy chips was established. Chippers are somewhat similar to forage harvesters in that feed rolls or a conveyor belt feeds material into a cutting device. The majority of chipper used for in-woods chipping use a rotating disk with

FIGURE 4.15
Examples of in-wood chipping equipment.

embedded knives to chip large material into smaller particles. A biomass chipper added to a conventional logging operation would be required to chip tree tops and limbs that are traditionally piled near the logging deck in tree-length operations because they have little to no value. This is an economical process because the residue material is being utilized and most of the costs associated with its production have already been covered by the higher value solid wood products. The energy chips are only required to pay for the operation of the chipper and any additional skidding or loading operations needed to feed the chipper.

The harvest and collection of biomass has received considerable amount of research and development interest from government, academic, and industrial organizations looking for low cost, highly effective means to gather biomass. New machinery markets are opening for equipment that is specifically designed for biomass collection; however, these machines are still in their early development and the economic conditions surrounding the operation of these machines are marginal. Modification kits for simultaneous collection of corn cobs and grain corn have been developed for grain combines. Also, bundling machines that collect woody residues from the forest floor then compress them into simulated synthetic logs are in operation in Scandinavia. As biomass logistics networks become more defined and biorefining operations define how they wish to receive biomass; harvest/collection systems will be optimized, economies of scale will dominate, and the costs will decrease.

4.3 Transportation

Many bioenergy stakeholders and logistics systems managers consider transportation as a potentially expensive and challenging aspect of the logistics system. The primary reason for this assumption is that transportation systems do not add value to the material; transportation only cuts into the potential profit. Scheduling the arrival and departure of shipments, investing capital in loading and unloading systems, and maintaining the feedstock quality standards during shipment present a significant challenge to a logistics manager. In the industrialized world, there are three major networks used for shipment of agricultural and forestry products: road, rail, and navigable bodies of water. Costs are generally fixed within a given system as a biorefinery would essentially pay a negotiated dollar per unit distance or dollar per unit volume price for the movement of goods into the factory. It is to the biorefinery's benefit to use the largest possible transportation unit for feedstock delivery as this will streamline material handling and reduce the number of deliveries into the facility. The primary factor creating variability within transportation costs is the expense associated with energy (i.e., fuel) to power the transportation system. The majority of the systems used for bulk transport of biomass rely on diesel engines as a power source. Fuel costs can account for up to 30% of the overall transportation expense; therefore, volatility in the petroleum prices can negatively impact the economics of a logistics system. If the largest possible transportation system for moving bulk material is used by a biorefinery, it is possible to spread these costs over a larger volume of feedstock.

Dollars per ton of material moved (i.e., $/ton) is the most important value when evaluating a logistics system. Traditionally, agricultural and forestry products have been sold or traded on a weight or volumetric basis. Equilibrating transportation system costs to these traditional commodity transaction units helps the land owner, crop producer, investors, and the biorefinery have a clear understanding of the impact of transportation on costs on profitability and expense allocation. Economies of scale will justify a switch to higher capacity transportation systems as distances increase. Consider the following scenario, a 2300 bushel shipment of grain has to travel 640 km. This volume of grain would require at least two semitruck trips, meanwhile this volume of grain would only fill half of a railcar. However, the railcar would cost nearly 30% less per bushel. If the volume was reduced to 1000 bushels, the truck would be considerably cheaper because of the severe underutilization of the train car's capacity and the payment to the rail transportation company.

Typically, the larger a biorefining operation the greater the transportation cost associated with the delivery of biomass. Larger biorefineries will require a larger service area to draw material into the plant. This problem will be compounded if much of the land use around the plant is tied into conservation or urban use. Also, a larger plant will receive a higher volume of trucks, train cars, or barges, which will require these entities to remain in a queue longer as they wait to unload their loaded material. This queue time is not free as workers are still paid, idling engines burn fuel, and equipment is generating hours that will add to maintenance and replacement fees. Typically, trucks serve as the lowest cost transportation system as it requires the least infrastructure and overhead to use. However, trucks can only move relatively small quantities of biomass and when haul distance increase the transportation costs will increase because the volume of material being transported is not sufficient. Rail and shipping are expensive propositions for short haul applications, but because of the considerable scale of these operations, these transportation costs associated with the system become competitive with increases in haul distance or volume of biomass transported. Rail and shipping transportation systems require relatively long haul distances to offset the considerable overhead associated with operating ports and rail yards.

Figure 4.16 provides an illustration of the cost and distance relationship between common modes of transportation used for biomass transport. The cost has been normalized on a dollars per ton basis. The *y*-intercepts of the lines on the chart represent the initial fixed costs or terminal fees associated

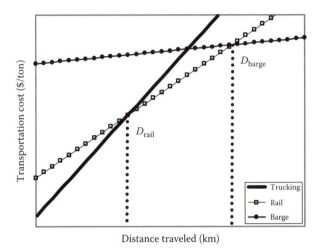

Distance traveled (km)

FIGURE 4.16
Economically favorable travel distances for truck, rail, and barge transportation systems.

with each mode. As the capacity of the transportation method increases, the terminal fee increases because the material handling facilities become far complex and specialized. The slope of the line indicates the mode's cost per kilometer travel. Trucking has the highest cost per kilometer; however, because it has relatively low material handling fees associated with the operation, it is economically viable for shorter distances. As the haul distance and volume of material transported increases, both the rail and barge systems are greater opportunities to distribute costs across entire payload. D_{rail} represents the travel distance where rail becomes more economical than trucking and D_{barge} marks the travel distance required for barges to become the lowest price transportation mode.

The selection of the optimal biomass transportation mode is based on many factors and some of these factors cannot be controlled or modified by the biorefinery. For instance, the transportation infrastructure surround the plant is an important parameter to analyze prior to selecting a biorefinery location. Quality of roads and bridges, access to a deep water port, or proximity to main line rail will dictate the transportation modes actually available to the plant. Also, the biomass refining operation may not have the capital to invest in specialized material-handling systems that may be required to unload trains, barges, or ships. Finally, the competitive environment in the region's transportation industry could drive costs down for certain modes of transportation. In the United States, cities such as Chicago and New Orleans have access to deep water, rivers, railroad, and the interstate highway system. Transportation companies must compete for customers because of the numerous transportation options available; therefore, transportation costs may be lower there than a region of the country with inferior transportation infrastructure.

4.3.1 Trucking

The use of motor vehicles to move food, feed, and fiber from the farm to a processing facility has long been the backbone of the agricultural transportation industry. On modern farms, the motor vehicle used for the movement of material consists of a truck and a trailer capable of hauling between 20 and 25 metric tons of material per load. Trucking is particularly attractive to a biorefinery operator because the truck is usually owned and operated by the feedstock producer. Therefore, the costs to own and operate the vehicle must be covered by the shipper. Also, truck transportation relies primarily on the use of roadways that have been paid for by public funds. This method of shipping involves very little investment for a biorefinery as long as they are not operating a fleet of trucks.

Trucking is also a much needed resource when intermodal transport systems are considered. Intermodal transportation involves the use of at least two modes of transportation between the origin and destination. If a docking facility, rail yards, or airport is located adjacent to the biorefinery, the trucking is usually required to move material either to or from the larger transportation hubs. In modern logistics networks, intermodal transportation is carried out with the use of shipping containers. The freight is loaded into a container at the origin and then the containers are handled within the transportation network. The freight is never handled by the shipping entities, only the containers. The ISO specification for an intermodal container is 2.438 m high by 2.438 m wide. The lengths of the containers do vary, but 5.8, 12.2, 14.6, and 16.15 m are standard lengths used today.

In addition to the intermodal container, there are numerous other truck and trailer combinations that are used to transport biomass. Figure 4.17

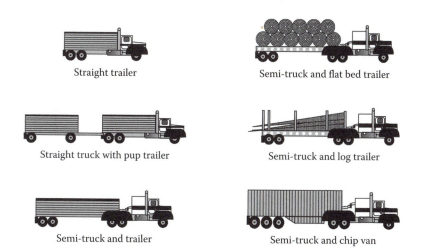

Straight trailer

Semi-truck and flat bed trailer

Straight truck with pup trailer

Semi-truck and log trailer

Semi-truck and trailer

Semi-truck and chip van

FIGURE 4.17
Examples of common truck and trailer combinations used for biomass transportation.

presents the prevailing modes of transporting biomass with a truck and trailer or truck and container system. There are a variety of truck and trailer combinations with a variety of dimension, payload capacity, and cost. Many of the critical dimensions required for truck are shown in Table 4.5. Other important dimensions to note are the maximum height and width dimensions for all trailers, 2.6 m wide and 4.11 m tall. These extreme dimensions are for trucks operating in North America; other regions of the world have unique truck and trailer dimension regulations. The majority of the trailers used to transport biomass are open top, which allows for quick loading operations of bulk material. These trailers may also be equipped with rapid unloading equipment such as hopper bottoms or hydraulic systems that will tilt the entire trailer.

The truck's capacity to haul material is bound by two parameters: weight and dimension. North American trucks are required to maintain a weight below 36.36 tons when traveling on public roads. This weight restriction includes the weight of the truck, trailer, and payload. This weight limit is usually accompanied by additional restrictions that insure the load is distributed evenly across the truck's axles. Additional height, width, and length restrictions are placed on the trailer as well. There are allowances for log trucks that will allow material to hang off the back of the trailer as long as the material is secure. Additional weight restrictions maybe imposed if

TABLE 4.5

Typical Dimensional Parameters Associated with Containers, Trailers, and Trucks Used to Transport Biomass on Road Networks

Trucks and Semitrailer Parameters

Truck Type	Weight
Straight truck chassis	7500–9500 kg
Semitruck	
Day cab	6000–7000 kg
Sleeper cab	7000–8500 kg

Trailer or Container Parameters

Type	Length	Weight	Capacity
Box[a]	3.66–7.62 m	750–1900 kg	12.5–28.5 m³
Hopper bottom trailer	6.70–12.80 m	1080–5017 kg	21–45 m³
Pup trailer	4.88–6.70 m	Box weight + 2000– 3000 kg for chassis	18–24 m³
Flatbed trailer	12.19–16.16 m	4000–5500 kg	2.40–2.60 m width
Chip van	12.80–13.72 m	4250–4850 kg	75–85 m³
Log trailer	12.20–13.72 m	4000–5000 kg	2.60 m wide 3.71 m bolster height

[a] The term box refers to the container mounted on a truck chassis to form a straight truck or the container mounted on an axle assembly to form a pup trailer.

the truck passes over bridges. Bridge weight restrictions may cause a truck's course to be rerouted, adding time and cost to the transportation expense.

The 36.36 ton weight restriction is not the only weight limit that is considered when loading a truck with biomass. Gross vehicle weight rating (GVWR) is a rating that defines the maximum amount of weight a road vehicle can carry while be able to operate legally within the public road network. GVWR includes the weight of the vehicle, fuel, people, and payload. Straight trucks are most susceptible to GVWR restrictions among the trucking systems illustrated in Figure 4.17. A single rear axle straight truck could have a GVWR rating as low as 14.85 tons and more than half of the allowable weight will be the truck chassis and grain box. If the straight truck was equipped with tandem rear axles (i.e., two axles at the rear of the truck), a 20 ton GWVR or more could be expected. Additional axles would increase the trucks carry capacity, but there would be additional expenses incurred when purchasing and maintaining the larger system.

Weight and dimension interactions are important for the selection of a trailer to haul biomass. The weight of both hopper-bottom grain trailers and the open-top boxes used on straight trucks increase as these containers get longer (Figure 4.18a). However, the increased length and weight is accompanied by a corresponding increase in capacity. Weight becomes an important component when considering straight trucks because the box weight will count against the GVWR and restrict the truck's payload. Weight is a minor consideration when purchasing a trailer for a semitruck because most modern trailers have been designed to minimize weight. The type of biomass and the capacity curve can be important when sizing a trailer.

The restrictions on weight and dimension require the biomass logistics network manager to consider how to optimize the trucking system to minimize costs. Because biomass is filled with void spaces, thereby making it light and fluffy, it is possible to fill the trailer's volume long before the weight limit is achieved. This is not an optimal shipping situation because biomass will be usually purchased on a unit weight basis. If the maximum weight is not achieved, then the shipping cost is not distributed across the highest possible load. For instance, if the biorefinery is 40 km one-way from a forestry biomass harvest site, the shipping charge would be $120 per trip (accounting for two-way travel), assuming a mileage rate of $1.50/km. If the truck's load was 12 ton instead for the maximum 36 ton, the shipping cost increases from a possible low value of $3.33/ton to a relatively expensive $10.00/ton.

The topic of densification will be covered later in this chapter. Densification is essentially the process to remove voids from the biomass to increase the material's density. However, there are densification methods, primarily pelletization, that create a product that may be too dense to fully utilize the trailer's full volume. If the weight criteria is achieved before the trailer is full, then the operator of the trailer is wasting money on ownership costs associated with a trailer that is too big. Generally, larger trailers will be more costly to own and operate versus a straight truck (i.e., a truck with the trailer

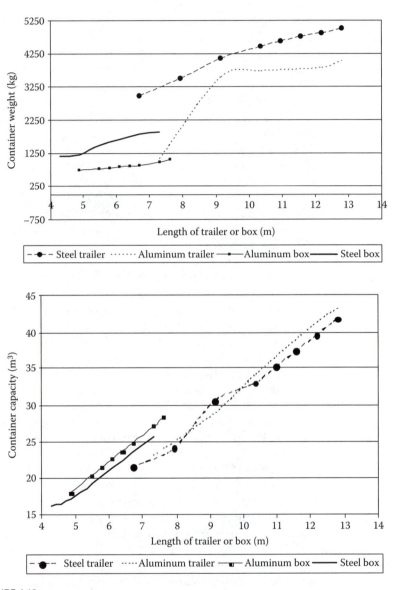

FIGURE 4.18
Comparison of weight and capacity values for open-top box containers and hopper bottom grain trailers constructed from aluminum and steel.

attached to the chassis) or a shorter trailer. It is important to use trailers that can be loaded to the maximum weight criteria and meet their volumetric capacity. Figure 4.19 shows the relationship between volumetric capacity (m³), maximum material density to achieve full weight, and trailer length for ISO intermodal container. Even the largest ISO intermodal container can

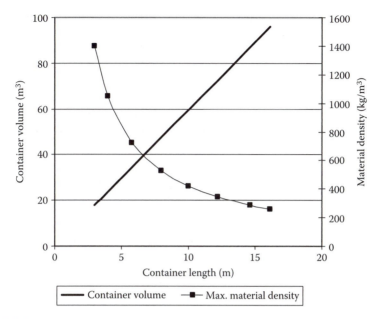

FIGURE 4.19

Relationship between ISO intermodal container length, the volumetric capacity of the container, and the maximum material bulk density that can fill the container without surpass weight restrictions.

be filled with biomass having a density of 260 kg/m³ and not surpass weight restrictions in North America. One of the concerns about biomass is that most uncompressed, chopped biomass materials have a bulk density well below the 260 kg/m³ threshold. Therefore, the truck will never achieve the maximum weight, and the biorefinery will essentially be charge freight to haul the air trapped in the void spaces between biomass particles.

Example 4.8:

An electrical generation company is interested in harvesting and pelletizing *Miscanthus* at farm sites prior to delivery to a power plant. The final pellet will have a bulk density of 650 kg/m³. Find the optimal aluminum trailer length for this operation.

Arbitrarily, a 12 m trailer is selected in Figure 4.17. The corresponding capacity (41 m³) and weight (3800 kg) were read from the graphs. A day cab equipped semitruck (6500 kg, Table 4.5) was assumed because moving material from fields to the biorefinery will not require the driver to sleep on the road.

The total weight of the semitruck-trailer system is 10,300 kg or 10.3 tons. The maximum weight allowed on the road is 36.36 tons, and there are no GVWR restrictions to consider.

36.36 tons–10.3 ton = 26.06 or 26,060 kg tons of weight capacity

26,060 kg ÷ 650 kg/m³ = 40.09 m³ of material capacity needed to achieve load

The 12 m trailer has a capacity of 41 m³, which is sufficient for meeting weight, volume, and space requirements. Having the trailer slightly larger than the required dimension is preferred as it allows additional room for dimensional variation in the material and the loading operation is not required to pack the container full to achieve the desired weight. However, a trailer that is significantly larger is not advisable as the loading operator might be tempted to overload the truck and trailer causing increased expenses, greater wear on the equipment, and the possibility of incurring fines from law enforcement.

Economic analysis of a truck is very similar to the analysis discussed for agricultural and forest machinery. The truck acts like a prime mover providing power to move an implement, in this case a trailer. The fixed and variable cost categories are essentially the same; however, many of the formula used to calculate estimates for specific cost components are abandoned. Instead of formulas, values must be estimated from sources including journal articles, trade publications, manufacturer information, and personnel experience. Fees related to maintaining licenses required to operated the truck on public roads are now included in the taxes, housing, and insurance category. Also, the cost to replace tires is an additional variable cost that must be considered. Because truck and trailer combinations can have 18 or more tires and their life is far less than the service life of the vehicle, it is important to quantify these costs.

Table 4.6 provides a breakdown of the variables required for estimating truck costs as well as the necessary modifications to apply the cost estimate equations discussed for agricultural and forestry machines. The cost variables (shown on the left side of the table) are based on direct estimates, there are no formulas used to calculate the values. Formulas for determining the actual cost component values are provided. Note that the letters in the formulas refer to the corresponding letter used to designate a cost variable. As mentioned earlier, there are modifications in the formulas previous developed for equipment cost approximation. The charge for annual tire replacement is a function of tire service life, tire replacement cost, the number of tires on the truck, and annual distance driven. Labor cost is primarily a function of the time spent driving the truck, which is found by dividing the annual distance driven by the average speed of the truck. Finally the fuel and lube costs are a function of the vehicle's fuel efficiency rating, which can be obtained from the manufacturer. The sum of total ownership and total operating costs is the total cost to operate the truck for a year. By diving this value by the annual volume of material transported or annual distance traveled, it is possible to acquire a value for dollars per ton and dollars per mile.

The cost estimation methods applied in Table 4.6 could also be used to calculate the costs associated with the operation of the trailer as well. The major differences would be much lower values for taxes/license fees, reduced repair costs, and the fuel, lubrication, and labor costs would be eliminated.

TABLE 4.6

Trucking Cost Estimation

Ownership Cost Variables			Ownership Costs	
A.	Purchase price	$85,000	Depreciation and interest	$= A*[(1-(C/A))/B+ ((1+(C/A))/2)*i]$ $= \$10,157$
B.	Economic life	10 years	Taxes, insurance, licenses	$3000
C.	Salvage value	$20,000	**Total ownership costs**	**$13,157**
D.	Taxes, insurance, license fees	$3000		
E.	Interest rate	7%		
Operating Cost Variables			**Operating Cost**	
F.	Annual distance driven	24,000 km	Annual repair costs	$1750
G.	Annual repair costs	$1750	Annual tire costs	$= (H*I*F)/J$ $= \$384$
H.	Number of tires	10	Fuel and lube cost	$= 1.1*K*(F/L)$ $= \$2511$
I.	Replacement cost per tire	$400	Labor costs	$= (F/M)*N*1.25$ $= \$6250$
J.	Life of tire	250,000 km	**Total operating costs**	**$10,895**
K.	Price of fuel	$0.92/L		
L.	Fuel efficiency	9.67 km/L		
M.	Average speed	72 km/h	**Total cost**	**$24,052/year or $1.00/ km**
N.	Driver labor rate	$15/h		

The fuel cost would only be considered if the trailer was a refrigerated unit or refer, which required the consumption of fuel for climate control purposes. Otherwise, the fuel and labor have been assigned to the truck. If the truck has a trailer or container attached physically to the chassis of the truck, then a trailer calculation is not necessary.

Example 4.9:

A crop producer is generating large rectangular bales, 1.22 m × 1.22 m × 2.44 m, with an average bulk density of 160 kg/m^3. The farm has purchased a day cab semitruck discussed in Table 4.6 and a flatbed trailer that holds 25 bales. The trailer will travel 50 km to a biorefinery approximately 120 times per year. Determine the dollars per kilometer and dollars per ton costs associated with operating this truck system. The trailer cost $38,500, has eight tires, a $10,000 salvage value, and requires two annual payment of $375 for repair/maintenance and insurance. Other values can be substituted from Table 4.6.

Solution:

From Table 4.6, the truck costs $1.00/km and from Table 4.7, the trailer costs an additional $0.17/km. This is a total transportation cost of $1.45/km. The biorefinery is 50 km one-way for a total round trip distance of 100 km and a cost of $145.

Each bale has a volume of 3.62 m³ for an average weight of 580 kg. A standard flatbed trailer has a deck height of 1.52 m restricting the stacking height of the bales to two levels. Therefore, the trailer is limited to 25 bales that would create an average load of 14.5 tons. This equates to a shipping cost of $10/ton.

4.3.2 Rail

Rail systems become important for the transportation of biomass when a biorefinery requires such a large volume of material that truck-based freight systems lose their initial competitive advantage. Trucks are cheaper to operate for short distances because of the low overhead invested in these systems. However, trucks have a relatively high cost per unit distance traveled and the overall cost per ton will increase because the payload capacity is not sufficient to distribute these costs. Rail provides a couple of advantages over trucking. First, the volume of material can be dramatically increased compared to trucking. Second, trains operate on rail systems owned by the freight carrier, this is a situation that is unique to railroads. All other methods of transportation either use a publicly financed system or natural resource

TABLE 4.7

Trailer Cost Estimation

Ownership Cost Variables		Ownership Costs	
A. Purchase price	$38,500	Depreciation and interest	$= A*[(1-(C/A))/B + ((1+(C/A))/2)*i]$ $= \$4548$
B. Economic life	10 years	Taxes, insurance, licenses	$375
C. Salvage value	$10,000	**Total ownership costs**	**$4923**
D. Taxes, insurance, license fees	$375		
E. Interest rate	7%		
Operating Cost Variables		**Operating Cost**	
F. Annual distance driven	12,000 km	Annual repair costs	$375
G. Annual repair costs	$375	Annual tire costs	$= (H*I*F)/J$ $= \$154$
H. Number of tires	8	**Total operating costs**	**$529**
I. Replacement cost per tire	$400		
J. Life of tire	250,000 km		
		Total cost	**$5452/year or $0.45/ km**

to travel. Because the freight carrier is interested in maximizing profits and efficiency, rail systems are well managed and the movement of material on rail is highly structured. This structure is allowed because the railroad carrier owns the trains, railroad, and typically the railcars too.

The disadvantages associated with rail transport are related to scale and the private ownership of the system. Because the capacity of the train is so large, considerable funds are required to construct infrastructure to load and unload railcars. Also, small segments of rail (i.e., spurs or sidings) linking the plant's loading/unloading facilities to the main line rail have to be constructed. Therefore, rail has a higher initial fixed cost because of the capital and infrastructure required to use rail systems. The privatized ownership of rail systems is a disadvantage in the sense there are no analytical methods to calculate the cost of moving material by rail. Fortunately, many of the large railroad freight carriers operating in North America provide cost estimates for shipping material to and from various rail hubs across the continent. These cost estimates can be accessed from internet websites for the individual rail companies. There can be variations in prices depending of fuel costs, demand for railcars, and the direction of travel. Railroads are typically set up to run along north-south or east-west routes. When a shipment needs to more in a southeast direction, there may be higher costs because the train is required to switch tracks to travel in the correct direction. Interaction with a rail yard and junction will incur higher costs for the transportation system.

A unit train or block train is a train that is used to move material from one point of origin to a single destination. Typically, unit trains only carry a single commodity; therefore, the railcars are usually the same type (i.e., tanker, flatbed, boxcar) if not completely identical. Unit trains are usually a predetermined length or number of railcars that is determined by the rail company that operates the railroad. In North America, the technical definition of a unit train refers to any train pulling more than 52 railcars loaded with the same commodity. However, due to economic and optimization factors, unit trains are generally 75 or 110 cars in length depending on the freight carrier. Because the unit train travels directly from the origin to the destination, shipping is fast, efficient, and runs on a tight schedule. The uniformity of the unit train system allows end users of the product being shipped to construct specialized material handling facilities that improve efficiency and economics.

A classic example of rail transport in the United States is the movement of corn used for animal feed, which is shipped from the midwestern United States to the southeastern United States in extremely high volumes. The southeastern United States is one of the leading livestock and poultry production regions in the world; however, due to marginal grain yields this region relies on grain imports to meet the demands of these animal feeding operations. A single railcar has a capacity of 90.72 tons, which is equivalent to 3500 bushels of corn. A unit train with 75 cars would move a volume of grain equivalent to more than 290 trucks. Some grain corn millers and refiners may use more than one unit train per week, and it becomes clear that unloading one or two scheduled trains per week is a more reliable and controlled situation compared to 600 pseudo-random truck arrivals. Consider Figure 4.20, which shows variations in transportation cost for corn based on transportation mode, the volume of grain filling the railcar, and distance traveled.

Notice that trucking remains competitive with the railroad traffic as long as the travel distance is 200 km or less. Also, if the railcar's payload is not fully utilized, the associated transportation costs cannot be distributed across an adequate volume or weight of grain. Therefore, in the most severe underutilization case (25% railcar loading), the rail system cost surpasses the trucking cost.

Rail systems are generally not considered a self-reliant mode of transportation because it is highly unlikely a biorefinery or rail carrier would be willing to invest capital to lay track for each individual farm supplying biomass to the refinery. Rail systems rely on trucking or shipping to move raw materials to a rail yard or depot where they are loaded into the railcars. The use of multiple transportation modes must be considered in the analysis, as the transportation cost to utilize a rail system may increase dramatically if long-distance trucking is required to bring material to the rail yard. Typically, rail and trucking systems are combined in what are known as satellite processing or satellite storage facilities. The satellite system works by placing small depots at multiple locations within the rail network servicing a biorefinery. The locations are strategically located to minimize the distances traveled by motor vehicles moving biomass from the field to the depot. The rail system will do the majority of the long-distance travel. Also, biomass is allowed to accumulate at these satellite facilities until the quantities are sufficient to dispatch a unit train; thereby,

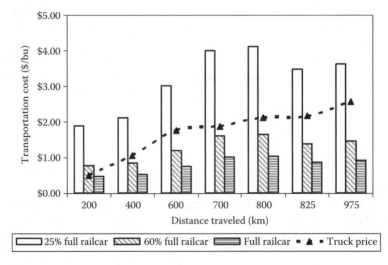

FIGURE 4.20
Comparison of truck and rail transportation costs for railcars filled to varying capacities.

improving the economics of this system. From a cost accounting standpoint, it will be necessary to consider rail, trucking, and the satellite facility's overhead costs. The pulp and paper industry has utilized satellite chipping facilities to process solid wood biomass into chips at locations far removed from the paper mill. This allows the procurement system to purchase woody fiber from a larger region, which insures adequate feedstock supply.

4.3.3 Maritime Shipping

Maritime shipping involves the use of navigable waterways or large bodies of water to move materials in bulk quantities to much greater distances than rail. Shipping typically involves the use of transoceanic ships that are capable of handling tens of thousands of tons of bulk material or the use of barges that travel along navigable rivers. As a reference point, the typical barge can carry the same weight of material as 15 railcars or approximately 1360 tons. Transoceanic shipping and barge transport is economical when the scale of the operation is considered; however, much like rail transportation, shipping is not a self-reliant transportation mode as both rail and road networks must be used to move materials to and from ports.

A unique aspect of maritime shipping is that there are times when it is the only means of economical transportation available for the bulk movement of goods and materials. Due to the expense of airborne freight, maritime shipping is the only means to ship materials between many continents. Maritime shipping is by far the largest capacity transportation used in the world today, and the costs associated with this scale are only justified if travel distances are long and the volume of biomass transported is high. The greatest disadvantages associated with maritime shipping are that it is the slowest transportation mode and this transportation system is possibly the least reliable. The reliability issues stemming from the probability of encountering a delay are increased because of the long travel distances involved. Also, many ports are nearly their material handling capacity so slight delays at a port facility can quickly spiral into considerable time delays when loading or unloading a ship.

There are numerous classifications of ships that can be employed in the movement of biomass. From a biomass transportation point-of-view, the shipping options come down to two general categories: bulk ships and container ships. A bulk freighter is designed to transport unpackaged material such as grain, coal, or iron ore in a series of cargo holds. The capacity of bulk freighters is specified in deadweight tonnage. A large grain bulk freighter has a DWT of approximately 76,000 tons. A ship with this DWT rating is known as a Panamax freighter because it has the maximum dimensions that allow clearance through the lock system on the Panama Canal. Another unit to express a ship's capacity is TEU or 20 ft equivalent unit. A TEU is equal to the cargo-carrying capacity of an ISO intermodal container 20 ft or 6.10 m in length. A Panamax ship has a 4000–4999 TEU capacity.

A container ship specializes in the transportation of ISO intermodal containers. These containers are the only freight that is handled by the ship during a voyage. Unlike bulk freighters, container ships usually run on fairly regular schedules. Bulk carriers operate on a principle known as tramping, this indicates the ship to operate on an "as-needed" basis. Bulk freighters are dispatched to ports when a client requests a ship. Bulk cargo accounts for the majority of the tonnage moved over the oceans; however, it only accounts for a small percentage of the total values of goods shipped by maritime vessels. Container ships are used to ship high-value materials across the ocean, but only carry 30% of the freight carried by maritime vessels.

There are many fees associated with the transportation of biomass on maritime vessels. First, there is a container rate that is the charge associated with renting or using an ISO intermodal container or in the case of a bulk freighter this is the rental fee for space on the ship. There are also currency adjustment factors due to the international movement of bulk and container ships. Port fees are paid to cover the costs associated with tug boat operations, material handling, docking, and dock worker salaries. There is a substantial fuel bill generated by crossing an ocean in a very large ship. Finally, arbitrary charges can be applied to the shipping costs for special needs such as ice breaking or electrical power outlet access to run refrigerator units on refrigerated units.

Typically, the customer booking a bulk freighter or container ship will be billed on a per day basis for use of the ship. Depending on demand for ships, a rate exceeding $50,000/day can be charged for use of a Panamax bulk freighter. However, if the ship is carrying 60,000 tons of materials, this corresponds to a shipping rate of $0.83 ton/day. The United States Department of Agriculture Agricultural Marketing Service (USDA AMS) provides weekly reports on transportation costs associated with the movement of grain. These reports are critical for assessing both transoceanic shipping and barge rates. Barge rates are adjusted daily, and these prices can be volatile if the supply of barges is restricted due to demand or natural disaster. Seasonal flooding can halt barge traffic on rivers; therefore, barges that are full of cargo cannot get to their destination to unload. The USDA AMS report has detailed information on barge traffic on the Mississippi River, barge availability, and cost per ton to move material along the river various distances. The price structure for maritime shipping is the most volatile of all transportation modes; therefore, predicting the associated cost is difficult. But this fact should not be too alarming because this mode of transportation is potentially the least important to a biorefining operation because one of the advantages for bioenergy systems is the reliance of locally available feedstocks.

4.3.4 GIS Analysis of Transportation Systems

The discussion on transportation systems has been limited to calculating various cost factors and making a decision regarding the optimal transportation mode based on the lowest cost per unit distance or per unit volume delivered. If more information is known about the transportation network in a region, it would be possible to conduct a more through transportation analysis. Information such as weight-restricted bridge locations, the speed limits associated with roads within the network, the number of intersections encountered by a truck, and the opportunity to carry freight on the return trip could be used to dramatically improve the operating efficiency of the supply chain. If the number of trucks arriving in a given period of time was better understood, weigh scales and truck unloading facilities could be designed to minimize truck queue and idle times. Also, the biorefinery could eliminate excess receiving capacity if truck arrivals could be effectively managed.

A geographic information system (GIS) is a powerful software tool that performs complex analysis of geospatial features and any attributes associated with these features. ESRI is one of the leading developers of GIS software, and they have developed a network analyst tool that can help logistics managers decide the shortest routes required for individual trucks, the area available to feed biomass into the plant based on transportation costs, and provide projections about the arrival frequency of trucks. GIS tools can be used to

1. Calculate an acceptable service area based on the time or costs required for a truck to travel from the farm to refining plant
2. Optimize trucking routes between farms and the biorefinery
3. Determine the trucking costs associated with moving material from various zones within a region so the biorefinery can make an informed decision regarding the transportation modes to employ

Figure 4.21 illustrates the use of GIS tools to determine trucking cost zones for feeding a pellet mill facility located in the deep water port in Wilmington, North Carolina.

The transportation costs were calculated based on the mileage and time required to travel from any address within the analyzed region to the Port of Wilmington. This analysis takes into account the infrastructure of the road

FIGURE 4.21
Using GIS tools to determine transportation cost zones based on existing infrastructure, speed limits, and roadway intersections.

network, speed limits, and number of intersections encountered along the way. Establishing the amount of material within a given transportation cost range of the plant will allow the operators to determine if there is enough biomass in the region to support the plant's operation. If the trucking costs escalate to the point that it is not an economical transport mode for biomass delivery, plant operators could investigate rail shipment from further out locations or study the viability of establishing satellite biomass-processing facilities, which would rely on rail and truck transportation. GIS would be a cost-effective tool to study the feasibility of employing one transportation mode or network versus another.

4.4 Storage

Biomass storage is a critical aspect of the logistics system because of the massive amount of material a biorefinery or energy generation plant must stockpile to insure continuous operation. Like transportation, biomass storage is an essential part of the supply chain, but it will do little to increase the value of the material. Storage facilities provide a place to accumulate biomass and minimize degradation while the material is held for later use. Because biomass is a biological material, it is subject to degradation due to microbial activity and climatic conditions. A properly designed and implemented storage system should minimize the degradation of the material while the material is stored. The relatively low bulk density of biomass may require considerable space requirements to store adequate biomass for continuous year-round operation, though it is required when selecting a particular storage method for biomass to minimize the footprint and structure required for storage. Typically, the harvest and collection systems used for a particular biomass resource will dictate the type of storage system that is needed.

Many of the storage systems currently utilized for agricultural and forestry commodities can be modified to accommodate other forms of biomass. The ability for most biomass resources to readily absorb water from the environment is a critical concern that biomass storage must address. Moisture content control depends on when the crop was harvested and the exact storage conditions needed to maintain the required biomass component (i.e., sugars, carbohydrates, or cellulose) in a useable form. In addition to moisture, oxygen concentrations, temperature, and nutritional parameters of the biomass can be stabilized and controlled using properly designed storage structures. Aerobic versus anaerobic storage consideration, cooling material to slow microbial activity, and attempting to starve or poison microorganisms by promoting conditions that either produce harmful compounds or introduce growth-inhibiting compounds (i.e., spray on chemical) to the biomass are all considerations for designing biomass storage structures.

Perhaps the simplest method that can be used to store biomass is to leave it exposed to the elements. This type of storage would not be acceptable for plant material that is susceptible to dry matter loss as a result of microbial activity or degradation from the environment. Biomass left outdoors unsheltered may be blown away by wind as extremely dry, brittle material breaks down or excessive moisture can create a situation that fosters microorganism growth. However, round bales of forage, grass, and hay material as well as solid wood harvested from a forest have been successfully stored outdoors without shelter. Round bales form a protective crust around the outer edge that sheds rain and snow fall. This crust can account for 10%–60% of the bales dry matter depending on the depth of the crust and the size of the round bale. The conversion of this "crust" material to a biofuel product may not yield the same conversion efficiency as the unaffected material in the bale. The flat edges of a round bale are highly susceptible to moisture intrusion because they do not develop a protective barrier. Also, unsheltered round bale storage should include the use of either a well-drained ground condition, such as gravel, or some type of barrier between the ground and the bale, such as a plastic sheet. If protection from ground moisture is not provided, there is an increased probability that dry matter loss will occur in the bottom portion of a round bale. Therefore, the only cost associated with the unsheltered round bale storage would be any ground moisture barrier and drainage structures.

Like round bales, solid woody biomass, primarily tree trucks, are stored outdoors under unsheltered conditions. However, often these logs are stored under high moisture conditions to help prevent degradation. Historically, logs were stored by floating them in ponds prior to use in a sawmill. The log ponds have been replaced with the use of sprinkler systems that provide enough water to saturate large log piles storage at a wood yard. Saturating the woody fiber with water provides dimensional stability for a log. If the log were to shrink and crack as it dries, then the log would be more susceptible to attack from insects. The saturated woody fiber is also more resistant to attack from mold, fungus, and bacteria because the water limits the movement of air (oxygen) through a wood pile and keeps the wood's moisture above the optimal moisture levels that are favorable for microbial growth. Sometimes logs used for solid wood products and paper production are treated with additional chemicals that provide additional protection against microorganisms and insects. Because many bioenergy conversion processes will rely on microbial activity to convert biomass into a more valuable energy form; the application of these preservatives should be carefully monitored to insure their presence or residues will inhibit the activity of beneficial microorganisms.

Biomass bales present an interesting challenge for a biomass conversion operation as they are in many ways the optimal form from transportation and material handing standpoint, yet both square and round bales present a unique set of storage challenges. Dry matter loss is inevitable when bales are considered, as material handling, weather conditions, microbial activity, and other pests (i.e., insects, rats, mice) will consume solid material. Square bales are particularly demanding because of their ability to readily absorb water, which leads to spoilage and material loss problems. Unlike round bales, which orient straw in a direction that provides moisture with a pathway around the bale's core, square bales have straw oriented in a way that will actually carry water into the center of the bale. Therefore, square bales require sheltered storage. Round bales can be stored under shelter as well to prevent the formation of a protective crust layer that might lead to excessive dry matter losses; however, because of the inefficiencies and void spaces created when stacking round bales storing under a roof is usually not economical.

Typically, the cost of a storage structure is converted to a dollars per ton figure, which has been a recurring term throughout this chapter. Normalizing the costs will allow a supply chain analyst to determine the total costs to produce and deliver a crop from the field to the start of processing operations. For storage structures, there are many references that will provide approximate costs required to construct a particular storage building on a per unit area ($/m²). For instance, a building with side walls can cost substantially more than a structure that just provides a roof over the biomass. A region's climate as well as the material's susceptibility to degradation will determine the extent of the structure as well as the complexity of the design. Based on the building's footprint or area covered, it is possible to determine the volume or weight of material that can be stored in the shelter. Using construction costs and storage capacity, it is possible to calculate the total investment ($/ton) required to operate the storage facility. The total investment value must be divided into annual installments so the storage system costs can be included in the overall economic model of the supply chain.

As with transportation systems and agricultural equipment, the annual cost to store the material can be found by breaking down the fixed and variable costs. The unique aspect of a building is essentially all costs can be treated as a fixed cost because the building is constantly exposed to the elements, protecting the contents of the building from degradation and harm. The economic factors associated with building ownership are depreciation, interest, taxes and insurance, and maintenance. Usually, each of these values is calculated as a percentage of the total investment costs, which was discussed in the preceding paragraph. Building depreciation is considered to be straight-line with a salvage value approaching zero at the end of the economic life. Building economic life is considerably longer than transportation or agricultural machinery. A building with a life of 25 years would have an annual depreciation component equivalent to 4% of the total investment. The interest rate can be plugged directly into the equation. The remaining maintenance, tax, and insurance costs can range between 2% and 7% of the total investment costs. A hay storage structure with an economic life of 25 years would have an annual cost between 10% and 17% of the total investment value.

Another way to protect both round and square bales is to wrap bales with some material that will provide protection from the elements. There are devices known as bale wrappers that rotate a bale on essentially a turntable while dispensing plastic wrap from a spool. The thickness of the plastic used and the number of plastic layers used to cover the bale are the primary parameters that must be decided. For each crop, there is an optimal number of plastic layers that will sufficiently protect the bale and this level of protection will not be significantly improved upon by adding additional layers. Proper application of the plastic layer will provide a barrier from moisture and air intrusion into the bale. Once any microbes in the bale consume the preexisting oxygen, the bale's temperature will stabilize. Bale wrapping should be done as soon as possible following the actual baling operation. The cost of the bale wrapper, the tractor needed to operate the bale wrapper, as well as the plastic consumed in the wrapping process must be considered when doing an economic analysis of bale wrapping. The main drawbacks to bale wrapping are there are increased material handling costs because bales are generally handled as individual units and they are relatively small in size. Also, if there is excessive moisture in the biomass or moisture is somehow introduced to the wrapped bale, it will pool at the bottom of the bale and lead to considerable dry matter loss.

Another form of biomass storage that is particularly useful for material that is valued for cellulosic material is ensiling chopped material. For many years, ensiling chopped forage material has been stored as animal feed using a process known as ensiling. This process relies on the production of lactic acid to lower the pH of the material to a point that will inhibit the growth of microorganism. Successful ensiling requires the biomass to be stored in an anaerobic state so bacteria that produce lactic acid will thrive. Ensiling is a relatively simple process to use; however, satisfactory results will only occur if the void spaces filled with air are removed from bulk biomass material. This can be accomplished by using the weight of the material or an external stimulus to pack the material tightly and force out the air. Another means of creating an anaerobic condition involves using plastic wrap to cover biomass that prevents air from entering the biomass once it is in place. Eventually the respiration of aerobic organisms contained in the biomass will use all the available oxygen, therefore creating an anaerobic condition.

As the material is packed, oxygen is present and the enzymes in the plant along with aerobic microorganisms will consume nutrients and sugars that are essential for acid production when anaerobic conditions are created. The aerobic microorganisms will also produce significant amounts of heat, which can lead to biomass loss or lead to spontaneous combustion of the material. Therefore, it is critical that processes required for ensiling are started as soon as possible. The speed and efficiency of field operations will impact the quality of ensiled biomass as exposure to the elements and aerobic microorganism will lead to inadequate moisture content and water-soluble carbohydrate (i.e., sugar) levels. The water-soluble carbohydrates are the source of energy for the anaerobic microorganism, and the level of available carbohydrate will determine how much acid will be produced by the organisms. The carbohydrate content is generally expressed as a percentage of dry matter. Carbohydrate content must be considered in accordance with the plant's ability to resist changes in pH or its buffer capacity. A crop that has a high buffering capacity will require more acid production to adjust the pH to a level that will no longer sustain microorganism activity. This increased acid production will require more carbohydrate consumption. Assuming there are sufficient available carbohydrate levels in the biomass, eventually the anaerobic bacteria will produce enough lactic acid that will prove toxic to themselves as well as any other microorganisms present in the biomass.

Ensiling typically makes use of either a horizontal or a vertical silo. Vertical silos rely on the weight of the material to compress and eliminate air pockets between particles. A horizontal or bunker silo would require additional stimuli to compress the material. Chopped, uncompressed material is fed into a bunker silo and then one or more heavy tractors are used to compress the material. Often additional ballast or weight is added to the tractor to improve the packing performance. Usually, multiple passes are required over the silage material to insure proper ensiling. The material packing requirement is significant and usually at least one tractor has 100% effort dedicated to this effort during the ensiling process. There are numerous guidelines and factsheets available from land-grant universities that discuss how to achieve proper ensiling densities. Ensiling can be achieved by packing silage on any hard surface; however this activity can be dangerous. A bunker silo has side walls that help contain material. If sidewalls are not provided, there is an increased risk of tractor overturns, especially when an operator is attempting to compress the material along the edges.

The Ritter process is a potential low-cost method to effectively store biomass and maintain the integrity of insoluble components contained in the material. The Ritter process involves saturating the biomass with water to radically increase the moisture content. This method has long been used in the pulp and paper industry to maintain a quality source of cellulose for paper production. Application of the Ritter process would involve liquefying the biomass in a slurry that is between 3% and 5% solid content. The slurry is then pumped and piled for storage. As the water drains from the pile, it is collected and circulated throughout the pile to maintain the moisture content of the material. The high moisture content keeps the material under anaerobic conditions, which promotes the formation of lactic acid as the soluble sugars and carbohydrates are consumed by microbes. The majority of the dry matter loss is the sugars consumed in lactic acid production. Also, the additional weight of the water helps densify the material, which will improve material handling operations and reduce the amount of space required to store the biomass.

The majority of the biomass storage systems discussed thus far work very well for cellulosic materials, and the primary focus of these systems is the perseveration of cellulose, hemicellulose, and lignin. However, many of these systems will not be adequate if soluble sugars and carbohydrates are the primary components a biorefining operation requires. Storage of starches is somewhat well understood as there is a large knowledge base focused on the storage of corn, soybeans, and cereal grains. More novel starch crops, such as sweet potatoes, do have unique storage issues that are currently being addressed by ongoing research and development issues. With forage and grass crops, damaging the crop through a crushing or conditioning operation is considered a critical process that will improve biomass storage. However, many starch and sugar crops will have significant dry matter loss if the material is handled too aggressively. Grains should be gently so kernels and seed will not crack. Careful handling is essential for sweet potatoes because damage to the outer peel will lead to dry matter loss.

Storage can be a limiting factor for potential bioenergy feedstocks as well. The inability to either maintain the integrity of the biomass or provide a sustainable year-round supply of material will reflect negatively on the crop's bioenergy potential. For instance, sweet sorghum provides a source of directly fermentable aqueous sugar that is relatively inexpensive to process into ethanol. However, this aqueous sugar is not only a source of subsistence for the alcohol-producing yeast, but it is also an appealing food source for numerous "pest" microorganisms. The ability to store sweet sorghum juice without the sugar being consumed by these detrimental organisms is a nontrivial task the storage system and process must address for the crop to be consider a biofuel feedstock. The final point to reinforce about storage systems is they do little to improve the quality or value of a feedstock; therefore, the cost and energy inputs into this activity should be kept to a minimum. Returning to the sorghum juice example, refrigeration or freezing is an appealing concept for storage. However, the energy intensiveness of these operations would negatively impact both cash flow and energy balance calculations.

4.5 Densification

A recurring theme surrounding biomass logistics networks is the fact that biomass is a relatively low-value material with a very low energy density. The low energy density is the result of two characteristics. First, biomass is a more light and fluffy material compared to other energy sources such as coal or petroleum. The second characteristic of biomass is that it generally has a high moisture content. Because the moisture content of the crop is high, additional processing is required to drive away the moisture. The energy required to dry the crop reduces the energy value of the material. To increase the energy density of biomass, there are several processing steps known as densification that can be employed to either increase bulk density or convert biomass into an energy-dense form.

Because of the low density, the crops are difficult to handle, have the potential to incur a material loss (i.e., blown away by the wind), and will not make full use of transportation space. Densification is the step in the logistics supply chain that looks to increase the biomass density to improve handling, storage, and transportation parameters. Usually, densification is achieved by applying a mechanical force to the biomass and effectively removing any voids in the material. Examples of physical densification include baling, loafing, ensiling, or pelletizing. Another form of densification is increasing the energy content per unit mass of the material. For example, if biomass is to be combusted for heat energy, the consumer of this biomass would like to receive material with as low moisture content as possible as they do not want to expend energy from the biomass to evaporate the water. Energy densification of this biomass would involve removing the water prior to delivery to the biorefinery. While this drying step adds a potentially costly step to the supply chain, it may be required to address a consumer's biomass requirement or the result could be a biomass product with superior value to a wetter or green product.

Pelletizing has long been considered the primary means to achieve densification of biological plant material. Pelletizing is a process in which biomass is broken down mechanically into small particles, and these particles are then compressed under high temperature and pressure. The outcome of this process is a pellet that has significantly increased the material's bulk density, dried the material, and stabilized the material's physical properties to slow degradation.

4.6 Problems

1. For a 225 kW, $235,000 grain combine, equipped with a 7.5 m cutting platform determine the following parameters:
   a. Field capacities, both theoretical and actual
   b. Salvage value assuming 350 h of annual use
   c. Average annual cost of ownership
   d. Cumulative repair costs
   e. Fuel and lube costs

   Use values from Tables 4.3 and 4.4 or make assumptions for labor, interest, and fuel rates.

2. A biomass refinery is considering purchasing equipment to mow switchgrass. The first option: a mower conditioner has a cutting width of 2.5 m and requires 45 kW to operate. The second mower conditioner under consideration cuts a 4.5 m swath and requires

105 kW to operate. Using a spreadsheet, calculate the machine rate for the following tractor-implement scenarios so you can make an informed purchasing decision. Determine the annual use of the tractor and the mower and the field capacity. The company plans on mowing 400 ha per year with this system.

Assumptions:

5% interest rate

15 year life for the tractor

7 year life for the rotary mower

The tractor will be used an additional 40% of the scheduled time for the mower

$12.75/h operator wage

Equipment/Implement Scenarios:

a. The company may purchase a 54 kW tractor (MSRP: $51,269) to pull the 2.5 m mower conditioner (MSRP: $18,617).

b. The company may purchase a 116 kW tractor (MSRP: $98,578) to pull the 4.5 m mower conditioner (MSRP: $35,753).

c. The third option found a used 2002 John Deere 7710 with 6000 h for $55,000. How about we go with this instead of the JD 7730? The 956 mower conditioner would still be purchased.

3. The mowing company is having a difficult time booking farmers to perform their contract mowing operation. Step through all three tractor scenarios from 200 to 2000 acres in 200 acre increments to determine how variations in acreage affect the machine operating cost. If a spreadsheet is developed correctly, only a few values will require modification. Plot the results.

4. A custom hired machine operations provide an opportunity to outsource farm equipment operations if the land base or expected crop yield is not sufficient to justify machine ownership. There are several sources that estimate custom hire rates on the internet. The $235,000 grain combine equipped with a 7.5 m cutting platform from Problem 1 could be purchased to harvest corn for an ethanol plant or the ethanol plant could hire a local farmer to complete the task for $60/ha. Using information in Tables 4.1 through 4.3, make a case for the correct machinery use decision. What are potential advantages and disadvantages of hiring out machinery operations?

5. A simple approximation for stover production is to equate total stover production to the total grain production. In the region interested in corn stover bioenergy conversion, the 5 year average grain yield is 10.5 Mg/ha. Corn stover is purchased from local farmers at a contracted rate of $22.50/Mg. The farmers in the area are having a difficult time making a decision about the best machine system to use to collect the stover. If direct baling recovers 38% of the stover, a rake

and bale operation recovers 52.5% of the stover, and a mow, rake, bale operation recovers 72% of the stover.

   a. Determine the maximum additional cost that can be spent to add an additional machine and still have an economic advantage to justify the machine.

   b. If there is a climatic event that drops grain yields to 7 Mg/ha, how are the prices affected?

   c. Based on knowledge of equipment and machine operations, discuss the possibility of adding operations to recover additional corn stover.

6. Compare the costs of shipping 1.2 m × 1.2 m × 2.4 m rectangular bales (density = 175 kg/m³) with the semitruck-trailer combination in Example 4.9 and the use of a $65,000 straight truck with a 6.5 m long flat bed, 8 ton capacity.

   a. Determine the equivalent annual distance travel to move the same material as the larger truck system.

   b. Calculate the expense to operate the truck on a dollars per ton basis and state which truck system is better.

   c. If the hauling distance was reduced from 50 to 10 km, how would the economic picture change?

7. A cellulosic ethanol plant receives two types of round bales; 60% of bales are 1.5 m × 1.2 m (diameter by length) and 40% of bales are 1.8 m × 1.5 m. The average bulk density of all bales is 105 kg/m³.

   a. Determine the number of bales required to operate the biorefinery 325 days/year if the daily consumption is 1500 tons/day.

   b. Determine the % of material loss if the crusted layer is 5, 10, and 15 cm deep.

   c. How does the value of Part (a) change if the average crusted layer depth of 7.5 cm is considered?

8. The cellulosic ethanol plant in Problem 5 thinks it might be a better idea to only accept large rectangular bales (1.4 m × 1.4 m × 2.75 m) for delivery. These large rectangular bales have an average bulk density of 182 kg/m³.

   a. How many square bales are needed to supply the plant's daily biomass consumption if the average daily consumption remains at 1500 tons/day.

   b. The biorefinery wishes to maintain a 15 day stockpile of large square bales. Based on this storage capacity, determine the sheltered area required to store the 15 day volume of bales. Assume the storage structure such that it provides 9 m of clearance height.

c. A building contractor has quoted a price of $70/m^3$ to build the required storage structure. Calculate the total investment required to operate the facility on a dollars per ton basis and then determine the annual cost per ton if the annual cost is estimated to be 13.5% of the total costs.

9. The cellulosic biomass plant from Problem 5 is now interested in wrapping 2000 larger round bales (1.8 m × 1.5 m) per year. The bale wrapper they may purchase cost $24,500 and after careful study it is determined 60 bales/h can be fed through the unit, it will cost $2.75/bale in plastic costs.

   a. Determine cost per bale to operate the bale wrapper. For this problem, use the coefficients from Table 4.2 for the forage wagon and any other assumptions can be estimated using Table 4.4.

   b. If the tractor cost $18.25/h to operate with an operator and additional laborer costs $15.00/h, what is the total per bale cost for the entire system (tractor, labor, wrapper, plastic)? What is the total cost associated with processing 2000 bales if the $30/ton payment to the biomass producer is also considered?

   c. How many unwrapped round bales could be purchased ($25/ton) and stored outdoors for the same amount of money spent in part b? Quantify the reduction in dry matter loss needed to justify the investment in a bale wrapper.

10. Based on the information presented in this chapter, what are the major considerations that must be accounted for when biomass logistics systems are considered?

11. Discuss the advantages and disadvantages of corn, woody fiber, and switchgrass from a logistics standpoint. Is one crop clearly superior to the other when logistics alone is considered?

12. Compare and contrast each of the steps or subsystems in the logistics supply chain focusing on expenses incurred, potential to increase costs, impact of feedstock quality, and ease of implementation/modification.

References

ASABE Standards. 2007. St. Joseph, MN: ASABE007.

Prewitt, R. M., M. D. Montross, S. A. Shearer, T. S. Stombaugh, S. F. Higgins, S. G. McNeill, and S. Sokhansanj. 2007. Corn stover availability and collection efficiency using typical hay equipment. *Transactions of the ASABE*, 50(3), 705–711.

Sokhansanj, S. and A. F. Turhollow. 2004. Biomass densification—Cubing operations and costs for corn stover. *Applied Engineering in Agriculture*, 20(4), 495–499.

5

Kinetics and Microbiology of Biological Processes

Jay J. Cheng

CONTENTS

5.1   Introduction of Biological Processes 135
5.2   Kinetics of Enzymatic Reactions 137
      5.2.1   Mickaelis–Menten Equation.. 137
      5.2.2   Inhibition Kinetic Models.. 140
              5.2.2.1   Substrate Inhibition 140
              5.2.2.2   Product Inhibition.. 141
              5.2.2.3   Competitive Inhibition.. 141
5.3   Microbiology of Biological Processes.. 142
      5.3.1   Microorganisms 142
      5.3.2   Microbial Growth 143
              5.3.2.1   Lag Phase 143
              5.3.2.2   Exponential Growth Phase... 144
              5.3.2.3   Stationary Phase... 147
              5.3.2.4   Death Phase 148
5.4   Problems... 148
References... 150

5.1 Introduction of Biological Processes

Biological conversion of biomass to renewable energy products involves biochemical reactions that convert organic substrates into products that can be utilized for energy production. For example, sugars, starch, and cellulose can be converted to bioethanol or biobutanol through a series of biochemical reactions and organic wastes can be converted to methane or hydrogen through anaerobic digestion. The biomass that can be utilized for biological conversion include sugar crops (sugarcane, sweet sorghum, and sugar beet), grains (corn, wheat, rice, barley, etc.), trees, agricultural residues (corn stover, wheat straw, etc.), grasses, and organic waste materials from municipalities, agriculture, and industries. A biological process usually consists of a series of elementary biochemical reactions catalyzed by enzymes, or enzymatic reactions. The overall process can be expressed as follows:

$$\text{Substrate} \xrightarrow{\text{Enzymes}} \text{Product} \tag{5.1}$$

Elementary reactions are the very basic chemical reactions and examples are as follows:

$$A \longrightarrow B \tag{5.2}$$

$$A + A \longrightarrow B \tag{5.3}$$

$$A + B \longrightarrow C \tag{5.4}$$

If the rate of an elementary reaction is defined as the accumulation of the reactant(s) (mole) per unit reactor volume per unit time, then

$$r_A = \frac{1}{V}\frac{dN_A}{dt} = \frac{dC_A}{dt} \tag{5.5}$$

where
  r_A is the rate of an elementary reaction (mol/L s)
  V is the reactor volume (L)
  N_A are the moles of reactant A (mol)
  T is the reaction time (s)
  C_A is the concentration of reactant A (mol/L)

The rates of the elementary reactions (Reactions 5.2 through 5.4) can be expressed respectively as follows:

$$r_A = -k_A C_A \tag{5.6}$$

$$r_A = -k_A C_A^2 \tag{5.7}$$

$$r_A = -k_A C_A C_B \tag{5.8}$$

where k_A is a kinetic constant.
  Based on these elementary reaction rate expressions, Reaction 5.2 is a first-order reaction, and Reactions 5.3 and 5.4 are second-order reactions.

5.2 Kinetics of Enzymatic Reactions

5.2.1 Mickaelis–Menten Equation

A typical enzymatic reaction can be presented as follows:

$$Substrate + Enzyme \longrightarrow Product + Enzyme$$

The enzyme serves as a catalyst in the reaction and remains unchanged when the reaction is completed. However, Reaction 5.8 is an overall reaction and can be broken down to the following elementary reactions:

$$S + E \longrightarrow E \bullet S \tag{5.9}$$

where
  S is the substrate
  E is the enzyme
  E • S is the enzyme–substrate complex

This is the first step of the enzymatic reaction in which the enzyme attaches to the substrate to form an enzyme–substrate complex. The enzyme–substrate complex is very unstable. It can be broken into the substrate and enzyme:

$$E \bullet S \longrightarrow S + E \tag{5.10}$$

which is the reversible reaction of Reaction 5.9. The enzyme–substrate complex can also be converted into the product and release the enzyme:

$$E \bullet S \longrightarrow P + E \tag{5.11}$$

where P is the product.

During the enzymatic reaction process, Reaction 5.11 is usually a rate-limiting elementary reaction. Its rate can be expressed as

$$r_3 = -k_3 [E \bullet S] \tag{5.12}$$

which is also considered as the overall enzymatic reaction rate,
where
  r_3 is the rate of Reaction 5.11
  k_3 is the kinetic constant of Reaction 5.11
  $[E \bullet S]$ is the concentration of E • S in the reactor

However, it is very difficult to obtain the concentration of E • S in the reactor because the enzyme–substrate complex is very unstable. For the same reason, it is safe to assume that the net rate of the enzyme–substrate complex (E•S) accumulation is zero at steady-state conditions. The accumulation of E•S involves the formation of E•S in Reaction 5.9 and its consumption in Reactions 5.10 and 5.11, so the net E•S accumulation rate can be expressed as

$$r_{E\bullet S} = k_1\,[S]\,[E] - k_2\,[E\bullet S] - k_3\,[E\bullet S] = 0 \tag{5.13}$$

where
$r_{E\bullet S}$ is the net E•S accumulation rate
k_1 is the kinetic constant of Reaction 5.9
[S] is the concentration of the substrate in the reactor
[E] is the concentration of the enzyme molecule in the reactor
k_2 is the kinetic constant of Reaction 5.10

The instant concentration of the enzyme molecule, [E], is also very difficult to obtain, but the initial or total enzyme concentration in the reactor, [E_t], is constant throughout the enzymatic reaction and

$$[E_t] = [E] + [E\bullet S] \quad \text{or} \quad [E] = [E_t] - [E\bullet S] \tag{5.14}$$

Substituting Equation 5.14 into Equation 5.13 results in the following:

$$k_1\,[S]\,([E_t] - [E\bullet S]) - k_2\,[E\bullet S] - k_3\,[E\bullet S] = 0 \tag{5.15}$$

Therefore,

$$[E\bullet S] = \frac{k_1[E_t][S]}{k_1[S] + k_2 + k_3} \tag{5.16}$$

Substituting Equation 5.15 into Equation 5.12, we will get the kinetic expression of the overall enzymatic reaction rate as follows:

$$r = r_3 - k_3\,[E\bullet S] = -\frac{k_3 k_1[E_t][S]}{k_1[S] + k_2 + k_3} \tag{5.17}$$

or

$$r = -\frac{k_3[E_t][S]}{[S] + \dfrac{k_2 + k_3}{k_1}} \tag{5.18}$$

or

$$-r = \frac{k[S]}{[S] + K_S} \tag{5.19}$$

where

r, substrate reaction rate
k, maximum substrate reaction rate
K_S, half-velocity constant

Equation 5.18 is also called the Mickaelis–Menten equation.

To understand the physical meaning of the kinetic parameters, k and K_S, Equation 5.18 can be graphically expressed as in Figure 5.1.

As shown in Figure 5.1, k is the maximum substrate utilization rate that the enzymatic reaction is approaching when the substrate concentration is very high; K_S is the substrate concentration when the substrate utilization rate $(-r)$ is at half of its maximum (k), so it is called the half-velocity constant.

When serving as catalysts for biochemical reactions, enzymes are very specific on substrates. In other words, a specific enzyme is working only on a specific organic compound or a certain group of organic compounds. The kinetic constants k and K_S for an enzymatic reaction are usually obtained through experiments. Linearization of a relationship between the variables is usually very helpful in determining the kinetic constants. To linearize the relationship, the Mickaelis–Menten equation can be rearranged as follows:

$$\frac{1}{-r} = \frac{K_S}{k}\frac{1}{[S]} + \frac{1}{k} \tag{5.20}$$

Equation 5.19 can be graphically expressed as in Figure 5.2.

As shown in Figure 5.2, a straight line will be obtained when $-1/r$ is plotted against $1/[S]$ from Equation 5.19. The intercept of the line to the vertical axis should be $1/k$ and the slope of the line would be K_S/k. For a specific enzymatic reaction, an experiment can be conducted with several different substrate concentrations. A reaction rate (r) can be measured with each substrate

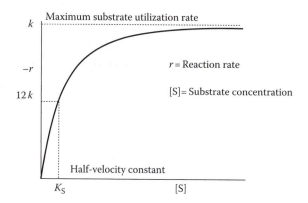

FIGURE 5.1
Graphical interpretation of the kinetic rate expression of an enzymatic reaction.

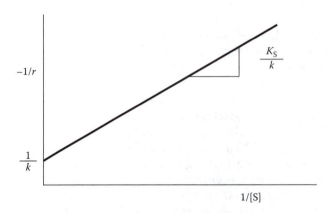

FIGURE 5.2
Linear expression of Mickaelis–Menten equation for enzymatic reactions.

concentration ([S]). A plot of $-1/r$ against $1/[S]$ should result in a straight line, so the kinetic constants k and K_S for the enzymatic reaction can be obtained.

5.2.2 Inhibition Kinetic Models

5.2.2.1 Substrate Inhibition

For some enzymatic reactions, the substrate may inhibit its own utilization when the substrate concentration is high. The Haldane kinetic model (Haldane, 1930) has been frequently used to describe the enzymatic utilization of inhibitory organic compounds. The Haldane model is also called the substrate inhibition model. It is based upon the hypothesis that the substrate can be combined with the enzyme–substrate complex to form a more complicated complex that inhibits the utilization of the substrate itself. The elementary enzymatic reactions can be expressed as follows:

$$E + S \underset{k_{-1}}{\overset{k_1}{\rightleftarrows}} E \bullet S \tag{5.21}$$

$$E \bullet S + S \underset{k_{-2}}{\overset{k_2}{\rightleftarrows}} S \bullet E \bullet S \tag{5.22}$$

$$E \bullet S \xrightarrow{k_3} S + E \tag{5.23}$$

where k_1, k_{-1}, k_2, k_{-2}, and k_3 are the reaction constants to the corresponding elementary biochemical reactions. The Haldane kinetic equation was developed upon these elementary enzymatic reactions with the following expression:

$$-r = \frac{k[S]}{K_S + [S] + \dfrac{[S]^2}{K_I}} \tag{5.24}$$

where K_I is the Haldane or substrate inhibition coefficient. In this model, K_s (the half-velocity constant) is equal to k_{-1}/k_1, which is the equilibrium constant of the reversible reaction (Reaction 5.21); K_I equals to k_{-2}/k_2, which is the equilibrium constant of the reversible reaction (Reaction 5.22).

5.2.2.2 Product Inhibition

In enzymatic reactions, products may inhibit the utilization of substrates, especially when the product concentration is high in the bioreactor. In an ethanol fermentation process, the product ethanol can be inhibitory to the fermentation microorganisms that convert glucose to ethanol. A kinetic model to describe the product inhibition in ethanol fermentation is shown in the following equation (Luong, 1985):

$$-r = \frac{k[S]}{K_S + [S]}\left[1 - \left(\frac{[P]}{[P]^*}\right)^n\right]$$ (5.25)

where
[P] is the inhibitory product concentration
[P]* is the critical product concentration above which microbial cells cease to grow
n is a constant

At low product concentrations, when $[P] \ll [P]^*$, Equation 5.24 becomes the Mickaelis–Menten equation.

5.2.2.3 Competitive Inhibition

When there is more than one substrate in a biological system, the substrates may compete against each other for the enzymes. For example, in the biological conversion of xylose from hydrolysis of hemicellulose to ethanol by *Pachysolen tannophilus*, minor sugars such as arabinose, rhamnose, and galactose that are by-products from the hydrolysis of hemicelluloses could competitively inhibit the microorganisms against xylose (Converti and Del Borghi, 1998). Chang et al. (1993) developed a model describing the competitive inhibition as follows:

$$-r = \frac{k[S_1]}{K_S\left(1 + \frac{[S_2]}{K_C}\right) + [S_1]}$$ (5.26)

where
$[S_1]$ is the concentration of primary substrate
$[S_2]$ is the concentration of competitive substrate
K_C is the competitive inhibition coefficient

5.3 Microbiology of Biological Processes

5.3.1 Microorganisms

Many biological processes involve microorganisms such as bacteria, fungi, and yeast. Fermentation of sugars to produce ethanol is usually carried out by yeast (e.g., *Sacchromyces cereviciae*) or bacteria. Biogas production from organic waste materials involves anaerobic bacteria through anaerobic digestion. Butanol is also a product of bacterial fermentation of sugars. Microbial cells are composed of water, proteins, carbohydrates, lipids, amino acids, and minerals. Water content in the microbial cells is approximately 80%. The contents of other materials are listed in Table 5.1.

The chemical elements that constitute the organic compounds of the microbial cells are mainly carbon (C), oxygen (O), nitrogen (N), hydrogen (H), phosphorus (P), and sulfur (S). The minerals in the microbial cells include potassium (K), sodium (Na), calcium (Ca), magnesium (Mg), chloride (Cl), iron (Fe), and trace minerals such as B, Al, V Mo, I, Si, F, Sn, Mn, Co, Cu, and Zn. The elemental chemical composition of the microbial cells is shown in Table 5.2.

In microbial biological processes, microorganisms need "food" to support their growth during the processes. Basically, the microorganisms need carbon source, energy source, and nutrients to reproduce cells and perform the metabolic functions in the biological processes. Microorganisms can obtain their carbon to support their cell reproduction from organic compounds or inorganic carbon dioxide. The microorganisms that take carbon from organic compounds to support their growth are called heterotrophs. The microorganisms that take carbon from carbon dioxide to support their growth are called autotrophs. Autotrophs usually have lower growth rates and yields of cell reproduction than heterotrophs because it takes more energy for the former to utilize inorganic carbon dioxide in the biosynthesis of microbial cells. Microorganisms can take energy for the synthesis of new cells or growth from light or biochemical reactions. Those microorganisms that are able to

TABLE 5.1

Typical Dry Matter Composition
of Microbial Cells

Constituent	% of Dry Weight
Proteins	55
Carbohydrates	12
Lipids	10
DNA and RNA	20
Amino acids	2
Minerals	1

TABLE 5.2

Typical Elemental Composition of Microbial Cells

Element	% of Dry Weight
C	50
O	20
N	14
H	8
P	3
S	1
K	1
Na	1
Ca	0.5
Mg	0.5
Cl	0.5
Fe	0.2
Trace elements	0.3

use light as energy source to support their growth are called phototrophs and they can be either autotrophs or heterotrophs. The microorganisms that have to obtain their energy to support their growth through chemical oxidation reactions are called chemotrophs. Chemotrophs can also be either autotrophs or heterotrophs. In addition to carbon and energy sources, microorganisms also need nutrients, organic nutrients and inorganic nutrients, to support their growth. Principal organic nutrients include amino acids and vitamins. Inorganic nutrients include macronutrients such as N, P, S, K, Na, Ca, Mg, and Cl and micronutrients such as B, Al, V Mo, I, Si, F, Sn, Mn, Co, Cu, and Zn.

5.3.2 Microbial Growth

In microbial biological processes, microorganisms utilize the substrate to support their own reproduction and to convert the substrate into product(s). Figure 5.3 illustrates the consumption of the substrate and the growth and death of microbial cells in a batch bioreactor. The microorganisms experience basically four phases in the batch bioreactor, that is, lag phase, exponential growth phase, stationary phase, and death phase (Figure 5.3).

5.3.2.1 Lag Phase

Initially when the substrate and microbial inoculum are added in a batch reactor, the microbial cells are placed in a new environment under different conditions such as temperature, pH, and/or ion strength. It takes some time

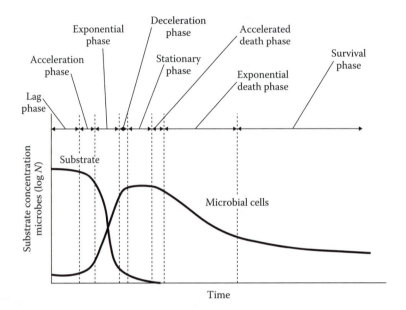

FIGURE 5.3
A typical microbial cell growth cycle in a batch bioreactor.

for the microorganisms to acclimate themselves to the new environment, which is called lag time. The length of lag time depends on individual type of microorganisms. During the lag phase, the microorganisms consume a small amount of the substrate to get energy and prepare themselves for growth.

5.3.2.2 Exponential Growth Phase

At the end of the lag phase, the microorganisms are acclimated to the new environment and ready to grow or start the exponential growth phase. The first part of the exponential growth phase is called the acceleration phase in which the microorganisms start to grow and substantially consume the substrate. During the acceleration phase, the growth rate of the microbial cells increases until it reaches the exponential growth rate, which is the start of the real exponential growth phase. In the exponential growth phase, the microorganisms have the highest growth rate and substrate consumption rate for the entire cycle of the cell growth in the batch reactor. As a result, biosynthesis of new cells and conversion of substrate to product(s) are very active.

The microbial growth or cell reproduction rate can be estimated through the following kinetic analysis. First, we need to introduce two terms: microbial cell generation time or double time (T) and cell generation (Z). T is the time it takes for the microbial cells to grow one generation or double their cell number. Z is the number of generations that the microbial cells have grown. Assume n is the number of cells at time t and n_0 is the initial cell number at time 0 as shown in Figure 5.4, we will have the following equation:

$$n = n_0\, 2^Z \tag{5.27}$$

or

$$\log n = \log n_0 + Z \log 2 \tag{5.28}$$

or

$$Z = (\log n - \log n_0)/\log 2 \tag{5.29}$$

Since

$$Z = t/T \tag{5.30}$$

Then

$$t/T = (\log n - \log n_0)/\log 2 \tag{5.31}$$

or

$$\frac{\log n - \log n_0}{t} = \frac{\log 2}{T} \tag{5.32}$$

If the lag phase is added into consideration and the time length of the lag phase is L as shown in Figure 5.5, then

$$Z = (t - L)/T = (\log n - \log n_0)/\log 2 \tag{5.33}$$

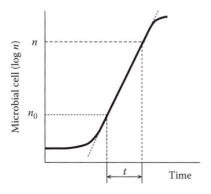

FIGURE 5.4
Growth of microbial cells during the exponential growth phase.

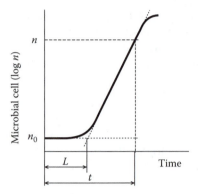

FIGURE 5.5
Growth of microbial cells during the lag and exponential growth phases.

or

$$\frac{\log n - \log n_0}{t - L} = \frac{\log 2}{T} \tag{5.34}$$

If the primary concern is in the exponential growth phase, we can go back to Equation 5.31, which can be converted to natural logarithm form as follows:

$$\frac{\ln n - \ln n_0}{2.303\, t} = \frac{\log 2}{T} \tag{5.35}$$

Therefore, the kinetic expression to describe the microbial cell growth during the exponential phase can be

$$\frac{d(\ln N)}{dt} = \frac{0.693}{T} = \mu \tag{5.36}$$

or

$$\frac{dN}{dt} = \mu N \tag{5.37}$$

where
  N is the microbial cell number
  μ is the specific growth rate of the microorganism

For the convenience, microbial cell biomass, X, is often used instead of cell number N, so

$$\frac{dX}{dt} = \mu X \tag{5.38}$$

When the microbial process approaches the end of the exponential growth phase, most of the substrate is consumed and the microbial cell growth slows down. The cell growth starts the deceleration phase. During this phase, the microbial cell growth rate can be limited by the substrate. Therefore, the microbial cell growth kinetics can be expressed with Monod model:

$$\mu = \frac{\mu_m[S]}{K_S + [S]} \tag{5.39}$$

where

μ_m is the maximum specific growth rate of the microorganism
[S] is the concentration of the substrate that limits the microbial cell growth
K_S is the saturation constant or half-velocity constant

It should be noticed that there is a similarity between the Mickaelis–Menten equation (Equation 5.18) and Equation 5.38. The former describes the substrate utilization rate, while the latter describes the microbial cell growth rate. Comparing the two equations indicates that the microbial growth rate is proportional to the substrate utilization rate, which can be expressed in the following relationship:

$$\frac{dX}{dt} = -Y \frac{d[S]}{dt} \tag{5.40}$$

where Y is the yield of microbial cell mass as a result of substrate consumption, or g cell mass generated per g of substrate consumed.

5.3.2.3 Stationary Phase

As most of the substrate is consumed at the end of the exponential growth phase, the microbial cell growth significantly slows down. At the same time, some old cells die. The microbial cell death rate can be expressed by the following equation:

$$\frac{dX}{dt} = -\lambda X \tag{5.41}$$

where λ is the specific death coefficient of the microorganism. When both cell growth and death are occurring at the same time in a bioreactor, the net cell growth rate is

$$\frac{dX}{dt} = (\mu - \lambda)X \tag{5.42}$$

As shown in Figure 5.3, the net cell growth during the stationary phase is zero or the microbial cell growth is equal to the death rate.

5.3.2.4 Death Phase

When the substrate is depleted in the bioreactor, the cell death rate increases and the microbial cells will start the accelerated death phase (Figure 5.3). The accelerated death phase is usually short. When the death rate reaches its maximum, the microbial cells start their exponential death phase and stay in this phase for a period of time. Finally, the cell death rate will slow down and the microbial cells are struggling for their survival during the survival phase. In this phase, the microorganisms are in the status of self-mainte-nance, which means they need a minimum energy supply to maintain their viability without biosynthesis or growth.

In summary, a microbial biological process in a batch reactor involves the substrate consumption and microbial cell growth. The microbial cell growth kinetics can be described as

$$\frac{dX}{dt} = \frac{\mu_m [S] X}{K_S + [S]} - \lambda X \tag{5.43}$$

The substrate consumption rate can be expressed as

$$\frac{d[S]}{dt} = -\frac{1}{Y}\frac{\mu_m [S] X}{K_S + [S]} = -\frac{k[S]X}{K_S + [S]} \tag{5.44}$$

and

$$k = \frac{\mu_m}{Y} \tag{5.45}$$

5.4 Problems

1.
    a. What is the kinetic expression of zero-order reaction?
    b. Derive the integrated form of the first- and second-order kinetic expressions of Equations 5.6 and 5.7, respectively.
    c. Draw graphical expressions of zero-, first-, and second-order kinetics and explain.

2. Substrate inhibition or Haldane inhibition was discussed in Section 5.2.2.1 and Haldane inhibition model was presented for the kinetics:

$$-r = \frac{k[S]}{K_S + [S] + \dfrac{[S]^2}{K_I}} \qquad (5.24)$$

where K_I is the Haldane or substrate inhibition coefficient. In this model, K_s (the half-velocity constant) is equal to k_{-1}/k_1, which is the equilibrium constant of the reversible reaction (Reaction 5.20); K_I equals to k_{-2}/k_2, which is the equilibrium constant of the reversible reaction (Reaction 5.21).

a. Use a similar approach for obtaining the Mickaelis–Menten equation to derive Haldane model or Equation 5.24.

b. Propose a method for analyzing steady-state data of r versus [S] with the objective of obtaining the values of the model constants.

3. In glucose-to-ethanol fermentation process, the growth of yeast cells may be inhibited by the product (ethanol), especially at a high ethanol concentration. The phenomenon is called product inhibition. A commonly used model for the kinetics of the yeast growth under product inhibition is as follows:

$$\mu = \frac{\mu_m [S]}{K_S + [S]} \left[1 - \left(\frac{[P]}{[P^*]} \right)^n \right]$$

where
  μ is the specific cell growth rate (g/d g cells)
  μ_m is the maximum cell growth rate (g/d g cells)
  [S] is the glucose concentration (g/L)
  K_S is the half-velocity constant (g/L)
  [P] is the ethanol concentration (g/L)
  [P*] is the critical ethanol concentration above which cells cease to grow (g/L)
  n is a constant

  Propose methods to determine the values of the model constants: μ_m, K_S, [P*], and n. Explain.

4. Autotrophic and heterotrophic microorganisms use different carbon sources to support their growth. Please explain the difference of autotrophs from heterotrophs in growth rates and yields of cell reproduction.

5. Why do microorganisms normally experience an initial lag phase when inoculated in a bioreactor? In a batch microbial cultivation process, the last death phase usually takes quite a long time. Explain what happens in the death phase.

References

Chang, M. K., Voice, T. C., and Criddle, C. S. (1993). Kinetics of competitive inhibition and cometabolism in the biodegradation of benzene, toluene, and *p*-xylene by two *Pseudomonas* isolates. *Biotechnology and Bioengineering*, 41(11), 1057–1065.

Converti, A. and Del Borghi, M. (1998). Inhibition of the fermentation of oak hemicellulose acid-hydrolysate by minor sugars. *Journal of Biotechnology*, 64(2–3), 211–218.

Haldane, J. B. S. (1930). *Enzymes*. Longmans, London.

Luong, J. H. T. (1985). Kinetics of ethanol inhibition in alcohol fermentation. *Biotechnology and Bioengineering*, 27(3), 280–285.

6

Anaerobic Digestion for Biogas Production

Jay J. Cheng

CONTENTS

6.1   Introduction 152
6.2   Anaerobic Process for Biogas Production 153
6.3   Temperature Effect on Anaerobic Digestion............................ 157
    6.3.1   Psychrophilic Anaerobic Digestion............................. 157
    6.3.2   Mesophilic Anaerobic Process 158
    6.3.3   Thermophilic Anaerobic Process.................................. 161
6.4   Hydrogen Effect on Anaerobic Digestion 163
6.5   pH Effect on Anaerobic Digestion.. 163
6.6   Toxicity/Inhibition.. 167
    6.6.1   Ammonia Inhibition... 167
    6.6.2   Sulfate/Sulfide Inhibition 169
    6.6.3   Metal Inhibition 170
    6.6.4   Organic Compound Inhibition 175
6.7   Anaerobic Digestion Kinetics 176
6.8   Modeling Anaerobic Digestion Processes.................................. 179
6.9   Anaerobic Digesters 182
    6.9.1   Suspended-Growth Anaerobic Digesters................... 183
        6.9.1.1   Design Equation 184
        6.9.1.2   Complete-Mix Digester 186
        6.9.1.3   Baffled Anaerobic Digesters 187
    6.9.2   Attached-Growth Anaerobic Digesters 188
        6.9.2.1   Packed-Bed Anaerobic Digesters.................. 188
        6.9.2.2   Expended-Bed Anaerobic Digesters............. 189
    6.9.3   Other Anaerobic Digesters 190
        6.9.3.1   Up-Flow Anaerobic Sludge Blanket
                  Digesters 190
        6.9.3.2   Anaerobic Sequencing Batch Reactors......... 192
        6.9.3.3   Covered Anaerobic Lagoon 193
6.10 Anaerobic Digestion for Hydrogen Production 194

6.11 Biogas Purification 196
    6.11.1 H_2S Scrubbing from Biogas 197
    6.11.2 CO_2 Scrubbing from Biogas 199
6.12 Problems 200
References 202

6.1 Introduction

Biogas is usually produced from anaerobic digestion of organic waste materials. It is a gas mixture of mainly methane and carbon dioxide with minor amounts of nitrogen, hydrogen, hydrogen sulfide, ammonia, and oxygen. A typical composition of biogas is shown in Table 6.1.

In an anaerobic digestion of organic waste materials, a consortium of anaerobic bacteria converts the organic compounds into biogas under oxygen-free environmental conditions. The major combustible component of biogas is methane, which can be easily burned with oxygen or air to produce water and carbon dioxide as well as heat:

$$CH_4 + 2O_2 \rightarrow CO_2 + 2H_2O + 192 \text{ kcal/mol} \tag{6.1}$$

Methane is a clean energy source and a complete combustion of $1\,m^3$ of methane produces heat of 8570 kcal. Biogas can be directly burned for heat production. It can also be combusted in engines for electricity production.

The first anaerobic digester for biogas production was built in France in 1860, and it was converted from a settling basin to a sewage system. In 1925, a heated anaerobic digester was designed and built in Germany. The first anaerobic digester in the United States was established in 1926, and the digester was operated with a temperature control system. Significant effort was spent on anaerobic digestion of organic waste materials for biogas production in Europe and United States after the World War II. However, relatively cheap petroleum has prevented anaerobic digestion for biogas production as an energy-generation technology from becoming a main energy industry although the anaerobic digestion research has made a lot of progress. In the mid-1970s, anaerobic digestion for biogas production received a lot of attention because of the "oil crisis." As a result, many industrial- and farm-scale anaerobic digesters were built in the United States and Europe to produce biogas for energy generation. Most of the digesters were converting municipal, agricultural, and industrial organic wastes into biogas under mesophilic (around 35°C) conditions and the biogas was utilized for electricity and heat production. More than

TABLE 6.1

Typical Composition of Biogas

Component	%
Methane (CH_4)	50–80
Carbon dioxide (CO_2)	20–40
Nitrogen (N_2)	0–5
Hydrogen (H_2)	0–1.0
Hydrogen sulfide (H_2S)	0.05–1.0
Ammonia (NH_3)	0.02–0.5
Oxygen (O_2)	0–0.5

5 million small-scale anaerobic digesters were built in China. Most of them were household scale and operated under psychrophilic (ambient temperature) conditions. Biogas produced from the digesters was mainly utilized for cooking, lighting, and heating. However, only a few anaerobic digesters in the United States and Europe were still operated for energy production after the "oil crisis." Many of them were gone with the crisis.

In recent years, biogas production has again attracted attention worldwide as a renewable energy source because of sustainability and environmental concerns of the fossil fuel. More than 4000 biogas plants were in operation in the European Union by the end of 2005. The plants were producing biogas through anaerobic digestion from different waste materials including landfill waste, urban sewage, industrial organic waste, and agricultural waste. The annual energy production from the biogas was equivalent to approximately 2.3 million tons of petroleum or crude oil. Many new anaerobic digesters have also been built in the United States to convert landfill, municipal, animal, and industrial wastes into biogas for energy production in the new century.

Nevertheless, anaerobic digestion, as an environmental protection technology, has been widely used throughout the world for the treatment of municipal wastewater, especially the sludge generated during the wastewater treatment. The technology has also been used for the treatment of high-strength industrial wastewaters such as food-processing, brewery, winery, dairy, and swine wastewaters. The main advantages of anaerobic technology compared to aerobic processes that are commonly used for municipal wastewater treatment are its low energy requirement and operational cost. In addition, anaerobic digestion converts the organics in wastes into biogas that can be utilized for energy generation. Recovery of energy from waste treatment would offset the cost of the digester operation. A disadvantage of anaerobic digestion compared to aerobic treatment is that the growth of anaerobic bacteria is very slow so a larger volume or longer hydraulic retention time (HRT) is required for the anaerobic digesters. To improve the efficiency of the anaerobic digestion, researchers and engineers have developed high-performance anaerobic reactors such as anaerobic packed-bed (APB) reactor, anaerobic fluidized-bed (AFB) reactor, and up-flow anaerobic sludge blanket (UASB) reactor. In these anaerobic processes, concentrated bacteria are either attached to the media or in the biological granules in the reactors. HRT is uncoupled with the solid residence time (SRT), and HRT can be significantly reduced compared to the conventional suspended-growth anaerobic digestion process.

6.2 Anaerobic Process for Biogas Production

Anaerobic digestion to convert organic compounds into biogas is a complex process that involves a series of microbial metabolism. The process can be divided into four main steps: hydrolysis, acidogenesis, acetogenesis, and methanogenesis. Each of the four steps involves different biochemical reactions with different substrates and microorganisms.

Hydrolysis: Original organic waste materials usually contain mainly large-molecule compounds, such as carbohydrates, proteins, lipids, and celluloses. These organic compounds are hydrolyzed by facultative (active in both environments with and without oxygen) and obligate (active only in the environment without oxygen) anaerobic bacteria to mainly smaller molecules, such as sugars, fatty acids, amino acids, and peptides, as well as a small amount of acetic acid, hydrogen, and carbon dioxide. The energy reserved in the original large-molecule organic compounds is redistributed into 5% in hydrogen, 20% in acetic acid, and 75% in the smaller molecule organic compounds during the hydrolysis.

Acidogenesis: The sugars, fatty acids, amino acids, and peptides are fermented by the anaerobic bacteria to volatile fatty acids (VFAs) such as propionic and butyric acids during the acidogenesis. Similar to the hydrolysis process, acidogenesis also produces a small amount of acetic acid, hydrogen, and carbon dioxide. In acidogenesis, 10% energy is released in the form of hydrogen, 35% in acetic acid, and the rest reserved in the VFA.

Acetogenesis: The VFAs are completely degraded into acetic acid, hydrogen, and carbon dioxide during acetogenesis. At this step, 17% of the energy is transferred to acetic acid and 13% to hydrogen.

Methanogenesis: The whole anaerobic digestion process is complete when both hydrogen and acetic acid are converted to methane during the methanogenesis. The conversion of hydrogen to methane involves a biochemical reaction of hydrogen and carbon dioxide to form methane:

$$4H_2 + CO_2 \rightarrow CH_4 + 2H_2O \qquad (6.2)$$

The bacteria that catalyze the biochemical reaction are obligate hydrogen-utilizing anaerobes. The conversion of acetic acid to methane is a degradation of acetic acid into methane and carbon dioxide by obligate acetate-utilizing anaerobic bacteria:

$$CH_3COOH \rightarrow CH_4 + CO_2 \qquad (6.3)$$

The anaerobic process to convert organic compounds in waste materials into methane and carbon dioxide is schematically shown in Figure 6.1.

Anaerobic microorganisms: A consortium of anaerobic bacteria are involved in the anaerobic digestion process to convert organic compounds into biogas. The microorganisms involved in the hydrolysis, acidogenesis, and acetogenesis are composed of both facultative and obligate anaerobic bacteria such as *Clostridium* spp., *Peptococcus anaerobes*, *Lactobacillus*, *Actinomyces*, and *Escherichia coli*. These microorganisms are fast-growing and flexible to pH, temperature, and inhibitory chemicals, compared to methane-producing

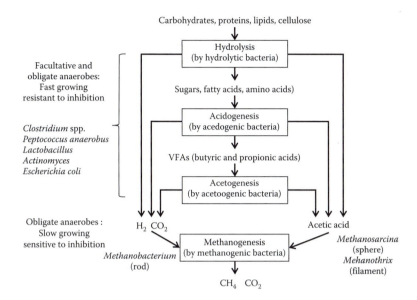

FIGURE 6.1
A schematic diagram of anaerobic degradation of organic compounds to biogas (methane and carbon dioxide).

bacteria or methanogens. They can grow in a pH range of 4.5–8.0 and are quite resistant to inhibition. The hydrogen-producing bacteria are usually the obligate anaerobic *Clostridium* spp. On the other hand, the methane-producing bacteria or methanogens are slow-growing, obligate anaerobes. They are also sensitive to pH change and inhibition. The hydrogen-utilizing methanogens are normally *Methanobacteria* (rod). The acetate-utilizing methanogens include *Methanosarcina* (sphere) and *Mehanothrix* (filament). Their pH range is from 6.5 to 8.0.

Substrates for methanogenesis: In addition to hydrogen and acetic acid, formic acid, methanol, and methylamine can also be the direct substrates for methane production through methanogenesis. The biochemical reactions are as follows:

$$4HCOOH \rightarrow CH_4 + 3CO_2 + 2H_2O \qquad (6.4)$$

$$4CH_3OH \rightarrow 3CH_4 + CO_2 + 2H_2O \qquad (6.5)$$

$$4(CH_3)_3N \rightarrow 9CH_4 + 3CO_2 + 6H_2O + 4NH_3 \qquad (6.6)$$

Stoichiometry of anaerobic reactions: Quantitative relationship of an organic compound with its products after anaerobic digestion can be obtained through mass balance. Large-molecule organic compounds are usually composed of C, H, O, N, and S. The compounds are converted with water to methane, carbon dioxide, ammonia, and hydrogen sulfide. A general form of the overall biochemical reactions can be expressed as

$$C_aH_bO_cN_dS_e + fH_2O \rightarrow gCH_4 + hCO_2 + iNH_3 + jH_2S \tag{6.7}$$

$$\text{A mass balance on N results in } i = d \tag{6.8}$$

$$\text{A mass balance on S results in } j = e \tag{6.9}$$

$$\text{A mass balance on O results in } 2h = c + f \tag{6.10}$$

$$\text{A mass balance on C results in } g + h = a \tag{6.11}$$

$$\text{A mass balance on H results in } 4g + 3i + 2j = b + 2f \tag{6.12}$$

The solution for the combination of Equations 6.8 through 6.12 gives:

$$f = (4a - b - 2c + 3d + 2e)/4 \tag{6.13}$$

$$g = (4a + b - 2c - 3d - 2e)/8 \tag{6.14}$$

$$h = (4a - b + 2c + 3d + 2e)/8 \tag{6.15}$$

Thus, a balanced equation for Reaction 6.7 should be

$$C_aH_bO_cN_dS_e + \left(a - \frac{b}{4} - \frac{c}{2} + \frac{3d}{4} + \frac{e}{2}\right)H_2O \rightarrow \left(\frac{a}{2} + \frac{b}{8} - \frac{c}{4} - \frac{3d}{8} - \frac{e}{4}\right)$$
$$\times CH_4 + \left(\frac{a}{2} - \frac{b}{8} + \frac{c}{4} + \frac{3d}{8} + \frac{e}{4}\right)CO_2 + dNH_3 + eH_2S \tag{6.16}$$

For the anaerobic digestion of a carbohydrate $C_aH_bO_c$ to biogas, the stoichiometric biochemical reaction can be simplified to

$$C_aH_bO_c + \left(a - \frac{b}{4} - \frac{c}{2}\right)H_2O \rightarrow \left(\frac{a}{2} + \frac{b}{8} - \frac{c}{4}\right)CH_4 + \left(\frac{a}{2} - \frac{b}{8} + \frac{c}{4}\right)CO_2 \tag{6.17}$$

Many organic waste materials contain a variety of chemical compounds that are sometimes difficult and unnecessary to be identified. Normally, biochemical oxygen demand (BOD) is used to represent all the chemical compounds in the waste materials. BOD is the amount of oxygen required to biochemically oxidize the organic compounds in the waste materials. For example, the BOD value of a wastewater containing $10\,g/L$ glucose can be obtained as follows:

$$C_6H_{12}O_6(\text{glucose}) + 6O_2 \rightarrow 6CO_2 + 6H_2O \tag{6.18}$$

The molecular weight of glucose and oxygen gas are 180 and 32, respectively. Thus, the BOD of $10 \, g/L$ glucose can be calculated with the ratio in Reaction 6.17 as $10.7 \, g$ BOD/L. In industrial practice, BOD values of certain waste materials are determined experimentally (APHA et al., 2005).

Chemical oxygen demand (COD) is also widely used to represent the organics in the waste materials. Actually, COD includes the oxygen demand from the oxidation of both organic compounds and reduced inorganic compounds. Usually, COD values for complex waste materials are also determined experimentally (APHA et al., 2005). For most municipal, food-processing, and agricultural waste materials, BOD is approximately 80% of the COD.

6.3 Temperature Effect on Anaerobic Digestion

Anaerobic digestion can be performed under different temperatures. It has generally three categories based on temperature: psychrophilic (low temperature), mesophilic (medium temperature), and thermophilic (high temperature) anaerobic digestion processes.

6.3.1 Psychrophilic Anaerobic Digestion

Psychrophilic anaerobic processes are normally operated at a temperature range of 10°C–25°C. They are usually applied in tropical and subtropical climate areas under ambient temperature without heating. These anaerobic systems are simple and relatively economical in construction and operation because no heat exchange units are involved. They can be used for treating a variety of waste materials containing high carbohydrates. However, there are disadvantages of psychrophilic anaerobic processes compared to mesophilic and thermophilic ones. The hydrolytic, acidogenic, acetogenic, and methanogenic reaction rates are low at low temperatures because the corresponding anaerobic bacterial activities are low. High residence time or large volume of digesters is necessary for efficient degradation of organics and biogas production.

Research efforts have been spent to improve the performance of anaerobic digestion at low or ambient temperatures. High-rate anaerobic reactors such as anaerobic sequencing batch reactors (ASBR), UASB reactors, and anaerobic filters (AF) have been studied for high efficiency of wastewater treatment (Hawkins et al., 2001; Alvarez et al., 2002; Bodik et al., 2002; del Pozo et al., 2002; Elmitwalli et al., 2002a,b; Ligero and Soto, 2002; Masse et al., 2002; Ong et al., 2002; Seghezzo et al., 2002).

UASB reactors have been widely used for the treatment of municipal and industrial wastewaters at different temperatures. UASB digesters were applied for the pretreatment of a high-solid domestic wastewater at a temperature between 17°C and 20°C (Alvarez et al., 2002) and for the treatment of settled sewage at ambient subtropical temperatures (16.5°C–21.6°C) in Argentina (Seghezzo et al., 2002). In the former case, 82%–85% of the suspended solids (SS) was hydrolyzed at an HRT of 2.9 h. In the latter case, granular sludge bed developed in the UASB reactor that had been inoculated with semidigested sewage sludge and the SRT was about 450 days. The conversion of the organic compounds reached 63.5% at an HRT of 6 h.

Animal slaughterhouse wastewaters usually contain high organics including fat. It is quite challenging to treat the wastewaters anaerobically under low temperatures. An ASBR was successfully used for the treatment of hog slaughterhouse wastewater containing pork fat in three-day cycles at 25°C (Masse et al., 2002). The pork fat particle degradation was mainly controlled by long-chain fatty acid oxidation and, to less extent, by neutral fat hydrolysis. del Pozo et al. (2002) studied the start-up process of an anaerobic fixed film reactor (AFFR) for the treatment of poultry slaughterhouse wastewater at low temperature (20°C–24°C). The anaerobic reactor was packed with vertically arranged PVC tubes. In spite of low temperature, the start-up process of the AFFR was successfully achieved in 74 days with a long acclimation phase followed by a low loaded growth phase.

Combination of two high-rate low-temperature anaerobic reactors could significantly enhance the efficiency of waste treatment and biogas production. An ASBR and up-flow anaerobic filter (UAF) filled with plastic insulating tubes were tested for municipal wastewater treatment at a temperature range of 9°C–23°C (Bodik et al., 2002). Average COD degradation efficiency of 56%–88% and 46%–92% were achieved for ASBR and UAF, respectively, at an HRT of 6–46 h. Elmitwalli et al. (2002a) investigated a two-step system consisting of an AF and an anaerobic hybrid (AH) reactor (a granular sludge bed at bottom half and a filled bed at the top half) for municipal wastewater treatment at a low temperature of 13°C. The medium used in both reactors was recycled polyurethane foam. High efficiencies of hydrolysis, acetogenesis, and methanogenesis were obtained in the system and as high as 71% COD degradation could be achieved in an HRT of 4 and 8 h in AF and AH reactors, respectively.

6.3.2 Mesophilic Anaerobic Process

Mesophilic anaerobic processes are the most commonly used ones for the treatment of a variety of industrial and municipal wastes for biogas production. They are usually operated at moderate temperatures of 30°C–37°C. Mesophilic anaerobic processes are popular because they are easy to start up, stable in performance, and relatively high in microbial activities. In mesophilic anaerobic processes with raised temperature, hydrolytic, acidogenic, acetogenic, and methanogenic reaction rates are much higher compared to psychrophilic processes. Recent research on mesophilic anaerobic processes has been focused on high-rate processes, codigestion of industrial and agricultural wastes, phased anaerobic digestion, and novel processes.

UASB reactors have been extensively studied for the treatment of food wastewaters such as fish processing, slaughterhouse, fruit cannery, brewing and winery wastewaters (Palenzuela-Rollon et al., 2002; Rodriguez-Martinez et al., 2002; Sigge et al., 2002; Wang et al., 2002). A common characteristic of these wastewaters was their high organic content. COD in the wastewaters was in the range of 4,000–12,800 mg/L. It would be very expensive to treat these wastewaters in aerobic processes such as activated sludge system. A common practice for treating high-strength food-processing wastewaters is to apply anaerobic system(s) for the degradation of most organics in the wastewaters and then use a polish system, usually an aerobic one, for further treatment of the anaerobic effluent. As a result of the anaerobic treatment, biogas is produced and can be utilized as an energy source. A UASB reactor treating a high-strength fish processing wastewater with high lipids at 30°C could achieve approximately 80% of COD and lipids conversion in the anaerobic process (Palenzuela-Rollon et al., 2002). A hybrid two-phase anaerobic system was studied for the treatment of food waste at 35°C (Wang et al., 2002). The food waste was from a university canteen, including residual fruits, vegetables, eggshell, and spoiled noodles. The system consisted of a solid fermentation reactor for acidification and a UASB reactor for methanogenesis. High efficiencies of COD and volatile solids (VS) conversion were achieved in the anaerobic system. It was found that leachate recycling in the acidification reactor was a fast way for effective acidification of the food waste. Conversion of VFA to methane by methanogenic bacteria in the UASB reactor was a key to maintain the system stability. Dilution effect of VFA in the methanogenic reactor made the pH control unnecessary. Sigge et al. (2002) tested a UASB reactor for the first-step treatment of fruit cannery and winery wastewaters at an HRT of 24 h and a temperature of 35°C. Over 90% COD was converted to biogas in the anaerobic reactor.

Tubular anaerobic digester and anaerobic sequencing batch reactor have also been studied for industrial wastewaters under mesophilic conditions (Ruiz et al., 2002; Bouallagui et al., 2003). Both HRT and solid loading affected the performance of the anaerobic reactor when a semicontinuously mixed tubular anaerobic digester was used for conversion of fruit and vegetable waste into biogas at 35°C (Bouallagui et al., 2003). Inhibition to the methanogenic bacteria was observed when HRT was less than 12 days. The reactor could perform well with the solid content in the influent up to 8%. Zero-order kinetics were found to be appropriate models to describe both acidification of the organic matter and methanization of the VFA during anaerobic treatment of winery wastewater in an anaerobic sequencing batch reactor at 35°C (Ruiz et al., 2002). The ASBR was operated at an HRT of 2.2 days, an organic loading rate of 8.6 g COD/L/day, and a specific organic loading rate of 0.96 g COD/L/day. The zero-order kinetics is most probably the results of the high substrate concentrations in both acidification and methanization processes. Under such a circumstance, Monod kinetic model can be simplified into zero-order model.

Many solid organic wastes such as domestic kitchen garbage and agricultural residues are anaerobically digestible. However, anaerobic digestion of these solid organic wastes alone usually results in poor digestion efficiency because they do not have sufficient nutrients and minerals to support bacterial growth. These nutrients and minerals are sufficiently present in sewage sludge and animal wastes. Therefore, a concept of codigestion of sewage sludge or animal wastes with kitchen garbage or agricultural residues has been studied for better treatment of the wastes for biogas production (Kaparaju et al., 2002; Komatsu et al., 2002). In a study on codigestion of kitchen garbage and sewage sludge in mesophilic anaerobic digesters, the sewage sludge:kitchen garbage ratio used was as high as 1:1.6 (Komatsu et al., 2002). Similar digestion efficiencies were achieved in these digesters compared to the digester treating sewage sludge alone. Steady-state operations were obtained in the digesters within 60 days of initiation. Some agricultural residues such as grass hay, crop straws, and stovers can be costly for environment-friendly management. These residues are usually anaerobically digestible for biogas production, but the digestion efficiencies are low. When the residues were codigested with cow manure in mesophilic anaerobic digesters, high efficiencies could be achieved (Kaparaju et al., 2002). The specific methane yields were 0.35, 0.26, and 0.21 m^3/kg VS for grass hay, oats, and clover, respectively.

Acetogenesis (including hydrolysis and acidogenesis) and methanogenesis, two major processes of anaerobic digestion, are usually performed in a single reactor. However, the two processes can be conducted separately in two-stage anaerobic reactors. In the latter case, both processes can be operated under their optimal conditions. Control of pH in the acidification reactor would determine the products of the acetogenesis. Under mesophilic conditions (35°C–37°C), productions of propionate and ethanol were favored at pH of 4.0–4.5; acetate, propionate, butyrate, and ethanol were the main products at pH of around 5.5; acetate and butyrate were the favored products at pH of 6.0–6.5; whereas acetate and propionate were the main products at pH of 8.0 (Yu and Fang, 2003). Dilution rate usually does not significantly affect the product formation in the acetogenic reactor.

Novel configurations of anaerobic systems have been studied for enhancing the performance of anaerobic processes (Han et al., 2002; Kim et al., 2002). Kim et al. (2002) developed a membrane-coupled anaerobic system for VFA fermentation and effective dissolved organics recovery from coagulated raw sewage. A ceramic micro-filtration membrane was used in the study. Solid retention time was maintained at 240 h by controlling the membrane filtration ratio, the volume of membrane permeate per day divided by volume of input coagulated raw sludge per day. VFA production increased with the increase in HRT until it reached 12 h at which a maximum VFA production was observed. Han et al. (2002) developed a multistep sequential batch two-phase anaerobic composting system to convert food waste into biogas. The system consisted of five leaching bed reactors for hydrolysis, acidification, and posttreatment, and a UASB reactor for methane production. The food waste was acidified in leaching bed reactors and soluble VFAs in the leachate were converted to methane in the UASB reactor. Solids after acidification in the leaching bed reactors were composted by blowing air through the beds for posttreatment. Rumen microorganisms were inoculated into the leaching bed reactors to enhance the degradation of cellulosic materials. A high efficiency of 85% VS conversion was achieved in this novel system.

6.3.3 Thermophilic Anaerobic Process

Thermophilic anaerobic bacteria grow well at elevated temperatures (50°C–65°C). The bacteria degrade organics in wastes at higher rates than those that grow at lower temperatures (psychrophilic and mesophilic). Thermophilic anaerobic digestion (TAnD) has been commonly applied for the treatment of animal, industrial, and municipal wastes in some European countries such as Denmark and the Netherlands. Because of their high digestion rates, thermophilic anaerobic digesters can be in a highly compact size with low HRT. Another major advantage of TAnD is its ability to kill most pathogenic bacteria and viruses because of a combined effect of high temperature and anaerobic environment (Shih, 1993). A big challenge for establishing a TAnD system is the slow and sometimes difficult start-up process. Some key thermophilic anaerobes necessary for a normal operation of a TAnD system are very low or sometimes missing in wastes. To develop a mature thermophilic anaerobic population, it is critical to provide optimal growing conditions for those thermophilic anaerobes.

With the practice of more strict regulations by U.S. Environmental Protection Agency on disinfection of biosolids such as digested sewage sludge before their application on crop land, thermophilic anaerobic process has recently attracted greater attention for waste management. A full-scale single-stage mesophilic anaerobic sewage sludge digester was successfully transformed to thermophilic operation (55°C) with a microbial management strategy at the Terminal Island Plant in Los Angeles, Calif., United States (Ahring et al., 2002). Mesophilic digested sewage sludge was used as the inoculum to start up the thermophilic process and the temperature in the digester was raised to 55°C within two days. The feeding strategy of the digester was based on the individual VFA concentrations in the digester with the specific methanogenic activity (SMA) of the culture in the digester. The strategy used to start up the thermophilic digester was to pulse feed the digester and at the same time monitor the development of VFAs to ensure that the culture was capable of handling the extra substrate added. SMA was another important parameter used to evaluate the capability of the thermophilic anaerobic culture for handling an extra load of feed. It took approximately three months for the thermophilic anaerobic digester to reach a stable operation with an HRT of approximately 12 days. The digester produced stabilized solids that contained less coliforms than the U.S. EPA requirement. Total solids and VS reductions in the thermophilic process increased by 9.8% and 9.0%, respectively, compared to the mesophilic process treating the same sludge.

Seed culture is very important in the start-up process of thermophilic anaerobic digesters. Waste activated sludge (WAS) from aeration systems was found a much better seed source than anaerobically (mesophilic) digested sludge (ADS) for starting up thermophilic anaerobic digesters (55°C) fed with acetate and propionate (Kim and Speece, 2002). WAS started to produce methane soon after acetate feeding without a lag time, while ADS had a lag time of 10 days. When the digesters were fed with propionate, WAS almost completely degraded propionate, while ADS consumed little propionate and produced little methane.

In addition to higher specific biogas production rate and much higher pathogen removal, the thermophilic anaerobic digester could result in much less odor and foaming problems than the mesophilic one (Zabranska et al., 2002). Hydrogen sulfide content in the thermophilic biogas can be approximately 30% less than that in the mesophilic biogas. The thermophilic biogas contains only lightweight volatile hydrocarbons and some volatile oxygenated organics but no other odorous sulfuric organic compounds in significant amount, while mesophilic biogas contains a significant amount of dimethylsufide, dimethyldisulfide, and methylpropyldisulfide that are strongly odorous compounds although their concentrations are low. The higher resistance of the thermophilic digester to foaming than the mesophilic one is because the former had a higher efficiency in destroying filamentous microorganisms.

When high-rate thermophilic UASB reactors are used for wastewater treatment, HRT can be much shortened because SRT is uncoupled from HRT. A good quality, well settleable granular sludge was cultivated and retained in a thermophilic UASB reactor treating methanol-containing wastewater at an HRT of 3.2 h, allowing an organic loading rate of 47 g COD/L/day (Paulo et al., 2002). Methanol removal rate was 93%, where 79% was converted to methane. Multistage thermophilic (55°C) UASB reactors were applied for the treatment of food-processing wastewater containing high lipids and proteins (Tagawa et al., 2002). The digester was operated at stable conditions for over 600 days with a high organic loading rate of 50 kg COD/m^3/day. COD conversion of 90% was achieved despite high protein and lipid levels in the feed wastewater. The presence of high lipids and proteins along with high concentrations of Mg and Ca ions in the raw wastewater could cause severe scum and/or insoluble substance formation within the UASB sludge bed, resulting in a low contact efficiency between substrate and sludge. Replacement of active microbial granules in the sludge bed by the insoluble lipids and proteins can cause deterioration of sludge methanogenic activity.

The characteristics of suspended-growth psychrophilic, mesophilic, and TAnD processes are summarized in Table 6.2.

TABLE 6.2

Characteristics of Suspended-Growth Psychrophilic, Mesophilic, and Thermophilic Anaerobic Digestion Processes

Anaerobic Process	Operating Temperature (°C)	Operating HRT (days)	Microbial Growth and Digestion Rates	Tolerance to Toxicity
Psychrophilic	10–25	>50	Low	High
Mesophilic	30–37	25–30	Medium	Medium
Thermophilic	50–60	10–15	High	Low

6.4 Hydrogen Effect on Anaerobic Digestion

Hydrogen gas is an important intermediate during the anaerobic digestion process. It is a major product of hydrolysis, acidogenesis, and acetogenesis. Hydrogen, at a high concentration, may exhibit product inhibition to the acetogenic reactions such as

$$CH_3CH_2COOH + 2H_2O \rightarrow CH_3COOH + CO_2 + 3H_2 \qquad (6.19)$$

$$CH_3CH_2OH + H_2O \rightarrow CH_3COOH + 2H_2 \qquad (6.20)$$

Hydrogen, at high concentration, may change the metabolic pathway of the anaerobic digestion of some organic compounds. For example, anaerobic digestion of ethanol to methane has two steps: the first step is the acetogenesis of ethanol to acetic acid and hydrogen (Reaction 6.19); the second one is the conversion of both acetic acid and hydrogen to methane. If the hydrogen concentration is high in the system, the two-step pathway could be changed. Hydrogen may react with acetic acid and carbon dioxide (normally abundant in anaerobic systems) to form propionic acid:

$$CH_3COOH + CO_2 + 3H_2 \rightarrow CH_3CH_2COOH + 2H_2O \qquad (6.21)$$

Reaction 6.20 is actually the reverse reaction of Reaction 6.18.

6.5 pH Effect on Anaerobic Digestion

The optimal pH for the growth of methanogenic bacteria is in the range of 6.5–8.0. Methane production is normally active in anaerobic digestion processes with the pH range. However, methanogenic activities may be inhibited or stopped if the pH is lower than 6.0. In that case, the anaerobic digestion system is in favor of hydrogen, instead of methane, production, which will be discussed in detail in a later section. The chemical compounds that may affect the pH in anaerobic digestion processes include organic acid, carbon dioxide, and ammonia. Organic acids could be components in the raw waste materials or metabolic intermediates in the anaerobic processes. Carbon dioxide and ammonia are by-products of anaerobic digestion for methane production. The former is mainly from the carbohydrates in the waste materials and the latter is mainly from proteins. Hydrogen sulfide is another common by-product of methane production. When the waste materials contain lipids, anaerobic digestion of the waste also produce phosphates. All these by-products could affect the pH of the aqueous phase in the anaerobic digestion. Almost all the phosphates stay in the aqueous phase in the anaerobic digestion processes:

$$H_3PO_4 \leftrightarrow H^+ + H_2PO_4^- \tag{6.22}$$

$$H_2PO_4^- \leftrightarrow H^+ + HPO_4^{2-} \tag{6.23}$$

$$HPO_4^{2-} \leftrightarrow H^+ + PO_4^{3-} \tag{6.24}$$

The chemical equilibrium of these three reactions can be respectively expressed as

$$K_{P,1} = \frac{[H^+][H_2PO_4^-]}{[H_3PO_4]} \tag{6.25}$$

$$K_{P,2} = \frac{[H^+][HPO_4^{2-}]}{[H_2PO_4^-]} \tag{6.26}$$

$$K_{P,3} = \frac{[H^+][PO_4^{3-}]}{[HPO_4^{2-}]} \tag{6.27}$$

On the other hand, carbon dioxide, ammonia, and hydrogen sulfide stay in both gaseous and aqueous phases of an anaerobic reactor. The phase equilibrium of the chemical compounds can be described with Henry's law:

$$[CO_2]_1 = H_{CO_2}P_{CO_2} \tag{6.28}$$

$$[NH_3]_1 = H_{NH_3}P_{NH_3} \tag{6.29}$$

$$[H_2S]_1 = H_{H_2S}P_{H_2S} \tag{6.30}$$

where $[CO_2]_l$, $[NH_3]_l$, and $[H_2S]_l$ are the concentrations of CO_2, NH_3, and H_2S in the aqueous phase, respectively; H_{CO_2}, H_{NH_3}, and H_{H_2S} are Henry's constants for CO_2, NH_3, and H_2S, respectively; P_{CO_2}, P_{NH_3}, and P_{H_2S} are the partial of CO_2, NH_3, and H_2S in the gaseous phase, respectively. When CO_2 is dissolved in water, it becomes carbonic acid, H_2CO_3, that can be ionized through the following processes:

$$H_2CO_3 \leftrightarrow H^+ + HCO_3^- \tag{6.31}$$

$$HCO_3^- \leftrightarrow H^+ + CO_3^{2-} \tag{6.32}$$

The chemical equilibrium of the carbonic acid dissociation reactions can be respectively expressed as

$$K_{CO_2,1} = \frac{[H^+][HCO_3^-]}{[H_2CO_3]} \tag{6.33}$$

$$K_{CO_2,2} = \frac{[H^+][CO_3^{2-}]}{[HCO_3^-]} \tag{6.34}$$

Ammonia is relatively soluble in water and can be readily ionized:

$$NH_{3,l} + H_2O \leftrightarrow NH_4^+ + OH^- \tag{6.35}$$

The chemical equilibrium can be expressed as

$$K_{NH_3} = \frac{[NH_4^+][OH^-]}{[NH_3]_l} \tag{6.36}$$

Similar to CO_2, H_2S is ionized in water in two steps:

$$H_2S \leftrightarrow H^+ + HS^- \tag{6.37}$$

$$HS^- \leftrightarrow H^+ + S^{2-} \tag{6.38}$$

The chemical equilibrium of the ionization reactions can be respectively expressed as

$$K_{H_2S,1} = \frac{[H^+][HS^-]}{[H_2S]_l} \tag{6.39}$$

$$K_{H_2S,2} = \frac{[H^+][S^{2-}]}{[HS^-]} \tag{6.40}$$

TABLE 6.3

Chemical Equilibrium Constants of CO_2, NH_3, H_2S, and Phosphates in Water at 25°C

Chemical Reaction	Equilibrium Constant (K) $-\log K = pK$
$H_2CO_3 \leftrightarrow H^+ + HCO_3^-$	6.3
$HCO_3^- \leftrightarrow H^+ + CO_3^{2-}$	10.3
$H_2S \leftrightarrow H^+ + HS^-$	7.1
$HS^- \leftrightarrow H^+ + S^{2-}$	14
$H_3PO_4 \leftrightarrow H^+ + H_2PO_4^-$	2.1
$H_2PO_4^- \leftrightarrow H^+ + HPO_4^{2-}$	7.2
$HPO_4^{2-} \leftrightarrow H^+ + PO_4^{3-}$	12.3
$NH_{3,l} + H_2O \leftrightarrow NH_4^+ + OH^-$	4.7

The chemical equilibrium constants and Henry's constants of CO_2, NH_3, H_2S, and phosphates are listed in Tables 6.3 and 6.4.

Ammonia is readily soluble in water, so the majority of the ammonia produced during the anaerobic digestion stays in the water in the form of ammonium ion, NH_4^+. The ammonia in the gaseous phase is usually very little, from 200 ppm to 1%, depending on the waste material being digested in the anaerobic digester. Ammonia concentration in the biogas can be estimated using Equations 6.29 and 6.36 if the ammonium (NH_4^+) concentration and pH in the digestate are known. Some waste materials contain proteins, lipids, and minerals including significant amount of magnesium (Mg). Anaerobic digestion of the waste materials results in ammonium, magnesium, and phosphate in the digestate that form magnesium ammonium phosphate ($MgNH_4PO_4$) or struvite that precipitates. Hydrated struvite is very hard, like a rock, but the dehydrated struvite looks like a powder and can be used as a slow-release fertilizer for crops.

Hydrogen sulfide is generated in anaerobic digestion of most waste materials. The H_2S concentration in the biogas depends on the characteristics of the waste materials, especially the content of heavy metals such as Cu, Zn, and Fe, and can be from 500 ppm to 2%. The heavy metals can react with sulfide ion to form hardly soluble metal sulfide such as CuS, ZnS, and FeS. H_2S concentration in the biogas can be estimated using Henry's law

TABLE 6.4

Henry's Constants of CO_2, NH_3, and H_2S in Water at 25°C

Gas	Henry's Constant, K_H (M/atm)
CO_2	0.035
NH_3	58
H_2S	0.1

Note: M/atm = Moles per liter per atm.

(Equation 6.30) and chemical equilibria (Equations 6.39 and 6.40) with the known or measured sulfide concentration and pH in the digestate.

Anaerobic digestion produces a significant amount of CO_2, usually 20%–40% in biogas. Some of the produced CO_2 dissolves in the water, causing a decrease of pH. To maintain the pH within the optimum range for the methanogenic bacteria (pH 6.5–8.0), it may be necessary to add alkalines such as $NaHCO_3$, Na_2CO_3, or lime (CaO). In industrial practice, an alkalinity of 1000–5000 mg/L needs to be maintained in an anaerobic digester to keep the pH within the optimal range. Alkalinity is the sum of alkaline concentrations in water and it can be expressed in anaerobic digestion as

$$\text{Alkalinity (mol equivalent/L)} = [HCO_3^-] + 2[CO_3^{2-}] + [OH^-] - [H^+] \qquad (6.41)$$

In industrial practice, equivalent mass of $CaCO_3$ is usually used for alkalinity. The conversion factor is obtained as follows:

The molecular weight of $CaCO_3$ is 100 g/mol; when 1 mol of $CaCO_3$ is completely dissolved in water, it generates 2 mol equivalent (meq) alkalinity; thus

$$\frac{100\,\text{g}\,CaCO_3/\text{mol}\ CaCO_3}{2\ \text{meq alkalinity/mol}\ CaCO_3} = 50\,\text{g}\,CaCO_3/\text{meq alkalinity} \qquad (6.42)$$

Increasing alkalinity in anaerobic digestion can be from the metabolism-generated alkalinity, for example, biodegradation of proteins releases NH_4^+. If necessary, alkalinity can be increased by adding chemicals such as CaO, MgO, $NaHCO_3$, Na_2CO_3, NaOH, and $NH_3 \bullet H_2O$.

6.6 Toxicity/Inhibition

Inhibitory chemical compounds at certain concentrations cause anaerobic digestion upset and failure. A chemical compound may be regarded as inhibitory when it stops the bacterial growth or causes an adverse shift of bacterial population. Indication of inhibition to anaerobic digestion is usually a decreasing or complete stop in methane production. Chemical compounds commonly found inhibitory to anaerobic digestion include ammonia, sulfide, metals, and some organic compounds. Chen et al. (2008) provided a comprehensive review of inhibition in anaerobic digestion processes, including inhibitory chemicals, mechanisms, and remedies.

6.6.1 Ammonia Inhibition

Ammonia is a by-product of anaerobic digestion of organic waste materials and generate through the biodegradation of proteins and urea. Ammonium ion (NH_4^+) and free ammonia are the main forms of inorganic nitrogen in the anaerobic digestate. It is generally believed that free ammonia is usually inhibitory to the anaerobic bacteria. The hydrophobic free ammonia molecules are permeable to cell membrane and may diffuse into the cells, causing proton imbalance, and/or potassium deficiency (Kroeker et al., 1979; de Baere et al., 1984; Sprott and Patel, 1986; Gallert et al., 1998). Methanogenic bacteria are much more sensitive to ammonia inhibition than are hydrolytic, acidogenic, and acetogenic bacteria (Kayhanian, 1994). Koster and Lettinga (1988) found that when the total ammonia nitrogen (TAN) (sum of both free ammonia and ammonium ion) concentration reached 4.0–5.7 g/L, acidogenic bacteria in the granular sludge were hardly affected while the methanogenic population lost 56.5% of its activity. The most sensitive methanogenic bacterium, *Methanospirillum hungatei*, could be inhibited by ammonia at the TAN concentration of 4.2 g/L (Jarrell et al., 1987).

Ammonia, at low concentrations (<200 mg/L), is usually beneficial to the anaerobic bacteria because they need N as an important nutrient to support their growth. It is generally believed that anaerobic digestion with TAN less than 1.5 g/L should be safe for a healthy methane production. The actual inhibitory concentration of TAN to a specific anaerobic digestion process depends on the organic compounds in the waste materials, bacterial inocula, pH, and temperature in the digester. pH affects the ratio of free ammonia to ammonium ion in anaerobic digestion even though the TAN may remain the same. Generally, a higher pH results in a higher free ammonia, while a lower pH associates with a higher ammonium ion at a certain TAN concentration. However, if the ammonia inhibition to the methanogenic bacteria is due to high pH, the inhibition usually results in an accumulation of VFAs, which leads to a decrease in pH and free ammonia concentration. The interaction of pH, free ammonia, and VFA may lead to a so-called inhibited steady state, a condition under which the anaerobic digestion process is "stably" running but with a lower methane yield (Angelidaki and Ahring, 1993). Presence of other cations such as Na^+, Ca^{2+}, and Mg^{2+} was found to be antagonistic to ammonia inhibition, a phenomenon in which the toxicity of one ion is decreased by the presence of other ion(s) (McCarty and McKinney, 1961; Braun et al., 1981; Hendriksen and Ahring, 1991). Another important factor that affects ammonia inhibition to the methanogenic bacterial growth is the acclimation. When the anaerobic digestion process is operated in a slowly increased ammonia concentration, the anaerobic bacteria experience an adaptation process to high ammonia concentration. The acclimated microorganisms can retain viability at the TAN concentrations far exceeding the initial inhibitory concentration.

When the ammonia concentration in the anaerobic digestion is much higher than the inhibitory concentration and detrimental to the anaerobic bacteria, the ammonia can be physically removed by stripping or chemical precipitation. Stripping is usually performed with air to drive the free ammonia out of the aqueous solution. Oxygen in the air may momentarily shock the obligate anaerobic bacteria, but their activities can be revived after a relatively short period of air stripping, especially for the waste materials with high BOD or COD. Chemical precipitation of ammonium ion can be achieved through the addition of magnesium and/or phosphate to form struvite precipitation, as discussed earlier in this chapter. The combination of ammonium removal and pH adjustment (keeping the pH around neutral) can significantly reduce free ammonia concentration in anaerobic digestion, thus alleviating the ammonia inhibition to the anaerobic microorganisms (mainly the methanogens).

6.6.2 Sulfate/Sulfide Inhibition

As discussed earlier in this chapter, sulfide is a by-product of biogas production during anaerobic digestion of organic waste materials. It is usually a result of sulfate reduction, which involves the reduction of sulfate that is a common constituent of many industrial wastes by sulfate-reducing bacteria (SRB) that are very common in the consortium of anaerobic bacteria. Sulfate reduction is performed by two major groups of SRB, that is, incomplete oxidizers that reduce organic compounds such as lactate to acetate and CO_2, and complete oxidizers that completely convert acetate to CO_2. The primary inhibition related to sulfide on methane production in anaerobic digestion is that the SRB are competing for the inorganic and organic substrates such as hydrogen, acetate, and VFA against mainly the acetogenic and methanogenic bacteria. The result of this inhibition is less methane yield. SRB generally do not compete against hydrolytic bacteria because the former cannot use biopolymers such as starch, glycogen, proteins, or lipids. Although a few strains of SRB have been reported to utilize sugars and amino acids as substrate to support their growth, SRB normally cannot rely on acidogenic substrates for their metabolism (Klemps et al., 1985; Min and Zinder, 1990; Hansen, 1993). SRB compete against acetogenic bacteria for substrates such as propionate, ethanol, and butyrate. In fact, sulfate-reducing biochemical reactions have a thermodynamic advantage over the corresponding acetogenic reactions. The hydrogen-oxidizing SRB have also shown their thermodynamic and kinetic advantages over hydrogenotrophic methanogens in anaerobic digestion. Hydrogen-oxidizing SRB have a lower hydrogen threshold concentration than hydrogenotrophic methanogens (Oude Elferink et al., 1994; Colleran et al., 1995). However, SRB usually require much lower oxidation–reduction potential (ORP) (<−450 mV) in their metabolism than most methanogens (−350–250 mV), which makes SRB very sensitive to ORP change.

On the other hand, sulfide, the product of sulfate reduction, is also toxic to methanogens. Many researchers believe that H_2S is the toxic form of sulfide since it can penetrate through the membrane of microbial cells. H_2S may form sulfide and disulfide cross-links between polypeptide chains, denaturing native proteins and interfering with the various coenzyme sulfide linkages and the assimilatory metabolism of sulfur (Conn et al., 1987;

Vogels et al., 1988). As discussed earlier in this chapter, the ratio of undisso-ciated H_2S to sulfide ion in anaerobic digestion changes substantially with the pH value. Therefore, the sulfide levels that exhibit inhibition to meth-ane production in anaerobic digestion is in the range of 100–800 mg/L dis-solved sulfide or approximately 50–400 mg/L undissociated H_2S (Parkin et al., 1990). Sulfur is also a required nutrient for most microorganisms. In fact, the sulfur content of methanogenic bacteria is normally higher than in many other anaerobic bacteria and the optimum sulfide concentration to support the growth of methanogens is between 1 and 25 mg/L (Scherer and Sahm, 1981).

Many industrial wastes, such as animal, food-processing, and mining wastes, contain a substantial amount of heavy metals such as Fe, Mn, Cu, Zn, Ni, and Co. These metal ions in the anaerobic digestate can react with sulfide to from insoluble metal sulfides, which precipitate into the anaerobic sludge. Solubility product is normally an indication of the solubility of a chemical compound in water. The expression of the solubility product is explained with the following example:

$$Fe^{2+} + S^{2-} \rightarrow Fe\,S\downarrow \tag{6.43}$$

The solubility product constant of FeS is defined as

$$SP_{FeS} = [Fe^{2+}][S^{2-}] \tag{6.44}$$

where
   SP_{FeS} is the solubility product constant of FeS
   $[Fe^{2+}]$ is the concentration of Fe^{2+} in the aqueous solution
   $[S^{2-}]$ is the concentration of S^{2-} in the aqueous solution

A list of solubility product constants for some heavy metal sulfides is shown in Table 6.5.

6.6.3 Metal Inhibition

Salts are commonly found in wastewaters such as animal, food-processing, and municipal wastewa-ters. They can be inhibitory to microorganisms at high levels because high salt concentrations cause the dehydration of microbial cells due to osmotic pressure (de Baere et al., 1984; Yerkes et al., 1997). Although salts are dissociated into both cations and anions in aqueous solutions, cations are usu-ally responsible for the inhibition (McCarty and McKinney, 1961).

TABLE 6.5

Solubility Product Constants of Metal Sulfide at 18°C

Metal Sulfide	Solubility Product Constant
FeS	3.7×10^{-19}
MnS	1.4×10^{-15}
CoS	3.0×10^{-26}
ZnS	1.2×10^{-23}
NiS	1.4×10^{-24}
CuS	8.5×10^{-45}

Note: The values are obtained from Weast (1975).

Cations that are commonly found in wastewaters include sodium, potassium, calcium, and magnesium. Heavy metals are also commonly found in wastewaters, for example, Fe, Cu, Zn, Ni, Co, Mn, Cr, Cd, and Pb. Most of these metals are required nutrients for the growth of anaerobic bacteria. However, at high levels they can be inhibitory or toxic to the bacteria.

Sodium: Sodium is a required nutrient for the growth of anaerobic bacteria, probably because of its role in the formation of adenosine triphosphate or in the oxidation of NADH (Dimroth and Thomer, 1989). The optimum level of the sodium concentration for mesophilic anaerobes in anaerobic digestion is 100–350 mg/L (McCarty, 1964; Patel and Roth, 1977). However, sodium can be inhibitory to the anaerobes when its concentration is very high. A variety of wastewaters contain a relatively high level of sodium, especially food-processing wastewaters. The inhibitory sensitivities of different anaerobic bacteria to sodium are different. In general, the hydrolytic, acidogenic, and acetogenic bacteria are more sensitive to sodium than methanogens. Among the VFA-utilizing bacteria, high sodium level is more toxic to propionate-utilizing anaerobic bacteria than the acetate-utilizing methanogens (Soto et al., 1993). When the sodium concentration reaches 3500–5500 mg/L, sodium exhibits moderate inhibition to methane production in anaerobic digestion (McCarty, 1964). As the sodium concentration increases to 8000 mg/L, the inhibition becomes very strong and may stop the methane production. Adaptation of the anaerobic bacteria to a slow increase of sodium concentration can significantly increase the sodium inhibition threshold. Methanogenic bacteria could survive at sodium concentration as high as 20 g/L when the methanogenic culture was acclimated to high sodium level of 12 g/L (Chen et al., 2003).

Potassium: Potassium is another required nutrient for the growth of anaerobic bacteria. Low level of potassium (<400 mg/L) can enhance methane production in anaerobic digestion. However, potassium can be inhibitory or toxic at high concentration because it leads to a passive influx of potassium ions that neutralize the membrane potential (Jarrell et al., 1984). High potassium level is commonly found in animal wastes. It can be as high as 4000 mg/L in swine waste. The potassium ion threshold for the inhibition to the methane production in anaerobic digestion generally decreases with the increase in temperature. In other words, the inhibition sensitivity of the anaerobic bacteria for the anaerobic digestion processes operated under different temperatures is in the following order from high to low: thermophilic > mesophilic > psychrophilic. Kugelman and McCarty (1964) found that 5.85 g K^+/L caused 50% inhibition of acetate-utilizing methanogens in a mesophilic anaerobic sludge digester, while Fernandez and Forster (1993, 1994) reported that potassium at a concentration of 1.2 g K^+/L inhibited the TAnD of simulated coffee wastes.

Acclimation of the anaerobic bacterial culture to a relatively high potassium concentration could substantially increase the potassium inhibition threshold on methane production in anaerobic digestion (Chen and Cheng,

2007). It was found that a thermophilic anaerobic culture in an anaerobic digester operated at 50°C had the inhibition threshold, beyond which methane production decreased significantly, of 3 g K^+/L. Accumulation of VFAs was observed during the decline of methane production. Propionic acid was the dominant fatty acid, indicating that the propionate-utilizing bacteria were more sensitive to potassium inhibition than acetate-utilizing methanogens. However, acclimation of the thermophilic anaerobic culture to high potassium concentrations of 6 and 9 g K^+/L resulted in a significant increase of the potassium inhibition threshold. Acclimation to a potassium concentration of 6 g K^+/L increased the tolerance of anaerobic inocula to potassium inhibition without significantly reducing the methanogenic activity. Inhibition threshold was increased from 3 g K^+/L for unacclimated inocula to 6 g K^+/L for inocula acclimated to 6 g/L of potassium. Acclimation of the anaerobic inocula to the potassium concentration 9 g K^+/L further increased the inhibition threshold to 7.5 g K^+/L.

The coexistence of other cations such as sodium, magnesium, and ammonium could mitigate the potassium toxicity, with sodium as the best mitigating cation (Kugelman and McCarty, 1964). Kugelman and McCarty (1964) also found that a combination of cations produced antagonism superior to that of a single cation. The combinations of sodium and calcium, and sodium, calcium, and ammonia have shown the best results of potassium inhibition antagonism.

Calcium: Calcium is also an essential nutrient for the growth of most microorganisms including anaerobic bacteria. However, excessive calcium in the anaerobic digestion process leads to precipitation of carbonate and phosphate, which may cause accumulation of insoluble inorganic solids in the digesters and pipes. Both carbonate and phosphate serve as buffers (HCO_3^-/CO_3^{2-} and $H_2PO_4^-/HPO_4^{2-}/PO_4^{3-}$) to contribute to the maintenance of pH within the optimum range for the anaerobic bacteria during the anaerobic digestion. The precipitation of carbonate and phosphate results in a loss of buffering capacity. Calcium carbonate precipitation could also impact the anaerobic bacterial activities, but the impact is complex with both benefits and drawbacks. The precipitated calcium carbonate could result in scaling of the bacterial biomass and the highly scaled biomass is less active because of mass transfer limitations. On the other hand, the precipitated calcium carbonate could serve as a core for the anaerobic bacteria to attach and grow to form a biofilm on the surface of the precipitates. As a result, the concentration of bacteria can be significantly increased for a higher rate of anaerobic digestion. The overall impact of calcium carbonate precipitation depends on individual anaerobic digestion conditions. Kugelman and McCarty (1964) found that the optimum Ca^{2+} concentration for the methanogenic bacteria using acetic acid as the substrate was 200 mg/L. When the Ca^{2+} concentration increased to 2500–4000 mg/L, moderate inhibition occurred. A strong inhibition was observed at the Ca^{2+} concentration of 8000 mg/L.

Magnesium: Magnesium, at low concentrations, is a necessary nutrient for many microorganisms to support their growth. It can stimulate the reproduction of microbial cells. The adequate Mg^{2+} level for the optimum growth of anaerobic bacteria is within a range of 300–720 mg/L (Harris, 1987; Xun et al., 1988; Ahring et al., 1991; Schmidt and Ahring, 1993). However, a high level of magnesium can be inhibitory and toxic to the anaerobic bacteria. The information about magnesium inhibition to anaerobic bacteria is very little in the literature, but it is generally believed that the inhibition threshold is around 1900 mg Mg^{2+}/L (Kugelman and Chin, 1971).

Aluminum: Aluminum is normally existent in some waste materials such as food-processing wastes. There is very limited information about aluminum inhibition to the anaerobic bacteria in the literature. Aluminum may cause inhibition to the anaerobic bacteria because of its competition against iron and manganese, which are required nutrients for most bacteria. Aluminum can also adhere to the microbial cell membrane or wall, causing its malfunction and thus negatively affecting the growth of the anaerobic bacteria (Cabirol et al., 2003). Cabirol et al. (2003) found that the specific activity of methanogenic and acetogenic bacteria decreased by 50% and 72%, respectively, after the anaerobic culture had been exposed to 1000 mg/L $Al(OH)_3$ for 59 days. Acclimation is helpful to enhance the tolerance of the anaerobic bacteria to aluminum inhibition. Jackson-Moss and Duncan (1991) found that the anaerobic bacteria could tolerate the aluminum level as high as 2500 mg Al^{3+}/L in anaerobic digestion after the acclimation of the culture.

Antagonism: The coexistence of some ions from dissolved salts, especially the cations, can significantly increase the inhibition threshold of an individual metal ion to the anaerobic microorganism. This phenomenon is called antagonism. For example, metal ions such as potassium or calcium have antagonistic effect on sodium in inhibiting the anaerobic bacteria. A combination of potassium and calcium significantly increases the antagonistic effect on sodium over that achieved by potassium alone (Kugelman and McCarty, 1964). Magnesium was also found to have antagonistic effect to sodium on the inhibition of anaerobic bacteria (Ahring et al., 1991).

Heavy metals: Heavy metals are commonly found in municipal, animal, food-processing, and feed-processing waste materials. The heavy metals that have been found to be particular concerns to anaerobic digestion include chromium, iron, cobalt, copper, zinc, cadmium, and nickel (Jin et al., 1998). A unique characteristic of the heavy metals is that they are not biodegradable and can accumulate to potentially toxic concentrations (Sterritt and Lester, 1980). The mechanism for the heavy metals to exhibit inhibitory or toxic effect on microorganisms is their disruption of enzyme function and structure by binding with thiol and other groups on protein molecules or by replacing naturally occurring metals in enzyme prosthetic groups (Vallee and Ulner, 1972).

Many enzymes that catalyze the biochemical reactions of anaerobic digestion have heavy metals as their components. Takashima and Speece (1989) studied 10 methanogenic bacterium strains and found that the heavy metal content in the bacterial cells was in the following order: $Fe \gg Zn \gg Ni > Co = Mo > Cu$. Some heavy metals, at low concentration, can be stimulatory to the growth of anaerobic microorganisms. The net effect of heavy metals to anaerobic digestion, stimulatory or inhibitory, depends on the total metal concentration, chemical forms of the metals, and process-related factors such as pH and ORP (Mosey et al., 1971; Lin and Chen, 1999; Zayed and Winter, 2000). In contrary to light metal (e.g., sodium and potassium) inhibition, methanogenic bacteria are generally believed to be more sensitive to heavy metal inhibition than the acidogenic bacteria (Zayed and Winter, 2000).

Heavy metals may be present in the aqueous solution of anaerobic diges-tion as dissolved ions, precipitated solids, adsorbed on microbial cells or inert particulates, or a part of the complex compounds formed with the metabolic intermediates or product. Among these metal forms, only soluble free metal ions may be inhibitory or toxic to the anaerobic bacteria. (Lawrence and McCarty, 1965; Mosey and Hughes, 1975; Oleszkiewicz and Sharma, 1990). The inhibitory level of heavy metals to the anaerobic bacteria varies from several to several hundred milligrams per liter, depending on the substrates, bacterium strains, and environmental conditions such as pH and tempera-ture. Bacterial biomass concentration in the anaerobic digestion is also an important factor. Generally, high bacterial biomass concentration could tol-erate higher heavy metal concentrations. Although most of the heavy metal inhibition data reported in the literature are in terms of heavy metal concen-tration, it would be more precise to express the inhibition in terms of the dos-age of the heavy metals per unit weight of bacterial biomass (Hickey et al., 1989). Normally, the relative sensitivity of acidogenic bacteria to heavy met-als is in the order of $Cu > Zn > Cr > Cd > Ni > Pb$, while the methanogenic bacteria is in the order of $Cd > Cu > Cr > Zn > Pb > Ni$ (Lin, 1992, 1993).

Many industrial and animal wastes contain a variety of heavy metals such as Cu, Zn, Fe, Ni, and Pb. These heavy metals can generate synergis-tic or antagonistic effects on anaerobic digestion to change the inhibition level of an individual heavy metal. The toxicity of most mixed heavy metals such as Cr–Cd, Cr–Pb, Cr–Cd–Pb, and Zn–Cu–Ni is synergistic (Lin, 1992). Ni has also shown to act synergistically in Ni–Cu, Ni–Mo–Co, and Ni–Hg mixtures to exhibit inhibition to anaerobic microorganisms (Babich and Stotzky, 1983). Some heavy metal mixtures have shown antagonistic effects of inhibition to anaerobic digestion. Ni has shown to act antagonistically in Ni–Cd, Ni–Zn, and Ni–Cu mixtures to decrease the toxicity of Cd, Zn, and Cu, respectively, to the anaerobic digestion (Babich and Stotzky, 1983; Ahring and Westermann, 1985).

The methods commonly used for the mitigation of heavy metal toxicity are precipitation, sorption, and chelation by inorganic and organic agents (Oleszkiewicz and Sharma, 1990). As discussed earlier in this chapter, sulfide can form precipitation with many heavy metals, including Fe, Mn, Co, Zn, Ni, and Cu, and is commonly used precipitation agent. A list of the solubility product constants for some metal sulfides is shown in Table 6.5. In fact, many waste materials contain a significant amount of sulfate that is reduced to sulfide during anaerobic digestion, which provides a readily available precipitation agent. Physical and chemical adsorptions can also alleviate heavy metal inhibition. Many heavy metals can be adsorbed on solid sludge in the anaerobic digester. Other adsorbents such as activate carbon, kaolin, and diatomite and waste materials such as compost and cellulose pulp waste can also be used for the mitigation of heavy metal inhibition. Chelation of heavy metals with organic ligands such as EDTA, PDA, NTA, aspartate, and citrate has also been used to fix the heavy metal ions in aqueous solutions (Babich and Stotzky, 1983), which can mitigate the heavy metal inhibition to the anaerobic digestion by reducing free heavy metal ions in the solution.

6.6.4 Organic Compound Inhibition

Organic compounds that have shown inhibition or toxicity to the anaerobic digestion include alkanes, alcohols, aldehydes, ethers, ketones, and carboxylic acids including long-chain fatty acids (LCFAs) as well as benzenes, phenols, halogenated aromatics, aliphatics and alcohols, nitro-aromatics, acrylates, amines, nitriles, amides, and pyridine and its derivatives (Chen et al., 2008). The mechanism of organic compound inhibition to the anaerobic digestion is that some organic compounds that are poorly soluble in water or adsorbed to the surface of sludge solids may accumulate to high levels in anaerobic digesters. The accumulation of apolar pollutants in bacterial membranes causes the membrane to swell and leak, disrupting ion gradients, and eventually causing cell lysis (Heipieper et al., 1994; Sikkema et al., 1994). The inhibition concentration of these organic compounds varies widely for specific toxicants, depending on biomass concentration, time of the microbial cells to the toxicant, cell age, feeding pattern, acclimation, and temperature (Yang and Speece, 1986). In fact, most of these organic compounds are biodegradable by the anaerobic microorganisms. They become toxic to the microorganisms when their concentration exceeds the inhibitory levels. Generally, the tolerance level of the anaerobic bacteria to toxicants increases with the increase in the microbial biomass concentration. Young active bacterial cells are also more resistant to the toxicity than the aged cells. Similar to other inhibitions, acclimation of the anaerobic culture to the specific toxicants can also improve the tolerance of the anaerobic bacteria to the toxicity.

Some animal and industrial waste materials such as food-processing and animal slaughterhouse wastes contain a high level of fatty materials. Anaerobic digestion of the fatty materials is often hampered because of the inhibitory effect of LCFAs. LCFAs have been reported to be inhibitory at low concentrations for Gram-positive but not Gram-negative microorganisms (Kabara et al., 1977). Methanogenesis can also be inhibited by LCFAs because the cell wall of the methanogens resembles that of Gram-positive bacteria (Zeikus, 1977). LCFAs also exhibit serious inhibition to anaerobic bacteria by adsorbing onto the cell wall/membrane, interfering with the transport or protective function of the bacteria (Rinzema et al., 1994). In addition, adsorption of a light layer of LCFAs to the microbial biomass leads to the flotation of the anaerobic culture and thus cause a washout of the culture (Rinzema et al., 1989). The LCFAs that are commonly found in anaerobic digestion include oleic acid, lauric acid, cyprylic acid, capric acid, and myristic acid. The IC50 (the toxicant concentration that causes 50% reduction in cumulative methane production over a fixed period of exposure time) of oleic acid and lauric acid is around 4.3 mM. Cyprylic acid is only slightly inhibitory (Koster and Cramer, 1987). The toxicity of capric acid and myristic acid can be enhanced at the presence of lauric acid. An important factor that affects the toxicity of LCFAs to the anaerobic bacteria is temperature. Thermophilic anaerobes are more sensitive to LCFAs inhibition than mesophilic anaerobes (Hwu and Lettinga, 1997).

Lignins and derivatives: Some agricultural residues such as wheat and rice straws can be anaerobically codigested with animal manure. The residues contain a significant amount of lignin, which has a complex structure including phenolic aromatics that has shown inhibition to anaerobic bacteria. Lignin derivatives with aldehyde groups or apolar substituents are highly toxic to methanogenic bacteria. The aromatic carboxylic acids are also mildly toxic to the anaerobes.

6.7 Anaerobic Digestion Kinetics

Anaerobic digestion process involves a series of biochemical reactions to convert the organic compounds into methane. As discussed earlier in this chapter, the process has four steps: hydrolysis, acidogenesis, acetogenesis, and methanogenesis. Each step involves specific substrates and anaerobic bacteria. The kinetic expressions for the microbial growth and substrate utilization in each step can be described using Equations 5.43 and 5.44, respectively:

$$\frac{dX}{dt} = \frac{\mu_m [S] X}{K_S + [S]} - \lambda X \qquad (5.43)$$

$$\frac{d[S]}{dt} = -\frac{1}{Y} \frac{\mu_m [S] X}{K_S + [S]} = -\frac{k [S] X}{K_S + [S]} \qquad (5.44)$$

The microbial biomass decay coefficient, k_d, is usually used for the specific death coefficient (λ) for the anaerobic bacteria in anaerobic digestion. Thus, Equation 6.42 can be rewritten as

$$\frac{dX}{dt} = \frac{\mu_m[S]X}{K_S + [S]} - k_d X = Y\frac{k[S]X}{K_S + [S]} - k_d X \tag{6.45}$$

The yield (Y) of anaerobic bacteria (ratio of microbial biomass growth to the substrate consumption) is approximately 0.06 g biomass/g BOD for municipal sludge, which is much lower than that of aerobic bacteria (around 0.6 g biomass/g BOD) because the energy release from an anaerobic fermentation is much lower than aerobic oxidation for the same organic compound. The following example is a comparison of anaerobic conversion of acetic acid to methane and carbon dioxide by methanogenic bacteria to aerobic oxidation of acetic acid to carbon dioxide:

Anaerobic methanogenesis:

$$CH_3COOH \rightarrow CH_4 + CO_2 \quad \Delta G = -12.59 \text{ kcal/mol} \tag{6.46}$$

Aerobic oxidation:

$$CH_3COOH + 2O_2 \rightarrow 2CO_2 + 2H_2O \quad \Delta G = -208.08 \text{ kcal/mol} \tag{6.47}$$

The decay coefficient for anaerobic bacteria is 0.03–0.04 day^{-1}.

The maximum specific substrate utilization rate (k) and the half-velocity constant (K_s) for the anaerobic bacteria varies with temperature. With the increase in temperature, k increases while K_s decreases, which is shown in Figure 6.2.

Typical values of k and K_s for the acetate-utilizing methanogesis at different temperatures are listed in Table 6.6.

The following equations are also widely used to estimate the maximum specific substrate utilization rate (k) and the half-velocity constant (K_s) for the anaerobic metabolism at different temperatures:

Lawrence and McCarty (1969) Arrhenius equation:

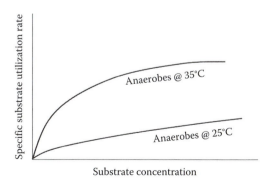

FIGURE 6.2
Kinetics of anaerobic metabolism at different temperatures.

TABLE 6.6

Kinetic Coefficients (Maximum Specific Substrate
Utilization Rate, k, and the Half-Velocity Constant,
K_s) for Acetate-Utilizing Methanogenesis at
Different Temperatures

Temperature (°C)	k (day⁻¹)	K_s (mg/L)
35	6.67	164
25	4.65	930
20	3.85	2130

$$\log \frac{K_{S,2}}{K_{S,1}} = 6980 \left(\frac{1}{T_2} - \frac{1}{T_2} \right) \qquad (6.48)$$

where $K_{S,1}$ and $K_{S,2}$ (mg/L) are the half-velocity constants at temperatures T_1 and T_2 (°K), respectively.

The van't Hoff–Arrhenius equation is

$$k_T = k_{20} \theta^{(T-20)} \qquad (6.49)$$

where k_T and k_{20} (day⁻¹) are the maximum specific substrate utilization rates at temperatures of T (°C) and 25°C; θ is the temperature-activity coefficient and its values are in the range of 1.02–1.25. The temperature-activity coefficient, θ, is used in the van't Hoff–Arrhenius equation because the sensitivities to temperature vary for different types of anaerobic bacteria. For example, methanogenic bacteria is more sensitive than the acetogenic bacteria for their growth and metabolism, as shown in Figure 6.3.

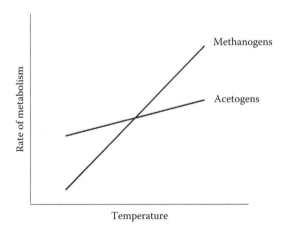

FIGURE 6.3
Sensitivities of methanogens and acetogens to the temperature effect on their metabolism.

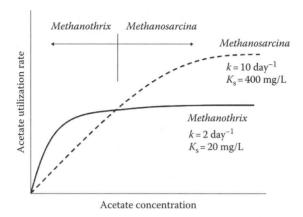

FIGURE 6.4

Competition of *Methanosarcina* and *Methanothrix* for acetate utilization during anaerobic digestion.

Different groups of microorganisms may act on the same substrate during anaerobic digestion, causing competition between the microbial groups. As shown in Figure 6.3, both *Methanothrix* and *Methanosarcina* utilize acetic acid as the substrate for methane production. However, their metabolic kinetics are different, as shown in Figure 6.4. *Methanosarcina* has higher values of the maximum specific utilization rate (k) and half-velocity constant (K_s) than *Methanothrix*. As a result, *Methanothrix* is dominant when acetate concentration is low in the anaerobic digestion, while *Methanosarcina* is dominant at high acetate concentration. When an anaerobic digester with suspended microbial biomass is operated at low acetate concentration for a long period of time, *Methanothrix* will be overwhelming and *Methanosarcina* could be washed out. Opposite situation could happen at high acetate concentrations.

6.8 Modeling Anaerobic Digestion Processes

To meet the needs of increased model application for full-scale plant design, operation, and optimization, a comprehensive IWA (International Water Association) Anaerobic Digestion Model No. 1 (ADM1) has been developed by Batstone et al. (2002). The ADM1 is a structured model with disintegration of hydrolysis, acidogenesis, acetogenesis, and methanogenesis steps including biochemical processes of (1) acidogenesis from sugars, (2) acidogenesis from amino acids, (3) acetogenesis from LCFA, (4) acetogenesis from propionate, (5) acetogenesis from butyrate and valerate, (6) aceticlastic methanogenesis, and (7) hydrogenotrophic methanogenesis. Substrate-based uptake Monod-type kinetics are used as the basis for all intracellular biochemical reactions. Death of biomass is represented by first-order kinetics, and dead biomass is maintained in the system as composite particulate material. Inhibition functions include pH (all groups), hydrogen (acetogenic groups), and free ammonia (aceticlastic methanogens). pH inhibition is implemented as an empirical equation, while hydrogen and free ammonia inhibitions are represented by noncompetitive functions. The other uptake-regulating functions are secondary Monod kinetics for inorganic nitrogen (ammonia and ammonium), to prevent growth when nitrogen is limited, and competitive uptake of butyrate and valerate by the single group that utilizes the two organic acids. Mechanisms included to describe the physicochemical processes are acid–base reactions and nonequilibrium liquid–gas transfer. Solid precipitation is not included in the model.

The ADM1 was applied to assess acid addition for pH control and avoidance of calcium carbonate ($CaCO_3$) precipitation in a UASB reactor treating paper mill wastewater (Batstone and Keller, 2002). The simulation work found, with a high degree of confidence, that acid dosing was neither economical for pH control nor had any real effect on the $CaCO_3$ levels in the reactor. The model was also used for assessment of the benefits of thermophilic (as opposed to mesophilic) operation for reduced ammonia inhibition, improved stability, and gas production in a solids digester at a gelatine production facility (Batstone and Keller, 2002). It was predicted that thermophilic operation could not attain either goal to a satisfactory extent.

Mathematical models have also been developed for anaerobic processes in specific digesters. Skiadas and Ahring (2002) have developed a new model for the anaerobic process in UASB reactors, based on cellular automata (CA). The liquid up-flow velocity and the superficial biogas velocity were believed to have considerable effect on the granular sludge in the UASB reactors, thereby acting as a selection pressure in the cultivation of the biomass. The new model combined differential equations with a CA system and was capable of quickly predicting the layer structure of the granules, and the granule diameter and the microbial compositions in the granules as functions of the operational parameters. The model is a combination of the biochemical reactions in the granule layers with the mass balances for a UASB digester:

$$\frac{dS_i^b}{dt} = D(S_i^{in} - S_i^b) + \frac{3BM}{rR} K_L^i (S_i^s - S_i^b) \tag{6.50}$$

where
  S_i^b is the concentration of the substrate i in the bulk solution
  t is the time of substrate utilization
  D is the dilution rate (the inverse of the reactor's HRT)
  S_i^{in} is the concentration of substrate i in the influent
  BM is the biomass concentration (as volatile suspended solids, VSS) in the
    reactor

r is the radius of a specific granule layer to the center of the granule
R is the radius of the granule
K_L^i is the mass transfer coefficient of substrate i
S_i^s is the concentration of the substrate i on the granule surface

The boundary conditions for Equation 6.48 are

$$\frac{\partial S_i}{\partial r} = 0 \qquad \text{at } r = 0 \tag{6.51}$$

$$-D_{ef,i} \frac{\partial S_i}{\partial r} = K_{L,i}(S_i^s - S_i^b) \qquad \text{at } r = R \tag{6.52}$$

where
$D_{ef,i}$ is the effective diffusivity of substrate i
S_i is the concentration of substrate i inside the granule
$K_{L,i}$ is the mass transfer coefficient for the substrate i

Sinha et al. (2002) developed an artificial neural network-based model to predict the steady-state performance of a UASB reactor treating high-strength wastewater. The model inputs were organic loading rate, HRT, and influent bicarbonate alkalinity. The output variables were one or more of the following: effluent substrate concentration, reactor bicarbonate alkalinity, reactor pH, reactor VFA concentration, average gas production rate, and percent methane content of the gas. Training of the neural network model was achieved using a large amount of experimentally obtained reactor performance data from the reactor mentioned above as the training set. Validation of the model was performed by using independent sets of performance data obtained from the same UASB reactor. Subsequently, simulations were performed using the validated neural network model to determine the impact of changes in parameters like influent COD concentration and HRT on the reactor performance. Simulation results were found to provide important insights into key variables that were responsible for influencing the working of the UASB reactor under varying input conditions.

A comprehensive mechanistic model was developed to simulate the constituent processes of a full-scale psychrophilic anaerobic digester treating swine wastewater (Fleming et al., 2002). The processes included in the simulation were the following: bulk fluid motion, sedimentation, bubble mixing, bubble entrainment, buoyant mixing, advection, biochemical reactions, internal heat transfer, and heat exchange with the environment, as shown in Figure 6.5. Three-year measured performance data for the digester, including weekly gas production rate, methane content, inlet and center temperature, volumetric flow rate, organics and nutrient concentrations in the influent and effluent, digester geometry, and hourly weather data, were used to validate

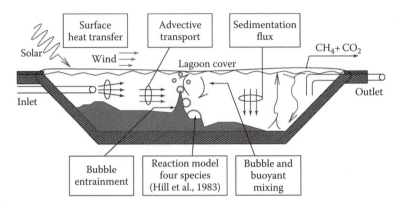

FIGURE 6.5
Model simulation of a full-scale ambient-temperature anaerobic digester converting swine waste into biogas. (Adapted from Fleming, J.G. et al., *Proceedings of the Animal Residuals 2002: Agricultural Animal Manure–Management, Policy and Technology*, Washington, DC, May 6–8, 2002.)

the model simultaneously. The validation results indicated that the model simulation fits the data very well.

Harmand et al. (2002) developed a nonlinear constrained model to predict the clogging of an anaerobic fixed-bed reactor. The mathematical model is a very useful tool in anaerobic reactor design, performance optimization and control, and insight understanding of the anaerobic processes. They also provide valuable information for further model development and technology transfer from research to industry.

6.9 Anaerobic Digesters

Anaerobic digesters are normally divided into two general categories: suspended-growth digesters and attached-growth or biofilm digesters, based on the nature of the presence of the bacterial cells in the digesters. In suspended-growth digesters, the microorganisms are suspended throughout in the digester. The HRT of the waste stream through the anaerobic digester is the same as the retention time of microbial cells or SRT in the digester. An adequate SRT is necessary for an efficient performance of an anaerobic digester because critical bacterial population can be washed out if the SRT is not long enough. Usually, methanogenic bacteria is the limiting factor for SRT in anaerobic digesters because of their low growth rate. However, hydrolytic bacteria can be the limiting microorganisms for SRT when the waste stream has a high level of organic solids. In the latter case, it probably takes long time to dissolve and degrade the organic solids by the hydrolytic bacteria.

In attached-growth anaerobic digesters, anaerobic bacteria are attached and grow on the surface of some supporting media such as plastic or ceramic packing materials. The bacteria can be immobilized on the surface of the supporting media and form biofilms. The benefits of the immobilization of the bacteria include (1) high microbial cell concentration in the digesters; (2) high SRT; (3) separation of SRT and HRT. High microbial cell concentration can result in high efficiency of anaerobic digestion. Separation of SRT from HRT makes it possible to design a relative low HRT or small digester with a high SRT to maintain a high conversion efficiency. High microbial cell concentration and high SRT increase the tolerance of anaerobic bacteria to toxic compounds. However, the attached-growth anaerobic digesters are usually sensitive to the organic solids in the waste stream because the solids cause clogging in the digesters. They are also more expensive to build and operate. A general comparison of suspended- and attached-growth anaerobic digesters is shown in Table 6.7.

6.9.1 Suspended-Growth Anaerobic Digesters

Based on the nature of waste stream flow through the digesters, suspended-growth anaerobic digesters can be classified as plug-flow reactors (PFR) and continuously stirred tank reactors (CSTR). In a PFR, the waste stream flows in the digester just like a pipe flow, as shown in Figure 6.6. There is almost no mixing of the aqueous materials in the direction of the flow in the PFR digester. Substrate concentration is the highest at the entrance of the digester and decreases along the digester as the substrate is degraded by the anaerobic bacteria. PFRs have a relatively high efficiency of digestion and work very well for the waste streams with stable substrate concentrations. However, if the substrate concentration significantly fluctuates in the influent waste stream, an occasionally very high concentration of substrate can inhibit the anaerobic bacteria in the digester, causing an upset or failure of the digester. In a CSTR, the substrate concentration is the same throughout the digester and in the effluent because of the continuous stirring as shown in Figure 6.7. In a CSTR, once the influent waste stream enters the digester, the substrate

TABLE 6.7

Comparison of Suspended- and Attached-Growth Anaerobic Digesters

Factor	Suspend-Growth	Attached-Growth
SRT	Low	High
Microbial cell concentration	Low	High
Digestion efficiency	Low	High
Tolerance to toxicity	Low	High
Tolerance to organic waste solids	High	Low
Cost of construction and operation	Low	High

Influent   Effluent

FIGURE 6.6
A plug-flow reactor (PFR) for anaerobic digestion of organic wastes.

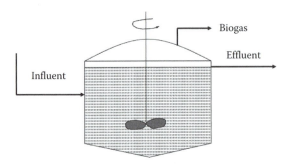

FIGURE 6.7
A continuously stirred tank reactor (CSTR) for anaerobic digestion.

concentration immediately drops to the same level as that in the effluent because the substrate is diluted in the digester. This phenomenon is called "dilution factor." Because of the dilution factor, CSTRs have a good capacity in handling the fluctuation of substrate concentration in the influent.

6.9.1.1 Design Equation

Mass balance for anaerobic digester: In general, an anaerobic digester involves waste materials entering the digester, substrate consumption in the digester, and effluent exiting the digester. A mass balance on the substrate can be expressed as the following:

Rate of accumulation = Rate of inflow − rate of outflow + rate of generation

or

$$\frac{dC}{dt}V = QC_0 - QC_e + r_c V \tag{6.53}$$

where
 C is the substrate concentration in the digester
 t is the time
 V is the digester volume
 Q is the waste flow rate
 C_0 is the substrate concentration in the influent
 C_e is the substrate concentration in the effluent
 r_c is the reaction rate of the substrate in the digester

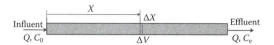

FIGURE 6.8
Mass balance for a plug-flow reactor (PFR) for anaerobic digestion.

In continuous flow digesters, there is no substrate accumulation, so Equation 6.51 becomes

$$0 = QC_0 - QC_e + r_c V \qquad (6.54)$$

PRF digesters: In a PFR digester, substrate concentration changes along the flow direction in the digester. The mass balance can be conducted as follows:

As shown in Figure 6.8, assume the substrate concentration at location X is C, a mass balance for a different volume of ΔV can be expressed as

$$QC - Q(C + \Delta C) + r_c \Delta V = 0 \qquad (6.55)$$

where
Q is the flow rate
C is the substrate concentration entering ΔV at location X
$C + \Delta C$ is the substrate concentration exiting ΔV at location $X + \Delta X$

An rearrangement of Equation 6.53 results in:

$$Q\Delta C = r_c \Delta V \qquad (6.56)$$

Assume the cross-sectional area of the PFR is A, Equation 5.99 becomes

$$Q\Delta C = r_c A \Delta X \qquad (6.57)$$

or

$$\frac{dC}{dX} = \frac{A}{Q} r_c \qquad (6.58)$$

which defines the relationship of the substrate concentration with the location in the digester. The volume of the PFR required to achieve the conversion efficiency is

$$V = \int_{C_0}^{C_e} \frac{Q}{r_c} dC \qquad (6.59)$$

CSTR digesters: Anaerobic digestion in a CSTR digester can be illustrated in Figure 6.9.

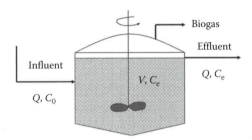

FIGURE 6.9
Anaerobic digestion in a continuously stirred tank reactor (CSTR).

A mass balance of the substrate in the digester can be expressed using Equation 6.52:

$$QC_0 - QC_e = -r_c V \tag{6.54}$$

Thus,

$$V = \frac{Q(C_0 - C_e)}{-r_c} \tag{6.60}$$

6.9.1.2 Complete-Mix Digester

Complete-mix digesters are the CSTR-type digesters, and they are the most commonly used digesters to convert organic wastes into biogas. The digesters include a continuously or semicontinuously mixing system to keep the materials uniformly distributed in the digesters. Complete-mix digesters are usually operated under mesophilic or thermophilic conditions with heat exchangers. Its advantages include:

- Relatively easy to build and operate
- Capable of digesting high solid waste streams
- High tolerance to the shock loading and toxicity because of the dilution factor

The major disadvantage of the complete-mix digesters is their relatively low efficiency or relatively high HRT required to perform the anaerobic digestion.

Mixing in the anaerobic digesters can be achieved through different means, including mechanical stirring, gas injection, and hydraulic mixing. Mechanical stirring was widely used in early anaerobic digesters. It provides a good mixing in anaerobic digesters, especially when the waste materials contain a high solid. The main drawback of the mechanical stirring is the maintenance of the moving parts to prevent gas leaking from the digester, which could be costly. Gas injection mixing is to use a portion of the biogas produced in the anaerobic digester, compress the biogas, and inject the gas at the bottom of the digester. The injected gas bubbles move through the mixed liquor from the bottom to the top of the digester. The gas injection provides a reasonably good mixing, especially in the vertical direction. The biogas bubbles normally do not interfere with the growth of the anaerobic bacteria in the digester because the biogas is generated in the digester. However, gas injection does not work well with high-solid waste streams. Gas pipe and compressor maintenance is also a concern. Hydraulic mixing is to recirculate the digestate from the top to the bottom of the digester. It involves a recirculation pump and a water distributor at the bottom of the digester. Recirculation in the digester can provide a very good mixing if the right flow rate is employed. The recirculated digestate usually do not affect the microbial growth in the digester because of the same reason as the injected biogas, unless the recirculation flow rate is too high and generate a very high shear force beyond the tolerance of the bacteria. Similar to the gas injection, recirculation does not provide a good mixing for high-solid waste streams. Solids in the recirculated digestate may cause clogging problems to the water distributor (nozzles and or draft tubes).

6.9.1.3 Baffled Anaerobic Digesters

A baffled anaerobic digester is a digester with baffles installed in the digester that guide the flow in the digester and make the flow like a plug flow. Figure 6.10 shows the wastewater flow in an baffled anaerobic digester. As a PFR-type digester, baffled anaerobic digesters have relatively high digestion efficiency. They are quite flexible with regard to the solid content in waste materials to be digested in the digesters. The main drawback of the conventional baffled anaerobic digesters is their low capacity of handling shock loadings of the influent waste streams. In a stably operated baffled anaerobic digester, a substantial increase of organic concentration over a short period of time or shock loading of the influent waste stream can cause inhibition of bacteria and failure of the digester. To avoid the negative impact of shock loading to the anaerobic culture in the baffled anaerobic digesters, a recirculation of some effluent to the influent can dilute the shock loading and prevent the inhibition (Figure 6.11).

FIGURE 6.10
Wastewater flow in a baffled anaerobic digester.

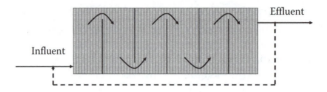

FIGURE 6.11
Modified baffled anaerobic digester with effluent recirculation to prevent inhibition of bacteria by shock loading.

6.9.2 Attached-Growth Anaerobic Digesters

Attached-growth or biofilm anaerobic digesters utilize inert media or biomass carriers for microorganisms to attach on the medium surface and to grow, which usually results in a high microbial cell concentrations and long SRTs in the digesters. The physical attachment prevents microorganisms from washout, which allows the digesters to be operated at high liquid velocities or short HRTs that would normally washout suspended cells. However, the continued growth and accumulation of microbial cells without control could cause internal mass transfer deficiency or even clogging in the biofilm digesters. Biofilm anaerobic digesters have been widely used for the treatment of wastewaters, especially low-strength wastewaters, because of their high performance with high microbial concentrations and SRTs. To convert many low-strength wastewaters with low concentration of organic compounds into biogas, a huge volume is necessary if suspended-growth anaerobic digesters are used because this type of digesters needs a long HRT (e.g., 20–30 days under mesophilic conditions). However, if biofilm anaerobic digesters are used for the same wastewaters, the volume of the digesters can be tremendously reduced because the HRT can be reduced to as low as a few hours. The commonly used biofilm anaerobic digesters include packed-bed and expended-bed anaerobic digesters.

6.9.2.1 Packed-Bed Anaerobic Digesters

Packed-bed anaerobic digesters are also called anaerobic biofilters. The digesters are usually packed with plastic or ceramic media for the bacteria to attach and grow on, as shown in Figure 6.12. The media bed is fixed in a packed-bed anaerobic digester. A good medium for the anaerobic biofilters usually has a high surface area and high porosity when packed in a digester. A high surface area provides a large platform for the anaerobic bacteria to attach and grow, which results in a high concentration in microbial cells in the digesters. High porosity enables a good flow of the wastewaters

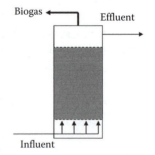

FIGURE 6.12
An up-flow packed-bed anaerobic digester or anaerobic biofilter.

through the digesters. During the digestion in an anaerobic biofilter, bacteria attached on the surface of the media grow while converting the organic compounds into biogas. The growth of the bacteria makes the biofilms thicker, which occupy the void space among the media. The lesser the void space, the shorter the HRT. When the void space is too little for the wastewater to flow or to result in a very short HRT that is not enough for an adequate digestion of organic compounds, a backwash is necessary to remove some bacteria from the digester. Backwash is normally conducted with the digester effluent passing through the biofilter at a very high velocity in the opposite direction to the flow of the wastewater during the normal digestion operation. The high-velocity water sloughs off most of the attached bacteria in the biofilms but still leaves a significant number of bacteria on the media to perform the biochemical reactions so the anaerobic biofilter can go back to the normal operation. The frequency of the backwash depends mainly on the concentration of organic compounds in the wastewater and the characteristics of the media (surface area and porosity). Usually, more frequent backwash is necessary for the anaerobic biofilter treating the wastewater with a relatively high concentration of organic compounds, which causes a high growth rate of the anaerobic bacteria. Because of their vulnerability to solid clogging in the packed-bed, anaerobic biofilters are usually not used for treating the wastewaters with high solid content. They are also not suitable for the treatment of concentrated wastewaters with high concentrations of organic compounds because high level of organics promotes the growth of anaerobic bacteria and therefore make it necessary to backwash the biofilter at high frequency. Anaerobic biofilters are very efficient for the treatment of low-strength wastewaters and HRT can be as low as a few hours. The very high SRT in the digesters makes the anaerobic culture tolerant to both shock organic loading and toxicities. The flow direction of an anaerobic biofilter is flexible, and can be either up-flow or down-flow.

6.9.2.2 Expended-Bed Anaerobic Digesters

The invention of the expended-bed anaerobic digester is an improvement of the packed-bed anaerobic digester to prevent clogging and avoid backwash. Different from the media structure in the packed-bed anaerobic digesters, the media bed is not fixed but expended in an expended-bed digester, which enables the movement of the media inside the digester. Figure 6.13 shows a schematic diagram of an expended-bed anaerobic digester. The media bed in the digester remains expended by a high velocity of upward flow of the mixed liquor in the digester, which is normally achieved by recirculation of the mixed liquor. Light media

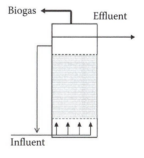

FIGURE 6.13
An expended-bed anaerobic digester.

such as plastic rings, granular activated carbons, and sands are usually used for the expended-bed anaerobic digesters. The expended-bed anaerobic digesters have almost all the advantage of the packed-bed anaerobic digesters, such as high digestion efficiency, high microbial cell concentration, high SRT, low HRT, and high tolerance to organic shock loading and toxicity. In addition, the expended-bed anaerobic digesters usually do not have clogging problems caused by excessive microbial cells and do not need backwash. The movement of the media inside the digester prevents the bacterial biofilms from becoming too thick that may cause clogging or need backwash. The main drawback of the expended-bed anaerobic digesters is its cost. It is relatively expensive to construct and operate. Keeping the media bed constantly expended requires a lot of energy.

6.9.3 Other Anaerobic Digesters

6.9.3.1 Up-Flow Anaerobic Sludge Blanket Digesters

6.9.3.1.1 Description of UASB Digester

UASB digester was invented by Lettinga et al. (1980, 1983) in early 1980s to convert high-carbohydrate wastewater into biogas. A schematic diagram of a typical UASB digester is shown in Figure 6.14. A unique characteristic of the UASB digester is the granule formation in the digester. The formation of anaerobic granules starts with the inorganic solids on which the anaerobic bacteria form coagulate and grow. It takes a few months to form mature anaerobic granules that are normally 1–3 mm in diameter. The granules are actually clusters of anaerobic bacteria, which can result in a very high concentration of bacterial cells (50–100 g/L). Similar to the

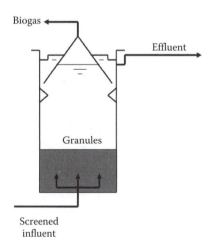

FIGURE 6.14
A schematics of a typical up-flow anaerobic sludge blanket (UASB) digester.

attached-growth culture, granular anaerobic culture has the following advantages:

- High SRT in the digesters
- High digestion efficiency with over 90% COD conversion
- Capable of treating high organic loading (12–20 kg COD/m³·d at 30°C–35°C)
- Low HRT (as low as 4–8 h)

UASB digesters work very well on high-carbohydrate wastewaters but not the high-protein wastewaters because the former is favorable for the formation of anaerobic granules. The optimum conditions for granule formation and effective digestion in UASB digesters are listed in Table 6.8.

6.9.3.1.2 Design of UASB Digesters

To design a UASB digester for the treatment of certain wastewater for biogas production, an adequate volume is necessary for efficient digestion and proper height needed to maintain an optimum up-flow velocity. An effective volume of a UASB digester can be estimated using the following equation:

$$V_n = QS_0/L_{org} \tag{6.61}$$

where
V_n is the effective liquid volume of the digester or volume occupied by sludge blanket and active microbial cells, m^3
Q is the influent flow rate, m^3/h
S_0 is the influent COD, kg/m^3
L_{org} is the organic loading rate, $kg\ COD/m^3 \cdot d$

Considering an effectiveness factor of anaerobic digestion, E, the total liquid volume of the UASB digester should be:

$$V_L = V_n/E \tag{6.62}$$

where $E = 0.8$–0.9, depending on the characteristics of the wastewater.

TABLE 6.8

Optimum Operational Conditions for UASB Digesters

Parameter	Optimum Value(s)
pH	Around 7.0
COD:N:P ratio in wastewater	300:5:1 at start-up
	600:5:1 at steady-state operation
Up-flow velocity, m/h	1.0–3.0 for wastewater with 100% soluble COD
	1.0–1.25 for wastewater with partially soluble COD

If the optimum up-flow velocity is v (m/h), then the cross-sectional area of the UASB digester should be:

$$A = Q/v \qquad (6.63)$$

Thus, the required liquid height of the UASB digester should be:

$$H_L = V_L/A \qquad (6.64)$$

It is necessary to leave some head space (usually 2.5–3.0 m) in the UASB digester to collect biogas. Therefore, the total height of the UASB digester should be:

$$H_T = H_L + H_G \qquad (6.65)$$

where H_G is the reactor height for biogas collection.

6.9.3.2 Anaerobic Sequencing Batch Reactors

When a batch reactor is used for anaerobic digestion of organic wastes, the process usually involves four steps: loading, reaction, settling, and decant. Loading is the process in which anaerobic culture and waste materials are transferred into the batch reactor. Loading time is relatively short. In the reaction step, biochemical reactions including hydrolysis, acidogenesis, acetogenesis, and methanogenesis take place in the reactor to convert the organics into biogas. This step takes a relatively long time, for example, 15–30 days for mesophilic culture, and continuous mixing is usually provided in the reactor. After the completion of the conversion of organic wastes to biogas, mixing is stopped to let the microbial cells settle down, which is called the settling step. The settling process normally takes 10–30 min, depending on the anaerobic culture. When the microbial cells are settled at the bottom of the batch reactor, the supernatant is drained or decanted as effluent of the reactor in the decant step. The settled microbial cells are used as inocula for the next batch operation in the reactor. Each batch cycle takes 15–30 days for mesophilic cultures, depending on the concentration of the organics in the wastes. The advantages of the batch digester include its high efficiency of anaerobic digestion and stability of the culture. It may take a while to develop the anaerobic culture for the waste stream in the first batch, but the culture stays fairly stable after its establishment in the following batches. The main disadvantage of the batch digester is that it takes a quite long time between the batches, while most waste streams are generated continuously or semicontinuously. To avoid the disadvantage, ASBR were invented to receive waste materials semicontinuously. An ASBR system usually contains several batch reactors, as shown in Figure 6.15.

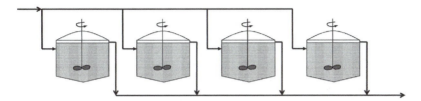

FIGURE 6.15
An anaerobic sequencing batch reactor (ASBR) system for anaerobic digestion.

The reactors can be connected in parallel (Figure 6.15) or series, making the ASBR system very flexible in organic loading. The bacteria concentration in each individual batch reactor can be very high, which result in a high digestion efficiency and high tolerance to shock loading and toxicity. In practice, the batch reactors receive waste materials at different times, allowing a semi-continuous feed of the waste to the ASBR system as a whole. Since an ASBR system involves several batch reactors operated at different steps, operation and control are quite complicated and usually require automation.

6.9.3.3 Covered Anaerobic Lagoon

Anaerobic lagoons are widely used for the treatment of animal wastewaters and municipal wastewater from small communities in warm climate areas. They are usually earthen structures with liners (clay or plastic) at the bottom to prevent wastewater leaking to soil and groundwater. As open systems, anaerobic lagoons are operated at different ambient temperatures. HRTs vary from 60 to 180 days. Although many anaerobic lagoons provide adequate treatment of the wastewaters, they generate serious environmental concerns about emissions of methane, ammonia, and odor (mainly from hydrogen sulfide and volatile organic acids). Covered anaerobic lagoons take the advantages of the anaerobic lagoon without emission problem. They are actually low-rate anaerobic digesters, as shown in Figure 6.16. Covered anaerobic lagoons are normally operated under ambient temperatures in tropical and subtropical areas with HRTs of 50–70 days, depending on the average annual temperature. Psychrophilic anaerobic bacteria are dominant in covered lagoons for most of the time. As shown in Figure 6.16, wastewater enters the covered anaerobic lagoon on one end of the lagoon while the effluent exits from the other end. Three layers are usually formed in the aqueous phase during the anaerobic digestion. A mixed liquor of organic materials and suspended microbial cells is in the middle layer where active metabolism takes place. As the organic compounds are degraded by the anaerobic bacteria to produce biogas, the bacteria also grow themselves. The old bacteria settle down to form a thick anaerobic sludge. The top layer in the lagoon is normally a supernatant with little solids. Usually, no mechanical mixing is provided in covered anaerobic lagoons, but the rising of the biogas bubbles

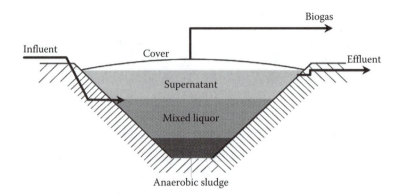

FIGURE 6.16
A covered anaerobic lagoon for wastewater treatment and biogas production.

generated in the anaerobic digestion process provides some mixing in the mixed liquor (Fleming et al., 2002). Effluent exiting the covered anaerobic lagoon is from the supernatant with very little solids, so the anaerobic bacteria stay in the lagoon for a very long time, which results in a high SRT. High SRT and HRT make the covered anaerobic lagoon very flexible in organic loading and solid content in the wastewater, and increase the tolerance of the anaerobic bacteria to the toxicities. High SRT also results in increased endogenous decay of the old bacteria, which releases nutrients that can be utilized by the active bacteria. As a result, the net yield of anaerobic bacteria in the covered anaerobic lagoons is significantly lower than that in many other anaerobic digesters. The most important advantage of the covered anaerobic lagoons is that they are economical to build and operate. A main disadvantage is their relatively large volume and surface area.

6.10 Anaerobic Digestion for Hydrogen Production

As discussed earlier in this chapter, anaerobic fermentation of organic wastes for methane production involves hydrolysis, acidogenesis, acetogenesis, and methanogenesis (Figure 6.1). Hydrogen gas is an important intermediate produced during the hydrolysis, acidogenesis, and acetogenesis. Hydrogen can be converted together with carbon dioxide to methane by H_2^- utilizing methanogenic bacteria. If the activity of the hydrogen-consuming methanogens is suppressed, the anaerobic fermentation process will exhibit significant capacity of converting organic wastes into hydrogen gas. Hydrogen gas is a clean energy source. When hydrogen gas is burned to produce energy, the only by-product is water. Hydrogen production from anaerobic digestion of organic waste materials has been extensively studied in the last decade.

Hydrogen-producing bacteria include *Clostridium* species, *Enterobacteriaceae*, and *Streptococcaceae* (Fang et al., 2002). Methods for promoting hydrogen production in the anaerobic digestion include pH adjustment, inhibition of hydrogen consuming methanogens, heat shock treatment, and SRT control.

pH adjustment: The optimum pH for H_2-utilizing methanogenic bacteria is around neutral (6.5–8.0). However, hydrogen-producing bacteria can tolerate a low pH (<6.5) or high pH (>9.0). If the pH is maintained at acidic (<6.5) or basic (>9.0) conditions in an anaerobic digester for organic waste digestion, methanogenic activities will be inhibited and hydrogen production will be favorable. Acidic conditions are more commonly used to create a favorable environment for hydrogen production. Chen et al. (2002) used the enrichment technique and obtained a hydrogen-producing anaerobic bacterium species from the biological sludge produced by wastewater treatment. The hydrogen-production potential of the sludge with acid or base enrichment was 200 and 333 times enhanced when the enrichment pH was 3 and 10, respectively. Voolapalli and Stuckey (2001) also observed high hydrogen production in an anaerobic reactor with acidically (pH<6.5) adapted inocula. Lay (2000) studied hydrogen production from anaerobic fermentation of starch and found that high hydrogen was produced within a pH range of 4.7–5.7.

Inhibition of H_2-utilizing methanogens: Inhibition of H_2-utilizing methanogens has also been studied for enhanced hydrogen production (Ueno et al., 1995; Zhu and Béland, 2006). Commonly used inhibitors include acetylene, 2-bromoethanesulfonic acid (BESA), and iodopropane. The acetate-utilizing methanogens are much more sensitive to the BESA inhibition than the H_2-utilizing methanogens. The activity of the former can be totally inhibited by 1 mmol BESA, while it takes 50 mmol BESA to completely inhibit the H_2-utilizing methanogenesis. Zhu and Béland (2006) found that the methanogenic activity could be inhibited by either 10 mmol BESA or 10 mmol iodopropane.

Heat shock treatment: Heat shock could also inhibit methanogenesis. When a liquid bacterial culture is exposed to a high temperature (>75°C), most of the bacteria cannot survive except some spore forming bacteria. Once the temperature is decreased back to the normal conditions, the spores should be able to produce viable growing cells. *Clostridium*, a spore forming bacterial species, has shown its potential to generate hydrogen gas in anaerobic processes (Brosseau and Zajic, 1982; Ueno et al., 1995). The *Clostridium* species are obligate anaerobic heterotrophs that produce hydrogen using the activities of the pyruvate-ferredoxin-oxidoreductase and hydrogenase enzymes and have an optimal pH range of 6.5–7.0 (Grady et al., 1999). van Ginkel and Sung (2001) used mixed bacterial cultures from a compost pile, a potato field, and a soybean field and heat shocked (baked) the cultures at 104°C for 2 h to inhibit the hydrogen-consuming methanogens and to enrich spore-forming, hydrogen-producing acidogens. A hydrogen production rate of 74.7 ml H_2/L·h was achieved at a pH of 5.5 and a substrate concentration of 7.5 g COD/L with a conversion efficiency of 38.9 mL H_2/(g COD/L).

SRT control: Methanogenic bacteria are slow-growing microorganisms and require a long SRT (e.g., 20–30 days under mesophilic conditions) to survive and perform methane production. When the SRT is reduced to hours or less than two days in an anaerobic digester, the methanogens will be washed out and hydrogen-producing activity will be dominant in the digester. In a two-phase configuration of anaerobic fermentation, the first phase is usually operated at a low SRT (2 h to 2 days), which makes it difficult for the methanogens to proliferate (Speece, 1996). Therefore, hydrogen gas production through the hydrolysis, acidogenesis, and acetogenesis is dominant in the first phase. The by-product of hydrogen production, mainly acetic acid, is converted to methane by acetate-utilizing methanogens in the second phase.

6.11 Biogas Purification

Biogas contains mainly 50%–80% methane and 20%–40% carbon dioxide as well as some minor gas components such as hydrogen sulfide (0.05%–1%) (Table 6.1). Among the components, methane is the only one that can generate energy through combustion. CO_2 does not change during the combustion, but its existence dilutes the energy density of the methane and consumes energy during the combustion for the increase of the temperature. Hydrogen sulfide is oxidized into sulfur dioxide and water, which forms sulfuric acid. Sulfuric acid is a very strong acid and corrosive. Extremely acidic electrolytes can rapidly dissolve metal parts of the combustion engines, especially at high temperatures (Figure 6.17).

Hydrogen sulfide itself can also be corrosive to the metals. It can form its own electrolyte and absorb directly onto the metal to cause corrosion. If

FIGURE 6.17
Metal corrosion caused by sulfuric acid. (Courtesy of Jerry Martin.)

the hydrogen sulfide concentration is very low, the corrosion will be slow. However, if the concentration of hydrogen sulfide in the biogas is greater than 100 ppm, it will cause fine pits on metals as shown in Figure 6.18.

The concentration of hydrogen sulfide in biogas depends on the waste materials treated in anaerobic digesters. Biogas generated from the anaerobic digestion of concentrated animal wastes usually contains a high level of hydrogen sulfide, as shown in Table 6.9.

Biogas can be upgraded to natural gas (>99% methane) if CO_2 and H_2S are removed. As natural gas, purified biogas can then be sent to the natural gas lines for households or used as fuel for the automobiles that utilize natural gas.

6.11.1 H_2S Scrubbing from Biogas

There are currently several methods to remove H_2S from the biogas, including precipitation, physical adsorption, chemical absorption, and biological oxidation.

Precipitation: As discussed earlier in this chapter, when H_2S is produced during the anaerobic digestion, it is dissociated in water. Many metal ions (Fe^{2+}, Fe^{3+}, Cu^{2+}, Zn^{2+}, Mn^{2+}, Ni^{2+}) can react with sulfide ion to form precipitation. Among these metals, the most economical choice is iron. Commonly used chemicals for sulfide precipitation and therefore removal of H_2S include iron chloride, iron phosphate, and iron oxide. The precipitants stay

FIGURE 6.18
Pitted anoxic corrosion caused by hydrogen sulfide. (Courtesy of Jerry Martin.)

TABLE 6.9

Hydrogen Sulfide Concentration in Biogas
Generated from Different Organic Wastes

Waste Material	H_2S in Biogas (ppm)
Swine manure	600–4000
Cattle manure	600–7000
Landfill wastes	0–2000

in the anaerobic sludge in the digester. Although precipitation can remove most of the hydrogen sulfide, there is usually still some left in the biogas.

Physical adsorption: H_2S can be adsorbed on porous media such as activated carbon and zeolites. The adsorption process is a clean process without any by-products. Although adsorption can remove almost all the H_2S in the biogas, it is a very expensive process because H_2S adsorption on activated carbon or zeolites is irreversible so it takes a lot of adsorbent.

Chemical absorption: Similar to precipitation, H_2S reacts with chemicals such as Fe $(OH)_2$, $FeCl_3$ to form solids and water, so H_2S can be removed from the biogas. The chemical reactions are as follows:

$$Fe(OH)_2 + H_2S \rightarrow FeS + 2H_2O \tag{6.66}$$

$$2FeCl_3 + 3H_2S \rightarrow Fe_2S_3 + 6H^+ + 6Cl^- \tag{6.67}$$

In practice, Fe $(OH)_2$ or $FeCl_3$ sponges are installed in biogas pipeline. When the wet biogas passes through the sponges, H_2S is removed from the biogas through the above chemical reactions. The chemical absorbents need to be replaced when their H_2S-absorbing capacity is exhausted.

Biological oxidation: When H_2S is burned or oxidized at high temperature, the product is sulfur dioxide. However, when H_2S is oxidized under mild conditions, incomplete oxidation could occur with sulfur as the product, as shown in Reaction 6.66. This can happen in biological

$$2H_2S + O_2 \rightarrow 2S + 2H_2O \tag{6.68}$$

oxidation of H_2S by aerobic bacteria. In practice, the biological oxidation is usually achieved by passing the biogas through moisturized biofilters or fixed film bioscrubbers with attached H_2S-oxidizing aerobic bacteria. Prior to the biogas entering the biofilters or bioscrubbers, a small amount of air (usually 2%–4% of the biogas volume) is added to the biogas to provide oxygen for the oxidation. The biological oxidation has been proven to be a very economical process for removing H_2S from biogas.

6.11.2 CO_2 Scrubbing from Biogas

Carbon dioxide can be removed from the biogas through physical absorption, physical adsorption, chemical absorption, and chemical conversion.

Physical absorption: As discussed earlier in this chapter, carbon dioxide dissolves in water:

$$CO_2 + H_2O \rightarrow H_2CO_3 \tag{6.69}$$

The relationship between CO_2 in gas and H_2CO_3 in water is determined by Henry's law. At high pressure, the phase equilibrium between CO_2 in gas and H_2CO_3 in water will be favorable toward the dissolution of CO_2 into the water. In industrial practice, pressurized water is usually used in scrubbing towers to remove CO_2 from the biogas. When the biogas passes through a CO_2 scrubbing tower filled with pressured water, most of the CO_2 will dissolve in water. H_2S can also be removed by the pressurized water based on the same principle. CO_2 removal efficiency in the scrubbing towers filled with pressured water can be as high as 99%. The quality of the water used in the scrubbing towers is flexible and treated municipal wastewater or grey water can be used for the purpose.

Physical adsorption: Activated carbon and zeolites can also be used for the adsorption of carbon dioxide from biogas. The adsorption is reversible, so the used activated carbon and zeolites can be regenerated for reuse. The regeneration is usually conducted by heating the used activated carbon or zeolites. Since H_2S can be irreversibly adsorbed on activated carbon and zeolites, H_2S needs to be removed from the biogas before the adsorbents are used for CO_2 removal from the biogas.

Chemical absorption: CO_2 reacts with bases such as NaOH, KOH, and $Ca(OH)_2$, which provides another option to remove CO_2 from biogas:

$$CO_2 + NaOH \rightarrow NaHCO_3 \tag{6.70}$$

$$CO_2 + KOH \rightarrow KHCO_3 \tag{6.71}$$

$$CO_2 + Ca(OH)_2 \rightarrow CaCO_3 + H_2O \tag{6.72}$$

Among the above bases, $Ca(OH)_2$ or lime is the most economical one for CO_2 removal from biogas.

Chemical conversion: CO_2, combined with hydrogen gas, can be chemically converted into methane at high temperature and high pressure:

$$CO_2 + 4H_2 \rightarrow CH_4 + 2H_2O \tag{6.73}$$

The process is commonly used in fertilizer industry, but it is rarely used to remove carbon dioxide from biogas because it is an expensive process and it involves the use of hydrogen gas.

6.12 Problems

1. Figure 6.4 shows a competition of *Methanosarcina* and *Methanothrix* for acetate utilization during anaerobic digestion. According to the figure, *Methanothrix* is dominant at low acetate concentrations, while *Methanosarcina* is dominant at high acetate concentration. At what acetate concentration would the two groups of anaerobic bacteria have the same substrate utilization rate? Can you keep the digester operated in a way that both groups of bacteria actively utilize the acetate at the same rate in a real practice? Explain.

2. What are the nutrient requirements for anaerobic digestion? Why do anaerobic bacteria normally grow at a much lower rate than aerobic bacteria?

3. An industrial wastewater has a daily average flow rate of $1000\,m^3/day$ and $4000\,mg/L$ of an organic substance with the following approximate composition: $C_{50}H_{75}O_{20}N_5S$. The organic waste is processed in an anaerobic digester for biogas production and the biodegradation efficiency is 95%. Determine:

    a. The alkalinity production in kilograms per day as $CaCO_3$

    b. Methane production rate in the anaerobic digester

    c. Approximate fraction of CH_4, CO_2, and H_2S in the biogas

4. For the anaerobic digestion of the industrial wastewater described in Problem 2, $10\,M$ ferric chloride ($FeCl_3$) solution is used to remove H_2S from the biogas produced in the digester.

    a. How much is the ferric chloride solution needed per week?

    b. If there are also iron, copper, and zinc ions existing in the wastewater at the concentration of 80, 60, and $50\,mg/L$, respectively, how much is the ferric chloride solution needed per week to remove the H_2S from the biogas?

5. The wastewater from an ethanol production facility contains 0.2% (w/w) ethanol. The wastewater flow is 250,000 gpd (gallons per day).

    a. What is the concentration of COD (chemical oxygen demand) in the wastewater if other organic compounds can be neglected?

    b. If the wastewater is treated in an anaerobic digester and 80% of the waste ethanol is converted to methane, what is the methane production rate in the digester?

    c. What are the approximate compositions of the biogas produced in the anaerobic digester?

    d. If the waste ethanol is utilized for hydrogen production and the conversion efficiency is 90%, what is the approximate hydrogen production rate?

6. A wastewater from a fermentation facility contains 800 mg/L acetate and 1200 mg/L ethanol. A suspended-growth anaerobic digester is used to convert the wastewater into biogas. The target conversion efficiency for both acetate and ethanol is 95%. In the anaerobic digestion process, methanogenesis is the limiting step for the conversion and the kinetic constants for the methanogenic bacteria at 35°C are

Maximum substrate utilization rate $k = 4 \, d^{-1}$;

Half-velocity constant $K_s = 40 \, mg \, COD/L$;

Yield $Y = 0.06 \, g$ biomass/g COD.

a. Assume that bacterial decay can be neglected, what is the appropriate HRT for the anaerobic digester to achieve the conversion at 35°C?

b. If the temperature is increased to 50°C, what is the appropriate HRT for the anaerobic digester to achieve the same conversion efficiency?

7. A food-processing wastewater has a daily average flow rate of 100 m³/day and 40,000 mg/L COD. An equivalent formula of the organic compounds in the wastewater can be expressed as $C_{10}H_{19}O_3N_2$. The wastewater is utilized for biogas production in a suspended-growth anaerobic digester at 35°C with a HRT of 25 days and the conversion efficiency is 95%.

a. What is the ammonium concentration in the mixed liquor of the digester?

b. Propose a method to remove ammonium from the mixed liquor to avoid inhibition and quantify the dosage of any reagents.

8. An anaerobic digestion process will be installed for biogas production from an animal wastewater. The wastewater has a biodegradable COD (chemical oxygen demand) concentration of 7000 and 5000 mg/L VSS with 70% of the VSS biodegradable. Briefly critique the compatibility of the following anaerobic processes for conversion of the wastewater to biogas and describe the potential impact of the influent solids on the reactor operation and performance.

Anaerobic processes:

UASB reactor

Anaerobic expanded-bed reactor

Anaerobic baffled reactor

Up-flow APB reactor

Covered anaerobic lagoon

9. How much is the annual organic kitchen waste produced in the households of the United States? Propose a strategy to utilize the

organic waste for biogas production. Assume the efficiency of converting the waste to biogas is 95%.

a. What is the approximate annual biogas production rate if all the organic kitchen waste is utilized for biogas production?

b. How much energy would be produced from the biogas annually?

c. How much natural gas would be saved if all the biogas would be utilized for energy production?

References

Ahring, B. K.; Alatriste-Mondragon, F.; Westermann, P.; Mah, R. A. (1991) Effects of cations on *Methanosarcina thermophila* TM-1 growing on moderate concentrations of acetate: Production of single cells. *Appl. Microbiol. Biotechnol.* 35, 686–689.

Ahring, B. K.; Mladenovska, Z.; Iranpour, R.; Westermann, P. (2002) State of the art and future perspectives of thermophilic anaerobic digestion. *Water Sci. Technol.* 46 (10), 293.

Ahring, B. K.; Westermann, P. (1985) Sensitivity of thermophilic methanogenic becteria to heavy metals. *Curr. Microbiol.* 12, 273–276.

Alvarez, J. A.; Zapico, C. A.; Gomez, M.; Presas, J.; Soto, M. (2002) Anaerobic hydrolysis of a municipal wastewater in a pilot scale digester. *Proceedings of the IWA 3rd World Water Congress* [CD-ROM]; Melbourne, Australia, April 7–12; International Water Association: London.

Angelidaki, I.; Ahring, B. K. (1993) Thermophilic digestion of livestock waste: The effect of ammonia. *Appl. Microbiol. Biotechnol.* 38, 560–564.

APHA/AWWA/WEF. (2005) Standard methods for the examination of water and wastewater, 21st edn. American Public Health Association/American Water Works Association/Water Environment Federation, Washington, DC.

Babich, H.; Stotzky, G. (1983) Toxicity of nickel to microbes: Environmental aspects. *Adv. Appl. Microbiol.* 29, 195–295.

Batstone, D. J.; Keller, J. (2002) Industrial applications of the IWA anaerobic digestion model No. 1 (ADM1). *Proceedings of the IWA 3rd World Water Congress* [CD-ROM]; Melbourne, Australia, April 7–12; International Water Association: London.

Batstone, D. J.; Keller, J.; Angelidaki, I.; Kalyuzhnyi, S. V.; Pavlostathis, S. G.; Rozzi, A.; Sanders, W. T. M.; Siegrist, H.; Vavilin, V. A. (2002) Anaerobic digestion model No. 1 (ADM1). IWA Publishing, London.

Bodik, I.; Herdova, B.; Drtil, M. (2002) The use of upflow anaerobic filter and AnSBR for wastewater treatment at ambient temperature. *Water Res.* 36, 1084.

Bouallagui, H.; Cheikh, R. B.; Marouani, L.; Hamdi, M. (2003) Mesophilic biogas production from fruit and vegetable waste in a tubular digester. *Bioresource Technol.* 86, 85.

Braun, B.; Huber, P.; Meyrath, J. (1981) Ammonia toxicity in liquid piggery manure digestion. *Biotechnol. Lett.* 3, 159–164.

Brosseau, J. D.; Zajic, J. E. (1982) Continuous microbial-production of hydrogen gas. *Int. J. Hydrogen Energ.* 7 (8), 623–628.

Cabirol, N.; Barragán, E. J.; Durán, A.; Noyola, A. (2003) Effect of aluminum and sulphate on anaerobic digestion of sludge from wastewater enhanced primary treatment. *Water Sci. Technol.* 48 (6), 235–240.

Chen, W.-H.; Han, S.-K.; Sung, S. (2003) Sodium inhibition of thermophilic methanogens. *J. Environ. Eng.* 129 (6), 506–512.

Chen, Y.; Cheng, J. J. (2007) Effect of potassium inhibition on the thermophilic anaerobic digestion of swine waste. *Water Environ. Res.* 79 (6), 667–674.

Chen, Y.; Cheng, J. J.; Creamer, K. S. (2008) Inhibition of anaerobic digestion process: A review. *Bioresource Technol.* 99(10), 4044–4064.

Chen, C. C.; Lin, C. Y.; Lin, M. C. (2002) Acid–base enrichment enhances anaerobic hydrogen production process. *Appl. Microbiol. Biot.* 58 (2), 224–228.

Colleran, E.; Finnegan, S.; Lens, P. (1995) Anaerobic treatment of sulphate-containing waste streams. *Antonie van Leeuwenhoek* 67, 29–46.

Conn, E. E.; Stumpf, P. K.; Bruening, G.; Doi, R. H. (1987) *Outlines of Biochemistry.* John Wiley and Sons, New York.

de Baere, L. A.; Devocht, M.; van Assche, P.; Verstraete, W. (1984) Influence of high NaCl and NH$_4$Cl salt levels on methanogenic associations. *Water Res.* 18, 543–548.

del Pozo, R.; Diez, V.; Salazar, G. (2002) Start-up of a pilot-scale anaerobic fixed film reactor at low temperature treating slaughterhouse wastewater. *Water Sci. Technol.* 46 (4–5), 215.

Dimroth, P.; Thomer, A. (1989) A primary respiratory Na$^+$ pump of an anaerobic bacterium: the Na$^+$-dependent NADH: Quinone oxidoreductase of *Klebsiella pneumoniae*. *Arch. Microbiol.* 151, 439–444.

Elmitwalli, T. A.; Oahn, K. L. T.; Zeeman, G.; Lettinga, G. (2002a) Treatment of domestic sewage in a two-step anaerobic filter/anaerobic hybrid system at low temperature. *Water Res.* 36, 2225.

Elmitwalli, T. A.; Sklyar, V.; Zeeman, G.; Lettinga, G. (2002b) Low temperature pretreatment of domestic sewage in an anaerobic hybrid or an anaerobic filter reactor. *Bioresource Technol.* 82, 233.

Fang, H. H. P.; Liu, H.; Zhang, T. (2002) Characterization of a hydrogen-producing granular sludge. *Biotechnol. Bioeng.* 78 (1), 44–52.

Fernandez, N.; Forster, C. F. (1993) A study of the operation of mesophilic and thermophilic anaerobic filters treating a synthetic coffee waste. *Bioresource Technol.* 45, 223–227.

Fernandez, N.; Forster, C. F. (1994) The anaerobic digestion of simulated coffee waste using thermophilic and mesophilic upflow filters. *Process Safety Environ. Protect.* 72(B1), 15–20.

Fleming, J. G.; Johnson, R. R.; Cheng, J. (2002) Validation of a three dimensional covered lagoon simulation. *Proceedings of the Animal Residuals 2002: Agricultural Animal Manure–Management, Policy and Technology;* Washington, DC, May 6–8.

Gallert, C.; Bauer, S.; Winter, J. (1998) Effect of ammonia on the anaerobic degradation of protein by a mesophilic and thermophilic biowaste population. *Appl. Microbiol. Biotechnol.* 50, 495–501.

Grady, C. P. L. Jr.; Daigger, G. T.; Lim, H. C. 1999. *Biological Waste Water Treatment.* Marcel Dekker, New York.

Han, S. K.; Shin, H. S.; Song, Y. C.; Lee, C. Y.; Kim, S. H. (2002) Novel anaerobic process for the recovery of methane and compost from food waste. *Water Sci. Technol.* 45 (10), 313.

Hansen, T. A. (1993) Carbon metabolism of sulfate-reducing bacteria. In: *The Sulfate-Reducing Bacteria: Contemporary Perspectives,* Odom, J. M.; Rivers-Singleton, J. R. (eds.), Springer-Verlag, New York, pp. 21–40.

Harmand, J.; Miens, F.; Conte, T.; Gras, P.; Buffiere, P.; Steyer, J. P. (2002) Model based prediction of the clogging of an anaerobic fixed bed reactor. *Water Sci. Technol.* 45 (4–5), 255.

Harris, J. E. (1987) Spontaneous disaggregation of *Mathanosaricina mazei* S-6 and its use in the development of genetic techniques for *Methanosarcina* spp. *Appl. Environ. Microbiol.* 53 (10), 2500–2504.

Hawkins, G. L.; Raman, D. R.; Burns, R. T.; Yoder, R. E.; Cross, T. L. (2001) Enhancing dairy lagoon performance with high-rate anaerobic digesters. *T. ASAE.* 44 (6), 1825.

Heipieper, H. J.; Weber, F. J.; Sikkema, J.; Kewelch, H.; de Bont, J. A. M. (1994) Mechanisms of resistance of whole cells to toxic organic solvents. *Trends Biotechnol.* 12, 409–415.

Hendriksen, H. V.; Ahring, B. K. (1991) Effects of ammonia on growth and morphology of thermophilic hydrogen-oxidizing methanogenic bacteria. *FEMS Microb. Ecol.* 85, 241–246.

Hickey, R. F.; Vanderwielen, J.; Switzenbaum, M. S. (1989) The effect of heavy metals on methane production and hydrogen and carbon monoxide levels during batch anaerobic sludge digestion. *Water Res.* 23, 207–219.

Hill, D. T. (1983) Design parameters and operating characteristics of animal waste anaerobic digestion systems—Swine and poultry. *Agric. Waste,* 5 (3), 157–178.

Hwu, C.-S.; Lettinga, G. (1997) Acute toxicity of oleate to acetate-utilizing methanogens in mesophilic and thermophilic anaerobic sludges. *Enzyme Microb. Technol.* 21, 297–301.

Jackson-Moss, C. A.; Duncan, J. R. (1991) The effect of aluminum on anaerobic digestion. *Biotechnol. Lett.* 13 (2), 143–148.

Jarrell, K. F.; Saulnier, M.; Ley, A. (1987) Inhibition of methanogenesis in pure cultures by ammonia, fatty acids, and heavy metals, and protection against heavy metal toxicity by sewage sludge. *Can. J. Microbiol.* 33, 551–555.

Jarrell, K. F.; Sprott, G. D.; Matheson, A. T. (1984) Intracellular potassium concentration and relative acidity of the ribosomal proteins of methanogenic bacteria. *Can. J. Microbiol.* 30, 663–668.

Jin, P.; Bhattacharya, S. K.; Williama, C. J.; Zhang, H. (1998) Effects of sulfide addition on copper inhibition in methanogenic systems. *Water Res.* 32, 977–988.

Kabara, J. J.; Vrable, R.; Lie Ken Jie, M. S. F. (1977) Antimicrobial lipids: Natural and synthetic fatty acids and monoglycerides. *Lipids* 12, 753–759.

Kaparaju, P.; Luostarinen, S.; Kalmari, E.; Kalmari, J.; Rintala, J. (2002) Co-digestion of energy crops and industrial confectionery by-products with cow manure: Batch-scale and farm-scale evaluation. *Water Sci. Technol.* 45 (10), 275.

Kayhanian, M. (1994) Ammonia inhibition in high-solids biogasification: An overview and practical solutions. *Environ. Technol.* 20, 355–365.

Kim, J. O.; Somiya, I.; Shin, E. B.; Bae, W.; Kim, S. K.; Kim, R. H. (2002) Application of membrane-coupled anaerobic volatile fatty acids fermentor for dissolved organics recovery from coagulated raw sludge. *Water Sci. Technol.* 45 (12), 167.

Kim, M.; Speece, R. E. (2002) Aerobic waste activated sludge (WAS) for start-up seed of mesophilic and thermophilic anaerobic digestion. *Water Res.* 36, 3860.

Klemps, R.; Cypionka, H.; Widdel, F.; Pfennig, N. (1985) Growth with hydrogen, and further physiological characteristics of *Desulfotomaculum* species. *Archives Microb.* 143, 203–208.

Komatsu, T.; Kikuta, T.; Momonoi, K. (2002) Methane production from municipal solid organic waste by anaerobic codigestion with sewage sludge. *Proceedings of the IWA 3rd World Water Congress* [CD-ROM]; Melbourne, Australia, April 7–12; International Water Association: London.

Koster, I. W.; Cramer, A. (1987) Inhibition of methanogenesis from acetate in granular sludge by long-chain fatty acids. *Appl. Environ. Microbiol.* 53 (2), 403–409.

Koster, I. W.; Lettinga, G. (1988) Anaerobic digestion at extreme ammonia concentrations. *Biol. Wastes* 25, 51–59.

Kroeker, E. J.; Schulte, D. D.; Sparling, A. B.; Lapp, H. M. (1979) Anaerobic treatment process stability. *J. Water Pollut. Control Fed.* 51, 718–727.

Kugelman, I. J.; Chin, K. K. (1971) Toxicity, synergism, and antagonism in anaerobic waste treatment processes. In *Anaerobic biological Treatment Processes*, American Chemical Society Advances in Chemistry Series 105: 55–90.

Kugelman, I. J.; McCarty, P. L. (1964) Cation toxicity and stimulation in anaerobic waste treatment. *J. Water Pollut. Control. Fed.* 37, 97–116.

Lawrence, A. W.; McCarty, P. L. (1965) The role of sulfide in preventing heavy metal toxicity on anaerobic treatment. *J. Water Pollut. Control Fed.* 37, 392–405.

Lawrence, A. W.; McCarty, P. L. (1969) Kinetics of methane fermentation in anaerobic treatment. *J. Water Pollut. Control Fed.*, 41 (2p2), R1.

Lay, J. J. (2000) Modeling and optimization of anaerobic digested sludge converting starch to hydrogen. *Biotechnol. Bioeng.* 68 (3), 269–278.

Lettinga, G.; van Velsen, A. F. M.; Hobma, S. W.; de Zeeuw, W. J.; Klapwijk, A. (1980) Use of the upflow sludge blanket (USB) reactor concept for biological wastewater treatment. *Biotech. Bioeng.* 22, 699–734.

Lettinga, G.; Roersma, R.; Grin, P. (1983) Anaerobic treatment of domestic sewage using a granular sludge bed UASB-reactor. *Biotech. Bioeng.* 25, 1701–1723.

Ligero, P.; Soto, M. (2002) Sludge granulation during anaerobic treatment of prehydrolysed domestic wastewater. *Proceedings of the IWA 3rd World Water Congress* [CD-ROM]; Melbourne, Australia, April 7–12; International Water Association: London.

Lin, C. Y. (1992) Effect of heavy metals on volatile fatty acid degradation in anaerobic digestion. *Water Res.* 26, 177–183.

Lin, C. Y. (1993) Effect of heavy metals on acidogenesis in anaerobic digestion. *Water Res.* 27, 147–152.

Lin, C. Y.; Chen, C. C. (1999) Effect of heavy metals on the methanogenic UASB granule. *Water Res.* 33, 409–416.

Masse, L.; Masse, D. I.; Kennedy, K. J.; Chou, S. P. (2002) Neutral fat hydrolysis and long-chain fatty acid oxidation during anaerobic digestion of slaughterhouse wastewater. *Biotechnol. Bioeng.* 79 (1), 43.

McCarty, P. L. (1964) Anaerobic waste treatment fundamentals. *Public Works* 95 (9), 107–112; (10), 123–126; (11), 91–94; (12), 95–99.

McCarty, P. L.; McKinney, R. (1961) Salt toxicity in anaerobic digestion. *J. Water Pollut. Control Fed.* 33, 399–415.

Min, H.; Zinder, S. H. (1990) Isolation and characterization of a thermophilic sulfate-reducing bacterium *Desulfotomaculum thermoacetoxidans* sp. nov. *Archives Microbiol.* 153, 399–404.

Mosey, F. E.; Hughes, D. A. (1975) The toxicity of heavy metal ions to anaerobic digestion. *Water Pollut. Control* 74, 18–39.

Mosey, F. E.; Swanwick, J. D.; Hughes, D. A. (1971) Factors affecting the availability of heavy metals to inhibit anaerobic digestion. *Water Pollut. Control* 70, 668–679.

Oleszkiewicz, J. A.; Sharma, V. K. (1990) Stimulation and inhibition of anaerobic process by heavy metals—A review. *Biol. Wastes* 31, 45–67.

Ong, S. L.; Hu, J. Y.; Ng, W. J.; Lu, Z. R. (2002) Granulation enhancement in anaerobic sequencing batch reactor operation. *J. Environ. Eng.* 128 (4), 387.

Oude Elferink, S. J. W. H.; Visser, A.; Hulshoff Pol, L. W.; Stams, A. J. M. (1994) Sulphate reduction in methanogenic bioreactors. *FEMS Microbiol. Rev.* 15, 119–136.

Palenzuela-Rollon, A.; Zeeman, G.; Lubberding, H. J.; Lettinga, G.; Alaerts, G. J. (2002) Treatment of fish processing wastewater in a one- or two-step upflow anaerobic sludge blanket (UASB) reactor. *Water Sci. Technol.* 45 (10), 207.

Parkin, G. F.; Lynch, N. A.; Kuo, W.; Van Keuren, E. L.; Bhattacharya, S. K. (1990) Interaction between sulfate reducers and methanogens fed acetate and propionate. *Res. J. Water Pollut. Control Fed.* 62, 780–788.

Patel, G. B.; Roth, L. A. (1977) Effect of sodium chloride on growth and methane production of methanogens. *Can. J. Microbiol.* 23, 893–897.

Paulo, P. L.; Jiang, B.; Roest, K.; van Lier, J. B.; Lettinga, G. (2002) Start-up of a thermophilic methanol-fed UASB reactor: Change in sludge characteristics. *Water Sci. Technol.* 45 (10), 145.

Rinzema, A.; Alphenaar, A.; Lettinga, G. (1989) The effect of lauric acid shock loads on the biological and physical performance of granular sludge in UASB reactors digesting acetate. *J. Chem. Tech. Biotechnol.* 46, 257–266.

Rinzema, A.; Boone, M.; van Knippenberg, K.; Lettinga, G. (1994) Bactericidal effect of long chain fatty acids in anaerobic digestion. *Water Environ. Res.* 66, 40–49.

Rodriguez-Martiínez, J.; Rodriguez-Garza, I.; Pedraza-Flores, E.; Balagurusamy, N.; Sosa-Santillan, G.; Garza-Garc, Y. (2002) Kinetics of anaerobic treatment of slaughterhouse wastewater in batch and upflow anaerobic sludge blanket reactor. *Bioresource Technol.* 85, 235.

Ruiz, C.; Torrijos, M.; Sousbie, P.; Lebrato-Martinez, J.; Moletta, R.; Delgenes, J. P. (2002) Treatment of winery wastewater by an anaerobic sequencing batch reactor. *Water Sci. Technol.* 45 (10), 219.

Scherer, P.; Sahm, H. (1981) Influence of sulfur-containing-compounds on the growth of *Methanosarcina Barkeri* in a defined medium. *Eur. J. App. Micro. Biotech.* 12, 28–35.

Schmidt, J. E.; Ahring, B. K. (1993) Effects of magnesium on thermophilic acetate-degrading granules in upflow anaerobic sludge blanket (UASB) reactors. *Enzyme Microbiol. Technol.* 15, 304–310.

Seghezzo, L.; Guerra, R. G.; Gonzalez, S. M.; Trupiano, A. P.; Figueroa, M. E.; Cuevas, C. M.; Zeeman, G.; Lettinga, G. (2002) Removal efficiency and methanogenic activity profiles in a pilot-scale UASB reactor treating settled sewage at moderate temperatures. *Water Sci. Technol.* 45 (10), 243.

Shih, J. C. H. (1993) Recent development in poultry waste digestion and feather utilization—A review. *Poultry Sci.,* 72, 1617–1620.

Sigge, G. O.; Britz, T. J.; Fourie, P. C.; Barnardt, C. A.; Strydom, R. (2002) Combining UASB technology and advanced oxidation processes (AOPs) to treat food processing wastewaters. *Water Sci. Technol.* 45 (10), 335.

Sikkema, J.; De Bont, J. A. M.; Poolman, B. (1994) Interactions of cyclic hydrocarbons with biological membrandes. *J. Biol. Chem.* 26, 8022–8028.

Sinha, S.; Bose, P.; Jawed, M.; John, S.; Tare, V. (2002) Application of neural network for simulation of upflow anaerobic sludge blanket (UASB) reactor performance. *Biotechnol. Bioeng.* 77 (7), 806.

Skiadas, I. V.; Ahring, B. K. (2002) A new model for anaerobic processes of up-flow anaerobic sludge blanket reactors based on cellular automata. *Water Sci. Technol.* 45 (10), 87.

Soto, M.; Mendéz, R.; Lema, J. M. (1993) Methanogenic and non-methanogenic activity tests: Theoretical basis and experimental setup. *Water Res.* 27, 1361–1376.

Speece, R. (1996) *Anaerobic Biotechnology for Industrial Wastewaters.* Archae Press, Nashville, TN.

Sprott, G. D.; Patel, G. B. (1986) Ammonia toxicity in pure cultures of methanogenic bacteria. *System. Appl. Microbiol.* 7, 358–363.

Sterritt, R. M.; Lester, J. N. (1980) Interaction of heavy metals with bacteria. *Sci. Tot. Environ.* 14(1), 5–17.

Tagawa, T.; Takahashi, H.; Sekiguchi, Y.; Ohashi, A.; Harada, H. (2002) Pilot-plant study on anaerobic treatment of a lipid- and protein-rich food industrial wastewater by a thermophilic multi-staged UASB reactor. *Water Sci. Technol.* 45 (10), 225.

Takashima, M.; Speece, R. E. (1989) Mineral nutrient requirements for high rate methane fermentation of acetate at low SRT. In *Proc. Ann. Water Pollut. Control Fed. Confer.*; Dallas Texas, October 5. Water Pollution Control Federation, Washington, DC.

Ueno, Y.; Kawai, T.; Sato, S., et al. (1995) Biological production of hydrogen from cellulose by natural anaerobic microflora. *J. Ferment. Bioeng.* 79 (4), 395–397.

Vallee, B. L.; Ulner, D. D. (1972) Biochemical effects of mercury, cadmium, and lead. *Annu. Rev. Biochem.* 41, 91–128.

van Ginkel, S. and Sung, S. (2001) Biohydrogen production as a function of ph and substrate concentration. *Environ. Sci. Technol.* 35, 4726–4730.

Vogels, G. D.; Kejtjens, J. T.; van der Drift, C. (1988) Biochemistry of methane production. In *Biology of Anaerobic Microorganisms.* Zehnder, A. J. B. (ed.). John Wiley and Sons, New York.

Voolapalli, R. K.; Stuckey, D. C. (2001) Hydrogen production in anaerobic reactors during shock loads—Influence of formate production and H-2 kinetics. *Water Res.* 35 (7), 1831–1841.

Wang, J. Y.; Xu, H. L.; Tay, J. H. (2002) A hybrid two-phase system for anaerobic digestion of food waste. *Water Sci. Technol.* 45 (10), 159.

Weast R. C. (1975) *CRC Handbook of Chemistry and Physics.* 55th edn. CRC Press, Inc. Cleveland, OH.

Xun, L.; Boone, D. R.; Mah, R. A. (1988) Control of the life cycle of Methanosarcina mazei S-6 by manipulation of growth conditions. *Appl. Environ. Microbiol.* 54(8), 2064–2068.

Yang, J.; Speece, R. E. (1986) The effects of chloroform toxicity on methane fermentation. *Water Res.* 20, 1273–1279.

Yerkes, D. W.; Boonyakitombut, S.; Speece, R. E. (1997) Antagonism of sodium toxicity by the compatible solute betaine in anaerobic methanogenic systems. *Water Sci. Technol.* 37 (6–7), 15–24.

Yu, H. Q.; Fang, H. H. P. (2003) Acidogenesis of gelatin-rich wastewater in an upflow anaerobic reactor: Influence of pH and temperature. *Water Res.* 37, 55.

Zabranska, J.; Dohanyos, M.; Jenicek, P.; Zaplatilkova, P.; Kutil, J. (2002) The contribution of thermophilic anaerobic digestion to the stable operation of wastewater sludge treatment. *Water Sci. Technol.* 45 (4–5), 447.

Zayed, G.; Winter, J. (2000) Inhibition of methane production fro whey by heavy metals-protective effect of sulfide. *Appl. Microbiol. Biotechnol.* 53, 726–731.

Zeikus, J. G. (1977) The biology of methanogenic bacteria. *Bacteriol. Rev.* 41, 514–541.

Zhu, H.; Béland, M. (2006) Evaluation of alternative methods of preparing hydrogen producing seeds from digested wastewater sludge. *This article is not included in your organization's subscription. However, you may be able to access this article under your organization's agreement with Elsevier International Journal of Hydrogen Energy,* 31 (14), 1980–1988.

7

Biological Process for Ethanol Production

Jay J. Cheng

CONTENTS

7.1 History of Alcohol Fermentation and Ethanol as Fuel 210
7.2 Bioethanol Production Process 212
7.3 Saccharification and Hydrolysis for Fermentable
    Sugar Production 214
    7.3.1 Saccharification of Starch 214
           7.3.1.1 Starch 217
           7.3.1.2 Liquefaction 218
           7.3.1.3 Saccharification 220
    7.3.2 Hydrolysis of Lignocellulose 220
           7.3.2.1 Lignocellulose 220
           7.3.3.2 Pretreatment 223
           7.3.3.3 Enzymatic Hydrolysis 229
7.4 Fermentation Process 234
    7.4.1 Enzymatic Reactions in Fermentation 235
           7.4.1.1 EMP Pathway 235
           7.4.1.2 Other Products 240
    7.4.2 Yeast Microbiology 241
    7.4.3 Fermenters 246
7.5 Ethanol Purification 247
    7.5.1 Fractionation 247
           7.5.1.1 Phase Equilibrium 248
           7.5.1.2 Fractionation to Separate Ethanol from
                    Water 252
    7.5.2 Dehydration 261
7.6 By-Products 263
    7.6.1 Distiller's Dried Grains with Solubles 263
    7.6.2 Carbon Dioxide 263
    7.6.3 Other By-Products 263
7.7 Problems 263
References 267

TABLE 7.1

Basic Properties of Ethanol

Molecular formula	CH_3CH_2OH
Molecular weight (g/mol)	46.07
Density at 20°C (kg/L)	0.7893
Viscosity at 20°C (mPa·s (cP))	1.201
Freezing or melting point	−114.3°C, 159 K, or −174°F
Boiling point	78.4°C, 352 K, or 173°F

Source: Weast, R.C., *Handbook of Chemistry and Physics*, 53rd edn. CRC Press, Cleveland, OH, 1972.

Ethanol is also called ethyl alcohol. Pure ethanol is a colorless, volatile, flammable liquid. It is also a major component in alcoholic beverages such as liquor, wine, and beer. Ethanol is volatile and the mixture of ethanol vapor with air can be explosive when the volume fraction of ethanol vapor is in the range of 3.3%–19%. Liquid ethanol can be dissolved in water at any ethanol:water ratios. Ethanol is a constitutional isomer of dimethyl ether (CH_3–O–CH_3) and often abbreviated as EtOH with Et representing C_2H_5–. The basic properties of pure ethanol are listed in Table 7.1.

As a fuel, ethanol can be burned to form carbon dioxide and water, releasing energy:

$$C_2H_5OH(g) + 3O_2(g) \rightarrow 2CO_2(g) + 3H_2O \text{ (L)} \quad (\Delta G = -1409 \text{ kJ/mol}) \quad (7.1)$$

7.1 History of Alcohol Fermentation and Ethanol as Fuel

Fermentation of ethanol and alcoholic beverages has a very long history. It is believed that the earliest alcoholic beverages appeared in the Chalcolithic Era in the Indus valley civilization about 12,000 years ago. Distilled alcoholic beverage was made from rice meal, wheat, sugar cane, grapes, and other fruits. The alcoholic drink was popular among the Kshatriya warriors and the peasant population.

The earlier evidence of alcohol in China dates back to about 9000 years ago. The ancient Chinese produced wine by fermenting rice, honey, and fruit. Wine was considered as a spiritual food rather than a material food, and widely used in memorial ceremonies, offering sacrifices to gods or their ancestors, pledging resolution before going into battle, celebrating victory, before feuding and official executions, for taking an oath of allegiance, while attending the ceremonies of birth, marriage, reunions, departures, death, and festival banquets.

The earliest wine production in ancient Egypt was about 6000 years ago. Brewing appeared shortly after that. Alcoholic beverages were very important at that time. Both beer and wine were deified and offered to gods as well as used for pleasure, nutrition, medicine, ritual, remuneration, and funerary purposes.

The art of winemaking reached the Hellenic peninsula in ancient Greece about 4000 years ago. The ancient Greeks made their first alcoholic beverage or wine by fermenting honey with water. Wine drinking was incorporated into religious rituals, became important in hospitality, was used for medicinal purposes, and was also popular in daily meals.

Although alcohol fermentation started for the production of alcoholic beverages, ethanol has been widely used in food, medicine, and the chemical industry as a scent, flavoring agent, and/or solvent. It has also been used as clean fuel for internal combustion engines as an alternative to gasoline. Ethanol was used as fuel for the first time in the United States to light the lamps in 1840. However, the federal alcohol tax on industrial alcohol made ethanol too expensive to be used as fuel. The first automobile using ethanol as fuel in the United States, Ford Motor T, was designed and built by Ford Motor Company in 1908, right after the alcohol tax was repealed in 1906. Alcoholic liquor sale was prohibited in the United States in the 1920s and the sellers of fuel ethanol were accused of being allied with liquor sellers, which drove the fuel ethanol out of the market. The first industrial-scale fuel ethanol plant in the United States was built by the U.S. Army in Omaha, Nebraska in the 1940s. The ethanol plant produced fuel for the army and provided ethanol for regional fuel blending.

Ethanol has attracted significant attention as an alternative fuel in the last two decades. In 1990, Clean Air Act Amendments require oxygenated fuels used for automobiles. As a result, methyl tertiary butyl ether (MTBE) and ethanol were commonly used as the additive oxygenates in automobile fuel. However, MTBE was found to leak from automobile fuel tanks to groundwater in many areas, which caused contamination of drinking water sources. Many states have banned the use of MTBE as an additive oxygenate in fuel since 1999. Instead, ethanol has replaced MTBE as additive oxygenate in automobile fuel. On the other hand, the U.S. government has installed regulations to promote the use of ethanol as alternative fuel in order to protect the environment and reduce the nation's dependence on imported foreign oil. In 1992, the U.S. Congress passed The Energy Policy Act of 1992 to reduce the nation's dependence of imported foreign petroleum by requiring the automobile fleets to acquire alternative fuel vehicles that are capable of operating on renewable fuels. The Energy Policy Act of 2005 sets a target of consuming 7.5 billion gal of bioethanol as fuel by 2012, which is approximately 5% of the projected annual gasoline consumption in 2012. In 2007, the Congress passed another energy-related bill, The Energy Independence and Security Act of 2007, which requires that the national production of biofuels (mainly bioethanol) increase to 9 billion gal in 2008, 15.2 billion gal in 2012, and 36 billion gal in 2022.

7.2 Bioethanol Production Process

Bioethanol can be produced through biological processes from different natural materials: sugar-, starch-, and lignocellulose-based materials. The biological processes for bioethanol production from different materials are illustrated in Figure 7.1.

Sugar-platform bioethanol production: The main natural materials that can be converted to bioethanol through the sugar-platform processes include sugarcane, sugar beet, and sweet sorghum. These materials are rich in sugars, mainly sucrose, that are readily fermented by yeasts or bacteria to bioethanol. Releasing of sugars from the materials can be achieved through simple mechanical extraction or squeezing. Sugarcane grows in warm temperate to tropical regions and accumulates biomass and sugars at a very high rate. Brazil is the largest sugarcane-producing country in the world, with an annual production of 514,080,000 tons in 2008. The average raw sugarcane production rate is about 74 tons per hectare in Brazil. The stalks, after separated from the leaves, are about 77% of the raw sugarcane. Each ton of sugarcane stalks yields 740 kg of juice (135 kg of sucrose and 605 kg of water) and 260 kg of moist bagasse (moisture content of around 50%). The juice is used for bioethanol production through fermentation and the bagasse is usually dried and burned to provide energy for the distillation of the fermentation beer. Brazil produced 5.96 billion gal of bioethanol as a renewable fuel from sugarcane in 2007. Sugar beet is a plant whose root contains a high concentration of sucrose. It normally grows in cool temperatures and produces a large (1–2 kg) storage root whose dry mass is 15%–20% sucrose by weight during its growing season. Sugar beet is widely grown in the European Union, Russia, and North America. Sweet sorghum is a sorghum that has a high

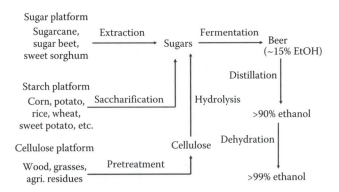

FIGURE 7.1
Biological processes for bioethanol production from different renewable materials.

sugar (sucrose) content and grows well under dry and warm conditions. It is widely grown in the United States, mainly for syrup production.

Starch-platform bioethanol production: The biomaterials that can be converted to ethanol through the starch-platform processes include starch-rich grains (corn, wheat, rice, barley, and grain sorghum), potatoes, and sweet potatoes. The starch content is in the range of 60%–75% in the grains, and 10%–30% in potatoes and sweet potatoes. Conversion of starch to ethanol involves one additional step, saccharification of starch to produce fermentable sugars (mainly glucose), compared to the sugar-to-ethanol process. The saccharification of starch is normally performed through enzymatic reactions catalyzed by amylases. The main product of the saccharification is glucose, which is readily fermented to ethanol by yeasts or bacteria. Among the starch-rich materials, corn is widely used for commercial fuel ethanol production in the United States, China, and Europe. The United States produced 6.49 billion gal of fuel ethanol from corn in 2007.

Cellulose-platform bioethanol production: Lignocellulosic materials such as wood, grasses, and agricultural residues can also be used for bioethanol production because they have a high content of cellulose and hemicelluloses. However, the conversion of lignocellulosic materials to ethanol is much more difficult than that of sugar-rich or starch-rich materials. The former involves three steps in conversion process: pretreatment, hydrolysis, and fermentation. Pretreatment is necessary because of the unique compact structure of lignocellulosic materials. The major components of lignocellulosic materials include cellulose, hemicelluloses, and lignin. The cellulose and hemicelluloses are tightly tangled together while the lignin firmly wraps them as a protection wall, which makes the structure of lignocellulosic materials very tough to biodegrade. The pretreatment is to remove lignin from the material and to loosen its structure, so the cellulose and hemicelluloses are accessible for hydrolysis to produce fermentable sugars. The complex processes for the conversion of lignocellulosic materials to ethanol make it very expensive economically compared to the sugar- and starch-platform bioethanol production. Although there is no commercial bioethanol production from lignocellulosic materials yet, the cellulose platform has a great potential because the lignocellulosic materials are so abundant in the world and the technologies are advancing very rapidly to lower the costs of the process.

Ethanol purification: Ethanol content in the beer or broth after the fermentation is completed is usually 10%–15%. To separate ethanol from the fermentation beer, distillation and dehydration need to be performed to obtain almost pure ethanol (>99%) as fuel. Distillation is a thermodynamic process based on the phase equilibrium theory. However, distillation can only produce ethanol with a purity of around 90% (by weight) because ethanol and water form an azeotropic mixture at ethanol concentration higher than 93% (by weight) (details will be discussed in Section 7.5.1). A further dehydration, usually through physical adsorption, to remove the water from the distillate is necessary to produce anhydrous fuel ethanol.

7.3 Saccharification and Hydrolysis for Fermentable Sugar Production

7.3.1 Saccharification of Starch

Corn, also called maize, is the most widely grown crop in the Americas due to its high yields of both grain and starch. It is also a popular crop in Asia and Europe. Corn is widely used for commercial ethanol production in North America, Asia, and Europe. It is also widely used for food production such as corn meal, popcorn, corn flakes, and chicha (a drink made from corn). In the United States and Canada, corn is also a major ingredient in animal feed. The main composition of corn kernels is shown in Table 7.2.

Because of its high starch content, corn is a major source of carbohydrates in food industry. Corn oil is also one of the popular vegetable oils in food preparation. Corn starch is stored in the endosperm of the kernels, while most of the proteins and oil are in the embryos, which makes it possible to separate starch from oil and proteins. When corn starch is utilized for bioethanol production, the corn oil, proteins, and fiber can be valuable by-products that may offset the cost of ethanol production. There are two different technologies for the corn-to-bioethanol conversion: dry milling and wet milling processes.

A flow diagram of dry milling technology to produce bioethanol from corn is shown in Figure 7.2. Harvested corn usually contains a small amount of soil and sand, which needs to be removed in the cleaning process. Soil and sand can be removed from corn kernels with sieves. Dry corn kernels, after cleaning, are crushed by mechanical mills to very small corn particles or powder, which are then mixed with water and heated to around 60°C. At this temperature, the corn starch starts to dissolve in water, which is called mashing. Mashing process is relatively short, about 5–10 min. When the corn starch dissolves in water to form a mash, α-amylase is added into the mash and temperature is further increased to 70°C–80°C in the liquefaction process to degrade starch into smaller molecules such as oligosaccharides and dextrin. The liquefaction process takes about 2 h. An indication of complete liquefaction is a dramatic decrease of the viscosity of the liquid in the processing tank. When the liquefaction is complete, glucoamylase is added into the liquid to convert the oligosaccharides and dextrin into glucose in the saccharification

TABLE 7.2

Main Composition of Corn Kernels

Moisture content (%)	7–16
Starch (%)	60–70
Proteins (%)	8–10
Oil (%)	3–5
Fiber (%)	1.0–1.5
Minerals (%)	1.5–2.0

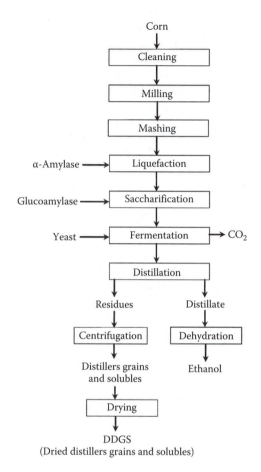

Corn
↓
Cleaning
↓
Milling
↓
Mashing
↓
α-Amylase → Liquefaction
↓
Glucoamylase → Saccharification
↓
Yeast → Fermentation → CO₂
↓
Distillation
↓ ↓
Residues Distillate
↓ ↓
Centrifugation Dehydration
↓ ↓
Distillers grains Ethanol
and solubles
↓
Drying
↓
DDGS
(Dried distillers grains and solubles)

FIGURE 7.2
Dry milling processes to produce bioethanol from corn.

process, which takes about 30 min. Saccharification is normally performed at 60°C–65°C, and iodine test is usually used to make sure that the saccharification is complete. Fermentation is the process in which glucose and other fermentable sugars are converted to ethanol by yeast that is added in the process. CO_2 is a by-product that can be used in soft drink manufacturing. The fermentation process takes 60–72 h and ethanol concentration in the fermentation beer is normally 10%–15% (w/w). The ethanol is concentrated to around 90% (w/w) in the distillate in the following distillation process. The distillate then goes through a dehydration process to remove the water and purified ethanol of over 99% (w/w) is obtained and is ready as biofuel. The residues from the distillation process contain proteins, amino acids, fiber, and water. The residues are centrifuged to remove the free water and then dried to remove the moisture; this by-product is called dried distillers grains and solubles (DDGS) and is sold as animal feed. The products from the dry milling processes are fuel ethanol, DDGS, and CO_2.

Fractionation of corn into value-added products by wet milling is an example of a biorefinery. The processes are illustrated in Figure 7.3. As shown in the figure, cleaned corn kernels are steeped in water to expand the corn structure. Corn oil is extracted and germ meal removed in the degermination process, and fiber and gluten are also separated from starch in the following processes. The separated starch can be used for various purposes such as in syrups and as a food ingredient. The following processes to convert starch to ethanol is almost the same as those in the dry milling processes except almost pure starch is used in the wet milling processes. The separated germ

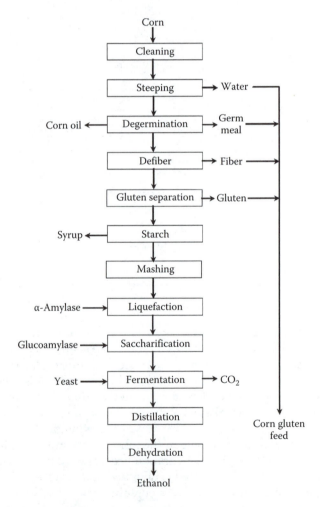

FIGURE 7.3
Wet milling processes to produce bioethanol from corn.

meal, fiber, and gluten can be used as animal feed supplement, while corn oil can be sold for cooking or food processing.

7.3.1.1 Starch

Starch is a polysaccharide that is composed of α-D-glucose units. Starch is commonly found in the seeds and roots of higher plants. Corn starch is in the endosperm of the corn kernels and stores chemical energy produced by photosynthesis. Starch consists of two different chemicals: amylose and amylopectin. Amylose is a long-chain polysaccharide of D-glucose units linked by α-1,4 glycosidic bonds (Figure 7.4). It is composed of 200–1000 glucose units. Amylose dissolves in water at 70°C–80°C. Iodine changes the color of amylose to blue. Amylopectin is a mixture of branched polysaccharides of D-glucose units linked by α-1,4 and α-1,6 glycosidic bonds (Figure 7.5). The degree of polymerization of amylopectin is much greater than that of

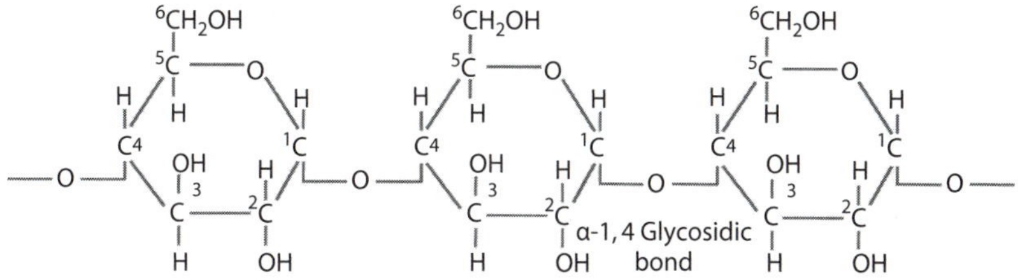

FIGURE 7.4
Basic structure of amylose.

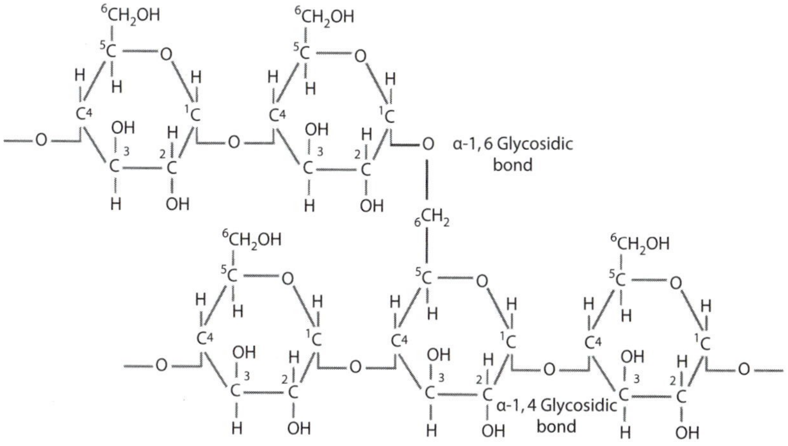

FIGURE 7.5
Basic structure of amylopectin.

amylose. Amylopectin is composed of 10,000–100,000 D-glucose residues. It is more difficult to dissolve in water and the temperature needs to be around 130°C for amylopectin to completely dissolve in water. Iodine turns amylopectin into purple color.

Starch is hardly soluble in water at room temperature. However, it can expand and be soluble at high temperature. When a mixture of starch and water is heated, the structure of starch will expand and the small-molecule water will penetrate into the structure of large-molecule starch. Viscosity of the mixed liquor is a good indication of the expansion and solubilization of starch in water at increasing temperatures (Figure 7.6). As shown in the figure, the viscosity of the mixture slightly decreases when the temperature increases from 20°C to 50°C. After the temperature reaches 50°C, there are significant starch expansion and water penetration, resulting in a sharp increase of the viscosity. Amylose can be completely solubilized in water at around 70°C, while a complete solubilization of amylopectin occurs when the temperature increases to 130°C. The viscosity reaches the peak at around 90°C with the expansion of amylopectin. A sharp decrease of viscosity occurs after 90°C when the solubilization of amylopectin is dominant until a complete solubilization of amylopectin at around 130°C.

7.3.1.2 Liquefaction

Liquefaction is the process in which starch is converted to oligosaccharides and dextrins. The process is catalyzed by α-amylase. α-Amylase can randomly hydrolyze α-1,4 glycosidic bonds inside the amylose and amylopectin structures. Although it cannot hydrolyze α-1,6 glycosidic bonds that link the branches in amylopectin, α-amylase can skip the α-1,6 glycosidic bonds and

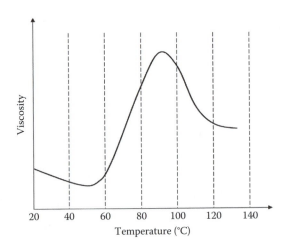

FIGURE 7.6
Viscosity change of starch–water mixture with the increase in temperature.

continue to hydrolyze the α-1,4 glycosidic bonds. Products of the hydrolysis of α-amylase on amylose are normally trisaccharides, disaccharides, and glucose. When α-amylase acts on amylopectin, the products also include dextrins that contain α-1,6 glycosidic bonds.

α-Amylase can be obtained from different sources including malt (e.g., barley), bacteria (e.g., *Bacillus subtilis*), and fungi (e.g., *Aspergillus* spp.). In brewing industry, α-amylase is mainly from malt of barley. However, commercial α-amylase is usually produced from bacteria or fungi due to the low cost. The optimum environmental conditions for α-amylases are slightly different, depending on the source. The α-amylases from bacteria prefer relatively higher temperature and pH, while those from fungi like lower temperature and pH, as shown in Table 7.3.

The sensitivity of α-amylase activity to pH depends on the temperature, but generally α-amylase maintains a high activity in a relatively wide range of pH. Figure 7.7 shows the relative activity of α-amylase from bacteria as a function of pH at different temperatures.

TABLE 7.3

Optimum Conditions of α-Amylases from Bacteria and Fungi

Optimum Condition	α-Amylase from Bacteria	α-Amylase from Fungi
Temperature (°C)	65–70	60–65
pH	6.0–7.5	5.0–6.5

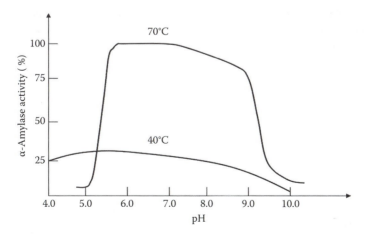

FIGURE 7.7
Relative activity of α-amylase from bacteria as a function of pH at different temperatures.

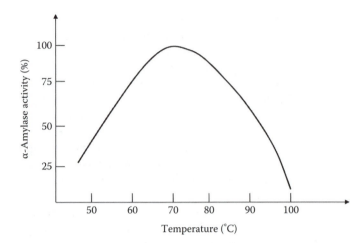

FIGURE 7.8
Relative activity of α-amylase from bacteria as a function of temperature.

α-Amylase usually has high sensitivity to temperature change. Figure 7.8 shows the relative activity of α-amylase from bacteria as a function of temperature. The α-amylase remains highly active only at temperatures around 70°C.

TABLE 7.4

Optimum Conditions of Glucoamylases for Saccharification

Temperature (°C)	58–60
pH	4.0–4.5

7.3.1.3 Saccharification

Saccharification is the process in which oligosaccharides and dextrins are hydrolyzed into glucose. The process is catalyzed by glucoamylase. Glucoamylase can hydrolyze α-1,4 glycosidic bonds of oligosaccharides from their nonreducing ends to produce glucose. It can also hydrolyze α-1,6 glycosidic bonds of dextrins at a lower rate to break the branches. Compared to α-amylase, glucoamylase prefers lower pH and temperature for hydrolysis. Table 7.4 shows typical optimum pH and temperature for glucoamylases.

Glucoamylase has a relative narrow pH range to maintain its maximum activity. Figure 7.9 shows the effect of pH on the glucoamylase activity at 60°C. Glucoamylase activity is also sensitive to temperature change, as shown in Figure 7.10.

7.3.2 Hydrolysis of Lignocellulose

7.3.2.1 Lignocellulose

With the rapidly increasing demand of fuel ethanol worldwide, it is not practical to dramatically increase ethanol production using the current

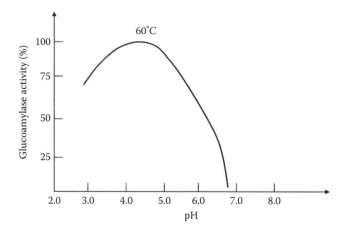

FIGURE 7.9
Effect of pH on the glucoamylase activity for saccharification at 60°C.

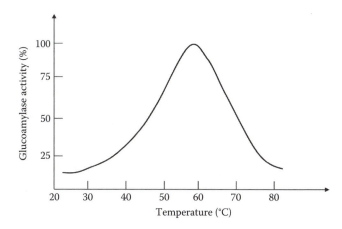

FIGURE 7.10
Effect of temperature on the glucoamylase activity for saccharification.

cornstarch- or sugarcane-based technologies because corn production for ethanol will compete for the limited agricultural land needed for food production and sugarcane production is limited by climate. A potential source to tremendously increase ethanol production is lignocellulosic materials such as grasses, woods, and agricultural residues because they are abundant almost all over the world. Although there is no commercial ethanol production plant using lignocellulosic materials at this time, extensive research has been done. The conversion of lignocelluloses to ethanol usually involves three steps: pretreatment, hydrolysis, and fermentation.

Pretreatment is necessary due to the unique structure of lignocellulosic materials. Lignocellulose contains mainly cellulose, hemicelluloses, and lignin. Cellulose is a long-chain homogenous polysaccharide of D-glucose units linked by β-1,4 glycosidic bonds (Figure 7.11). It is composed of over 10,000 glucose units. Hemicellulose is a complex, heterogeneous polymer of sugars and sugar derivatives that form a highly branched network. The monomers include hexoses (glucose, galactose, and mannose) and pentoses (xylose and arabinose). Its degree of polymerization is much smaller than cellulose, consisting of about 100–200 units. Figure 7.12 shows an example of the basic structure of hemicellulose arabinoxylan. Lignin is a very complex heterogeneous mixture of mainly phenolic compounds and their derivatives. It is a main component in plant cell walls and holds the cellulose and hemicellulose fibers together. Lignin also provides support to the plants. A general structure of lignocellulose is shown in Figure 7.13. In this structure, the basic long cellulose chains are connected to each other by hydrogen bonds. They are also tangled with hemicellulose molecules to form a highly crystalline structure that is protected by the lignin in plant cell wall. To utilize the cellulose and hemicellulose in the lignocellulosic materials, pretreatment is necessary to remove lignin and to break the crystalline structure.

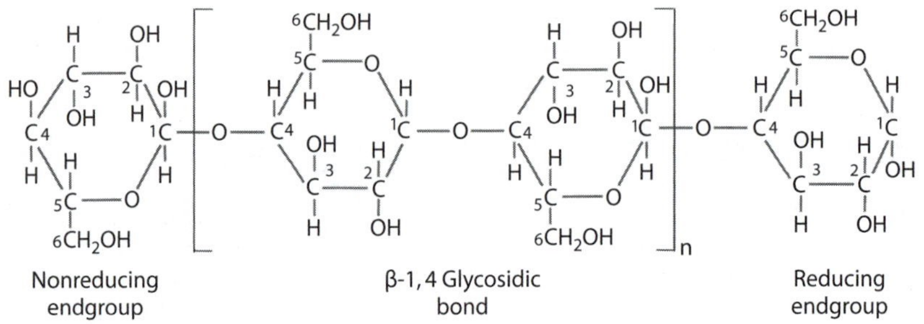
Nonreducing endgroup
β-1, 4 Glycosidic bond
Reducing endgroup

FIGURE 7.11
Basic structure of cellulose.

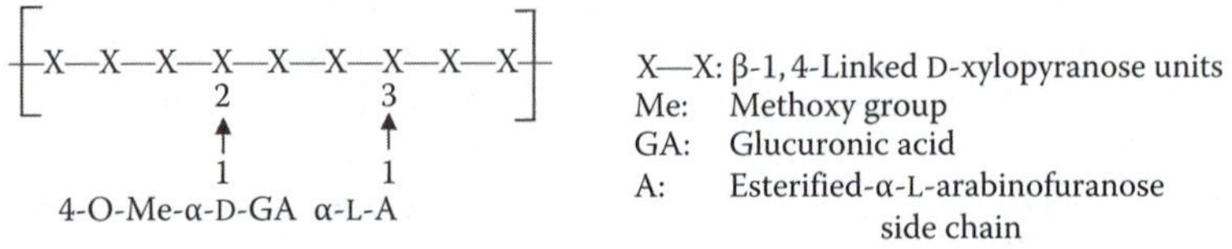
X—X—X—X—X—X—X—X—X
4-O-Me-α-D-GA α-L-A
X—X: β-1, 4-Linked D-xylopyranose units
Me: Methoxy group
GA: Glucuronic acid
A: Esterified-α-L-arabinofuranose side chain

FIGURE 7.12
Basic structure of arabinoxylan, an example of hemicellulose.

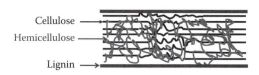

Cellulose →
Hemicellulose →
Lignin →

FIGURE 7.13
General structure of lignocellulosic materials.

7.3.3.2 Pretreatment

The purpose of the pretreatment is to remove lignin from the lignocellulose, reduce cellulose crystallinity, and increase the porosity of the material, so the enzymes can access their substrates (cellulose and hemicellulose) in the following enzymatic hydrolysis to produce fermentable sugars. Thus, pretreatment must meet with the following requirements: (1) improve the formation of sugars or the ability to subsequently form sugars by enzymatic hydrolysis; (2) avoid the degradation or loss of carbohydrate; (3) avoid the formation of by-products inhibitory to the subsequent hydrolysis and fermentation processes; and (4) be cost-effective. Extensive investigations of pretreatment technologies have been conducted in the last three decades, including physical, chemical, and biological processes for lignocellulosic materials (Sun and Cheng, 2002).

7.3.3.2.1 Physical Pretreatment

Mechanical comminution: Lignocellulosic materials can be comminuted by a combination of chipping, grinding, and milling to reduce cellulose crytallinity. The size of the materials is usually 1–3 cm after chipping and 0.2–2 mm after milling or grinding. Vibratory ball milling was found to be more effective in breaking down the cellulose crystallinity of spruce and aspen chips and improving the digestibility of the biomass than ordinary ball milling (Millet et al., 1976). Power requirement of mechanical comminution of lignocellulosic materials depends on the final particle size and the biomass characteristics (Cadoche and López, 1989). A comparison is shown in Table 7.5.

Steam explosion (autohydrolysis): Steam explosion is one of the most commonly used methods for the pretreatment of lignocellulosic materials (McMillan, 1994). In this method, chipped biomass is treated with high-pressure saturated steam and then the pressure is swiftly released, which makes the materials undergo an explosive decompression. Steam explosion is typically carried out at temperature of 160°C–260°C (corresponding pressure 0.69–4.83 MPa) for a period of time and then exposed to atmospheric pressure. It causes hemicellulose degradation and lignin transformation due to high temperature, thus increases the potential of cellulose hydrolysis. A 90% efficiency of enzymatic hydrolysis of poplar chips pretreated by

TABLE 7.5

Energy Requirement of Mechanical Comminution of
Lignocellulosic Agricultural Wastes with Different Size
Reduction

Lignocellulosic Materials	Final Size (mm)	Energy Consumption (kW h/ton)	
		Knife Mill	Hammer Mill
Hardwood	1.60	130	130
	2.54	80	120
	3.2	50	115
	6.35	25	95
Straw	1.60	7.5	42
	2.54	6.4	29
Corn stover	1.60	NA[a]	14
	3.20	20	9.6
	6.35	15	NA[a]
	9.5	3.2	NA[a]

Source: Data from Cadoche, L. and López, G.D., *Biol. Wastes*, 30, 153, 1989.
[a] NA: not available

steam explosion in 24 h was achieved, compared to only 15% hydrolysis of untreated chips (Grous et al., 1986). The factors that affect steam explosion pretreatment are residence time, temperature, chip size, and moisture content (Duff and Murray, 1996). Addition of H_2SO_4 (or SO_2) or carbon dioxide in steam explosion can effectively improve enzymatic hydrolysis, decrease the production of inhibitory compounds, and lead to more complete removal of hemicellulose (Morjanoff and Gray, 1987). The optimal conditions of steam explosion pretreatment of sugarcane bagasse were found to be the following: temperature, 220°C; residence time, 30 s; water-to-solids ratio, 2; and H_2SO_4 dose, 1.0 g H_2SO_4/100 g dry bagasse (Morjanoff and Gray, 1987). Sugar production was 65.1 g sugars/100 g starting bagasse after steam explosion pretreatment. The advantages of steam explosion pretreatment include low energy requirement compared to mechanical comminution and no recycling or environmental costs associated with chemical pretreatment. The conventional mechanical methods require 70% more energy than steam explosion to achieve the same size reduction. Steam explosion is recognized as one of the most cost-effective pretreatment processes for hardwoods and agricultural residues, but it is less effective for softwoods (Clark and Mackie, 1987). Limitations of steam explosion include destruction of a portion of xylan fraction, incomplete disruption of lignin–carbohydrate matrix, and generation of compounds that may be inhibitory to microorganisms used in downstream processes (Mackie et al., 1985). Because of the formation of degradation products that are inhibitory to microbial growth, enzymatic hydrolysis, and/or fermentation, steam-exploded biomass usually needs to be washed with water prior to enzymatic hydrolysis and subsequent fermentation. The water wash decreases the overall saccharification yields due to the removal of soluble sugars, such as those generated by hydrolysis of hemicellulose. Typically 20%–25% of initial dry matter is removed by water wash (Mes-Hartree et al., 1988).

Ammonia fiber explosion (AFEX): AFEX is another type of physical pretreatment in which lignocellulosic materials are exposed to liquid ammonia at high temperature for a period of time, and then flashed to a lower pressure. The concept of AFEX is similar to steam explosion. In a typical AFEX process, the dosage of liquid ammonia is 1–2 kg ammonia/kg dry biomass, temperature 90°C, and residence time 30 min. AFEX pretreatment can significantly improve the saccharification rates of various herbaceous crops and grasses. It can be used for the pretreatment of many lignocellulosic materials including alfalfa, wheat straw, wheat chaff (Mes-Hartree et al., 1988), barley straw, corn stover, rice straw (Vlasenko et al., 1997), municipal solid waste, softwood newspaper, kenaf newspaper (Holtzapple et al., 1992), coastal bermudagrass, switchgrass (Reshamwala et al., 1995), aspen chips (Tengerdy and Nagy, 1988), and bagasse (Holtzapple, 1991). In contrast to steam explosion, AFEX pretreatment does not significantly solubilize hemicellulose (Mes-Hartree et al., 1988; Vlasenko et al., 1997). It was reported that over 90% hydrolysis of cellulose and hemicellulose was obtained after AFEX pretreatment of bermudagrass and bagasse (Holtzapple et al., 1991). However, the AFEX process was not very effective for the lignocellulosic biomass with high lignin content such as newspaper (18%–30% lignin) and wood chips (25%–35% lignin). To reduce the cost and protect the environment, ammonia needs to be recycled after the pretreatment. Mes-Hartree (1988) compared the steam and ammonia pretreatment for enzymatic hydrolysis of aspenwood, wheat straw, wheat chaff, and alfalfa stems and found that steam explosion solubilized the hemicellulose, while AFEX did not. The composition of the materials after AFEX pretreatment was essentially the same as the original materials. The ammonia pretreatment did not produce inhibitors for the downstream biological processes, so water wash is not necessary (Dale et al., 1984; Mes-Hartree et al., 1988). It is also reported that AFEX pretreatment did not require small particle size for efficacy (Holtzapple et al., 1990). However, the cost of AFEX process is higher than steam explosion.

Carbon dioxide explosion: Similar to steam and ammonia explosion pretreatment, carbon dioxide explosion is also used for pretreatment of lignocellulosic materials. It was hypothesized that CO_2 would form carbonic acid and increase the hydrolysis rate. Dale and Moreira (1982) used this method to pretreat alfalfa (4 kg CO_2/kg fiber at the pressure of 5.62 MPa) and obtained 75% of theoretical glucose released during 24 h of the following enzymatic hydrolysis. The yields are relatively low compared to steam or ammonia explosion pretreatment, but high compared to the enzymatic hydrolysis without pretreatment. Zheng et al. (1998) compared carbon dioxide explosion with steam and ammonia explosion for pretreatment of recycled paper mix, sugarcane bagasse, and repulping waste of recycled paper, and found that carbon dioxide explosion was more cost-effective than ammonia explosion and did not cause the formation of inhibitory compounds that may happen in steam explosion.

Pyrolysis: Pyrolysis has also been used for pretreatment of lignocellulosic materials. When the materials are treated at temperature higher than 300°C, cellulose rapidly decomposes to produce gaseous and tarry compounds. Mild acid hydrolysis of the residues from pyrolysis pretreatment resulted in 80%–85% conversion of cellulose to reducing sugars with more than 50% glucose (Fan et al., 1987). Pretreatment of yellow poplar sawdust using liquid water at temperatures above 220°C under high pressure achieved 80%–90% conversion of cellulose in hydrolysis and 55% of theoretical ethanol yield in fermentation (Weil et al., 1997). When zinc chloride was added as a catalyst, the decomposition of pure cellulose could occur at lower temperature.

7.3.3.2.2 Chemical Pretreatment

Ozonolysis: Ozone could be used to degrade lignin and hemicellulose in many lignocellulosic materials such as wheat straw, bagasse, cotton straw, green hay, peanut, poplar, and pine. The degradation was essentially limited to lignin and hemicellulose, which were slightly attacked, whereas cellulose was hardly affected. The rate of enzymatic hydrolysis increased by a factor of 5 following 60% removal of the lignin from wheat straw in an ozone pretreatment (Vidal and Molinier, 1988). Enzymatic hydrolysis yield increased from 0% to 57% as the percentage of lignin decreased from 29% to 8% after an ozonolysis pretreatment of poplar sawdust (Vidal and Molinier, 1988). Ozonolysis pretreatment has the following advantages: (1) it effectively removes lignin; (2) it does not produce toxic residues for the downstream processes; and (3) the reactions are carried out at room temperature and pressure (Vidal and Molinier, 1988). However, a large amount of ozone is required in the process, making the process expensive.

Acid hydrolysis: Concentrated acids such as H_2SO_4 and HCl have been used to treat lignocellulosic materials. Although they are powerful agents for cellulose hydrolysis, the concentrated acids are toxic, corrosive, and hazardous in nature and require reactors that are resistant to corrosion. In addition, the cost of recovering the concentrated acids is very high, making the process too expensive for a large-scale pretreatment of lignocellulosic materials. Dilute acid hydrolysis has been successfully developed for pretreatment of lignocellulosic materials. The dilute sulfuric acid pretreatment could achieve high reaction rates and significantly improve cellulose hydrolysis (Esteghlalian et al., 1997). There are primarily two types of dilute acid pretreatment processes: high temperature ($T > 160°C$), continuous-flow process for low solids loading (5%–10% [w/w]) (Brennan et al., 1986; Converse et al., 1989), and lower temperature $(T < 160°C)$, batch process for high solids loading (10%–40% [w/w]) (Cahela et al., 1983; Esteghlalian et al., 1997). Although dilute acid pretreatment can significantly improve the cellulose hydrolysis, its cost is usually higher than some physical pretreatment processes such as steam explosion or AFEX. A neutralization of pH is necessary for the downstream enzymatic hydrolysis or fermentation processes.

Alkaline hydrolysis: Alkaline hydrolysis is another promising pretreatment method. The mechanism of alkaline pretreatment of lignocellulosic materials is believed to be the saponification reaction of intermolecular ester bonds cross-linking hemicellulose and cellulose or lignin (Figure 7.14). Alkaline pretreatment can also disrupt lignin structure and decrease crystallinity of cellulose and degree of sugar polymerization (Sun and Cheng, 2002). Unlike other physical or chemical pretreatment methods, the alkaline pretreatment process is simple and does not need much energy input. Compared with acid pretreatment, which is the other most commonly used chemical method, alkaline pretreatment has less sugar degradation and inhibitory compound (furan derivatives) formation. In addition, some caustic base such as lime can be recovered and/or regenerated. NaOH and lime are commonly used in alkaline pretreatment. NaOH is very efficient in removing lignin from lignocellulosic materials at a temperature of 100°C for 15–30 min (Wang et al., 2008; Xu et al., 2008). Lime pretreatment of switchgrass at low temperatures could significantly improve the sugar yield of the biomass in the following enzymatic hydrolysis, but the pretreatment took a much longer time (6 h) (Xu et al., 2008). Ammonia was also used for the pretreatment to remove lignin. Iyer et al. (1996) described the ammonia recycled percolation process (temperature 170°C, ammonia concentration 2.5%–20%, reaction time 1 h, flow rate 1 mL/min) for the pretreatment of corn cobs/stover mixture and switchgrass. The efficiency of delignification was 60%–80% for corn cobs and 65%–85% for switchgrass.

Oxidative delignification: Lignin biodegradation could be catalyzed by peroxidase enzyme with the presence of H_2O_2 (Azzam, 1989). The pretreatment of cane bagasse with hydrogen peroxide could greatly enhance its susceptibility to enzymatic hydrolysis. It was reported that about 50% lignin and

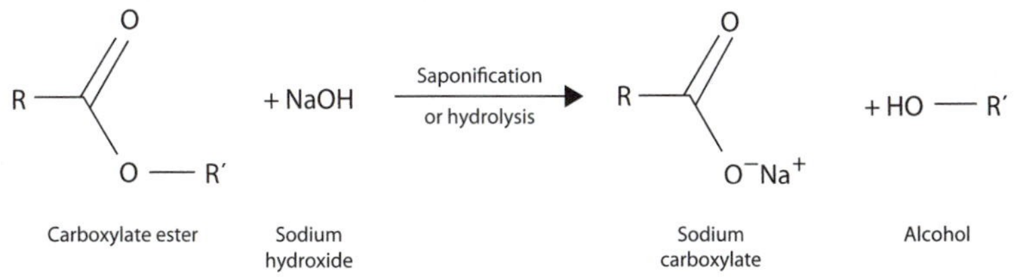

FIGURE 7.14
Saponification reaction in alkaline pretreatment of lignocellulosic materials.

most hemicellulose were solubilized by 2% H_2O_2 at 30°C within 8 h, and 95% efficiency of glucose production was achieved in the subsequent saccharification by cellulase at 45°C for 24 h (Azzam, 1989).

Organosolv pretreatment: In the organosolv process, an organic or aqueous organic solvent mixture with inorganic acid catalysts (HCl or H_2SO_4) is used to break the internal lignin and hemicellulose bonds. The organic solvents used in the process include methanol, ethanol, acetone, ethylene glycol, triethylene glycol, and tetrahydrofurfuryl alcohol (Chum et al., 1988; Thring et al., 1990). Organic acids such as oxalic, acetylsalicylic, and salicylic acid can also be used as catalysts in the organosolv process (Sarkanen, 1980). At high temperatures (above 185°C), the addition of catalyst was unnecessary for satisfactory delignification (Sarkanen, 1980; Aziz and Sarkanen, 1989). Usually, a high yield of xylose can be obtained with the addition of acid. Solvents used in the process need to be drained from the reactor, evaporated, condensed, and recycled to reduce the cost. Removal of solvents from the system is necessary because the solvents may be inhibitory to the growth of organisms, enzymatic hydrolysis, and fermentation.

7.3.3.2.3 Biological Pretreatment

In biological pretreatment processes, microbes such as brown-, white-, and soft-rot fungi are used to degrade lignin and hemicellulose in lignocellulosic materials (Schurz, 1978). Brown rots mainly attack cellulose, while white and soft rots attack both cellulose and lignin. White-rot fungi are the most effective basidiomycetes for biological pretreatment of lignocellulosic materials (Fan et al., 1987). Hatakka (1983) studied the pretreatment of wheat straw by 19 white-rot fungi and found that 35% of the straw was converted to reducing sugars by *Pleurotus ostreatus* in 5 weeks. Similar conversion was obtained in the pretreatment by *Phanerochaete sordida* 37 and *Pycnoporus cinnabarinus* 115 in 4 weeks. In order to prevent the loss of cellulose, cellulase-less mutant of *Sporotrichum pulverulentum* was developed for the degradation of lignin in wood chips (Ander and Eriksson, 1977). Akin et al. (1995) also reported the delignification of Bermuda grass by white-rot fungi. The biodegradation of Bermuda grass stems was improved by 29%–32% using *Ceriporiopsis subvermispora* and 63%–77% using *Cyathus stercoreus* after 6 weeks.

The white-rot fungus *Phanerochaete chrysosporium* produces lignin-degrading enzymes, lignin peroxidases and manganese-dependent peroxidases, during secondary metabolism in response to carbon or nitrogen limitation (Boominathan and Reddy, 1992). Both enzymes have been found in the extracellular filtrates of many white-rot fungi for the degradation of wood cell walls (Kirk and Farrell, 1987; Waldner et al., 1988). Other enzymes including polyphenol oxidases, laccases, H_2O_2-producing enzymes, and quinone-reducing enzymes can also degrade lignin (Blanchette, 1991). The advantages of biological pretreatment include low energy requirement and mild environmental conditions. However, the rate of hydrolysis in most biological pretreatment processes is very low.

7.3.3.3 Enzymatic Hydrolysis

Enzymatic hydrolysis of pretreated lignocellulosic biomass involves biochemical reactions that convert cellulose into glucose and hemicellulose into pentoses (xylose and arabinose) and hexoses (glucose, galactose, and mannose) (Figure 7.15). The conversion of cellulose and hemicellulose is catalyzed by cellulase and hemicellulase enzymes, respectively. The enzymes are highly specific (Béguin and Aubert, 1994). Utility cost of enzymatic hydrolysis is low compared to acid or alkaline hydrolysis because enzymatic hydrolysis is usually conducted at mild conditions (pH 4.8 and temperature 45°C–50°C) and does not have a corrosion problem (Duff and Murray, 1996).

Cellulases or β-(1–4) glycoside hydrolases are a mixture of several enzymes and at least three major groups of cellulases are involved in the hydrolysis of cellulose: endoglucanase (EG, endo-1,4-D-glucanohydrolase, or EC 3.2.1.4.) or cellulase, exoglucanase or cellobiohydrolase (CBH, 1,4-β-D-glucan cellobiohydrolase, or EC 3.2.1.91.), and β-glucosidase (EC 3.2.1.21). The joint hydrolysis of the three groups of enzymes completes the conversion of cellulose into glucose. The detailed processes for the bioconversion are illustrated in Figure 7.16. After the pretreatment, most of lignin is removed from the biomass, the

$$\underset{\text{Cellulose}}{(C_6H_{10}O_5)_n + n\,H_2O} \xrightarrow{\text{Cellulases}} \underset{\text{Glucose}}{n\,C_6H_{12}O_6}$$

$$\text{Hemicelluloses} \xrightarrow[\text{Hemicellulases}]{\text{Cellulases}} \text{Xylose, arabinoase glucose, galactose, mannose}$$

FIGURE 7.15
Enzymatic hydrolysis of cellulose and hemicelluloses to produce reducing sugars.

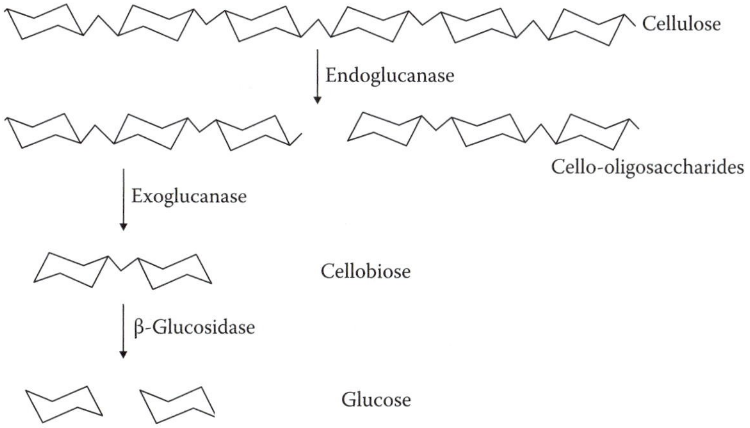

FIGURE 7.16
Enzymatic processes to convert cellulose into glucose.

crystallinity of the biomass is significantly reduced, and the porosity is substantially increased, which allows the enzymes to penetrate into the biomass and access the substrates. Endoglucanase randomly attacks regions of low crystallinity in the cellulose fiber and hydrolyze the β-(1,4) glycosidic bonds of cellulose to produce cello-oligosaccharides with free-chain ends. Exoglucanase or cellobiohydrolase can hydrolyze the β-(1,4) glycosidic bonds from the nonreducing ends of the cello-oligosaccharides to generate cellobiose, which is further hydrolyzed by β-glucosidase to glucose.

Commercial cellulase enzymes are usually a mixture of β-(1–4) glycoside hydrolases. The cellulase enzyme loadings in the hydrolysis of lignocellulose vary from 7 to 33 FPU (filter paper unit, defined as a micromole of reducing sugar as glucose produced by 1 mL of enzyme per minute)/gram substrate, depending on the type and concentration of substrates. Cellulase dosage of 10–15 FPU/g cellulose is often used in laboratory studies because it provides a hydrolysis profile with high levels of glucose yield in a reasonable time (48–72 h) at a reasonable enzyme cost (Gregg and Saddler, 1996). However, supplemental β-glucosidase greatly improve the hydrolysis of celluloses and the yield of glucose. Sun and Cheng (2004) investigated the effect of additional β-glucosidase on the enzymatic hydrolysis of coastal Bermuda grass. Significant amount of cellobiose accumulation was observed in the hydrolysis with cellulase loading of 5–15 FPU/g dry biomass. However, the addition of β-glucosidase at loading rates of 25 or 50 CBU (cellobiose unit, defined as a mole of cellobiose that is converted into glucose per minute with cellobiose as substrate)/g dry biomass prevented the cellobiose from accumulating, resulting in a more rapid hydrolysis with higher glucose yield (Figure 7.17).

7.3.3.3.1 Enzymes for Hydrolysis

Endoglucanase: Endoglucanase randomly attacks the β-(1, 4) glycosidic bonds of cellulose, resulting in a rapid decrease of the viscosity of the cellulose–water mixed liquor. The enzymes normally act on only amorphous cellulose not crystalline cellulose. Endoglucanase can be produced from both fungi and bacteria. Fungi that have been reported to produce cellulases include *Aspergillus* sp., *Humicola* sp., *Phanerochete chrysosporium*, and *Trichoderma* sp. The optimum conditions for the endoglucanase to perform at high activity depend on the source of the enzymes. Table 7.6 shows the optimum pH and temperature for the endoglucanases from different fungi. As shown in the table, the fungal endoglucanases prefer a slightly acidic environment at medium temperature. Although the optimum conditions for the fungal endoglucanases are within quite narrow ranges, the enzymes can survive in a much wider range of pH and temperature. The bacteria that can produce endoglucanases include both aerobes and anaerobes such as the species of *Clostridium, Bacillus, Thermomonospora, Ruminococcus, Pseudomonas,* and *Streptomyces*. The optimum pH and temperature for the endoglucanases from commonly used bacteria are shown in Table 7.7. As shown in the table, bacterial endoglucanases prefer neutral pH and relatively high temperature,

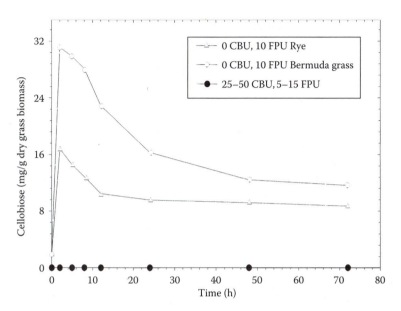

FIGURE 7.17

Cellobiose as an intermediate during the enzymatic hydrolysis of acid-pretreated rye and coastal Bermuda grass with different cellulase and β-glucosidase loadings (CBU and FPU are β-glucosidase and cellulase units, respectively, based on per gram of dry biomass). (Data from Sun, Y. and Cheng, J., T. *ASAE*, 47, 1, 343, 2004.)

TABLE 7.6

Optimum pH and Temperature for the Endoglucanases from Different Fungi

Source Fungi	Optimum pH	Optimum Temperature (°C)
Aspergillus sp.	4.0–5.0	45–70
Humicola sp.	4.5–5.5	50–65
Phanerochete chrysosporium	4.0–5.0	40–50
Trichoderma sp.	4.5–5.5	50–65

TABLE 7.7

Optimum pH and Temperature for the Endoglucanases from Bacteria

Source Bacteria	Optimum pH	Optimum Temperature (°C)
Bacillus sp.	5.0–7.0	60–70
Clostridium sp.	6.0–6.5	60–70
Pseudomonas sp.	7.0–8.0	60–70

compared to fungal endoglucanases. However, bacterial endoglucanases are more sensitive to pH change than the fungal endoglucanases.

Exoglucanase: Exoglucanases or cellobiohydrolases release cellobiose from the nonreducing ends of cello-oligosaccharides. Exoglucanases can hydrolyze both amorphous and crystalline celluloses, but generally do not hydrolyze substituted cellulose such as carboxymethyl cellulose. Exoglucanases are mainly from fungi such as the species of *Coniophora, Fusarium, Humicola, Penicillium,* and *Trichoderma.* Just like the fungal endoglucanases, fungal exoglucanases also prefer slightly acidic environment and medium temperature. The optimum pH and temperature for the fungal exoglucanases from different sources are shown in Table 7.8.

β-Glucosidase: β-Glucosidases degrade cellobiose into glucose that provides the source of energy and C to the host microorganisms during the enzyme production. They have very broad specificity to both glycon and aglycon substrates such as steroid β-glucosides and β-glucosylceramides of mammals, compared to endoglucanases and exoglucanases. β-Glucosidase can greatly improve the hydrolysis efficiency of cellulose by degrading cellobiose, the end-product and competitive inhibitor of endoglucanases and exoglucanases. β-Glucosidase can be produced from both fungi and bacteria. The commonly used fungi for β-glucosidase production include the species of *Aspergillus, Candida, Humicola, Penicillium, Saccharomyces,* and *Trichoderma.* The optimum conditions for the β-glucosidases from some of the above fungi are listed in Table 7.9. Bacteria that are commonly used for β-glucosidase production include *Clostridium* sp., *Ruminococcus* sp., and *Streptomyces* sp. The optimum conditions for the β-glucosidases from the bacteria are listed in Table 7.10.

Other enzymes: In addition to the three major groups of cellulase enzymes, there are also a number of ancillary enzymes that attack hemicelluloses, such as xylanase, β-xylosidase, glucuronidase, acetylesterase, galactomannanase, and glucomannanase (Duff and Murray, 1996). Among these enzymes, xylanase and β-xylosidase are usually responsible for the conversion of xylan

TABLE 7.8

Optimum pH and Temperature for the Endoglucanases from Different Fungi

Source Fungi	Optimum pH	Optimum Temperature (°C)
Coniophora sp.	5.0	50
Fusarium sp.	5.0	50
Humicola sp.	5.0	50
Penicillium sp.	4.5	60
Trichoderma sp.	5.0	60

TABLE 7.9

Optimum pH and Temperature for the
β-Glucosidase from Different Fungi

Source Fungi	Optimum pH	Optimum Temperature (°C)
Aspergillus sp.	4.5–5.0	65
Humicola sp.	5.0	50
Saccharomyces sp.	6.8	45
Trichoderma sp.	6.0–6.5	50

TABLE 7.10

Optimum pH and Temperature for the
β-Glucosidase from Different Bacteria

Source Bacteria	Optimum pH	Optimum Temperature (°C)
Clostridium sp.	6.0	65
Ruminococcus sp.	6.5	30–35
Streptomyces sp.	6.5	50

into xylose. Xylanase is an endo-acting enzyme and randomly attack β-(1,4) bonds between D-xylose residues of heteroxylans and xylo-oligosaccharides to release xylobiose and xylo-oligosaccharides. β-Xylosidase further degrades xylobiose and xylo-oligosaccharides into xylose, a five-carbon sugar.

7.3.3.3.2 Substrates

Substrate concentration is one of the main factors that affect the yield and initial rate of enzymatic hydrolysis of cellulose (Sun and Cheng, 2002). At low substrate levels, an increase of substrate concentration normally results in an increase of the yield and reaction rate of the hydrolysis (Cheung and Anderson, 1997). However, high substrate concentration can cause substrate inhibition, which substantially lowers the rate of the hydrolysis, and the extent of substrate inhibition depends on the ratio of total substrate to total enzyme (Huang and Penner, 1991; Penner and Liaw, 1994). Huang and Penner (1991) found that the substrate inhibition occurred when the ratio of the micro-crystalline substrate Avicel pH101 to the cellulase from *T. reesei* (grams of cellulose/FPU of enzyme) was greater than 5. Penner and Liaw (1994) reported that the optimum substrate to enzyme ratio was 1.25 g of the microcrystalline substrate Avicel pH105 per FPU of the cellulase from *T. reesei*.

7.3.3.3.3 Product Inhibition

Cellulase activity can be inhibited by cellobiose, as discussed earlier in this chapter. Glucose, to a lesser extent, can also inhibit the hydrolysis. Several methods have been developed to reduce the inhibition, including the use of high concentrations of enzymes, the supplementation of β-glucosidase during hydrolysis, and the removal of sugars during hydrolysis by ultrafiltration or simultaneous saccharification and fermentation (SSF) (Sun and Cheng, 2002). As discussed earlier, the supplementation of β-glucosidase in the hydrolysis can prevent cellobiose from accumulation and thus eliminate cellobiose inhibition. In SSF process, reducing sugars produced in cellulose hydrolysis or saccharification are simultaneously fermented to ethanol, which greatly reduces the product inhibition to the hydrolysis. Compared to the two-stage hydrolysis–fermentation process, SSF has the following advantages: (1) increase of hydrolysis rate by conversion of sugars that inhibit the cellulase activity, (2) lower enzyme requirement, (3) higher product yields, (4) lower requirements for sterile conditions since glucose is removed immediately and ethanol is produced, (5) shorter process time, and (6) less reactor volume because a single reactor is used (Sun and Cheng, 2002). However, ethanol may also exhibit inhibition to the cellulase activity in the SSF process. Wu and Lee (1997) found that cellulase lost 9%, 36%, and 64% of its original activity at ethanol concentrations of 9, 35, and 60 g/L, respectively, at 38°C during SSF process. The disadvantages that need to be considered for SSF include (1) incompatible temperature of hydrolysis and fermentation; (2) ethanol tolerance of microbes; and (3) inhibition of enzymes by ethanol (Sun and Cheng, 2002).

7.4 Fermentation Process

Fermentation of glucose for ethanol production can be carried out by either yeast or bacteria. Although bacterial fermentation may be used in some alcoholic liquor production, yeast fermentation is commonly used in fuel ethanol production. In yeast fermentation, the glucose solution obtained from starch saccharification or cellulose hydrolysis is cooled to around 32°C and acclimated yeast culture is added into the solution under aseptic conditions. Glucose in the solution penetrates into yeast cells and is converted by a group of enzymes created by yeast cells through a series of enzymatic reactions to eventually ethanol, CO_2, and energy. Some of the released energy and glucose are utilized by the yeast cells to support their growth during the fermentation. The rest of the energy becomes heat to the fermentation broth and may increase the temperature if not taken out of the system. Both ethanol and CO_2 penetrate out of yeast cells. CO_2 readily dissolves in water, but can be easily saturated in fermentation broth. The excess CO_2 bubbles out of the liquid and can be collected for food and soft drink preparation. Ethanol dissolves in water at any ratio and the CO_2 bubbling helps the transportation of ethanol from around the yeast cells to the bulk fermentation broth, avoiding the occurrence of high ethanol concentration in local areas that may be toxic to yeast cells.

In a batch ethanol fermentation system, initially yeast cell concentration is low and yeast growth is dominant. Glucose is mainly utilized to support the growth of yeast cells, so little ethanol and CO_2 are produced and the glucose conversion rate is relatively low. The length of initial stage depends on the yeast inoculation ratio and the fermentation temperature. At normal inoculation ratio of 5%–10% of the glucose solution at around 30°C, the initial stage takes approximately 6–8 h.

After the initial stage of the yeast growth, the yeast cell concentration has tremendously increased to normally over 10^8 cells/mL. The fermentation becomes very active, resulting in a rapid ethanol, CO_2, and energy production, which is indicated by vigorous bubbling and heat production. At this time, heat exchange or cooling needs to be provided to maintain the fermentation temperature around 30°C. Active fermentation normally lasts about 12 h, after which the fermentation activity slows down because less glucose is available. In this phase, the yeast slowly ferments the rest of the glucose to ethanol and CO_2.

During the slow fermentation period, the yeast cells do not grow any more, the biochemical reactions are limited by the substrate (glucose) concentration. Although there are still bubbles generated from the fermentation activity, the vigorousness of bubbling is much less than that in the active fermentation period. To complete the conversion of glucose and other fermentable sugars into ethanol and CO_2, it takes approximately 40–48 h for the slow fermentation period.

7.4.1 Enzymatic Reactions in Fermentation

The overall biochemical reactions to convert glucose to ethanol and CO_2 in yeast fermentation can be expressed as follows:

$$C_6H_{12}O_6 + 2ADP \xrightarrow{\text{Enzymes}} 2C_2H_5OH + 2CO_2 + 2ATP + 10.6 \text{ kJ} \qquad (7.2)$$

where ADP and ATP represent adenosine diphosphate and adenosine triphosphate, respectively. The process involves a series of enzymatic reactions carried out by the enzymes generated by yeast cells under anaerobic conditions. Yeast cells normally generate 20–30 types of enzymes, but only about 15 of them are involved in the fermentation of glucose to ethanol. The process can be divided into two major stages: conversion of glucose to pyruvate and fermentation of pyruvate to ethanol.

7.4.1.1 EMP Pathway

The conversion of glucose to pyruvate in yeast cells is through Embden-Meyerhof-Parnas (EMP) pathway, which is actually a common pathway under either anaerobic or aerobic conditions. The product of EMP pathway, pyruvate, can be further converted to different products such as acetaldehyde, ethanol, lactic acid, and carboxylic acids. Yeast cells normally convert pyruvate to ethanol via acetaldehyde under anaerobic conditions. The detailed processes to convert glucose to ethanol by yeast cells through anaerobic fermentation can be described in the following 13 steps (Jia et al., 2004):

1. Glucose enters yeast cells in which glucose reacts with ATP that provides phosphorus to form glucose-6-phosphate. The biochemical reaction is catalyzed by an enzyme called glucokinase. It is an energy-consuming biochemical reaction and the energy is provided from the conversion of ATP to ADP. This is a relatively slow reaction and could be one of rate-limiting reactions for the entire process.

$$ (7.3) $$

2. Under the catalysis of phosphoglucose isomerase, glucose-6-phosphate is isomerized to fructose-6-phosphate. This is a reversible reaction and the ratio of glucose-6-phosphate to fructose-6-phosphate is normally 7:3 at equilibrium. However, the reaction rate is very high.

$$ (7.4) $$

3. Fructose-6-phosphate, under the attack of fructose-1,6-bisphosphatase, takes a phosphorus from ATP to form fructose-1,6-bisphosphate. This is also an energy-consuming reaction and the energy is provided from the conversion of ATP to ADP. It is also a slow reaction and could be one of the rate-limiting biochemical reactions in the entire process.

$$ (7.5) $$

4. Fructose-1,6-bisphosphate, under the catalysis of fructose-1,6-bisphosphate adolase to break the bond connecting C_3 and C_4, is broken into two compounds: dihydroxyacetone phosphate and glyceraldehyde 3-phosphate. This is a reversible and energy-producing reaction. The two products of the enzymatic reaction are actually isomers, but only glyceraldehyde 3-phosphate is involved in the following EMP pathway to eventually produce ethanol.

Fructose-1,6-bisphosphate   Dihydroxyacetone phosphate   Glyceraldehyde 3-phosphate

$$(7.6)$$

5. Fortunately, the enzymes generated by the yeast cells include triose phosphate isomerase that can convert dihydroxyacetone phosphate to its isomer, glyceraldehyde 3-phosphate.

Dihydroxyacetone phosphate   Glyceraldehyde 3-phosphate

$$(7.7)$$

6. In the presence of NAD+ (nicotinamide adenine dinucleotide) and phosphorus, glyceraldehyde 3-phosphate is catalyzed by glyceraldehyde 3-phosphate dehydrogenase to release a hydrogen to NAD+ and take the phosphorus to form 1,3-diphosphoglycerate.

Dihydroxyacetone phosphate   Glyceraldehyde 3-phosphate dehydrogenase   1,3-Diphosphoglycerate

$$(7.8)$$

7. Under the catalysis of phosphoglycerate kinase assisted by Mg^{2+} and the presence of ADP, 1,3-diphosphoglycerate transfers a high energy phosphorus to ADP to form 3-phosphoglycerate.

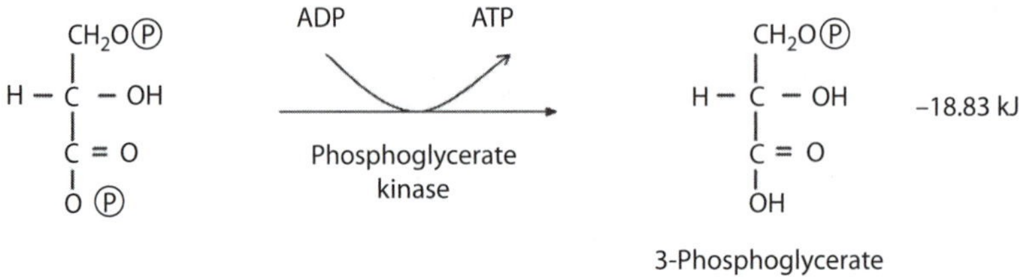

$$-18.83 \text{ kJ}$$

3-Phosphoglycerate

(7.9)

8. Catalyzed by glycerophosphate mutase, 3-phosphoglycerate is mutated into its isomer, 2-phosphoglycerate. It is a reversible reaction.

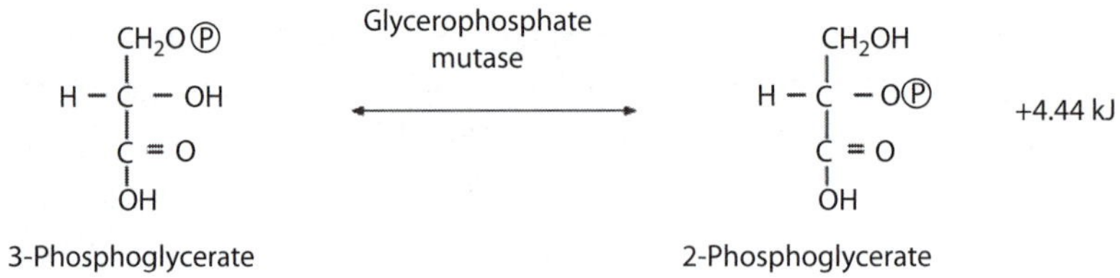

$$+4.44 \text{ kJ}$$

3-Phosphoglycerate   2-Phosphoglycerate

(7.10)

9. Catalyzed by enolase assisted by Mg^{2+}, 2-phosphoglycerate is dehydrated into 2-phosphoenolpyruvate. This is again a reversible reaction.

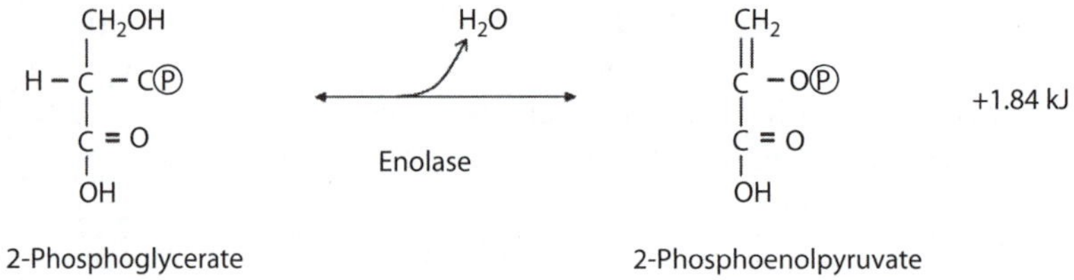

$$+1.84 \text{ kJ}$$

2-Phosphoglycerate   2-Phosphoenolpyruvate

(7.11)

10. 2-Phosphoenolpyruvate, catalyzed by pyruvate kinase assisted by Mg^{2+} and K^+, transfers a high energy phosphorus to ADP and form enolpyruvate. This reaction is another rate-limiting one of the entire process.

$$
\begin{array}{ccc}
\begin{array}{l} CH_2 \\ \| \\ C - O\textcircled{P} \\ | \\ C = O \\ | \\ OH \end{array}
&
\xrightarrow[\text{Pyruvate kinase}]{\text{ADP} \quad \text{ATP}}
&
\begin{array}{l} CH_2 \\ \| \\ C - OH \\ | \\ C = O \\ | \\ OH \end{array} \quad -6.28\ kJ
\\[2em]
\text{2-Phosphoenolpyruvate} & & \text{Enolpyruvate}
\end{array}
$$

(7.12)

11. Under the catalysis of pyruvate kinase, enolpyruvate is isomerized into pyruvate, a very important intermediate during the fermentation process.

$$
\begin{array}{ccc}
\begin{array}{l} CH_2 \\ \| \\ C - OH \\ | \\ C = O \\ | \\ OH \end{array}
&
\xrightarrow{\text{Pyruvate kinase}}
&
\begin{array}{l} CH_3 \\ | \\ C = O \\ | \\ C = O \\ | \\ OH \end{array} \quad +5.06\ kJ
\\[2em]
\text{Enolpyruvate} & & \text{Pyruvate}
\end{array}
$$

(7.13)

12. Pyruvate is catalyzed by pyruvate decarboxylase assisted by Mg^{2+} to form acetaldehyde and release a CO_2 through a decarboxylation reaction.

$$
\begin{array}{ccc}
\begin{array}{l} OH \\ | \\ C = O \\ | \\ C = O \\ | \\ CH_3 \end{array}
&
\xrightarrow[\text{Pyruvate decarboxylase}]{\quad CO_2 \quad}
&
\begin{array}{l} CH_3 - CH \\ \qquad\ \| \\ \qquad\ O \end{array}
\\[2em]
\text{Pyruvate} & & \text{Acetaldehyde}
\end{array}
$$

(7.14)

13. Finally, acetaldehyde is catalyzed by alcohol dehydrogenase in the presence of NADH and H^+, and reduced to ethanol, completing the fermentation process to convert glucose to ethanol.

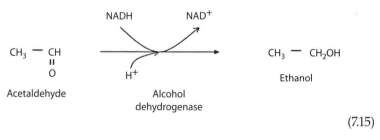

(7.15)

As mentioned earlier, pyruvate is a very important intermediate during the fermentation process. It can also be reduced to lactate in the presence of NADH and H⁺, which is catalyzed by lactate dehydrogenase.

$$
\begin{array}{ccc}
\text{Pyruvate} & \xrightarrow[\text{Lactate dehydrogenase}]{\text{NADH} \quad \text{NAD}^+ \quad \text{H}^+} & \text{Lactate}
\end{array}
$$

OH
|
C = O
|
C = O
|
CH₃

Pyruvate

OH
|
C = O
|
H — C — OH
|
CH₃

Lactate

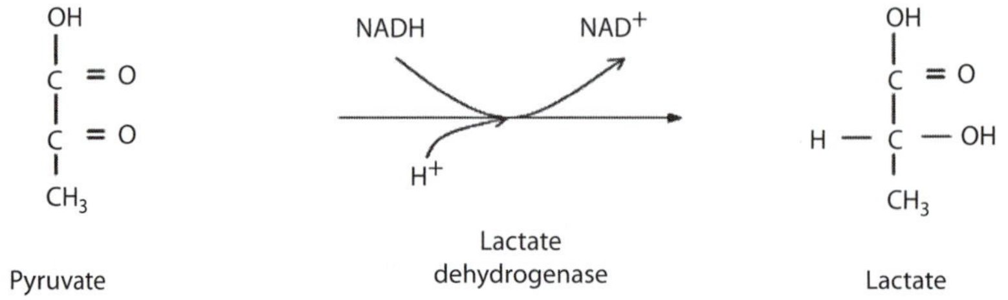

$$(7.16)$$

At high pH (>7.6), acetaldehyde can be catalyzed by acetaldehyde dehydrogenase to form acetate, in the presence of NADH and H⁺.

$$
\text{CH}_3 - \underset{\text{O}}{\overset{\text{CH}}{\parallel}} \; + \; \text{H}_2\text{O} \quad \xrightarrow[\text{Acetaldehyde dehydrogenase}]{\text{NAD}^+ \qquad \text{NADH}_2} \quad \text{CH}_3 - \underset{\text{O}}{\overset{\text{C}}{\parallel}} - \text{OH}
$$

Acetaldehyde

Acetate

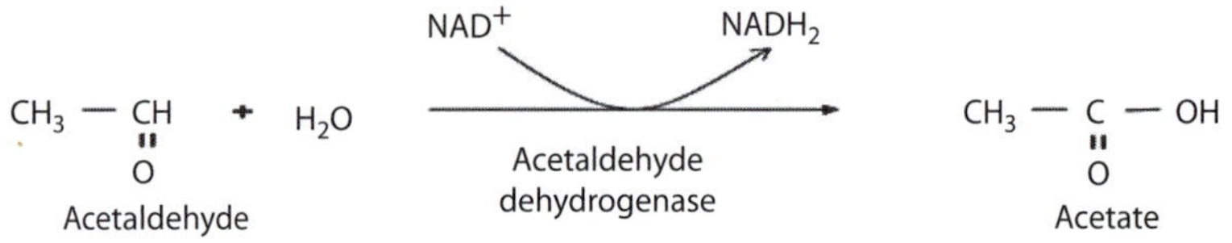

$$(7.17)$$

The formation of acetate can start the tricarboxylic acid (TCA) cycle for the production of different organic acids.

7.4.1.2 Other Products

The main products of the yeast fermentation are ethanol and carbon dioxide. However, there are many by-products generated in the fermentation processes, including other alcohols and organic acids in small amount.

Glycerol: Glycerol formation usually occurs at the initial stage of the fermentation when there is very limited acetaldehyde to take the hydrogen from NADH. A relatively high NADH favors the formation of glycerol through the following process:

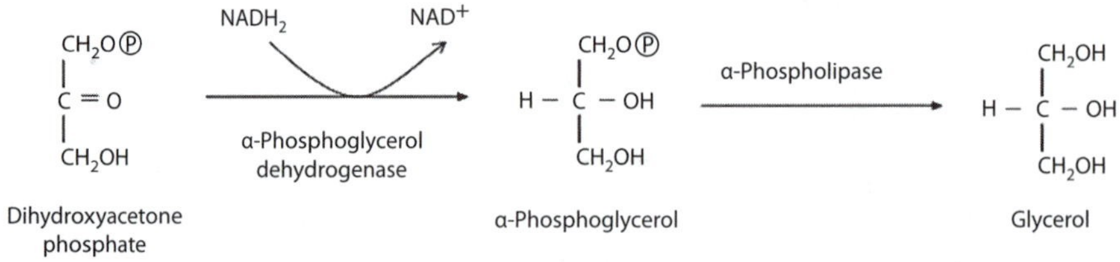

$$
\begin{array}{ccccc}
\text{CH}_2\text{O}\,\textcircled{P} & & \text{CH}_2\text{O}\,\textcircled{P} & & \text{CH}_2\text{OH} \\
| & \xrightarrow[\text{α-Phosphoglycerol dehydrogenase}]{\text{NADH}_2 \quad \text{NAD}^+} & | & \xrightarrow{\text{α-Phospholipase}} & | \\
\text{C} = \text{O} & & \text{H} - \text{C} - \text{OH} & & \text{H} - \text{C} - \text{OH} \\
| & & | & & | \\
\text{CH}_2\text{OH} & & \text{CH}_2\text{OH} & & \text{CH}_2\text{OH}
\end{array}
$$

Dihydroxyacetone phosphate

α-Phosphoglycerol

Glycerol

During the active fermentation, ethanol production is dominant, inhibiting the glycerol-formation reactions. Glycerol concentration is usually about 0.3% in the broth in a normal yeast fermentation.

Acetic acid: Acetic acid formation is usually a result of bacterial and oxygen contamination in the yeast fermentation. *Acetobacter* is a most common bacterium that causes contamination in the fermentation. It is an aerobic bacterium that oxidizes ethanol into acetic acid.

$$CH_3-CH_2OH+O_2 \xrightarrow{\quad Acetobacter \quad} CH_3-COOH+H_2O \qquad (7.18)$$

This oxidation reaction is very rapid if oxygen is abundant in the fermentation broth. Thus, maintaining the fermentation broth under anaerobic or oxygen-free conditions is important to prevent contamination. Anaerobic acetogenic bacteria can also oxidize ethanol to form acetic acid under the oxygen-free environment, but their activities can be inhibited by a relatively high concentration of ethanol produced during the active fermentation.

$$CH_3-CH_2OH+H_2O \xrightarrow{\quad Acetogenic\ bacteria \quad} CH_3-COOH+2H_2 \quad (7.19)$$

Butyric acid: Butyric acid is usually a result of contamination by anaerobic *Clostridium* species that convert glucose to butyric acid:

$$C_6H_{12}O_6 \xrightarrow{\quad Clostridia \quad} CH_3CH_2CH_2COOH+2CO_2+2H_2+61.44\ kJ \qquad (7.20)$$

The mechanism of butyric acid formation by *Clostridium* species is that the bacteria can convert pyruvate produced in fermentation to aceto-CoA in the presence of coenzyme A and then to acetophosphoric acid, which is easily converted to butyric acid.

Other alcohols: During the yeast fermentation, a small amount of other alcohols (more than two carbons) are also produced, in addition to the major alcohol, ethanol. These other alcohols include *n*-propanol, isopropanol, isobutanol, and isoamylol. It is believed that the alcohols are usually from the amino acids produced from the proteins in the raw material for fermentation. These alcohols dissolve in highly concentrated ethanol, but normally not in water. They are believed to contribute to the unique flavor of alcoholic beverages.

7.4.2 Yeast Microbiology

Yeasts are eukaryotic microorganisms classified in the kingdom Fungi, with about 1,500 species. Most yeasts are unicellular and reproduce asexually by budding, but a few grow by binary fission. They are widely used in bakeries and the production of alcoholic beverages such as beer, wine, and liquor. The yeast species *Saccharomyces cerevisiae* has been used in baking and fermenting

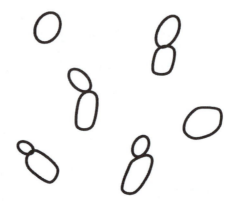

FIGURE 7.18
Shapes of *Saccharomyces cerevisiae* commonly used in ethanol fermentation.

alcoholic beverages for thousands of years. It is also widely used for the fermentation to produce industrial and fuel ethanol.

Yeast morphology: *Saccharomyces cerevisiae* is normally an egg-shaped yeast as shown in Figure 7.18. The size of the yeast cells varies depending on individual strains, but is typically in the range of 2.5–4.5 μm and 10.5–20 μm along the short and long axes, respectively. The unique shape and size of yeast cells can be used for the identification of individual yeast strains.

Yeast cell structure: The structure of yeast cells varies slightly depending on the species, but typically composed of cell wall, cell membrane, cytoplasm, nucleus, ribosomes, mitochondrion, volutin granules, glycogen granules, and lipid granules, as shown in Figure 7.19.

Yeast cell wall consists of approximately 90% carbohydrates and some proteins (mainly glucans, glycogen, chitin, and mannoproteins). It is about 150–250 nm in thickness and accounts for approximately 18%–25% of the total yeast cell dry weight. The main function of the cell wall is to protect the cell. Cell wall forms a protective layer over a relatively flimsy cell membrane. It is flexible, which allows rapid fluctuations in cell volume to respond to changes in the osmotic pressure of the external medium (Briggs et al., 2004). On the other hand, cell wall has a sufficient mechanical strength to prevent lysis when cells are subject to hypo-osmotic shock. Cell wall also sets the limits for the size of molecules that pass in and out of the cell.

Cell membrane is composed of mainly lipids and proteins. It serves as a barrier between the cell wall and cytoplasm and enclose the cytoplasm of the cell. Cell membrane selectively allows the solutes to pass in and out of the cell, preventing a free diffusion of the solutes. It is like a "filter" that allows substrates to penetrate to the inside of the yeast cell for enzymatic reactions and the products to pass through to the exterior environment.

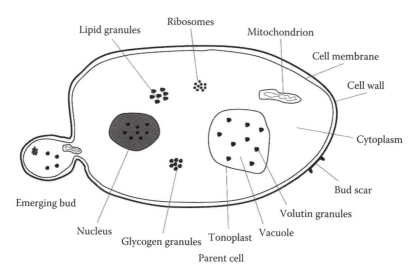

FIGURE 7.19
A diagram of yeast cell structure.

Cytoplasm is an aqueous colloidal liquid in which the metabolic enzymatic reactions take place. It is slightly acidic, with a pH around 5.2. Cytoplasm contains a variety of enzymes and metabolites as well as granules generated through different metabolic pathways. Lipid granules are usually formed during the initial aerobic growth of the yeast cells and store energy for the reproduction of yeast cells. Glycogen is synthesized and utilized during the metabolism in the cytoplasm of the yeast cell. Ribosomes contain high concentrations of RNA and proteins. Their function is to synthesize proteins from activated amino acids sequenced in response to the code in the molecules of messenger RNA. Ribosomes are present throughout the cytoplasm.

Nucleus contains a large portion of the genetic materials of the yeast cell. It is enclosed by a double membrane with pores. The main function of cell nucleus is the synthesis of DNA and chromosomes. The nucleus splits into two when the daughter cell is separating from the parent cell, so each individual cell has a complete set of chromosomes and DNA.

Vacuoles are often formed when the yeast cells are stressed, especially in starvation. They are bound by tonoplast, another membrane. Vacuoles serve as temporary stores for the metabolites, especially the nitrogen-containing metabolites. They contain several proteases that degrade proteins into amino acids. Vacuoles provide the cell with a mechanism to control metabolite concentrations in other cellular compartments. They also store inorganic phosphate in the form of polyphosphate linked by high-energy phosphoanhydride bonds.

Mitochondria serve the yeast cell with energy generation through oxidative phosphorylation to produce ATP, normally during the depressed growth of yeast cells. They do not have the energy generation role during the active fermentation. Mitochondria have two membranes as shown in Figure 7.19 and contain a self-replicating genome that accounts for a small portion of the cell genetic materials.

Yeast propagation: Yeast cell contains approximately 80% water and 20% dry matter. The composition of the dry matter includes 40%–45% proteins, 30%–35% carbohydrates, 6%–8% nucleic acids, 4%–5% lipids, and 5%–10% minerals. Yeasts are heterotrophic microorganisms that rely on biochemical reactions to generate energy to support their growth. They also need organic carbon and nutrients for reproduction. In the fermentation process with *Saccharomyces cerevisiae*, glucose serves as a carbon source for the yeast cells. A small amount of phosphorus nutrient is required during the initial stage of fermentation for yeast cell reproduction. The availability of phosphorus depends on the raw material for ethanol production. For example, corn syrup usually contains sufficient phosphorus for yeast cells. Nitrogen is another important nutrient and yeast cells require nitrogen in the form of ammonia or organic nitrogen. Nitrogen is normally insufficient in the raw materials and needs to be supplemented during the fermentation process. Common sources for nitrogen supplementation include $(NH_4)_2SO_4$, $(NH_4)_3PO_4$, and urea. Mineral nutrients required for yeast cell growth include macronutrients such as K, Na, Ca, Mg, and Cl and micronutrients such as B, Al, V Mo, I, Si, F, Sn, Mn, Co, Cu, and Zn. These nutrients are usually sufficient in the raw materials and water used for the fermentation. In addition, vitamins are also required to support the growth of yeast cells and they are usually from the raw materials but sometime need to be supplemented.

Preparation of yeast culture for industrial ethanol production involves laboratory cultivation and pilot-scale propagation. Original yeast strains are usually stored in the slope culture in test tubes in low-temperature (−80°C) freezer. A typical process for yeast culture preparation is shown in Figure 7.20. All the transfer and cultivation should be conducted under aseptic conditions to prevent any bacterial contamination. An original yeast culture from the freezer is usually thawed at room temperature and then transferred into a liquid medium in test tubes. Initially, yeast extract supplemented with peptone glucose (YEPG) medium is used to provide carbon, energy, and nutrients for yeast growth because YEPG is from yeast cells and has the same ingredients that are needed for growing yeast cells. The amplification or propagation of yeast culture is processed every 15–24 h with the medium amplification ratio of 1:10–25. In the late steps, the saccharification product, glucose, maltose, or sucrose solution, depending on the raw material used for ethanol production, is used to prepare the yeast culture so the yeast cells are acclimated to the substrate in the fermenter to avoid a lag phase of yeast cells during the fermentation process. To start the fermentation, the inocula of yeast culture is approximately 5%–10% of the substrate solution in the fermenter. Maintaining a high-level hygiene during the yeast

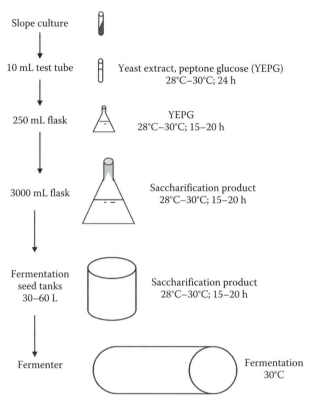

Slope culture

10 mL test tube — Yeast extract, peptone glucose (YEPG)
28°C–30°C; 24 h

250 mL flask — YEPG
28°C–30°C; 15–20 h

3000 mL flask — Saccharification product
28°C–30°C; 15–20 h

Fermentation
seed tanks
30–60 L — Saccharification product
28°C–30°C; 15–20 h

Fermenter — Fermentation
30°C

FIGURE 7.20
A typical yeast culture preparation for fermentation to produce ethanol.

culture preparation and fermentation is very important because ethanol vapor is everywhere in an ethanol production facility, which attracts ethanol-utilizing bacteria, especially acetobacters. Special cautions need to be made to prevent the acetobacters from entering the cultivation and fermentation systems.

Saccharomyces cerevisiae can grow in a wide range of temperatures, normally 5°C–38°C. However, the fermentation rate increases with the increase in temperature. In the production of alcoholic beverages, in addition to ethanol content, flavor is very important to the product, which requires the fermentation in a certain range of temperatures. For example, fermentation for brewing usually takes place around 10°C. In industrial fermentation for ethanol production, ethanol yield and fermentation rate are the main concerns, so fermentation is normally conducted at relative high temperature, around 30°C. The yeast can also grow at a wide range of pH, 2.0–8.0, but the optimum pH for the yeast growth is slightly acidic, 4.8–5.0.

Extensive efforts have been spent on exploring new yeast strains to improve the fermentation process. Some yeasts can tolerate temperatures as high as 50°C. Fermentation with regular yeasts generates a lot of heat, so heat exchangers need to be employed to cool down the fermentation broth and maintain a consistent temperature (around 30°C). However, if high-temperature yeasts are used for fermentation, the produced heat will be absorbed to maintain a high temperature of the broth and heat exchangers become unnecessary. Utilization of high temperature yeasts also makes it easier to adopt the SSF technology because the optimum fermentation temperature is much closer to the optimum saccharification temperature, compared to traditional fermentation. As discussed earlier in this chapter, ethanol at high concentrations may inhibit the microbial activities during fermentation. Thus, ethanol concentration is normally at 10%–15% (w/w) in the final fermentation broth or beer. Some yeast strains can tolerate higher ethanol concentration up to 20% (w/w), which makes it possible to produce a high ethanol broth with ethanol concentration of 18%–20%. High ethanol concentration in the fermentation beer will tremendously reduce the energy consumption during the following distillation process. Employment of active dry yeast can also improve the fermentation efficiency. Yeasts normally contain approximately 80% water. Under rapid vacuum drying at 50°C–60°C, water content of yeasts can be reduced to around 5% and the yeast cell concentration in the active dry yeast can be increased to 30–40 billion cells/g. Active dry yeasts are normally vacuum-packed to keep their activity and added to the fermentation tank to initiate the fermentation process, which can result in a high yeast cell concentration at the beginning of the fermentation and increase the fermentation rate.

7.4.3 Fermenters

There are mainly two types of fermenters, batch and continuous-flow fermenters, in industrial fermentation. Batch fermentation usually involves fermenter disinfection, loading of substrate solution, yeast culture, and nutrients, fermentation, and separation of yeasts from fermentation beer. The main advantages of the batch fermentation are that it is a simple system and is easy to control with little chance for bacterial contamination. Disinfection can be operated for each batch or once in a few batches, depending on the potential of contamination. In case microbial contamination happens in a batch fermentation, it will affect only that batch, limiting the loss to a minimum. Because of its advantage in contamination control, batch fermentation is overwhelmingly used in industrial ethanol production. The disadvantages of batch fermentation include a relatively low efficiency and constantly changing environment for the yeast cells. The fermentation rate can be high at high substrate concentrations during the active fermentation stage. However, the rate decreases significantly with the consumption of the substrates, especially at the time close to the end of fermentation. Because the substrates and their concentrations change all the time in a batch fermentation, the environment for yeast cell growth and metabolism changes constantly. As a result, yeast cell population experience different phases including acceleration growth, exponential growth, and stationary phases. During the exponential growth phase, the yeast cells are in their most active stage for ethanol production. However, at the stationary phase, by-products are also generated by yeast fermentation, in addition to ethanol.

Continuous flow fermentation system usually involves two stages: start-up and steady-state operation. During the start-up of a continuous flow fermenter, the fermenter is operated like a batch fermenter with the inoculum yeast, substrate solution, and nutrients to initiate the yeast cell growth and establish an active yeast culture in the fermenter. The start-up normally takes 12–18 h. Once an active yeast culture is established in the fermenter, substrate and nutrient solutions continuously enters the fermenter, while fermented liquor or fermentation beer continuously flows out of the fermenter under steady-state operation. Yeast cells can be either suspended or immobilized in the fermenter. In the former case, yeast cells flowing out with the fermentation beer can be separated and recycled back into the fermenter to increase the yeast concentration in the fermenter. The main advantages of the continuous flow fermentation are its high efficiency and stable environment for the yeast cells. The fermenter can be designed to keep the yeast cells at their exponential growth phase and produce ethanol at a high rate. The main disadvantage of continuous flow fermentation is the difficulty to control contamination because of the constant inflow and outflow. In case there is a microbial contamination, it is quite challenging to define the boundary of contamination because there is a time gap between the invading microorganisms entering the fermentation system and the identification of the contamination.

7.5 Ethanol Purification

At the end of the fermentation process, ethanol concentration in the fermentation beer is usually 10%–15% (w/w). As a biofuel, especially for automobiles, ethanol must be separated from water in the beer to obtain almost pure ethanol (>99%). The presence of water in fuel ethanol not only reduces the energy density of the fuel but can also cause damage to the engine in which fuel ethanol is combusted. The separation process usually takes two steps: fractionation and dehydration. Fractionation can produce an ethanol–water mixture with ethanol content as high as 90% (w/w) and dehydration removes the rest of the water to generate fuel-grade ethanol.

7.5.1 Fractionation

Fractionation or fractional distillation is a thermal physical process, based on phase equilibrium of the ethanol–water mixture. The process is performed in a fractionation column with plates or packing materials. Boiling ethanol–water mixture in the column generates an up-flow vapor with higher ethanol content and a down-flow liquid with lower ethanol content. Thus, the effluent collected from the top of the column has much higher ethanol content than that from the bottom of the column. To understand the mechanism, we need to start with fundamental phase equilibrium of mixed liquid.

7.5.1.1 Phase Equilibrium

Boiling point and dew point: Water, under normal atmospheric pressure, becomes water vapor when heated to 100°C, which is a boiling or bubble point for water. When a continued heating is provided to the mixture of water and water vapor, more water will become water vapor but the temperature remains at 100°C until all the water becomes vapor. After that, if more heating is provided to the vapor, the temperature rises again. On the other hand, when water vapor is cooled from a high temperature (>100°C) to 100°C, the first dew or water droplet will be formed. Further cooling of the water–vapor mixture will result in more water formation from the vapor but the temperature remains at 100°C until all the vapor becomes water. So, 100°C is also the dew point for water vapor. Similarly, pure ethanol has the same bubble and dew points of 78.5°C under normal atmospheric pressure. This is true for almost all single compound liquid. However, when two or more liquid compounds that can be dissolved in each other at any ratio (e.g., ethanol–water, ethanol–benzyne, ethanol–propanol, benzyne–toluene, acetone–butanol), their mixtures can fall into ideal solution or nonideal solution categories.

An ideal solution means that the mixture behaves just like a single compound liquid with the same bubble and dew points. For example, two compounds, A and B, form an ideal solution. As a single compound, A and B have the same bubble and dew points of 80°C and 115°C, respectively, as shown in Figure 7.21. When they form an ideal mixture solution, the mixture still has the same bubble and dew point at any given ratio of A:B. The temperature of the bubble (or dew) point depends on the ratio of A:B in the mixture and the relationship can be expressed with a bubble (or dew) point line that separates the liquid and vapor phases of the ideal mixture solution.

When two compounds, C and D, form a nonideal mixture solution, the bubble and dew point properties of the solution is different from a single compound. As an example, Figure 7.22 shows the liquid–vapor phase equilibrium of a nonideal solution composed of C and D. As a single compound, C and D have the bubble (or dew) points of 76°C and 113°C, respectively. However, when C and D are mixed together to form a solution, the bubble point and dew point of the solution are different at a given ratio of C:D. For example, a mixed solution of 50% (mole fraction) C and 50% (mole fraction) D has a bubble point of 87°C but the dew point is 102°C. If the nonideal solution is heated starting at 75°C, the temperature increases. When the temperature

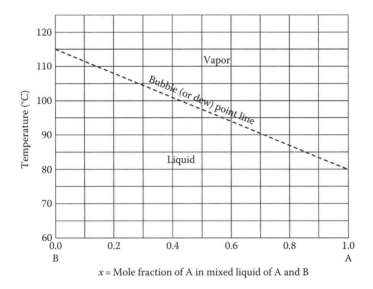

FIGURE 7.21
Bubble point or dew point line for an ideal mixed solution of A and B.

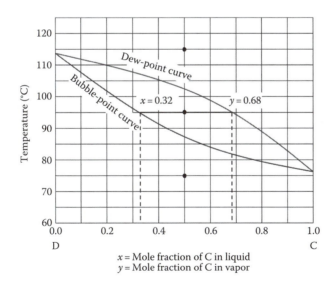

FIGURE 7.22
Phase equilibrium of a typical nonideal mixed solution of C and D.

is raised to 87.5°C, the solution starts to bubble or reaches the bubble point. Unlike a single compound liquid or ideal solution, the temperature of the liquid–vapor mixture keeps rising when more solution becomes vapor with further heating. However, the compositions of the two components in liquid and vapor are different and they are changing with the increase in temperature (e.g., the mole fraction of C in liquid and vapor is 0.32 and 0.68, respectively, at temperature of 95°C) until the temperature reaches the dew point of 102.5°C. At the dew point, all the solution becomes vapor and the composition comes back to 50% C. All the bubble points at different ratios of C:D form a bubble point curve and dew points form a dew point curve. Below the bubble point curve is the liquid phase and beyond the dew point curve is the vapor phase. Between the two curves is the mixture of liquid and vapor. The unique property of nonideal solution makes it possible to separate C and D through fractional distillation. As shown in Figure 7.22, if a mixed solution of C and D with 50% (mol/mol) C is heated to 95°C in a distiller, C content in the vapor phase will be 68% (mol/mol), which can be condensed to obtain a solution with 68% (mol/mol) C in a single stage distillation. Similarly, the solution with 68% (mol/mol) C can also be distilled to further increase the C content. After a series of distillers, the C content can be increased to close to 100%.

Azeotropic mixture: Some mixed solutions may have the characteristics of nonideal solutions within a certain range of molar ratios and behave like ideal solutions at other molar ratios. A typical example is the mixed solution of ethanol and benzene. The mixed solution is generally a nonideal solution except when the ethanol:benzene ratio is 1:1 (mole). Figure 7.23 shows the bubble and dew point curves of the ethanol–benzene mixture.

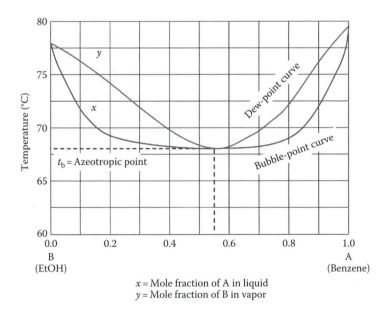

x = Mole fraction of A in liquid
y = Mole fraction of B in vapor

FIGURE 7.23
Phase equilibrium of an azeotropic nonideal solution of ethanol and benzene.

At the mole ratio of ethanol:benzene of 1:1, the mixture behaves like an ideal solution with the same bubble and dew points. The special mixture is called azeotropic mixture and the bubble or dew point temperature is called azeotropic point. To separate the components of a nonideal solution with azeotropic point(s), the concentration at the azeotropic point(s) is the limit for separation using ordinary fractional distillation.

Ethanol–water mixture is an azeotropic nonideal solution. Figure 7.24 shows the phase equilibrium of the ethanol–water solution. When the mixture contains ethanol of 0%–84% (mol/mol), the solution is a nonideal solution, as shown in the figure. However, at the ethanol mole fraction greater than 84% (93%, w/w), the phase equilibrium becomes an azeotropic zone in which the bubble point curve and dew point curve merge into one. For a given ethanol–water mixture below 84% (mol/mol) (e.g., 30% [mol/mol] ethanol), distillation can generate a vapor with higher ethanol content as shown in Figure 7.24. A series of distillation can concentrate ethanol to nearly 84% (mol/mol) or 93% (w/w), which is the maximum ethanol concentration than can be achieved using ordinary fractional distillation. In industrial practice, fractional distillation or fractionation to concentrate ethanol is usually performed in fractionating columns with plates. Each plate inside the column serves as a distillation stage. The question is how many plates are necessary to perform the fractionation to concentrate ethanol to a certain level?

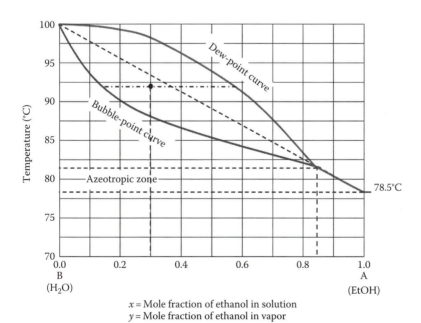

FIGURE 7.24
Phase equilibrium of ethanol–water solution.

7.5.1.2 *Fractionation to Separate Ethanol from Water*

Figure 7.25 shows a schematic diagram of a typical fractional distillation in a fractionating column. The column usually contains several or many plates inside, each of which serves as a distillation stage. The column is divided by the feeding location into two sections. The upper one from the feeding location up to the top of the column is called rectifying section; the lower one from the feeding location down to the bottom of the column is called stripping section. A boiler is used to generate boiling or bubbling solution at the bottom of the column. An ethanol–water mixture is fed to the process in the middle of the column. The solutions are boiling or are at the temperatures between the bubble and dew points through all plates, with vapor moving upward and liquid flowing downward. Theoretically, the vapor and liquid leaving each plate should be at equilibrium and ethanol content in the vapor is higher than in the liquid. Thus, ethanol content in the vapor increases as the vapor gets closer to the top of the column. The vapor stream at the top of the column, which has the highest ethanol content, is liquefied in a condenser with cooling water. A portion of the condensed liquid is sent back to the column to serve as the source of the down-flow liquid. The rest of the condensed liquid is the overhead product with high concentration of ethanol (usually close to 90% [w/w]). Another product of the fractionation is from

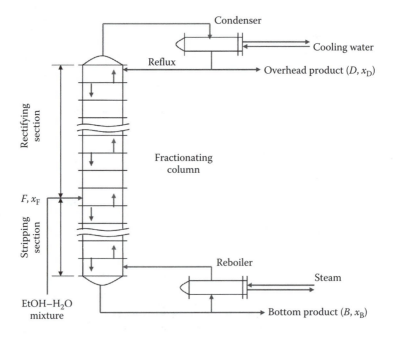

FIGURE 7.25
A typical fractionating or fractional distillation process.

the bottom of the column, so it is called the bottom product. Ethanol content in the down-flow liquid becomes lower and lower, as the liquid flows down toward the bottom of the column. Liquid flowing out of the bottom of the column usually contains around 1% (w/w) ethanol. A portion of the liquid is sent to a reboiler to boil the liquid into vapor, providing the source and heat of the up-flow vapor. The bottom product is mainly a hot water at a temperature around 95°C. The hot water is usually used to provide heat to the ethanol–water mixture fed to the fractionating column to raise its temperature to around the bubble point.

Mass balance: A mass balance for the total mass and ethanol around the fractionating column will give us

   Mass in-flow = Mass out-flow and Ethanol in-flow = Ethanol out-flow

or

$$F = D + B \tag{7.21}$$

and

$$F\, x_F = D\, x_D + B\, x_B \tag{7.22}$$

where F, D, and B are the flow rates of the feed (ethanol–water mixture), overhead product, or distillate, and bottom product or distillation residue, respectively. x_F, x_D, and x_B are ethanol concentrations in feed, distillate, and bottom product, respectively. Thus, the ratio of the overhead product (ethanol) and the bottom product (water) to the original feed can be respectively expressed as

$$\frac{D}{F} = \frac{x_F - x_B}{x_D - x_B} \tag{7.23}$$

$$\frac{B}{F} = \frac{x_D - x_F}{x_D - x_B} \tag{7.24}$$

Operating lines in fractionating column: A mass balance around the boundary from between the nth and $n + 1$th plates in the rectifying section of the fractionating column to the top of the column, as shown in Figure 7.26, yields

$$\text{Total mass balance: } V_{n+1} = D + L_n \tag{7.25}$$

$$\text{Ethanol mass balance: } V_{n+1}\, y_{n+1} = D\, x_D + L_n\, x_n \tag{7.26}$$

where
   V_{n+1} is the flow rate of the vapor leaving the $n + 1$th plate and moving to nth plate

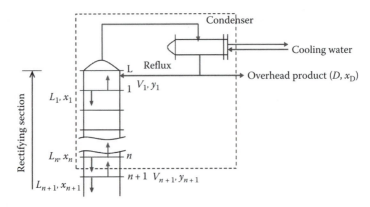

FIGURE 7.26
Mass balance boundary in the rectifying section of a fractionating column.

y_{n+1} is the ethanol concentration in vapor V_{n+1}
L_n is the flow rate of the liquid leaving the nth plate and moving to $n + 1$th
     plate
x_n is the ethanol concentration in liquid L_n

A relationship between y_{n+1} and x_n can be established from the two equations above:

$$y_{n+1} = \frac{L_n}{D+L_n} x_n + \frac{S x_D}{D+L_n} \tag{7.27}$$

It is a straight line on a phase equilibrium diagram plotted with ethanol concentration in vapor against the ethanol concentration in liquid, so Equation 7.27 is called the operating line for the rectifying section of a fractionation column.

As shown in Figures 7.24 and 7.25, the temperature in the fractionating column decreases from the bottom to the top. The temperature and ethanol concentrations in the vapor and liquid are different on each individual plate. The flow rates of the vapor and liquid are slightly different for each plate because the heat exchanges at different temperatures are different. The heat exchange involved on each individual plate include mainly latent heat (energy required to change the solution into vapor or energy release when the vapor becomes liquid), heat required to increase the temperature, and heat loss. This heat exchange determines the ratio of liquid to vapor leaving an individual plate. The latent heat for the ethanol–water mixture at certain temperature is usually much bigger than other forms of heat. Therefore, the following assumptions are made to simplify the analysis:

1. Other forms of heat can be neglected compared to latent heat.
2. Latent heat is the same for all the ethanol–water mixtures at different ratios inside the fractionating column.

With these two assumptions, we can approximate that the flow rates of the
liquid and vapor leaving each plate in the fractionating column are the same,
or $L_n = L_{n-1} = \cdots = L$ and $V_{n+1} = V_n = \cdots = V$. Thus, the operating line for
the rectifying section Equation 5.27 can be simplified as

$$y_{n+1} = \frac{L}{D+L} x_n + \frac{Dx_D}{D+L} \tag{7.28}$$

or simply

$$y = \frac{L}{D+L} x + \frac{Dx_D}{D+L} \tag{7.29}$$

If R_D is defined as a reflux ratio or $L{:}D$, then $R_D = L/D$. Equation 7.29 can also
be expressed as

$$y = \frac{R_D}{R_D+1} x + \frac{x_D}{R_D+1} \tag{7.30}$$

For the first plate from the top or top plate, the reflux liquid is from the condensed vapor from the same plate. Thus, the ethanol content in both liquid
and vapor is the same, which is x_D, or $y = x = x_D$.

Similarly, a mass balance can also be conducted around a boundary in the
stripping section of the fractionating column as shown in the following figure. A general form of the operating line in Figure 7.27 can be expressed as

$$y_{m+1} = \frac{L_m}{L_m - B} x_m + \frac{Bx_B}{L_m - B} \tag{7.31}$$

A simplified operating line for the stripping section with the same assumptions as discussed in the analysis of rectifying section can be expressed as

$$y = \frac{L_S}{L_S - B} x + \frac{Bx_B}{L_S - B} \tag{7.32}$$

where L_S is the liquid flow rate in the stripping section and it depends on the
thermal status of the feed. For example, if the feed is sent to the column at
the bubble point, then $L_S = L + F$. The temperature in the reboiler determines
the ethanol content in the bottom product, x_B, and the ethanol content in the
vapor from the reboiler back to the column, as shown in Figure 7.24.

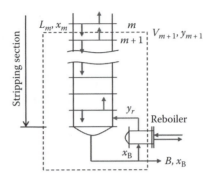

FIGURE 7.27
Mass balance boundary in the stripping section of a fractionating column.

Design of the fractionating column: To design a fractionating column, the most important thing is to figure out how many plates are necessary to perform a desired separation of ethanol from water. A theoretical number of plates required for a fractionation process can be obtained using the principles of phase equilibrium. Figure 7.24 shows the bubble and dew point curves of ethanol–water mixture. The data for that figure can be used to plot the relationship between the ethanol concentration or mole fraction in the vapor against that in the liquid, as shown in Figure 7.28. On the phase equilibrium curve, the ethanol–water mixture shows an azeotropic property in a range of ethanol mole fraction from 84% to 100% or 93% to 100% (w/w). This means

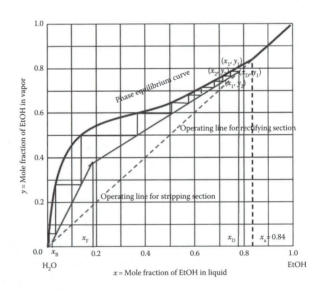

FIGURE 7.28
Relationship of ethanol fractions in liquid and vapor in a fractionating column.

that the theoretical maximum concentration of ethanol that can be obtained from fractionation is 84% (mol/mol) or 93% (w/w). Figure 7.28 also shows the operating lines in both rectifying and stripping sections of a fractionating column. The concentration difference between the phase equilibrium curve and the operating lines is the driving force to increase ethanol content on each theoretical plate. For the first plate from the top of the column or top plate, y_1 should be the same as x_D. Assume that the liquid leaving the first plate is in equilibrium with the vapor moving up from the same plate, then the ethanol concentration in the leaving liquid (x_1) can be determined graphically by drawing a horizontal line from the point (x_D, y_1, or x_D) to the intersection with the phase equilibrium curve (x_1, y_1). The relationship of x_1 with y_2, which is the ethanol concentration of the vapor on the second plate, is defined by the operating line for the rectifying section. Continuation of drawing the lines between the phase equilibrium curve and the operating line shows the 2nd, 3rd, ... and nth theoretical plate. It has to be noted that when the drawing is passing through the feed plate, the operating line for the stripping section needs to be used to determine the concentrations. The total numbers of the triangles between the phase equilibrium curve and the operating lines are the required theoretical number of plates to perform the fractionation of ethanol–water mixture to the desired ethanol product (x_D) and the bottom product (x_B).

Example:

A fermentation beer of 10,000 kg/h has an ethanol concentration of 16% (w/w). The beer needs to be fractionated in a column to get an overhead product with ethanol concentration of 90% (w/w). The ethanol concentration in the bottom product or residue is supposed to be less than 1.0% (w/w). Assume other components in the beer can be neglected. Before the bottom product is getting out of the system, it is cooled in a heat exchanger to heat up the feed to the fractionating column. The feed enters the column as a saturated liquid at the bubbling temperature. A reflux ratio of 2.5–1 mol of product is to be used. (1) Calculate the flow rate of the overhead product and the bottom product. (2) Determine the number of theoretical plates necessary in the fractionating column and the position of the feed plate.

Solution:

In this problem, $F = 10,000$ kg/h, $x_F = 16\%$ (w/w), $x_D = 90\%$ (w/w), $x_B = 1\%$ (w/w), $R_D = 2.5$.

1. The flow rate of the overhead product and the bottom product can be determined using mass balance: $F = D + B$ and $Fx_F = Dx_D + Bx_B$ or

   $D + B = 10,000$ kg/h and $10,000$ kg/h $\times$ 16% = 90% D + 1% B
   So, $D = 750$ kg/h
   and $B = 9250$ kg/h

2. To determine the number of theoretical plates necessary in the fractionating column and the position of the feed plate, we need to use the phase equilibrium of ethanol–water mixture, so we have to convert the weight concentrations into mole fractions. The molecular weight of ethanol and water is 46 g/mol and 18 g/mol, respectively.

   Therefore, $x_F = 16\%$ (w/w) = (0.16/46)/(0.16/46 + 0.84/18) = 0.069 mol/mol;
   $x_D = 90\%$ (w/w) = (0.9/46)/(0.9/46 + 0.1/18) = 0.78 mol/mol;
   $x_B = 1.0\%$ (w/w) = (0.01/46)/(0.01/46 + 0.99/18) = 0.004 mol/mol;
   Since $R_D = 2.5$ and $x_D = 0.78$ mol/mol, the operating line for the rectifying

   section should be $y = \dfrac{2.5}{2.5+1} x + \dfrac{0.78}{2.5+1} = 0.714\,x + 0.223.$

This operating line should start at the point (0.78, 0.78) with the interception to the y-axis of 0.223 (Figure 7.29). The feed with ethanol mole fraction of 0.069 enters the column at the bubble point, so the feeding line should be a vertical starting at (0.069, 0). The operating line for the stripping section can be determined using Equation 7.32:

$$y = \frac{L_S}{L_S - B} x + \frac{Bx_B}{L_S - B}$$

where $L_S = L + F = R_D \cdot D + F$ and B should be in mol/h. A shortcut to draw the operating line for the stripping section is to connect the intersection between the operating line for the rectifying section and the feeding line and the point (0.004, 0.004), because the intersection should also be shared by the operating line for the stripping section and the point (0.004, 0.004) representing the ethanol concentrations in the reboiler. A graphical analysis indicates that 10 theoretical plates are

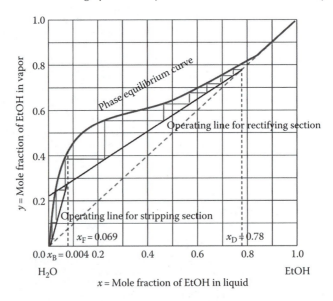

FIGURE 7.29
Graphical solution for the example.

required to perform the fractionation with the feed entering the seventh plate, as shown in Figure 7.29.

Feeding line depends on the status of the feed, saturated liquid at bubble point, saturated vapor at dew point, mixture of liquid and vapor, or cool liquid. In many cases for ethanol–water fractionation, the mixture is fed to the column as saturated liquid, which is a vertical line as shown in Figure 7.30.

Maximum and minimum reflux ratio (R_D): If all the liquid from the condenser returns to the column, the reflux ratio is the maximum ($+\infty$). This means there is no overhead product or $D = 0$. In this case, the operating line is overlapping with the diagonal line, as shown in Figure 7.30. This usually happens at the start-up of the fractionating column to speed up the start-up process. However, there is a minimum reflux ratio required for any possible fractionation. As discussed earlier, the driving force for fractionation is the difference of the ethanol concentrations between the operating lines and the phase equilibrium curve. If an operating line intersects with one point of the phase equilibrium curve, the number of theoretical plates for the fractionation will be infinite ($+\infty$). In this case, the reflux ratio is the minimum R_D, which is shown in Figure 7.30.

Fractionating column: Fractionating columns used for separating ethanol from water are divided into two categories: plate columns and packed column. Figure 7.31 shows typical liquid and vapor flows in a plate column. Plates with small openings are installed inside the column with a slight slope. Liquid flows on the slope of the plates and in the guided pipes between the

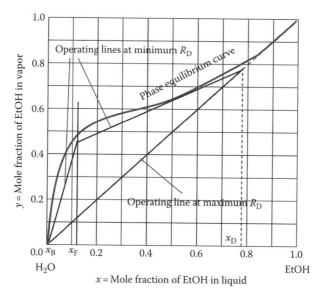

FIGURE 7.30
Maximum and minimum reflux ratios in fractionation.

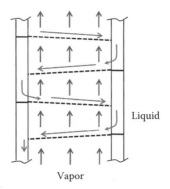

FIGURE 7.31
Typical liquid and vapor flow patterns inside a plate column.

plates. Vapor rises through the openings of the plates and bubbles through the liquid layers on the plates where the vapor and liquid contact each other. The efficiency of a plate column depends on the structures of the individual plate and the column and includes plate efficiency and local efficiency. In industrial practice, overall plate column efficiency is usually used for certain type of plates for the design of fractionating columns, which is approximately 50%–60%. The actual number of plates needed for a fractionating column can thus be estimated using the following equation:

$$\text{actual number of plates} = \text{theoretical number of plate/overall efficiency}$$

(7.33)

Another type of the fractionating column is the packed columns. Instead of plates, packing materials (plastic and ceramic) are installed inside the columns, as shown in Figure 7.32. Liquid flows down on the surface of the

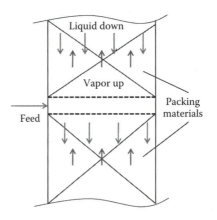

FIGURE 7.32
Typical structure of a packed fractionating column.

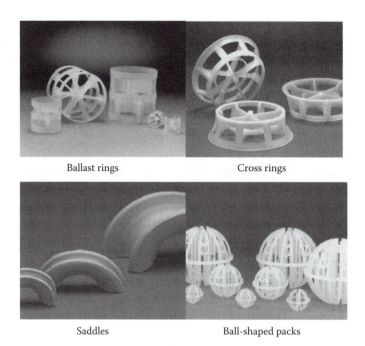

Ballast rings   Cross rings

Saddles   Ball-shaped packs

FIGURE 7.33
Shapes of commonly used packing materials for fractionating columns.

packing materials, while vapor rises through the void space in the column. A good material usually has a high surface area and porosity when placed in the column. Commonly used packing materials include ballast rings, cross rings, saddles, and ball-shapes packs. Their shapes are shown in Figure 7.33. To estimate the necessary height of a packing material in packed column, the equivalent height of a theoretical plate (EHTP) is usually determined experimentally. Thus, the necessary height of a certain packing material for a fractionating column can be estimated as

$$\text{Total height of packs} = \text{EHTP} \times \text{theoretical number of plates} \quad (7.34)$$

7.5.2 Dehydration

After the fractionation or fractional distillation process, the overhead product usually contains around 90% (w/w) ethanol and 10% (w/w) water. The water has to be completely removed to produce fuel ethanol or anhydrous ethanol (ethanol content > 99%), which is normally achieved in a dehydration process. There are different technologies for dehydration to generate anhydrous ethanol and molecular sieve separation is the most commonly used one in ethanol industry.

Molecular sieve adsorption: A fundamental principle of molecular sieve separation is based on the difference of the sizes of the molecules that need to be separated, ethanol and water in this case. A water molecule has a diameter of approximately 0.28 nm, while an ethanol molecule has a diameter of about 0.4 nm. Molecular sieves are synthesized materials with uniform micropores. The molecular sieves used in the dehydration to produce anhydrous ethanol usually have pores with a diameter of 0.3–0.35 nm. When the ethanol–water mixture is in contact with the molecular sieves under pressure, the smaller water molecules can enter the micropores and be adsorbed on the surface inside the pores, while the larger ethanol molecules stay outside of the pores, so the ethanol is separated from the water. In industrial practice, ethanol vapor from the fractionating column is usually pressurized to about 0.3 MPa and then passes through a dehydration column packed with molecular sieves. Water is adsorbed inside the micropores of the molecular sieves and anhydrous ethanol flows out of the column. When the adsorption capacity of the molecular sieves is saturated, the molecular sieves need to be regenerated for reuse in dehydration. Regeneration of the molecular sieves is normally conducted under vacuum conditions to release the water from the micropores. In fuel ethanol plants, multiple dehydration columns are usually used to generate anhydrous ethanol, with some columns in operation while others in regeneration. Molecular sieve technology has the following advantages: (1) It generates almost pure ethanol (ethanol content > 99.9%) without impurities. (2) The molecular sieves can be used, regenerated, and reused for many times and they keep their water adsorption capacity for several years. (3) It requires low energy input and environmentally friendly.

Dry corn powder adsoprtion: Dry corn powder can also be used for dehydration, which especially makes sense in ethanol plants with corn as feedstock. Dry corn powder can be very effective in adsorbing water from the distillation vapor out of the fractionating column at its dew point. The corn powder saturated with water can be regenerated by heating to 80°C–100°C and blowing air through. The saturated corn powder can also be sent to the saccharification tanks in the conversion process. The anhydrous ethanol from the dry corn powder adsorption system usually carries some impurities from the corn powder.

Supercritical CO_2 extraction: CO_2 at supercritical conditions (normally at the high pressure of about 80 atm) has a high density like a liquid but very low viscosity. It has a high capacity of dissolving organic compounds such as ethanol and butanol. Ethanol is much more soluble in supercritical CO_2 than in water. Thus, when the distillate from the fractionating column enters a supercritical CO_2 system, ethanol will be extracted by the liquid-like CO_2. When the pressure of the supercritical CO_2 saturated with ethanol is dropped, supercritical CO_2 will become a gas and anhydrous ethanol will be released. A supercritical CO_2 ethanol extraction system involves a pressurized vessel for the supercritical CO_2 to extract ethanol and a flashing vessel to lower the pressure and release anhydrous ethanol. This system usually has a high energy requirement to generate supercritical CO_2.

7.6 By-Products

7.6.1 Distiller's Dried Grains with Solubles

Distiller's dried grains with solubles (DDGS) is a main by-product generated from the dry-milled corn to ethanol process. Residues from the distillation process contain fibers, proteins, fats, amino acids, vitamins, and minerals, which are good ingredients in animal feed. The residues are usually centrifuged to remove the free water and become DDGS after drying. DDGS is normally sold as animal feed. Its generation in ethanol plants with corn as feedstock is very significant. Production of 1 kg of ethanol generates approximately 900 g DDGS. Depending on specific separation process, extra yeast cells after fermentation can be ended in the DDGS. They can also be singled out and dried separately to make yeast extract, a high-value by-product.

7.6.2 Carbon Dioxide

During the fermentation for ethanol production, almost the same amount of CO_2 is generated in the biological process. Theoretically, the production ratio of ethanol to CO_2 in fermentation is 92:88 g. The CO_2 gas is normally collected and pressurized in tanks and can be utilized in food, drink, and chemical industries. CO_2 is widely used in soft drinks and beer to supplement the carbonate in the drinks. It is also commonly used to fill the packs of vegetables and meat to keep them fresh. CO_2 can also be used as a raw material for the synthesis of methanol, formic acid, and urea. Other applications of CO_2 include its use as a medium in supercritical CO_2 extraction and in fire extinguishing equipment.

7.6.3 Other By-Products

In the wet mill process of converting corn into ethanol, corn oil is a by-product that can be sold to the food industry and household for food preparation. When sugarcane is used for ethanol production, the bagasse is usually burned to provide energy in the distillation process. Bagasse contains a high content of cellulose and hemicelluloses, which have a potential for additional ethanol production to improve the overall efficiency of the sugarcane-to-ethanol process.

7.7 Problems

1. How much energy will be produced when 15 gal of ethanol is burned? If ethanol is used as automobile fuel, how many gallons of gasoline can be replaced with 15 gal of ethanol?

2. Starch and cellulose both are polymers of glucose residues. When they are used as feedstock for ethanol production, starch and cellulose have to be converted to glucose, which is then fermented to ethanol. Compare starch with cellulose as feedstock for ethanol production and discuss their advantages and disadvantages.

3. Potatoes and sweet potatoes are excellent starch crops. Compare them with corn in starch production and as feedstock for ethanol production.

4. A pilot fuel ethanol plant will use switchgrass as feedstock. Please select an appropriate pretreatment process for the switchgrass and explain.

5. If poplar, a hardwood wood, is used as the feedstock for ethanol production, please select an appropriate pretreatment process for the switchgrass and explain.

6. The product of the yeast fermentation, ethanol, may be inhibitory to the yeast growth when the ethanol concentration in the fermentation broth is over 12% (w/w). Propose a method to mitigate the inhibition problem and explain.

7. An ethanol plant is being planned to produce 100,000 metric tons of fuel ethanol per year, using corn as the raw material.

   a. Design a conceptual process for the ethanol plant.

   b. Estimate how much corn is needed per year for the ethanol plant.

   c. Estimate how much cropland is required to produce enough corn for the ethanol plant.

   d. Estimate how much net energy can be produced through the corn-to-ethanol process.

   e. Estimate how much less green house gas emission can be achieved when the produced ethanol is consumed as automobile fuel in comparison with the consumption of gasoline with equivalent amount of energy production.

8. If both corn and corn stover are used as the raw materials for ethanol production in Problem 7.

   a. Design a conceptual process for the ethanol plant.

   b. Estimate how much cropland is required to produce enough corn and corn stover for the ethanol plant.

   c. Estimate how much net energy can be produced through the corn and corn stover-to-ethanol process.

   d. Estimate how much less green house gas emission can be achieved when the produced ethanol is consumed as automobile fuel in comparison with the consumption of gasoline with equivalent amount of energy production.

9. An ethanol plant is being planned to produce 100,000 metric tons of fuel ethanol per year, using switchgrass as the raw material.

   a. Design a conceptual process for the ethanol plant.

   b. Estimate how much switchgrass is needed per year for the ethanol plant.

   c. Estimate how much land is required to produce enough switchgrass for the ethanol plant.

   d. Estimate how much net energy can be produced through the switchgrass-to-ethanol process.

   e. Estimate how much less green house gas emission can be achieved when the produced ethanol is consumed as automobile fuel in comparison with the consumption of gasoline with equivalent amount of energy production.

10. An ethanol plant is being planned to produce 100,000 metric tons of fuel ethanol per year, using poplar wood as the raw material.

    a. Design a conceptual process for the ethanol plant.

    b. Estimate how much poplar wood is needed per year for the ethanol plant.

    c. Estimate how much forest land is required to produce enough poplar wood for the ethanol plant.

    d. Estimate how much net energy can be produced through the wood-to-ethanol process.

    e. Estimate how much less green house gas emission can be achieved when the produced ethanol is consumed as automobile fuel in comparison with the consumption of gasoline with equivalent amount of energy production.

11. Lignocellulosic materials contain a significant amount of lignin, usually 15%–35% (w/w). Utilization of lignin is an important factor that affects the economics of fuel ethanol production from lignocellulosic materials. Identify the possible applications of lignin and explain.

12. A fermentation broth of 10,000 kg/h has an ethanol concentration of 16% (wt). The beer needs to be fractionated in a column to get an overhead product with ethanol concentration of 90% (wt). The ethanol concentration in the bottom product is supposed to be less than 1.0% (wt). Assume that other components in the beer can be neglected. Before the bottom product is getting out of the system, it is cooled in a heat exchanger to heat up the feed to the fractionating column. The feed enters the column as a saturated liquid at the bubbling temperature. A reflux ratio of 2.5 to 1 mol of product is to be used. Figure 7.34 shows a phase equilibrium diagram of ethanol–water mixture. The molecular weight of ethanol and water is 46 and 18 g/mol, respectively.

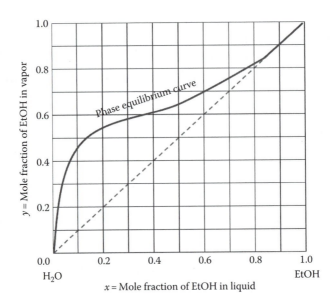

FIGURE 7.34
Phase equilibrium diagram of ethanol–water mixture.

    a. Calculate the flow rate of the overhead product and the bottom product.

    b. Determine the number of theoretical plates necessary in the fractionating column and the position of the feed plate.

13. A fermentation broth of 10,000 kg/h has an ethanol concentration of 16% (wt). The beer needs to be fractionated in a column to get an overhead product with ethanol concentration of 90% (wt). The ethanol concentration in the bottom product is supposed to be less than 1.0% (wt). Assume that other components in the beer can be neglected. Before the bottom product is getting out of the system, it is cooled in a heat exchanger to heat up the feed to the fractionating column. The feed enters the column as a saturated liquid at the bubbling temperature. Figure 7.34 shows a phase equilibrium diagram of ethanol–water mixture. The molecular weight of ethanol and water is 46 and 18 g/mole, respectively.

    a. Determine the minimum reflux ratio for the fractionating column.

    b. If the fractionating column is operated with a reflux ratio that is two (2) times of the minimum reflux ratio, determine the number of theoretical plates necessary in the fractionating column and the position of the feed plate.

14. Estimate the cost of ethanol production (in dollar per gallon of ethanol) for Problem 9.

15. Estimate the cost of ethanol production (in dollar per gallon of ethanol) for Problem 10.

References

Akin, D.E., Rigsby, L.L., Sethuraman, A., Morrison, W.H.-III., Gamble, G.R., and Eriksson, K.E.L. (1995) Alterations in structure, chemistry, and biodegradability of grass lignocellulose treated with the white rot fungi Ceriporiopsis subvermispora ad Cyathus stercoreus. *Appl. Environ. Microb.*, 61, 1591–1598.

Ander, P. and Eriksson, K.-E. (1977) Selective degradation of wood components by white-rot fungi. *Physiol. Plant.*, 41, 239–248.

Aziz, S. and Sarkanen, K. (1989) Organosolv pulping—A review. *Tappi J.*, 72, 169–175.

Azzam, A.M. (1989) Pretreatment of cane bagasse with alkaline hydrogen peroxide for enzymatic hydrolysis of cellulose and ethanol fermentation. *J. Environ. Sci. Heal. B*, 24(4), 421–433.

Béguin, P. and Aubert, J.-P. (1994) The biological degradation of cellulose. *FEMS Microbiol. Rev.*, 13, 25–58.

Blanchette, R.A. (1991) Delignification by wood-decay fungi. *Annu. Rev. Phytopathol.*, 29, 381–398.

Boominathan, K. and Reddy, C.A. (1992) cAMP-mediated differential regulation of lignin peroxidase and manganese-dependent peroxidase production in the white-rot basidiomycete Phanerochaete chrysosporium. *P. Natl. Acad. Sci. (USA)*, 89(12), 5586–5590.

Brennan, A.H., Hoagland, W., and Schell, D.J. (1986) High temperature acid hydrolysis of biomass using an engineering-scale plug flow reactor: Result of low solids testing. *Biotechnol. Bioeng. Symp.*, 17, 53–70.

Briggs, D.E., Boulton, C.A., Brookes, P.A., and Stevens, R. (2004) *Brewing Science and Practice*. CRC Press LLC, Boca Raton, FL.

Cadoche, L. and López, G.D. (1989) Assessment of size reduction as a preliminary step in the production of ethanol from lignocellulosic wastes. *Biol. Wastes*, 30, 153–157.

Cahela, D.R., Lee, Y.Y., and Chambers, R.P. (1983) Modeling of percolation process in hemicellulose hydrolysis. *Biotechnol. Bioeng.*, 25, 3–17.

Cheung, S.W. and Anderson, B.C. (1997) Laboratory investigation of ethanol production from municipal primary wastewater. *Bioresource Technol.*, 59, 81–96.

Chum, H.L., Johnsoon, D.K., and Black, S. (1988) Organosolv pretreatment for enzymatic hydrolysis of poplars: 1. enzyme hydrolysis of cellulosic residues. *Biotechnol. Bioeng.*, 31, 643–649.

Clark, T.A. and Mackie, K.L. (1987) Steam explosion of the soft-wood *Pinus radiata* with sulphur dioxide addition. I. Process optimization. *J. Wood Chem. Technol.*, 7, 373–403.

Converse, A.O., Kwarteng, I.K., Grethlein, H.E., and Ooshima, H. (1989) Kinetics of thermochemical pretreatment of lignocellulosic materials. *Appl. Biochem. Biotechnol.*, 20/21, 63–78.

Dale, B.E. and Moreira, M.J. (1982) A freeze-explosion technique for increasing cellulose hydrolysis. *Biotechnol. Bioeng. Symp.*, 12, 31–43.

Dale, B.E., Henk, L.L., and Shiang, M. (1984) Fermentation of lignocellulosic materials treated by ammonia freeze explosion. *Dev. Ind. Microbiol.*, 26.

Duff, S.J.B. and Murray, W.D. (1996) Bioconversion of forest products industry waste cellulosics to fuel ethanol: A review. *Bioresource Technol.*, 55, 1–33.

Esteghlalian, A., Hashimoto, A.G., Fenske, J.J., and Penner, M.H. (1997) Modeling and optimization of the dilute-sulfuric-acid pretreatment of corn stover, poplar and switchgrass. *Bioresource Technol.*, 59, 129–136.

Fan, L.T., Gharpuray, M.M., and Lee, Y.-H. (1987) Cellulose hydrolysis. *Biotechnology Monographs*, Springer-Verlag, Berlin, Heidelberg.

Gregg, D.J. and Saddler, J.N. (1996) Factors affecting cellulose hydrolysis and the potential of enzyme recycle to enhance the efficiency of an integrated wood to ethanol process. *Biotechnol. Bioeng.*, 51, 375–383.

Grous, W.R., Converse, A.O., and Grethlein, H.E. (1986) Effect of steam explosion pretreatment on pore size and enzymatic hydrolysis of poplar. *Enzyme Microb. Tech.*, 8, 274–280.

Hatakka, A.I. (1983) Pretreatment of wheat straw by white-rot fungi for enzymatic saccharification of cellulose. *Appl. Microbiol. Biot.* 18, 350–357.

Holtzapple, M.T., Jun, J.-H., Ashok, G., Patibandla, S.L., and Dale, B.E. (1990). Ammonia Fiber Explosion (AFEX) Pretreatment of Lignocellulosic Wastes. American Institute of Chemical Engineers National Meeting, Chicago, IL.

Holtzapple, M.T., Jun, J.-H., Ashok, G., Patibandla, S.L., and Dale, B.E. (1991) The ammonia freeze explosion (AFEX) process: A practical lignocellulose pretreatment. *Appl. Biochem. Biotech.*, 28/29, 59–74.

Holtzapple, M.T., Lundeen, J.E., and Sturgis, R. (1992) Pretreatment of lignocellulosic municipal solid waste by ammonia fiber explosion (AFEX). *Appl. Biochem. Biotech.*, 34/35, 5–21.

Huang, X.L. and Penner, M.H. (1991) Apparent substrate inhibition of the Trichoderma reesei cellulase system. *J. Agric. Food Chem.* 39, 2096–2100.

Iyer, P.V., Wu, Z.-W., Kim, S.B., and Lee, Y.Y. (1996) Ammonia recycled percolation process for pretreatment of herbaceous biomass. *Appl. Biochem. Biotech.*, 57/58, 121–132.

Jia, S., Li, S., and Wu, G. (2004) *Ethanol Technology-New Edition*. Chemical Industry Press, Beijing.

Kirk, T.K., and Farrell, R.L. (1987) Enzymatic combustion: The microbial degradation of lignin. *Annu. Rev. Microbiol.* 41, 465–505.

Mackie, K.L., Brownell, H.H., West, K.L., and Saddler, J.N. (1985) Effect of sulphur dioxide and sulphuric acid on steam explosion of aspenwood. *J. Wood Chem. Technol.*, 5, 405–425.

McMillan, J.D. (1994) Pretreatment of lignocellulosic biomass. *Enzymatic Conversion of Biomass for Fuels Production*, American Chemical Society, Washington, DC, 292–324.

Mes-Hartree, M., Dale, B.E., and Craig, W.K. (1988) Comparison of steam and ammonia pretreatment for enzymatic hydrolysis of cellulose. *Appl. Microbiol. Biotech.*, 29, 462–468.

Millet, M.A., Baker, A.J., and Scatter, L.D. (1976) Physical and chemical pretreatment for enhancing cellulose saccharification. *Biotech. Bioeng. Symp.*, 6, 125–153.

Morjanoff, P.J. and Gray, P.P. (1987) Optimization of steam explosion as method for increasing susceptibility of sugarcane bagasse to enzymatic saccharification. *Biotechnol. Bioeng.*, 29, 733–741.

Penner, M.H. and Liaw, E.-T. (1994) Kinetic consequences of high ratios of substrate to enzyme saccharification systems based on Trichoderma cellulase. In: *Enzymatic Conversion of Biomass for Fuels Production*, Himmel, M.E., Baker, J.O., Overend, R.P. (Eds.), American Chemical Society, Washington, DC, pp. 363–371.

Reshamwala, S., Shawky, B.T., and Dale, B.E. (1995) Ethanol production from enzymatic hydrolysates of AFEX-treated coastal bermudagrass and switchgrass. *Appl. Biochem. Biotech.*, 51/52, 43–55.

Sarkanen, K.V. (1980) Acid-catalyzed delignification of lignocellulosics in organic solvents. *Prog. Biomass Convers.*, 2, 127–144.

Schurz, J. (1978) *Bioconversion of Cellulosic Substances into Energy Chemicals and Microbial Protein Symposium Proceedings*. BERC IIT Delhi, New Delhi.

Sun, Y. and Cheng, J. (2002) Hydrolysis of lignocellulosic materials for ethanol production: A review. *Bioresource Technol.*, 83, 1–11.

Sun, Y. and Cheng, J. (2004) Enzymatic hydrolysis of rye straw and bermudagrass using cellulases supplemented with β-glucosidase. *T. ASAE.* 47(1), 343–349.

Tengerdy, R.P. and Nagy, J.G. (1988) Increasing the feed value of forestry waste by ammonia freeze explosion treatment. *Biol. Wastes*, 25, 149–153.

Thring, R.W., Chorent, E., and Overend, R. (1990) Recovery of a solvolytic lignin: Effects of spent liquor/acid volume ration, acid concentration and temperature. *Biomass*, 23, 289–305.

Vidal, P.F. and Molinier, J. (1988) Ozonolysis of lignin–improvement of *in vitro* digestibility of poplar sawdust. *Biomass*, 16, 1–17.

Vlasenko, E. Yu., Ding, H., Labavitch, J.M., and Shoemaker, S.P. (1997) Enzymatic hydrolysis of pretreated rice straw. *Bioresource Technol.*, 59, 109–119.

Waldner, R., Leisola, M.S.A., Sun, Y., Cheng, J., and Fiechter, A. (1988) Comparison of ligninolytic activities of selected fungi. *Appl. Microbiol. Biot.*, 29, 400–407.

Wang, Z., Keshwani, D.R., Redding, A.P., and Cheng, J.J. (2008) Alkaline Pretreatment of Coastal Bermudagrass for Bioethanol Production. *Proceedings of the ASABE Annual International Meeting*, June 29–July 2, 2008, Providence, RI (Paper No. 084013).

Weast, R.C. (1972) *Handbook of Chemistry and Physics*. 53rd edn. CRC Press, Cleveland, OH.

Weil, J., Sarikaya, A., and Rau, S.-H. (1997) Pretreatment of yellow poplar sawdust by pressure cooking in water. *Appl. Biochem. Biotech.*, 68, 21–40.

Wu, Z. and Lee, Y.Y. (1997) Inhibition of the enzymatic hydrolysis of cellulose by ethanol. *Biotech. Lett.*, 19, 977–979.

Xu, J., Cheng, J.J., Sharma-Shivappa, R.R., and Burns, J.C. (2008) Lime Pretreatment of Switchgrass for Bioethanol Production. *Proceedings of the ASABE Annual International Meeting*, June 29–July 2, 2008, Providence, Rhode Island (Paper No. 083998).

Zheng, Y.Z., Lin, H.M., and Tsao, G.T. (1998) Pretreatment for cellulose hydrolysis by carbon dioxide explosion. *Biotechnol. Prog.*, 14, 890–896.

8

Biological Process for Butanol Production

Jian-Hang Zhu and Fangxiao Yang

CONTENTS

8.1 History of Butanol Production and Its Application 272
    8.1.1 Butanol as Fuel 272
    8.1.2 History of Biobutanol Production 273
    8.1.3 Butanol as a Commodity Chemical..275
8.2 Feedstocks for Biobutanol Production...279
    8.2.1 Starch 280
    8.2.2 Molasses 280
    8.2.3 Lignocellulose 280
8.3 Upstream Processes for Feedstocks 281
8.4 Microbial Cultures...282
    8.4.1 Metabolism and Pathway 283
    8.4.2 Enzymes and Genes Involved in Biobutanol Fermentation 286
        8.4.2.1 Thiolase 286
        8.4.2.2 3-Hydroxybutyryl-CoA Dehydrogenase....................287
        8.4.2.3 Crotonase 287
        8.4.2.4 Butyryl-CoA Dehydrogenase....................................288
        8.4.2.5 Acetoacetate: Acetate/Butyrate CoA-Transferase.......288
        8.4.2.6 Acetoacetate Decarboxylase 289
        8.4.2.7 Aldehyde Dehydrogenase 289
        8.4.2.8 Alcohol Dehydrogenase..290
        8.4.2.9 Operons for Biobutanol Production 291
    8.4.3 Genomes of Biobutanol-Producing Clostridia........................293
8.5 Fermentation Technology 294
    8.5.1 Batch Process 295
    8.5.2 Fed-Batch Fermentation 297
    8.5.3 Continuous Fermentation..299
    8.5.4 Immobilized Cell Bioreactors...302
8.6 Biobutanol Recovery..302
    8.6.1 Distillation 304
    8.6.2 Gas Stripping...304
    8.6.3 Liquid–Liquid Extraction 304
    8.6.4 RSEMC Technology for Solvent Extraction..............................306

           8.6.4.1   Solvent Toxicity...308
           8.6.4.2   Butanol Partition 312
           8.6.4.3   Energy Consumption Calculation..............................314
           8.6.4.4   Extraction Kinetics..315
     8.6.5   Pervaporation 316
     8.6.6   Reverse Osmosis 318
     8.6.7   Perstraction 318
     8.6.8   Integration Process for Biobutanol Fermentation 319
8.7   Development Potential for Biobutanol Production..............................320
     8.7.1   Issues for Biological Process of Biobutanol Production 320
     8.7.2   New Technologies Being Developed 321
     8.7.3   Biobutanol Fermentation Waste and By-Products 323
8.8   Commercialization and Economic Evaluation 323
     8.8.1   Technology...323
     8.8.2   Economics 324
8.9   Problems..329
Acknowledgment...330
References..331

8.1 History of Butanol Production and Its Application

8.1.1 Butanol as Fuel

Butanol (biobutanol, if derived from biosources), an aliphatic saturated alcohol, is an industrial commodity, currently produced from petrochemical feedstock, with a worldwide capacity of 5 million metric tons and an average selling price of $0.9/kg ($3.4/gal, as of December 2006). Its primary industrial use is as a solvent. Although butanol is not currently used as a biofuel, it has a number of properties that make it extremely attractive. A comparison of critical fuel parameters of butanol with gasoline and ethanol clearly shows that it indeed represents a better alternative over ethanol (see Table 8.1).

TABLE 8.1

Comparison of Fuel Characteristics of Gasoline and Bioalcohols

Fuel	Energy Density	Air–Fuel Ratio	Heat of Vaporization	Research Octane Test	Motor Octane Test
Gasoline	32.0 MJ/L	14.6	0.36 MJ/kg	91–99	81–89
Butanol	29.2 MJ/L	11.2	0.43 MJ/kg	96	78
Ethanol	19.6 MJ/L	9.0	0.92 MJ/kg	130	96
Methanol	16.0 MJ/L	6.5	1.20 MJ/kg	136	104

Source: Yang, F.X. and Biermann, M., Butanol production with Montana wheat straw (Montana Board of Research and Commercialization Technology MBRCT), Contract No. 08-56, 2007–2009, 2007.

From the fuel properties shown in Table 8.1, butanol appears superior to ethanol due to the similarity of its fuel characteristics with gasoline, non-corrosive nature, and unrestricted miscibility with conventional fossil fuel. Butanol can be easily integrated into the existing internal combustion engine by blending with conventional fuels like gasoline or diesel, without any modification. Moreover, it offers similar energy density to gasoline, has low vapor pressure, and is easy to store, handle, and transport via pipelines. Compared to other liquid fuels such as gasoline and ethanol, biobutanol has many advantages as follows:

1. It contains 25% more energy (Btus) than ethanol, since it has four carbons versus two in ethanol.
2. It can be produced biologically from the materials such as corn, grasses, leaves, and agricultural residues.
3. It is safer to handle due to its lower evaporation when compared to gasoline and ethanol.
4. It can be used in the existing engines in pure form or blended in any ratio with gasoline, while ethanol can be blended only up to 85%.
5. It can be transported through the existing pipelines due to its low corrosive properties.
6. It is a green fuel without SO_x, NO_x, or carbon monoxide (CO) emissions when burned in internal combustion engines.

Use of butanol as automobile fuel has only been reported recently. In 2005, David Ramey drove a passenger car fueled with pure biobutanol across the United States, showing the feasibility of using biobutanol as transportation fuel (Environmental Energy Inc., 2005). Although the volume of butanol consumption was approximately 9% higher than that of gasoline, emissions of CO, hydrocarbons, and NO_x decreased enormously.

8.1.2 History of Biobutanol Production

The history of biological butanol production dates back to the nineteenth century. As early as in 1862, the well-known French scientist Louis Pasteur reported this alcohol to be a fermentation product of "Vibrion butyrique," which did not represent a pure culture but presumably contained *Clostridia* of the *Clostridium butyricum* or *Clostridium acetobutylicum* type. The first pure culture of such an organism was probably isolated by Albert Fitz in Strasbourg, who published a series of papers on *Bacillus butylicus*. This anaerobic, spore-forming bacterium fermented glycerol, mannitol, and sucrose to butanol, butyrate, CO_2, and hydrogen (plus small amounts of acetate, ethanol, lactate, and propanediol), produced typical clostridial forms, and was killed by higher concentrations of butanol. These pioneering studies paved the way for isolation of further solvent-forming bacteria around the turn to the twentieth century.

Numerous microbiologists were involved, among them the famous "Martinus Beijerinck" and "Sergei Winogradsky." At that time, the genus name *Clostridium* was a mere morphological description, meaning small spindle. Thus, it is not surprising that a variety of different designations were used, including *Granulobacter saccharobutyricum*, *Amylobacter butylicus*, and *Bacillus orthobutylicus*. Meanwhile, it is known that mostly *Clostridia* (strictly or moderately anaerobic, spore-forming, Gram-positive bacteria, incapable of dissimilatory sulfate reduction) perform butanol fermentation. Well-characterized and extensively used and investigated butanol-producing bacteria are *C. acetobutylicum*, *C. beijerinckii*, *C. saccharoperbutylacetonicum*, and *C. saccharobutylicum*. A few other microorganisms that form butanol as a major fermentation product include *Butyribacterium methylotrophicum* belonging to the taxonomic kingdom bacteria and *Hyperthermus butylicus* that is an archaeon.

Microbial solvent formation also made its way to industrial use. The so-called ABE (acetone–butanol–ethanol) fermentation has become the second largest industrial biotechnological process in volume, next to ethanol fermentation. The story dates back to the beginning of the twentieth century, when a number of scientists tried to produce rubber artificially (Durre and Bahl, 1996; Gabriel, 1928; Jones and Woods, 1986). The idea was to use isoprene for polymerization, which could be produced from isoamyl alcohol. An alternative was butadiene, derived from butanol (Durre, 2007). An English company, Strange and Graham, had employed Fernbach and Schoen from the Institute Pasteur in Paris and Perkins and Weizmann from Manchester University to work on that project. In 1911, Fernbach and Strange applied for two patents describing fermentation of various organic materials to mainly amyl, butyl, and ethyl alcohols by a mixed culture as well as recovery of the products. Weizmann left the research group in 1912, but continued to work in this field. In his memoirs, he described his success of isolating a new culture, producing large amounts of a liquid that smelled like amyl alcohol. In fact, it turned out to be a mixture of acetone and butanol. Ironically, the price of natural rubber started to decrease during those years, due to increasing production from plantations in Asia. However, the disregarded by-product acetone helped the ABE fermentation process to an industrial breakthrough. In November 1914, shortly after the start of World War I, two British cruisers (Good Hope, flagship of Admiral Sir Christopher Cradock, and Monmouth) were sunk off Coronel in Chile. They were part of a fleet getting into combat with the German East Asia squadron, commanded by Admiral Maximilian von Spee. It was reported that the British ammunition suffered from bad preparation, resulting in shells that plopped harmlessly into the water far short of enemy ships. For preparation of cordite, the propellant of cartridges and shells, acetone was required in large quantities, and there was a great shortage of chemically produced acetone in England at the outbreak of the war. Weizmann applied for a patent in 1915 and obtained a patent in the United States in 1919 (Weizmann, 1919), describing production of acetone and alcohols from starchy material by a mixed culture, mainly consisting of what later became known as *C. acetobutylicum*. Advantages over the Fernbach strain were higher productivity with respect to acetone and the ability to use a variety of starch-containing substances. This process guaranteed a constant supply of acetone in Great Britain, Canada, and the United States during the war. Weizmann denied personal honors from the British government, but, being a member of the Zionist movement, made clear that he favored to repatriate Jewish people in Palestine. Without doubt, this influenced the British Foreign Secretary and led eventually to the famous Balfour declaration of 1917. Several years later, Chaim Weizmann became the first president of the newly founded State of Israel.

There was almost no use of butanol during the war, but the substance had been stored in immense vats (one later used as a swimming pool). After the war, many ABE facilities had been closed because of the sharp decrease of acetone demand. Upon introduction of prohibition in the United States in 1920, there was an immediate shortage of amyl alcohol, a by-product of alcoholic fermentation. Amyl alcohol had been used to produce amyl acetate, a solvent for quick-drying lacquers, needed by the growing automobile industry in large amounts. Butanol was proven to be a perfect replacement, which started the production of butyl acetate. As the consequence, all ABE production plants closed after the armistice were reopened and new ones were built. Many countries used the fermentation to fulfill their industrial needs of butanol, and up to 1950 approximately two thirds of world's butanol supply stemmed from the biological process. The largest plant was located in Peoria, IL, and had a capacity of 96 fermenters with a volume of 50,000 gal or 189,000 L each. Facilities in South Africa, the former Soviet Union (using lignocellulose hydrolysates as a substrate), and China (using continuous culture technology) were smaller, but operated until the 1980s (South Africa and the former Soviet Union) and 2004 (China). The decline of the butanol fermentation was caused by increasing substrate (molasses) costs and much cheaper butanol from crude oil refinery processes. The research and development on butanol production via biological processes were still ongoing until mid-1990s (see Table 8.2—a list of patent application) (EEI, 2005). Research activity on ABE has been resumed in the wake of crude oil price skyrocket in the recent years.

Butanol was also temporarily produced by the aldol condensation of acetaldehyde, which can be obtained by oxidation of ethanol, to crotonaldehyde followed by hydrogenation. In the late 1940s, the production of butanol via the hydroformulation synthesis using propene and syngas as feedstock took the place of the ABE and aldol condensation process. In this process, butyraldehyde, a primary product of the hydroformulation reaction of propene, is converted via hydrogenation into butanol (Figure 8.1).

8.1.3 Butanol as a Commodity Chemical

Until 2005, butanol was mainly considered as a chemical precursor for the production of acrylate and methacrylate esters, glycol ethers, butyl acetate,

TABLE 8.2

Biobutanol Fermentation Patents

Name	Year	Title of the Patent	Patent No. (United States)
A. Fernbach	1912	Manufacture and production of a nutrient medium for use in fermentation processes	1,044,446
A. Fernbach	1912	Manufacture and production of nutrient medium for use in fermentation processes	1,044,447
J. Scheckenbach	1914	Process of manufacturing fuel-oil	1,118,238
J. Northrup	1919	Process for the production of acetone	1,293,172
C. Weizmann	1919	Production of acetone and alcohol by bacteriological processes	1,315,585
C. Weizmann	1920	Fermentation process for the production of acetone and butyl alcohol	1,329,214
G. Horton	1922	Process of producing butyl alcohol and acetone by fermentation	1,427,595
C. Weizmann	1922	Fermentation process for the production of acetone and butyl alcohol	1,437,697
G. Robinson	1924	Production of butyl alcohol and acetone by the fermentation of molasses	1,510,526
G. Freiberg	1925	Process for producing acetone and butyl alcohol	1,537,597
L. Waters	1925	Fermentation process	1,546,694
E. Ricard	1925	Manufacture of acetone and butyl alcohol by fermentation	1,550,746
E. Halford	1925	Process for the manufacture of alcohols and acetone	1,550,928
D. Legg	1926	Process for the production of butyl alcohol and acetone	1,582,408
E. Pike	1928	Process of producing acetone and butyl alcohol by fermentation	1,655,435
E. R. Weyer	1928	Fermentation process	1,696,022
A. Fernbach	1932	Acetono-butylic fermentation process	1,854,895
F. Gerretsen	1932	Biological production of butyl alcohol and acetone	1,858,808
C. Haner	1932	Process of producing butyl alcohol and acetone and ethyl alcohol	1,885,096
H. Brougham	1933	Production of butyl alcohol and acetone by fermentation	1,898,961
A. Izsak	1933	Process of producing butyl alcohol	1,908,361
D. Legg	1933	Production of butyl alcohol and acetone by fermentation	1,913,164
H. Wertheim	1933	Process for the simultaneous production of butyl alcohol and acetone by fermentation	1,917,676
H. Wertheim	1933	Process for producing butyl alcohol and acetone by fermentation	1,917,677
H. Hutchinson	1933	Production of butyl alcohol and acetone by fermentation	1,928,379

TABLE 8.2 (continued)

Biobutanol Fermentation Patents

Name	Year	Title of the Patent	Patent No. (United States)
J. Loughlin	1935	Production of butyl alcohol and acetone by fermentation	1,992,921
A. Izsak	1935	Process for producing butyl alcohol and acetone	2,016,112
W. McCutchan	1935	Production of butyl alcohol by fermentation	2,023,087
D. Legg	1935	Production of butyl alcohol by fermentation	2,023,368
H. Stiles	1935	Butyl-acetonic fermentation process	2,023,374
C. Arzberger	1936	Process for the production of butyl alcohol by fermentation	2,050,219
D. Legg	1936	Process for controlling hydrogen ion concentration of butyl alcohol fermentation mashes	2,063,449
J. Woodruff	1937	Process of producing butyl alcohol	2,089,522
D. Legg	1937	Process of producing butyl alcohol	2,089,562
J. Loughlin	1937	Manufacture of butyl alcohol, acetone, and isopropyl alcohol	2,096,377
H. Stiles	1937	Fermentation of amylaceous materials	2,098,199
H. Stiles	1937	Butyl-acetonic fermentation process	2,098,200
E. McCoy	1938	Production of butyl alcohol and acetone by fermentation	2,110,109
D. Legg	1938	Butyl alcohol fermentation	2,132,358
H. Hall	1939	Fermentation process for solvent manufacture	2,147,487
A. Freising	1939	Production of biobutanol and acetone by the fermentation of whey	2,166,047
F. Hildebrandt	1939	Fermentation process	2,169,244
F. Hildebrandt	1939	Butyl acetone fermentation process	2,169,246
N. Cunningham	1978	Apparatus for production alcohol from grains of starch materials	4,108,052
S. Levy	1984	Continuous process for producing N-biobutanol employing anaerobic fermentation	4,424,275
R. Heady	1985	Production of biobutanol by a continuous fermentation process	4,520,104
S. Bergstrom	1985	Production of biobutanol by fermentation in the presence of cocultures of *Clostridium*	4,539,293
C. Lemme	1985	Strain of *C. acetobutylicum* and process for its preparation	4,521,516
R. Datta	1985	Production of biobutanol by fermentation in the presence of carbon monoxide	4,560,658
R. Datta	1987	Utilization of zylan and corn fiber for direct fermentation by *C. acetobutylicum*	4,649,112

(continued)

TABLE 8.2 (continued)

Biobutanol Fermentation Patents

Name	Year	Title of the Patent	Patent No. (United States)
G. Husted	1988	Method for producing biobutanol by fermentation	4,777,135
M. Hermann	1988	Production of *C. acetobutylicum* mutants of high biobutanol and acetone productivity, the resultant mutants, and the use of these mutants in the joint production of biobutanol and acetone	4,757,010
A. D. Telles	1989	Continuous process of optimized fermentation for the production of alcohol	4,889,805
M. Czyko	1989	Continuous process for preparing organic acids by fermentation	4,882,277
J. H. Hsieh	1990	Continuous fermentation process for aromatic hydrocarbon bioconversions	4,968,612
R. Driscoll	1991	Process for the separation of biobutanol and butoxyacetaldehyde	4,986,885
D. Tedder	1991	Alcohol recovery by continuous fermentation	5,036,005
D. Glassner	1991	Process for the fermentation production of acetone, biobutanol, and ethanol	5,063,156
M. Jain	1993	Mutant strain of *C. acetobutylicum* and process for making biobutanol	5,192,673
L. Robbins	1994	Method for removing dissolved immiscible organics from an aqueous medium at ambient temperatures	5,294,303

Source: Adapted from Environmental Energy Inc., http://www.butanol.com/, 2005.

butylamines, and amino resins. Their use is manifold: production of adhesives/scalants, alkaloids, antibiotics, camphor, deicing fluid, dental products, detergents, elastomers, electronics, emulsifiers, eye makeup, fibers, flocculants, flotation aids (e.g., butyl xanthate), hard-surface cleaners, hormones and vitamins, hydraulic and brake fluids, industrial coatings, lipsticks, nail care products, paints, paint thinners, perfumes, pesticides, plastics, printing ink, resins, safety glass, shaving and personal hygiene products, surface coatings, super absorbents, synthetic fruit flavoring, textiles, mobile phases in paper and thin-layer chromatography, oil additive, and leather and paper polishers (Durre, 2007).

Recently, oil price increase resulted in a revival of research activities on a number of fermentations for biofuels, including butanol, ethanol, and 2,3-butanediol, in order to reduce the dependence on petroleum. DuPont and BP reopened their industrial ABE fermentation facilities in Great Britain in 2007, to produce butanol for use as a biofuel in the United Kingdom (DuPont, 2006). Fermentation plants are also restarted in China, and new ones built in

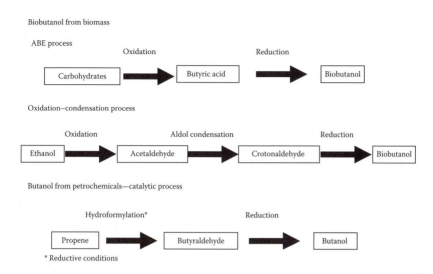

Biobutanol from biomass

ABE process

Oxidation

Reduction

Carbohydrates → Butyric acid → Biobutanol

Oxidation–condensation process

Oxidation

Aldol condensation

Reduction

Ethanol → Acetaldehyde → Crotonaldehyde → Biobutanol

Butanol from petrochemicals—catalytic process

Hydroformylation*

Reduction

Propene → Butyraldehyde → Butanol

* Reductive conditions

FIGURE 8.1
Past and current industrial processes for butanol production.

Brazil. Research projects in butanol fermentation are ongoing in industries as well as universities. Bright prospects for the biobutanol production are really coming toward us (Ezeji et al., 2004).

8.2 Feedstocks for Biobutanol Production

The primary raw materials used in ABE fermentation were corn and molasses. Corn was primarily used during the era of the Weizmann process, which was based on *C. acetobutylicum*, whereas molasses was used when other bacteria replaced *C. acetobutylicum* in the industrial fermentation. Besides corn and molasses, other starchy substrates were also used in industrial ABE fermentation outside North America. For example, the raw materials for biobutanol production in Taiwan between 1947 and 1957 included sweet potato, casava, and wheat starch. Potato waste, including industrial potato residues, low-grade potatoes, and spoiled potatoes, and other waste materials such as cheese whey, hydrolyzed lignocellulosic and hemicellulosic materials, palm oil mill effluent, apple pomace, and soy molasses can also be used as substrates for butanol production (Jones and Woods, 1989).

In the past few years, several carbon sources such as glucose, corn starch, molasses, and soy molasses have been utilized for laboratory biobutanol production. As substrate cost has a dramatic influence on butanol price, the use of agricultural residues has been recently focused for biobutanol production with *C. beijerinckii* BA101 (Qureshi and Blaschek, 1999). In particular, research has been focused on the use of corn stover, corn fiber, and fiber-rich distillers dried grains and solubles (DDGS) as substrates for butanol fermentation.

8.2.1 Starch

Starch is widely used in ABE fermentation due to its abundance and cheapness. Moreover, starch can be hydrolyzed stepwise by the excretive enzymes produced by the strains to overcome the disadvantage of glucose rapid assimilation. Although starch was a major substrate for industrial solvent production, little is known about the amylolytic system in *C. acetobutylicum* or *C. beijerinckii*, and a consistent picture of the regulation of starch hydrolysis and metabolism is yet to emerge. In recent years, with grain prices soaring, the food crisis worldwide began to appear. In order to avoid robbing food from hungry people and increasing cost due to the price rise of starch, alternative feedstock has been sought for biobutanol produciton (Qureshi, 2001; Walton and Martin, 1979).

8.2.2 Molasses

Brazil, a South American country with the majority of its territory in tropical climate, is rich in the sugarcane production. Molasses, a by-product of cane mill, is the mother liquor after sugar crystallization in the process of sugar refinery. Molasses was the most widely used raw material for ABE fermentation. Some saccharolytic microorganisms are capable of fermenting molasses directly under suitable conditions and producing almost full yields of the solvents. Molasses contains sugars (50%–70%), inorganic salts (some of them are added to aid in the recovery of crystalline sucrose), and vitamins, which makes it an excellent raw material for fermentation (Mitchell, 1998). As a by-product of cane mill, molasses is also reasonable in cost. Molasses can also be obtained from the sugar refinery of sugar beet. Generally, molasses is inadequate in nitrogen and phosphorus content for efficient fermentation. Therefore, nitrogen and phosphorus sources need to be added into the medium for ABE fermentation. In industrial production, ammonia is commonly used to adjust the pH value and as nitrogen source. Superphosphate and P_2O_5 were usually added as the phosphorus source. Batch-to-batch variations in the composition of molasses could cause wide variation in ABE yields and nutrient requirements for fermentation.

8.2.3 Lignocellulose

Lignocellulosic materials have a wide range of sources, including agricultural residues (e.g., corn stover and wheat straw), wood and forestry residues (e.g., sawdust, tinnings, and mill wastes), energy crops, and some industrial wastes. Lignocellulosic materials provide a low-cost and abundant resource for the production of biofuels and chemicals. Lignocellulose is composed of three major components: cellulose, hemicellulose, and lignin. Among them, the cellulose and hemicellulose, through the microbiological or chemical methods, may be transformed into sugars, which could be used in bioethanol or biobutanol production (Chandra et al., 2007; Fan and Beardmore, 1981; Luukkonen et al., 2001).

8.3 Upstream Processes for Feedstocks

The raw materials mentioned previously may require upstream processing, depending upon the starting materials and the processes being used for fermentation. Upstream processes could include both pretreatment and hydrolysis. Molasses requires dilution and removal of insoluble solids; whey permeate requires concentration, possibly by reverse osmosis (RO); corn requires removing bran, germ, and oil, milling; sieving; centrifugation; and saccharification. Table 8.3 shows the upstream process steps that may be required for the production of butanol from various feedstocks. Cellulosic

TABLE 8.3

Summary of Upstream Processes for Various Feedstocks

Corn
Wet milling (germ removal): steeping, grinding, fine grinding, sieving, and centrifugation
Dry milling: grinding, mash cooking

Molasses
Dilution and sediment removal by centrifugation

Whey Permeate
For batch fermentation: no upstream processing
For integrated fermentation and product removal process: lactose concentration possibly by reverse osmosis

Soy Molasses
Dilution, removal of sediment, and supplementation with additional carbon source

Corn Stover
Steam explosion, acid or enzymatic hydrolysis, and possible removal of inhibitors

Potato
Cooking, mashing, sediment removal, and dilution

Agricultural Residues
Steam explosion, hydrolysis by dilute acid or enzymatic methods

biomass, a promising feedstock for biobutanol production, requires pre-treatment and hydrolysis either by acids or by enzymes. The acid hydrolysis method is comparatively faster than the enzymatic methods. However, acid hydrolysate contains sugar degradation products that may inhibit the fermentation. Removal of these inhibitors is essential for good productivity in the fermentation.

8.4 Microbial Cultures

In a typical ABE fermentation, the microorganisms first produce a mixture of organic acids, for example, butyric, lactic acid, and acetic acid. This metabolic phase is called acidogenesis (see Figure 8.2).

When the pH of the culture reaches a value below 5 due to the increased organic acid production, the microorganisms shift its physiology to a different pathway to produce acetone, butanol, and ethanol. This metabolic phase has been dubbed as solventogenesis.

There are a number of cultures capable of producing significant amounts of solvents from various carbohydrates. A systematic study of all these cultures showed that these cultures belong to strains of four *Clostridium* species,

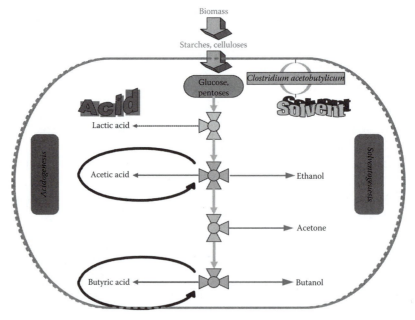

FIGURE 8.2
Cartoon of metabolic pathways for *Clostridia* fermentation.

TABLE 8.4

Three Groups of Clostridia with Different Product Patterns

Product Pattern	Microbial Cultures
Butanol, acetone, and ethanol	*C. acetobutylicum, C. beijerinckii* (some strains), *C. puniceum, C. pasteurianum, C. saccharoperbutylacetonicum, C. saccharobutylicum*
Butanol, isopropanol, acetone, and ethanol	*C. aurantibutyricum, C. beijerinckii* (some strains), *C. toanum*
Butanol and ethanol	*C. tetanomorphum, C. thermosaccharolyticum*

Source: Adapted from Rogers, P., *Prokaryotes*, 1, 511, 2006.

that is, *C. acetobutylicum, C. beijerinckii, C. saccharobutylicum,* and *C. saccharoperbutylacetonicum,* via the analysis of DNA–DNA reassociation as well as the physiological traits. Several other characterized species of *Clostridium* could also produce butanol but in lower concentrations. According to the pattern of their products, they can be divided into three groups (Table 8.4) (Rogers, 2006). Among these strains, *C. acetobutylicum* and *C. beijerinckii* are the two primary species that have been investigated for the production of butanol. The strain used for the production of solvents in large-scale plant in South Africa was *C. acetobutylicum*. Unfortunately, the plant was shut down due to molasses shortages. The other strains used in the past include *C. acetobutylicum* ATCC 824, *C. acetobutylicum* NRRL B643, *C. acetobutylicum* B18, *C. beijerinckii* 8052, *C. beijerinckii* BA101, and *C. beijerinckii* LMD27.6 (Qureshi, 2005).

Significant research efforts have been made to genetically improve some of the existing cultures or isolate new solventogenic strains. Montoya et al. (2001) isolated strains that produced 24.2–29.1 g/L total solvents. *C. beijerinckii* BA101 is a strain developed from *C. beijerinckii* 8052. *C. berijerinckii* BA101 has been reported to accumulate 23.5–33.0 g/L solvents. Attempts have been made to enhance solvent production capability of *C. acetobutylicum* ATCC 824. The newly developed strain was reported to produce 25.6–30.0 g/L solvents (Qureshi, 2005). *C. acetobutylicum* can use a wide range of carbohydrates from pentose to starch for growth and is strongly proteolytic, while the medium pH should be controlled within a narrow range. Moreover, as an obligate anaerobe, *C. acetobutylicum* is relatively insensitive to O_2. The oxygen tolerance of *C. acetobutylicum* may have contributed to its commercial success. A number of reports suggest that *C. beijerinckii* might have greater potential for the industrial production of solvents than does the previously sequenced *C. acetobutylcium* ATCC 824 type strain because the former has a broader substrate range and pH optimum for growth and solvent production (Rogers, 2006).

8.4.1 Metabolism and Pathway

There are generally two phases involved in the batch fermentation for butanol production, that is, the acidogenic phase followed by solventogenic phase. In

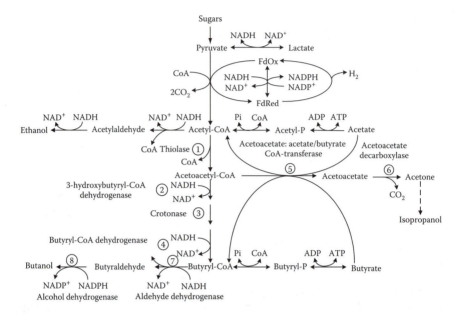

FIGURE 8.3
Pathways of acids and solvents production (the numbers represent the enzymes catalyzing the steps: (1) thiolase; (2) 3-hydroxybutyryl-CoA dehydrogenase; (3) crotonase; (4) butyryl-CoA dehydrogenase; (5) acetoacetate: acetate/butyrate CoA-transferase; (6) acetoacetate decarboxylase; (7) aldehyde dehydrogenase; (8) alcohol dehydrogenase.

addition to butanol, end products of *Clostridium* fermentation include ethanol, acetone, and isopropanol. Their metabolic pathways, including enzymes, have been summarized in Figure 8.3 (Rogers, 2006). Most metabolic data are available for *C. acetobutylicum*, which typically performs butyric acid fermentation, with acetate, butyrate, CO_2, and H_2 as major products (Figure 8.3). Butyrate is often produced about twice as much as acetate. Some ethanol and acetone are formed constitutively. Glucose transported into the cell is mediated by a phosphoenol pyruvate–dependent phosphotransferase system. Hexoses could be further metabolized via glycolysis and pentoses via the pentose phosphate pathway. At the end of the exponential growth of the bacterium, a major metabolic switch takes place in *C. acetobutylicum*. The organism slows down in acid production, and takes up excreted acetate and butyrate and converts them into the solvents, that is, acetone and butanol (approximately twice more butanol than acetone) (Durre, 2007).

The first enzyme, required for solventogenesis, is an acetoacetyl-CoA: acetate/butyrate-CoA transferase (CoA transferase, CtfA/B) that converts reinternalized butyrate and to a lesser extent also acetate into butyryl-CoA and acetyl-CoA. The latter is recycled to acetoacetyl-CoA, the starting point of the reaction. The corresponding enzyme, however, a thiolase (ThlA) is also needed for butyrate formation. Acetoacetate, one of the products of the

CoA transferase action, is converted into acetone and CO_2 by acetoacetate decarboxylase (Adc), a reaction required to "pull" the thermodynamically unfavorable butyryl-CoA formation. This can also be achieved using different substrates, resulting in massively increased butanol production over acetone. It is noteworthy that Adc occurs in two different forms in the cell, possibly indicating a regulation via protein modification. Finally, butyryl-CoA is reduced to butyraldehyde and butanol by several enzymes. AdhE is a bifunctional butyraldehyde/butanol dehydrogenase and catalyzes initiation of solventogenesis, whereas BdhB, a butanol dehydrogenase, takes over later and is responsible for the massive production. BdhA is an alcohol dehydrogenase that uses acetaldehyde and butyraldehyde almost equally well, assuming its physiological role to be just a sink for surplus reducing equivalents and resulting in minor ethanol and butanol formation. AdhE2, another bifunctional butyraldehyde/butanol dehydrogenase that is only formed when *C. acetobutylicum* is growing in reduced substrates (e.g., a mixture of glucose and glycerol), could lead to so-called alcohologenic fermentation (only butanol and ethanol are formed, but no acetone). This reaction can also be induced by adding the redox dye methyl viologen. All of the aforementioned enzymes could be regulated at the level of their genes, some of them by additional mechanisms as well (Durre, 2007).

In principle, there are two different options for the metabolic engineering of butanol-producing microorganisms: (1) the genes required for butanol formation can be cloned into a desired production organism and (2) a butanol producer such as *C. acetobutylicum* can be mutated in a way that it only produces butanol (homobutanol producer) plus some CO_2 and H_2 (Durre, 2007).

The first possibility would be suitable for biotechnology workhorses such as *Escherichia coli*. Since this microorganism does not possess the ability to synthesize C_4 compounds, the respective genes would have to come from *C. acetobutylicum* as well. A problem might become the expression of oxygen-sensitive essential clostridial enzymes, such as *AdhE*. Such an approach also does not take into account tolerance of the envisaged production strain against butanol. As already mentioned before, this alcohol damages membranes and certain membrane proteins. This seems to be less of a problem in *C. acetobutylicum*. There have already been a number of attempts to improve solvent tolerance in this organism (Durre, 2007).

Another approach is the over-expression of cyclopropane fatty acid synthase. Tolerance could indeed be improved, but butanol production was decreased. However, even without such targeted constructs, a number of other mutants have produced butanol in concentrations considered impossible a few years ago. The current best performer (238 mM butanol) is an *ORF5*-negative strain, over-expressing *AdhE*. Thus, it may be wise to start with *C. acetobutylicum* to construct a tailor-made production strain. Acetate and butyrate production could be inactivated by mutations in the phosphotransacetylase and phosphotransbutyrylase genes or the corresponding kinase genes, respectively. Lactate formation can be inhibited by inactivating the lactate dehydrogenase gene, acetone synthesis by knockout of *Adc*, and acetoin production by a mutation in the 2-acetolactate synthase gene. A gene knockout procedure has recently been developed for the *Clostridia* (Chai, 2008). Finally, there is still another approach. A patent from the Fraunhofer-Institute in Stuttgart, Germany, describes butanol formation from butane by means of a specific monooxygenase (Fraunhofer IGB, 2008). Over-expression of the respective gene might result in a different, yet suitable recombinant production strain (Durre, 2007).

8.4.2 Enzymes and Genes Involved in Biobutanol Fermentation

As shown in Figure 8.3, acetyl-CoA is an important branching point in the whole pathway. So the enzymes (enzyme 1–4, 7, and 8) catalyzing the process beginning from acetyl-CoA to butanol will be discussed. In addition, the enzymes concerning acetone production (enzymes 5 and 6) are significant to the butanol production and therefore also discussed.

8.4.2.1 Thiolase

Thiolase (acetyl-CoA: acetyl-CoA C-acetyltransferase, E.C. 2.3.1.19, enzyme 1 in Figure 8.3) catalyzes the condensation of two acetyl-CoA molecules into an acetoacetyl-CoA, which is the precursor for acetone, isopropanol, butanol, and butyrate. Thiolase has been purified and characterized from *Clostridium pasteurianum* and *C. acetobutylicum* ATCC 824. The native enzyme of *C. acetobutylicum* ATCC 824 consists of four 44,000-molecular-weight subunits. Unlike that from *C. pasteurianum*, the thiolase of *C. acetobutylicum* ATCC 824 keeps high relative activity throughout the physiological pH range of 5.5–7.0, indicating that the change in internal pH is not an important factor in regulating the activity of thiolase. In the condensation direction, coenzyme A, ATP, and butyryl-CoA are inhibitors of the enzyme. Its kinetic binding mechanism is Ping-Pong mechanism. The K_m values for acetyl-CoA, sulfhydryl-CoA, and acetoacetyl-CoA are 0.27 mM at 30°C and pH 7.4, 0.0048 and 0.032 mM at 30°C and pH 8.0, respectively. Another difference of this enzyme to other bacterial ones is that this thiolase is relatively stable in the presence of 5, 5′-dithiobis (2-nitrobenzoic acid) (DTNB) (Wiesenborn et al., 1988).

There are two genes of thiolase in the genome of *C. acetobutylicum* ATCC 824 (Rogers, 2006). Gene CAC2873 was located on the chromosome, whereas another gene, CAP0078, was found on the plasmid pSOL1. Thiolase purified from *C. acetobutylicum* ATCC 824 previously is encoded by CAC2873. There are also two thiolase genes, thlA and thlB, in the genome of *C. acetobutylicum* DSM 792. They all have been cloned and sequenced. The region preceding the transcriptional start site of thlA exhibits high similarity to the σ70 consensus promoter sequence of Gram-positive and -negative bacteria. Gene thlA is transcribed to a high level in acid- and solvent-producing cells, but gene thlB shows a very low expression. Gene thlA could be organized as a monocistron, whereas thlB form an operon with two adjacent genes, thlR and thlC. ThlR shows significant homology to regulatory proteins belonging to the TerR/AcrR family of transcriptional regulators. ThlR may be a transcriptional repressor of thlB operon expression. thlA was proved to be involved in the acid and butanol production, but the physiological function of thlB remains unclear and needs to be elucidated further (Winzer et al., 2000).

8.4.2.2 3-Hydroxybutyryl-CoA Dehydrogenase

3-Hydroxybutyryl-CoA dehydrogenase (β-hydroxybutyryl-CoA dehydrogenase, enzyme 2 in Figure 8.3) catalyzes the reduction of acetoacetyl-CoA to 3-hydroxybutyryl-CoA. It has been purified from *Clostridium beijerinckii* NRRL B593 with a native Mr of 213 kDa and a subunit Mr of 30.8 kDa. This enzyme is specific for the (S)-enantiomer of 3-hydroxybutyryl-CoA. Its Michaelis constants for NADH and acetoacetyl-CoA are 8.6 and 14 mM, respectively. And the maximum catalytic velocity is 540 μmol/min/mg for the reduction of acetoacetyl-CoA with the addition of NADH as the co-substrate. Although NADPH could also be added as a co-substrate, the k_{cat}/K_m for the NADPH-linked reaction is much higher than that for the NADH-linked reaction and the NADH-linked activity is less sensitive to changes in pH than the NADPH-linked one. The enzyme can be inhibited by acetoacetyl-coenzyme A at concentrations as low as 20 μM in the presence of 9.5 μM NADH, but this inhibition could be relieved by the increase of NADH concentration, showing that a modulating mechanism for the enzyme exists (Colby and Chen, 1992).

Structural genes coding 3-hydroxybutyryl-CoA dehydrogenase (*hbd*) from three *Clostridium* species have been sequenced. The lengths of deduced polypeptides and their calculated molecular weight are 282 and 30,500 (*C. acetobutylicum* ATCC 824), 281 and 30,167 (*C. beijerinckii* NRRL B593) (AF494018), 282 and 31,435 (*C. saccharobutylicum* NCP262 [formerly *C. acetobutylicum* P262]), respectively. The *hdb* gene of *C. acetobutylicum* ATCC 824 was designated CAC2708 on the annotated genome sequence. In addition, it is part of the *bcs* (butyryl-CoA synthesis) operon (Rogers, 2006).

8.4.2.3 Crotonase

Crotonase (enzyme 3 in Figure 8.3) catalyzes the dehydration of 3-hydroxybutyryl-CoA to form crotonyl-CoA. It has been purified from *C. acetobutylicum*. The native enzyme has a molecular weight of 158,000 ± 3,000 and consists of four 40,000-molecular-weight subunits. The enzyme is specific for short-chain fatty acyl-CoA substrates (C4 and C6) and inhibited by high concentrations of crotonyl-CoA. In addition, it requires a complete coenzyme A thioester substrate for efficient catalysis (Waterson et al., 1972).

Predicted genes for crotonases (*crt*) were identified in *C. acetobutylicum* ATCC 824 and *C. beijerinckii* NRRL B593. The gene from *C. acetobutylicum*

ATCC 824 is designated CAC2712 and encodes a polypeptide of 261 amino acids. The gene from *C. beijerinckii* NRRL B593 also encodes a polypeptide of 261 amino acids. They occupy a similar location in the *bcs* (butyryl-CoA synthesis) operon (Rogers, 2006).

8.4.2.4 Butyryl-CoA Dehydrogenase

Butyryl-CoA dehydrogenase (enzyme 4 in Figure 8.3) catalyzes the reduction of crotonyl-CoA to butyryl-CoA. This enzyme has not been purified from butanol-producing *Clostridia*. Studies using cell-free extracts suggest that the immediate electron donor for the enzyme is not NADH or NADPH. The enzyme consists of 379 amino acid residues deduced from its structural gene (*bcd*) no matter from *C. acetobutylicum* ATCC824 or *C. beijerinckii* NRRL B593. Its amino acid sequence has high levels of homology with various acyl-CoA dehydrogenases. Two *etf* genes reside downstream of *bcd*, encoding proteins showing high homology with α and β subunits of the electron transfer flavoprotein. This finding suggests that the enzyme interacts with electron transfer flavoprotein in its redox function.

Genes for butyryl-CoA dehydrogenases (*bcd*) from *C. acetobutylicum* and *C. beijerinckii* have been identified based on the conserved amino acid sequence. They are all located in the *bcs* operon. In the *bcs* (butyryl-CoA synthesis) operon of *C. acetobutylicum* ATCC 824, between *bcd* and *hbd*, there are two genes (*etfA*, *etfB*) coding for subunits of an electron-transferring flavoprotein (ETF). These two genes occur in the same location of *bcs* operon of *C. beijerinckii* NRRL B593, while they precede the *hbd* gene in *C. saccharobutylicum* NCP262 (formerly *C. acetobutylicum* P262) (Rogers, 2006).

8.4.2.5 Acetoacetate: Acetate/Butyrate CoA-Transferase

CoA-transferase (enzyme 5 in Figure 8.3) catalyzes the reversible transfer of the coenzyme A moiety between acetoacetate-CoA and butyrate or acetate. When cells enter the solventogenic phase, preformed butyrate can be catalyzed to butyryl-CoA by the enzyme and butyryl-CoA will be used to produce butanol. So the enzyme plays an important role in the reutilization of preformed butyrate. It has been purified from *C. acetobutylicum* ATCC 824. The native enzyme is a heterotetramer with subunit molecular weights of about 23,000 and 25,000. In order to preserve its activity *in vitro*, the presence of high concentrations of ammonium sulfate (0.5–0.75 M) and glycerol (15%–20%, v/v) at pH 7 is required. Its kinetic binding mechanism follows Ping-Pong mechanism. Its K_m value at pH 7.5 and 30°C for butyrate is 660 mM, while the K_m values for acetoacetyl-CoA range from 21 to 56 μM with acetate or butyrate as the co-substrate, respectively. The K_m value for butyrate is high relative to its intracellular concentration, suggesting *in vivo* the enzyme activity is sensitive to changes in the concentration of butyrate. The butyrate conversion reaction *in vitro* is inhibited by the physiological level of butanol.

This may be another factor in the *in vivo* regulation of enzyme activity (Wiesenborn et al., 1989).

Genes encoding acetoacetate decarboxylase (*adc*) have been identified in *C. acetobutylicum, C. beijerinckii* NRRL B592 and B593. In the genome of *C. acetobutylicum* ATCC 824 and *C. acetobutylicum* DSM 792, *adc* forms a single operon. CoA-transferase genes (*ctfA, ctfB*) reside in another operon upstream of the *adc* operon. These two operons are arranged convergently. The intervening region separating the two operons is characterized by an inverted repeat that forms a stem-loop structure functioning as a Rho-independent transcription terminator in both directions. However, *adc, ctfA,* and *ctfB* are organized as one operon in *C. beijerinckii* NRRL B592 and B593 (Rogers, 2006).

8.4.2.6 Acetoacetate Decarboxylase

Acetoacetate decarboxylase (ADC, EC 4.1.1.4, enzyme 6 in Figure 8.3) catalyzes the decarboxylation of acetoacetate to acetone and CO_2. It is a crucial enzyme for butanol production because the reaction is irreversible and the consumption of acetoacetate promotes the CoA-transferase (enzyme 5) to continue to channel preformed butyrate into butanol production. The enzyme has been purified from *C. acetobutylicum* and well characterized with respect to molecular mass, subunit structure, mechanism of reaction, and amino acid composition (Westheimer, 1969).

8.4.2.7 Aldehyde Dehydrogenase

Aldehyde dehydrogenase (ALDH, enzyme 7 in Figure 8.3) catalyzes butyryl-CoA to form butyraldehyde for the production of butanol. ALDHs have been purified from *Clostridium saccharobutylicum* NRRL B643 (formerly *C. acetobutylicum* NRRL B643), *C. beijerinckii* NRRL B592, and NRRL B593.

ALDH of *C. saccharobutylicum* NRRL B643 is responsible for the synthesis of both butyraldehyde and acetaldehyde to produce butanol and ethanol, respectively, but it is clearly separated from butanol dehydrogenase activity. The native enzyme is a 115 kDa homodimer with 56 kDa subunits. Its kinetic constants have been determined in both the forward and the reverse directions. In the reverse direction, both the V_{max} and the apparent affinity for butyraldehyde are significantly higher than they are for acetaldehyde. In the forward direction, the V_{max} for butyryl-CoA is fivefold of that for acetyl-CoA (Palosaari and Rogers, 1988).

ALDH of *C. beijerinckii* NRRL B592 has been purified under anaerobic conditions, showing a native molecular weight (Mr) of 100 kDa and a subunit Mr of 55 kDa. Purified ALDH has no alcohol dehydrogenase activity. The enzyme also catalyzes the formation of both acetaldehyde and butyraldehyde, but it is more effective for the production of butyraldehyde than for acetaldehyde. It can use either NADP(H) or NAD(H) as the coenzyme, but the K_m value for NAD(H) is much lower than that for NADP(H). Kinetic data indicate that its kinetic mechanism is Ping-Pong mechanism. The enzyme is more stable in Tris buffer than in phosphate buffer. The apparent optimum pH for the forward reaction (physiological direction) is between 6.5 and 7.0. The ratio of NAD(H)/NADP(H) linked activities increases with the decrease of pH. The enzyme is sensitive to O_2 and can be protected by dithiothreitol (Yan and Chen, 1990).

ALDH from *C. beijerinckii* NRRL B593 is similar to that of NRRL B592 in which it also has no alcohol dehydrogenase activity and is more active with butyraldehyde than with acetaldehyde, but it is NAD(H) specific (Toth et al., 1999).

ALDHs have not been purified from *C. acetobutylicum* ATCC 824, but their genes (*adhE, adhE2*) have been identified and expressed. The *adhE* encodes a 96.5 kDa protein that exhibits 56% amino acid homology to the trifunctional protein encoded by *adhE* from *Escherichia coli*. The N-terminal amino acids show homology to aldehyde dehydrogenases from bacteria, fungi, plants, and mammals, while the C-terminal amino acids show homology to alcohol dehydrogenases from bacteria (including *Clostridia*) and yeast. The enzyme expressed from *adhE* shows activities of NADH-dependent butanol dehydrogenase, NAD-dependent acetaldehyde dehydrogenase, butyraldehyde dehydrogenase, and a weak activity of DADH-dependent ethanol dehydrogenase (Nair et al., 1994).

The adhE2 encodes a 94.4 kDa protein that occurs in alcohologenic cultures and not in solventogenic cultures. The *adhE2* has been expressed in *E. coli* as a strep-tag fusion protein demonstrating NADH-dependent butyraldehyde dehydrogenase and butanol dehydrogenase activities. This is the second ALDH identified in *C. acetobutylicum* ATCC 824, and it is the first example of a bacterium with two ALDHs (Fontaine et al., 2002).

The gene (*ald*) encoding aldehyde dehydrogenase has been cloned and sequenced from *C. beijerinckii* NRRL B592 and B593. The gene *ald* encodes a polypeptide of 468 amino acids, with a predicted molecular weight of 51,312 (B592) or 51,353 (B593). Gene ald is also a part of the solvent production operon of *C. beijerinckii*.

The aldehyde dehydrogenase of *C. acetobutylicum* ATCC 824 is a fused aldehyde-alcohol dehydrogenase. Its gene, designated *aad/adhE*, has been cloned and sequenced from *C. acetobutylicum* ATCC 824 and DSM 792. The gene was located on the plasmid pSOL1. There is another gene (*adhE2*) for aldehyde dehydrogenase located on the plasmid pSOL1. The gene encodes a polypeptide of 858 amino acids that has a 66.1% identity with the *adhE* of *C. acetobutylicum* ATCC 824 (Rogers, 2006).

8.4.2.8 Alcohol Dehydrogenase

Alcohol dehydrogenase (ADH, enzyme 8 in Figure 8.3) catalyzes the final reaction of the butanol pathway, that is, butyraldehyde to butanol. ADHs have been purified from *C. acetobutylicum* ATCC 824, *C. beijerinckii* NRRL

B592, *C. beijerinckii* NRRL B593, and *C. beijerinckii* NESTE 255. There are two butanol dehydrogenases with different cofactor requirements (NADH/ NADPH) and different pH ranges in *C. acetobutylicum* ATCC 824. One of them, the NADH-dependent butanol dehydrogenase has been purified and characterized. Zn^{2+} is needed to obtain significant recovery during the purification. The native enzyme is a dimer with a molecular mass of 82 ± 2 kDa, consisting of two subunits with subunit molecular mass of 42 kDa. Inhibition studies with NADH and butanol suggest that the enzyme follows an ordered bi-bi mechanism. Kinetic constants are 4.86/s, 0.18 and 16 mM for K_{cat}, K_{NADH}, and $K_{butyraldehyde}$, respectively. Its activity in the reverse direction is 50-fold lower than that in the forward direction. The enzyme has a higher activity with longer-chained aldehydes and is inhibited by metabolites containing an adenine moiety (Welch et al., 1989).

ADHs purified from *C. beijerinckii* NRRL B593 and *C. beijerinckii* NESTE 255 are primary–secondary ADHs, which are coincident with isopropanol production. They have similar structural and kinetic properties. Each has a native molecular weight of between 90,000 and 100,000 and a subunit molecular weight of between 38,000 and 40,000. They are NADP (H) dependent but with a low NAD⁺-linked activity. They are equally active to reduce aldehydes and 2-ketones, but the oxidation activity with primary alcohols is much lower than that with secondary alcohols (Ismaiel et al., 1993).

C. acetobutylicum ATCC 824 has two genes (bdhA, bdhB) that are contiguous but are transcribed separately. The bdhA and bdhB genes code for polypeptides of 389 and 390 amino acids, respectively, which have a 72.9% identity. The ADH of *C. beijerinckii* NRRL B593 is a primary–secondary ADH. Its gene has been cloned and sequenced. *C. beijerinckii* NRRL B592, which does not produce isopropanol, has three ADHs. They are homo- or hetero-dimers subunits encoded by adhA and adhB genes. Polypeptides encoded by these two genes differ only by 13 amino acids (13%). These genes are not contiguous and unrelated flanking sequences (Rogers, 2006).

8.4.2.9 Operons for Biobutanol Production

The genome of *C. acetobutylicum* consists of a 3.94 Mbp chromosome and a 192 kbp plasmid (pSOL1). Genes encoding solventogenic enzymes are located on both elements. The megaplasmid harbors three operons, two of which are organized in a cluster. The *sol* operon consists of the genes *orf*L (encoding a peptide with still unknown function), *adh*E (encoding the bifunctional butyraldehyde/butanol dehydrogenase, the gene is sometimes also designated as *aad*), *ctf*A, and *ctf*B (encoding the two subunits of the CoA transferase). Transcription of the operon stops at a typical *rho*-independent terminator, which serves the same function for the downstream location, but convergently transcribed monocistronic *ad* operon, encoding the acetoacetate decarboxylase. This organization is unique within the solventogenic

Clostridia, as all other species contain a *sol* operon starting with a butyralde-hyde dehydrogenase gene (*ald* or *bld*), followed by *ctf*A, *ctf*B, and *adc*. In *C. ace-tobutylicum*, the organization in two different operons explains the ability of the organism to change the acetone/butanol production ratio. The *sol* operon of *C. acetobutylicum* is transcribed from a single, inducible promoter, which shows typical A-dependent motifs. The transcript is processed after its for-mation, representing another stage of regulation. A binding site for Spo0A~P, the master transcription factor of sporulation, is located at the upstream of the control region (in reverse orientation). This represents the regulatory link between solventogenesis and sporulation. The suggestion that an upstream located gene, *orf*5, served as a repressor (*sol*R) of the *sol* operon, turned out to be an error. However, the existence of additional regulators is very likely. It is also important to note that the *adh*E part of the transcript is translated into both the complete bifunctional enzyme and a separated C-terminal butanol dehydrogenase domain. It may be that this counteracts instability problems. The promoter strength of the *sol* promoter is ~2 orders of magnitude lower than those of the *adc* and *bdh*B operons, possibly reflecting the physiological function of *Adh*E in the initiation of solventogenesis (and implying higher stability of the other encoded enzymes) (Durre, 2007).

The *adh*E2 gene is also located on the megaplasmid. It forms a monocis-tronic operon and two transcriptional start points have been determined. As one of these also seems to be an mRNA processing site (very little homology of deduced "promoter" sequences to the consensus) and a Spo0A~P binding site (0A box) precedes the proper promoter, the regulation seems to be very similar to the *sol* operon (Durre, 2007).

The other solventogenesis genes reside on the chromosome. *bdh*A and *bdh*B are organized in contiguous, monocistronic operons. Both promoters seem to be A-dependent, the *bdh*A-controlling one having only ~6% of the strength of the *bdh*B one. As expected, upstream of the *bdh*B promoter a 0A box is found, as the gene product is important for butanol formation. However, such a motif is present before the *bdh*A control region as well (two 0A boxes), although under usual experimental conditions this promoter turned out to be constitutive. That could indicate the involvement of additional regulators or point to dan-gerous conclusions drawn from *in silico* analysis (Durre, 2007).

As mentioned before, thiolase is involved in both acidogenesis and sol-ventogenesis. Two respective genes are present in *C. acetobutylicum*, *thl*A and *thl*B, of which the former encodes the metabolically active enzyme. Gene *thl*B is part of a polycistronic operon, which is expressed at very low level, and the physiological relevance of its gene products is not yet known. However, the monocistronic *thl*A operon shows a so far unique expression pattern in solventogenic *Clostridia*: massive transcription during acidogenesis, drastic decrease at initiation of the shift, and again massive transcription during solventogenesis. The two different expression phases are explained by the need for this enzyme in both pathways. The regulator(s) responsible for this control is (are) still unknown (Durre, 2007).

Two global approaches for investigation of regulatory links inside a cell are proteome analysis (Sullivan, 2006) and transcriptional profiling (DNA microarrays). Both have been started for *C. acetobutylicum*. Comparing acidogenic and solventogenic cells, several proteins were found to increase under the latter conditions. Among them were only one true solventogenic enzyme, *adc* (in two modifications, see before), several stress proteins DnaK, GroEL, Hsp18 (also in two modifications), PdxY (possibly quenching singlet oxygen), and three proteins catalyzing serine biosynthesis and incorporation into proteins. The physiological relevance of the latter phenomenon is unclear. Another study compared the wild-type and *spo0A*-negative and over-expression mutants. Stress proteins again were found to be affected, protein modifications again were noticed, and six proteins were missing in the *spo0A* knockout mutant. While proteome analysis still suffers in part from the fact that a number of proteins cannot be resolved on the gels in use, DNA microarrays cover the complete transcription. The group of Papoutsakis (Northwestern University, Evanston, IL) has performed pioneering studies with *C. acetobutylicum* in this field (Alsaker et al., 2004). Microarray analysis has been used to study a *spo0A*-knockout and a degenerated strain, over-expression of the *spo0A* gene, the effects on antisense RNA down-regulation of *ctfB*, over-expression of *groESL*, butanol stress and tolerance, and early sporulation and stationary phase events. These experiments yielded a vast amount of data, confirming in part earlier results obtained by other methods as well as resulting in unexpected findings. Mining, however, has just begun, and we are still far from understanding the system biology of *C. acetobutylicum*. Nevertheless, first comparisons of physiology and sporulation between Bacilli and *Clostridia* became possible (Durre, 2007).

An open question that still exists is the nature of the signal(s) responsible for initiation of solventogenesis. In continuous culture, a very reliable method to induce the shift under controlled conditions, a surplus of glucose, a limiting concentration of phosphate, and a pH value less than 4.3 are required. Approximately neutral pH values are possible if higher concentrations of acetate and/or butyrate are added. In addition to salt concentration, temperature also plays an important role. All these parameters affect DNA topology, so that the DNA might act as a sensor for environmental changes and even trigger transcription. Support for this theory stems from experiments showing transcription of the *sol* operon to be directly dependent on DNA relaxation. Intracellular metabolites, whose concentration is changing during the shift to solventogenesis, also might be involved in signaling. Several groups suggested butyryl-CoA, butyryl phosphate, NAD(P)H, and/or ATP/ADP (Durre, 2007).

8.4.3 Genomes of Biobutanol-Producing Clostridia

Genetic maps are useful tools in genome analyses. Physical and genetic maps of chromosomes of *C. acetobutylicum* ATCC 824 (Cornillot et al., 1997),

C. beijerinckii NCIMB 8052 (Wilkinson and Young, 1995), and *C. saccharobutylicum* NCP 262 (Keis et al., 2001) have been constructed. Furthermore, the whole genome sequence of *C. acetobutylicum* ATCC 824 has also been determined (Palmer Rogers, 2006).

The genome of *C. acetobutylicum* ATCC 824 includes a chromosome of about 4 Mb and a megaplasmid of about 200 kb. The chromosome is 3,940,880 bp in length with a total of 3,740 polypeptide-encoding ORFs and 107 stable RNA genes having been identified and accounting for 88% of the chromosome DNA. The megaplasmid, pSOL1, is 192,000 bp in length and appears to encode 178 polypeptides. The genes required for butanol production, aad/ adhE, ctfA, ctfB, and adc, occur as two converging operons on the pSOL1 plasmid. The pSOL1 plasmid also contains the second gene for aldehyde-ADH, adhE2. Other genes involved in butanol production, including thl, hbd, crt, bcd, and bdh, are present on the chromosome.

C. beijerinckii NCIMB 8052 has a single, circular, 6.7 Mb chromosome. Its combined physical and genetic map has been constructed by using a combination of cloned DNA fragments as hybridization probes and a bank of strains harboring insertions of the conjugative transposon Tn1545. Several genes concerned with solventogenic fermentation are found at different locations. Genes encoding thiolase, acetoacetate decarboxylase, CoA transferase, and possible butanol dehydrogenase are located on the right arm of the chromosome, whereas genes coding phosphate transbutyrylase, butyrate kinase, and 3-hydroxybutyryl-CoA dehydrogenase lie on the other arm of the chromosome.

The physical map of *C. saccharobutylicum* NCP 262 was constructed using pulsed-field gel electrophoresis (PFGE) and hybridization techniques. The size of the circular genome is about 5.3 Mb. Thirty-nine restriction sites were positioned on the map resulting in a mean resolution of approximately 140 kb. The locations of 28 genes were determined. The gene of the iron-only hydrogenase (hydA) and the metronidazole susceptibility (sum) gene, which are involved in controlling electron flow during butanol fermentation, were found to be near the replication origin. The genes involved in the acid- and solvent-producing pathway are clustered in a small region, representing 2.5% of the genome. They include the alcohol dehydrogenase gene (adh1), the butyryl-CoA synthesis (BCS) operon genes, crt, bcd, etfAB, and hbd, and butyrate operon genes.

8.5 Fermentation Technology

ABE fermentation process is obviously divided into two stages as mentioned above with the name of acidogenesis and solventogenesis, respectively (see Figure 8.2). Up to now, the signal of switch from acidogenesis to solventogenesis remains unclear. In batch fermentation, it was thought that environmental parameters, including the low culture pH, the high concentration of butyric acid, and the increased growth-limiting concentration of phosphate and/or sulfate, appeared at the end of bacteria exponential growth, lead to solvent production. It has been revealed that the difference between intracellular and extracellular environmental conditions plays a critical role in activating the transcription of solvent production genes, before the starting of solvent production, although the molecular mechanism is still unclear (Durre et al., 2002). In addition, specific regulatory proteins might be of great importance in the transcriptional activation of solvent-production genes. Different carboxylic acids influence transcriptional activation of gene expression, indicating that the signal conduction of solvent production is complex.

As for *C. acetobutylicum* DSM 1731 (and DSM 792), it appears necessary for solvent production at pH of 5.5 or lower in laboratory, and the optimal pH is 4.3. Meanwhile, in batch or continuous fermentations, the limitation of phosphate could enhance solvents production, showing the synchronism with the falling of pH (Bahl et al., 1982). The onset of solvent production of *C. acetobutylicum* ATCC 824, a *buk*-gene mutant, starts during the exponential stage when external butyrate concentration is at low level (4 mM), indicating that high concentration of butyrate might not be the onset trigger of the solvent production. Because of the low level of butyrate kinase (*buk*), it was suggested that the accumulation of butyryl-phosphate or butyryl-CoA might be the signal for the inducement of solvent production. When *buk* gene was inactivated by antisense RNA, butyrate production could be enhanced. When the concentration of butyrate, acetoacetate, and acetate was controlled at the ranges of 6–20, 12–20, and 8–33 mM, respectively, solvent production could be initiated, and this induction could be strengthened by the addition of another C1–C4 straight-chain acids (Harris et al., 2000; Husemann and Papoutsakis, 1988). While *C. beijerinckii* strain NRRL B592 (George and Chen, 1983) and strain NCIMB 8052 (Holt et al., 1984) switch to solvent production could be accelerated by the addition of butyrate and acetate together, either butyrate or acetate alone has little effect.

ABE fermentation in the past has employed many process modes such as batch, fed-batch, and continuous operations. Some processes use combination of these unit operations. Following paragraphs are brief summary and descriptions of these operations and their characteristics.

8.5.1 Batch Process

Batch fermentation is the simplest mode of operation, and is often used in the laboratory to obtain substantial quantities of cells or product for further analysis. A batch fermentation is a closed system, where all of the nutrients required for the organism's growth and product formation are contained within the vessel at the start of the fermentation process. The vessel can take the form of a shake flask, single use disposable system, or, for tighter control of parameters such as oxygen transfer, pH, and agitation, a bioreactor can be used (Figure 8.4).

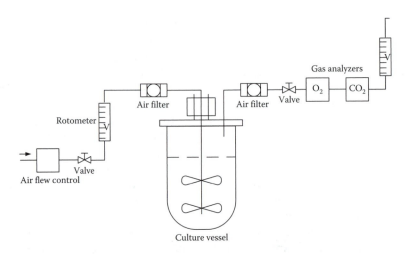

FIGURE 8.4

A diagram of a simple batch fermentation process. The system is "closed," containing all the nutrients required by the organism prior to inoculation, except for the gaseous requirement, which is continuously added to, and removed from, the reactor via sterilizing-grade hydrophobic filters.

Historically, these processes had involved non-sterile systems with self-selecting or natural inoculants. However, nowadays nearly all fermentation processes involve inoculation of a selected and specially bred strain of microbe, plant, or animal cell into a sterile medium held within a sterile fermenter vessel. After medium sterilization, the organism is inoculated into the vessel and allowed to grow. The fermentation is terminated when one or more of the following has been reached: (1) microbial growth has stopped due to the depletion of the nutrients or the accumulation of toxic compounds; (2) after a fixed predetermined period of time; and (3) the concentration of desired product has been achieved.

When cells are grown in a batch culture, they will typically proceed through a number of distinct phases, as shown in Figure 5.3. The lag phase, which may or may not be present, is described as "little or no growth at the beginning of the fermentation due to the physiochemical equilibrium between the microorganism and the environment following inoculation." The lag phase can be time-consuming and costly and so it is highly desirable to minimize this phase. This can be achieved by growing the inoculum in comparable medium to the bioreactor and under similar growth conditions (pH, temperature, etc.). A minimum of 5% by volume inoculum of exponentially growing cells should also be used for inoculation. Once the cells have adapted to the new conditions of growth, they enter the exponential phase. Then, key to minimizing the length of the lag phase lies in making sure that the culture being transferred undergoes the minimum levels of stress possible. In practical terms, this implies keeping the environments in the two fermentation systems as similar as possible. In reality, this is sometimes very difficult, for example, a late exponential stage shake-flask culture will typically exist in an environment where substrate levels are reduced, oxygen levels low, and carbon dioxide levels elevated; pH may also be very different from the process start point. Inoculation of fermenters to minimize lag phase is something of a compromise and often involves empiricism and experience of that particular culture character.

Nutrient depletion and the formation of inhibitors (typically excreted products such as ethanol, lactic acid, acetic acid, methanol, and aromatic compounds) have the effect of decelerating cell growth, and the cells then enter the stationary phase where the rate of cell growth equals that of cell death. Eventually, the cells enter death phase and this is characterized by a drop in optical density and biomass levels in most cultures.

The batch culture growth curve gives a good indication of when to stop the fermentation. Growth-associated products (primary metabolites) are produced during the exponential phase with their formation decreasing when growth ceases. Typically, the rate of product formation directly relates to the rate of growth. The fermentation can be terminated at the end of the exponential growth phase before the cell enters stationary phase. This growth phase is sometimes referred to as the trophophase. Nongrowth-associated products (e.g., classic secondary metabolites) have a negligible rate of formation during active cell growth.

The most commonly used industrial process for butanol production is batch fermentation with large fermenters ranging in capacity from 200,000 to 800,000 L and 8%–10% corn mash (sterilized) as medium. The solvent concentration that can be produced in a batch process depends upon the culture used. ABE concentrations in the order of 25–33 g/L can be produced in a batch process depending upon the culture used. Using newly developed *C. beijerinckii* BA101, an ABE yield of 0.45 could be obtained.

The traditional batch AB fermentation process has some limitations. Butanol is highly toxic to bacteria at quite low concentrations, which means that the level of solvents obtained in the final fermentation broth were only of the order of 2% maximum. Therefore, some technologies had to be introduced to avoid the toxicity of butanol to bacteria to increase the production of solvents and decrease the fermentation cost. The butanol fermentation process using *C. acetobutylicum* or *C. beijerinckii* is more complicated than the ethanol fermentation, and end-product inhibition at low butanol concentrations is a serious problem. The total solvent (acetone, butanol, and ethanol) concentration rarely exceeds 20 g/L, and this means that only low sugar concentrations can be fermented. This results in large volumes for downstream processing and wastewater treatment.

8.5.2 Fed-Batch Fermentation

Fed-batch culture is similar to batch culture, and most fed-batches begin life with a straightforward batch phase (Figure 8.5). However, unlike batch

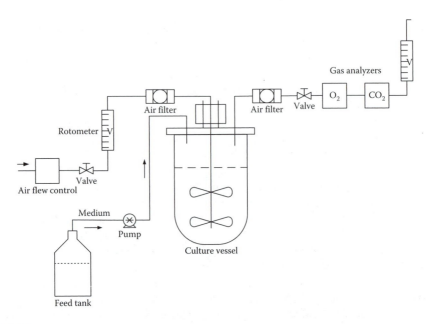

FIGURE 8.5
A simple fed-batch fermentation set-up. Feeding begins at a predetermined point during the process, often at the end of the exponential growth phase. The feed consists of a super-concentrated version of the process medium, allowing control of the growth rate and carbon flux through the system.

fermentation, these cultures do not operate as closed systems. At a given point during the fed-batch process, one or more substrates, nutrients, and/or inducers are introduced into the bioreactor. Fed-batch cultures can be run in different ways, for example, at a fixed volume where at a certain time point, a portion of the fermenter contents (consisting of spent medium, cells, product, and unused nutrients) is drawn off and replaced with an equal volume of fresh medium and nutrients (withdraw and fill), or at a variable volume where nothing is removed from the bioreactor during the time course of the process, with the cells and product remaining within the vessel until the end of the fermentation period, and the addition of fresh medium and nutrients having the effect of increasing the culture volume. This feeding strategy allows the organism to grow at the desired specific growth rate, minimizing the production of unwanted by-products, and allowing the achievement of high cell densities and product concentrations. The addition of the feed can be over a short or long period, starting immediately after inoculation or at a predetermined point during the run. The feeding strategy can be continuous over a long period of time or incremental, with the addition of fixed volumes at given time points (Figure 8.5). All are determined either from past fermentation data allowing the process operator to permit the same predetermined feed, or from the organism's physiology and the concentrations of key metabolites within the fermentation broth. In summary, fed-batch can be used to minimize or prevent flow of nutrients to waste products, and by this means to extend the productive phase of the process.

The fed-batch fermentation is an industrial technique that is applied to the processes where a high substrate concentration is toxic to the culture. In such a situation, the bioreactor is initiated in a batch mode with low substrate concentration and low volume of medium (usually less than half the volume of the fermenter). As the substrate is utilized by the culture, it is replaced by adding a concentrated substrate solution at a slow rate, thereby keeping the substrate concentration in the fermenter below a level toxic to the culture. When using this approach, the culture volume increases over time. The culture is harvested when liquid volume is approximately 75% of the volume of reactor. Because butanol is toxic to *C. acetobutylicum* or *C. beijerinckii* cells, the fed-batch fermentation technique cannot be applied in this case unless one of the novel product recovery techniques is applied for simultaneous separation of the product.

8.5.3 Continuous Fermentation

Historically, continuous culture techniques have not been widely used in laboratory scale, but are more common in industry where these techniques are used for such processes as vinegar production, wastewater treatment, ethanol production, and single cell protein production. In the laboratory, these techniques have been used increasingly to study the growth and physiology of microorganisms. Continuous cultivation is a method of prolonging the exponential phase of an organism in batch culture, while maintaining an environment that has less fluctuation in nutrients, cell number, or biomass. This is known as "steady state." The organisms are fed with fresh nutrients, and spent medium and cells are removed from the system at the same rate. This ensures that several factors remain constant throughout the fermentation, such as, culture volume, biomass, or cell number, product and substrate concentrations, as well as the physical parameters of the system such as pH, temperature, and dissolved oxygen.

Several control techniques can be used with continuous culture. The most commonly used continuous culture technique, the chemostat, operates on the basis of growth being restricted by the availability of a limiting substrate, while the turbidostat is operated under no limitations. Figure 8.6 shows a schematic representation of a typical chemostat. The feed medium contains an excess of all but one of the nutrients required for growth of the culture. The supply of the nutrient that is not in excess therefore determines growth rate of the microorganism.

At steady state, the flow of medium into the bioreactor equals the flow of spent medium and cells out of the bioreactor. There is therefore a mass balance across the system. The biomass, normally calculated as dry cell weight and denoted as grams per liter, remains constant because the addition of

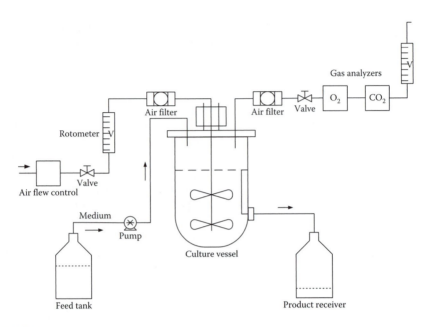

FIGURE 8.6

A continuous fermentation set-up. Feed is pumped into the reactor, increasing the volume within the reactor. At the same time, the overflow weir allows the excess volume to flow into the product tank. This ensures that the volume within the reactor remains constant.

fresh medium allows the formation of new biomass, and this is balanced by the loss of biomass in the spent medium outflow. When running a chemostat culture, it is imperative that the condition of steady state is calculated and monitored; this means that it is necessary to calculate dilution rate, specific growth rate, yield of product on substrate, etc., and a number of equations have been derived in order to do this. Here, we simply deal with the most basic of these relationships that anyone seeking to run chemostats should have an awareness of.

The dilution rate (D) describes the relationship between the flow of medium into the bioreactor (F) and culture volume within the bioreactor (V):

$$D = \frac{F}{V} \tag{8.1}$$

Flow rate is expressed in liters per hour and volume is expressed in liters, therefore dilution rate is expressed in units per hour (h^{-1}). Residence time (t) is the inverse of dilution rate and is also related to the reactor volume and flow rate:

$$t = \frac{V}{F} \tag{8.2}$$

Residence time is measured in hours. In general, the continuous culture must go through four or five residence times before it can be considered to be at steady state. When a reactor is perfectly mixed, all nutrients and cells in the bioreactor are equally distributed throughout the vessel. A small-scale stirred tank bioreactor with moderate concentrations of cells present is assumed to be perfectly mixed and this assumption is used in the derivation of the continuous culture equations. Therefore, the following statements are assumed to be correct throughout a chemostat at steady state:

$$\frac{dX}{dt} = 0 \qquad (8.3)$$

and

$$\frac{dS}{dt} = 0 \qquad (8.4)$$

that is, change in biomass (X) over time (t) is zero and change in substrate concentration (S) over time (t) is zero, that is, no net accumulation of cell mass or substrate. Some systems never actually reach a real steady state, but a pseudo-steady state. This is typically where the concentrations of key analytes oscillate around a single value.

The specific growth rate of a culture at steady state is set by the dilution rate (i.e., $\mu - D$), which is in turn determined by the rate of flow of nutrient solution to the culture. However, when the dilution rate is greater than the maximum specific growth rate (μ_{max}) of the microorganism, the result is "washout." This can be calculated using the biomass equation for the continuous reactor:

$$\frac{dX}{dt} = \mu X - DX \qquad (8.5)$$

So when DX is greater than μX, dX/dt becomes negative, that is, the dilution rate, or the rate at which fresh medium is added to and spent medium is removed from the system, exceeds the maximum specific growth rate of the organism, resulting in a decrease in the number of cells in the bioreactor over time. The number of cells in the bioreactor will therefore eventually become zero (washout). Most chemostat cultures become progressively more unstable as the dilution rate approaches the critical dilution rate (above which washout occurs). Hence, physiological and other studies involving this type of continuous culture system should not be operated near this region, as the results may be highly variable. Likewise, production systems using chemostats rarely run near the critical dilution rate because of instability problems and the risk of inadvertent washout occurring. This obviously limits biomass productivity since it imposes a ceiling on the rate of culture growth.

The continuous cultivation may be used to improve reactor productivity. Because of high efficiency of fermentation facility usage, a continuous process is superior to the batch fermentation. Since the early 1980s, many researches have been conducted the continuous fermentation characteristic description and the parameters control. On the basis of the conclusions of these experiments, it was suggested that butanol production was not induced by single growth-limiting factor in a chemostat. When limited glucose, nitrogen, or magnesium was added in a chemostat, it is difficult to maintain steady-state solvent production or low solvent production, whereas, butanol production was promoted when limited phosphate or sulfate was added. In a single stage continuous system, high reactor productivity may be obtained. However, this occurs at the expense of low product concentration when compared to that achieved in a batch process. In a single stage continuous reactor, 15.9 g/L of total solvents was produced at a dilution rate of 0.1/h resulting in a productivity of 1.5 g/L/h. When the dilution rate was increased to 0.22/h, the productivity was increased to 2.55 g/L/h, while the product concentration decreased to 12 g/L. Two or more multistage continuous fermentation systems have been investigated to reduce productivity fluctuations and increase solvent concentration in the product stream. This is done by allowing growth, acid production, and solvent production to occur in separate bioreactors. Bahl et al. (1982) reported an ABE concentration of 18.2 g/L using *C. acetobutylicum* DSM 1731 in a two-stage system, which is comparable to that in a batch reactor.

8.5.4 Immobilized Cell Bioreactors

Cell density is critical to the fermentation productivity. Immobilized bioreactor has several advantages: first and foremost, high cell density could be achieved in immobilized cell systems, to raise the production efficiency, including improved reaction rates and simplicity of operation. Second, the structure of bioreactor may be relatively simple. When the fermentation finished, the bacteria could be easily separated from the fluid, making the downstream process simple and cheap. Immobilized bioreactor could improve reactor productivity as high as 6.5–15.8 g/L/h, as compared to <0.50 g/L/h in batch reactors.

8.6 Biobutanol Recovery

High butanol recovery cost is the main economic bottleneck for biobutanol production. The cost of downstream process is very sensitive to butanol concentration in the broth. While in the fermentation broth, butanol is often less than 20 g/L. Apart from the traditional distillation technology, several alternative downstream separation and purification technologies, including membrane-based systems (RO, pervaporation, perstraction, and membrane evaporation), *in situ* gas stripping and liquid–liquid extraction have been investigated. There is still no answer to the question of which technology is most suitable (Table 8.5).

TABLE 8.5

Novel Product-Recovery Techniques for Biobutanol Production

Method	Principle	Advantage	Disadvantage
Gas stripping	Heating of effluent, purging with gas, condensation of solvent/water vapors	Simple to perform, low chance of clogging or fouling	Low selectivity, no complete removal of solvents, more energy required compared to membrane-based processes
Liquid–liquid extraction	Contact of water-immiscible solvent with fermentation broth, recovery of dissolved acetone/butanol/isopropanol by distillation	High capacity, high selectivity, low chance of clogging or fouling	Expensive to perform, possible forming of emulsions
Membrane evaporation	Selective diffusion of solvents across a porous membrane, recovery of evaporated vapors by applying vacuum and condensation	Smaller membrane area required compared to pervaporation, simple to perform	Possible clogging or fouling
Perstraction	Similar to liquid–liquid extraction, with a membrane separating fermentation broth and extractant	High selectivity, simple to perform	Large membrane area required, possible clogging or fouling
Pervaporation	Selective diffusion of solvents across a nonporous membrane, recovery of evaporated vapors by applying vacuum or sweep gas	High selectivity compared to membrane evaporation, simple to perform	Lower membrane flux compared to membrane evaporation, possible clogging or fouling
Reverse osmosis	High-pressure separation of dilute aqueous solution into a concentrated one and pure water by use of a semipermeable membrane, distillation of concentrated solution	Lower costs than conventional distillation	Possible clogging or fouling

Sources: Durre, P. *Appl. Microboil. Biotechnol.*, 49, 639, 1998; Data based on reports by Ennis, B.M. et al., *Process Biochem.*, 21, 131, 1986; Gapes, J.R. et al., *Appl. Environ. Microbiol.*, 62, 3210, 1996; Groot, W.J. et al., *Process Biochem.*, 27, 61, 1992; Maddox, I.S. et al., Utilization of whey by clostridia and process technology, in Woods, D.R. (Ed.), *The Clostridia and Biotechnology*, Butterworth-Heinemann, Stoneham, MA, 1993.

8.6.1 Distillation

The fundamental principles of distillation have been discussed previously in Chapter 7. Distillation has played a dominant role in separation technology owing to its simplicity and universal applicability. However, it is not a high-energy-efficient process. The cost of recovering butanol by distillation is high due to the low concentration of butanol in the broth. Phillips and Humphrey (1985) evaluated the economics of butanol recovery from broth using distillation and demonstrated that energy savings by a factor of several orders of magnitude may be achieved if the concentration of butanol is increased from 10 to 40 g/L. At a 10 g/L feed butanol concentration, the ratio of tons of oil used for fuel to tons of 100% recovered butanol is 1.5, while at a 40 g/L feed butanol concentration, this ratio is 0.25, suggesting that a tremendous amount of energy could be saved with increased butanol concentration in the broth.

8.6.2 Gas Stripping

Gas stripping is a convenient method for butanol purification from the broth, following the gas–water partitioning principle known as Henry's law. In order to increase effectiveness, a portion of the sales gas (or other available gas) can be injected directly into the fermentor to aid in stripping. Gases capture butanol during the period of its bubbling in the fermentor. Once gases with butanol go through the condenser, butanol can be condensed and collected in a receiver. Moreover, the gases can be recycled back to the fermentor for next round of stripping.

Figure 8.7 is a graphical illustration of a typical process of butanol recovery by gas stripping (Ezeji et al., 2003). A considerable improvement in the productivity and yield can be obtained, up to 200% and 118%, respectively, in comparison with the nonintegrated process, due to the removal of inhibition of butanol by keeping butanol concentration at low levels. During gas stripping, acids are not removed from the fermentation broth. Hence, gas-stripping stimulates the assimilation of acids into solvents.

8.6.3 Liquid–Liquid Extraction

Liquid–liquid extraction is another technique that can be used to remove ABE from the fermentation broth. Liquid–liquid extraction is a mass transfer operation in which a liquid solution (the feed) is contacted with an immiscible or nearly immiscible liquid (solvent) that exhibits preferential affinity or selectivity toward one or more of the components in the feed. Two streams result from this contact: the extract, which is the solvent-rich solution containing the desired extracted solute, and the raffinate, the residual feed solution containing little solute. Some of the requirements for extractive fermentation are (Qureshi, 2005)

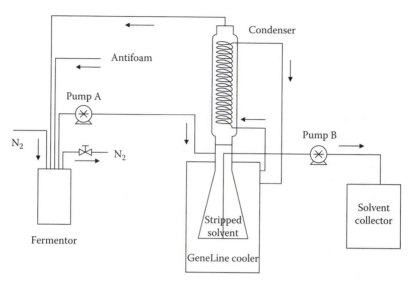

FIGURE 8.7
A schematic diagram of biobutanol production and *in situ* recovery by gas stripping. Pump A: gas recycle pump; Pump B: condensed solvent removal pump.

1. Nontoxic to the producing organism
2. High partition coefficient for the fermentation products
3. Immiscible and non-emulsion forming with the fermentation broth
4. Inexpensive and easily available
5. Can be sterilized and does not pose health hazards

Liquid–liquid extraction is an important energy-saving process for the purification of butanol from the fermentation broth by treating the broth with a non-miscible solvent in which butanol has preferential partition. As the extractant and broth are immiscible, the extractant can easily be separated from the broth after the extraction of butanol, still leaving nutrients in the broth. Extraction can successfully be used for *in situ* alcohol recovery in butanol fermentations to increase the substrate conversion.

An early application of extractive butanol fermentation was described by Wang et al. (1979) who used corn oil as an extraction solvent and reported an increase in biobutanol production in repeated batch fermentations. Since that time, there have been many reports on the use of various extraction solvents for extractive butanol fermentation (Maddox, 1989). An advantage of extraction over other recovery methods may be the high capacity of the solvent and the high selectivity of the butanol/water separation. Thirty-six chemicals were tested for the distribution coefficient for butanol, the selectivity of butanol/water separation, and the toxicity to Clostridia. Convenient extractants were found in the group of esters with high molecular mass. The common extractant for butanol extraction is oleyl alcohol because of its low toxicity to cell and good extraction selectivity (Ezeji et al., 2004, 2006). Roffler et al. (1987) were successful in producing biobutanol in an extractive fed-batch fermentation using a sugar solution containing 339 g/L glucose.

Extractant toxicity is a major problem with extractive fermentations. In order to avoid the toxicity problem brought in by the extraction solvent, some investigators have used a membrane to separate the extraction solvent from the cell culture (Traxler et al., 1985; Eckert and Schugerl, 1987). The fermentation broth was circulated through the membrane and the cells were returned to the fermentor, while the permeate was extracted with extractant to remove the butanol. In this system, a productivity of 3.08 g/L/h was achieved.

Another approach for reducing the toxicity and improving the partition coefficient has been to mix a high partition coefficient, high toxicity extractant with a low partition coefficient, and low toxicity extractant (Evans and Wang, 1988). The resultant mixture is an extractant with a high partition coefficient and low toxicity. At this time, research is underway in the author's laboratory on the extractive recovery of butanol from the fermentation broth of *C. beijerinckii* BA 101. In integrated extractive fermentations, oleyl alcohol forms an emulsion when agitated at high speed to improve removal rate of butanol.

8.6.4 RSEMC Technology for Solvent Extraction

Resodyn Corporation (Butte, Montana) has developed a technology called "The ResonantSonic Enhanced Mixer and Coalescer (RSEMC) as an Advanced Solvent Extraction" and applied the technology to butanol fermentation process (Yang, 2008). The technology is based on a resonant bar member as mixing vehicle to speed up the contact of two immiscible liquids. It has been demonstrated that the liquid droplets produced with RSEMC are much more uniform than those obtained by impeller agitated mixing. The uniform droplets are coalescing faster than those with broad size distribution. Therefore, shorter settling time is required to separate the two liquids.

The approach by Resodyn Corporation for butanol recovery from fermentation system (batch operation mode) was to use organic solvent to extract the product from microbial cell free fermentation broth. The cell was removed by filtration and only cell free culture media was in contact with organic solvent. Butanol lean culture media was fed back to fermentation system and merged with filtrated microbial cells. Fresh concentrated nutrient media was added to make up the fermentation requirement for the cells. Concurrent flow extraction mode for continuous flow operation to remove butanol and recycle the fermentation broth back to solventogenesis fermentor to achieve higher conversion of butyrate and other organic acids was proposed. Figure 8.8 has shown the conceptual design for the RSEMC extraction process. Extractants such as oleyl alcohol and benzyl benzoate were screened for their efficiency of removing butanol from aqueous media. The toxicity

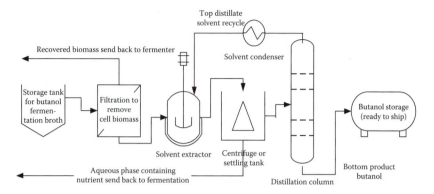

FIGURE 8.8
ResonantSonic enhanced mixer and coalescer (RSEMC) solvent extraction process to recover butanol from fermentation broth.

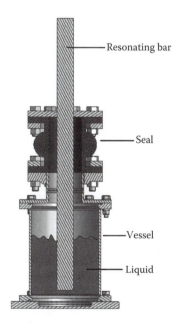

FIGURE 8.9
A bench-scale, high-intensity, low-frequency ResonantSonics acoustic mixer/coalescer for butanol recovery from fermentation broth.

of selected organic solvents on *C. acetobutylicum* and other solventogenic microbes was also evaluated. Experiments were carried out with a bench-scale RSEMC, as shown in Figures 8.9 and 8.10. The rate of butanol transfer from aqueous media to organic phase was determined by analyzing butanol concentration in organic phase at exit of the reactor with bar resonant mixing time as the variable. This kinetic behavior was used to judge the efficiency

FIGURE 8.10
Experimental setup. The vessel and RS mixer/coalesce are seen mounted to the platform of the stand. The sample plate is mounted to the left upright. The vacuum pump for sample collection is located on the cart on the lower left.

of the RSEMC. The transfer rate for the system with impeller as agitation, or mixing vehicle, was also determined with the same vessel. Both real-world and simulated solutions were used to compare the outcome.

8.6.4.1 Solvent Toxicity

One of key considerations in solvent selection is the influence of the solvent on microorganism growth and the effect on butanol production. The concentration of target solvent is usually its saturation concentrations in butanol fermentation media at a typical temperature of 37°C.

Kollerup (1986) reported the toxicity, predicted partition coefficient, and cost of some possible solvents. The solvents have been evaluated as candidates for extractive fermentation of ethanol using *Saccharomyces cerevisiae* (Kollerup and Daugulis, 1985). Potential candidate solvents for butanol recovery from fermentation broth are listed in Table 8.6.

Microbial screen: Publicly available *C. acetobutylicum* strains can be obtained from American Type Culture Collection (ATCC) and cultured as described in the literature (Huang et al., 2004) for cell mass growth observation and

TABLE 8.6

List of Potential Solvents That Can Be Used for Butanol Extraction

Solvent	Boiling Point (°C)	Density (g/mL)	Cell Growth Rank[a]	Butanol Partition Coefficient (P)	
				25°C	35°C
Ethanol	78.3	0.79			
Butanol	117.6	0.81			
Vinyl bromide	15.8	1.49	N/A	2.95 (10°C)	—
Oleic acid methyl ester	218.5 at 20 mmHg	0.87	4	0.47	0.58
Decanol	230	0.83	2	5.85	6.21
Oleyl alcohol	207 at 13 mmHg	0.85	5	3.26	3.60
Dodecanol	261	0.83	3	3.82	4.05
Hexane	69	0.66	1	0.79	0.52
Isooctane	99.2	0.69	1	0.42	0.38
Octane	126	0.70	2	0.37	0.32
Decane	174	0.73	2	0.31	0.28
Hexadecane	287	0.77	2	0.27	0.25
Kerosene	200–250	0.82	3	0.25	0.25
Poly(propylene glycol) 1200	>188	1.01	5	3.92	4.80
Benzyl benzoate	323	1.11	3	1.54	2.31
Trioctylamine	366	0.81	5	1.15	1.78

[a] Cell growth is for yeast (Kollerup 1986). Cell growth: 0 = no growth, 5 = full growth.

TABLE 8.7

Results of Microbial Screen in Terms of Cell Mass and Butanol Production

Strain	Cell Mass at 56 h (OD$_{600}$)	Butanol Conc. (g/L)	Butyric Acid Conc. (g/L)	Glucose Residue (g/L)
ATCC 824	0.5	5.6	0.2	39.6
ATCC 4259	2.7	9.8	0.3	34.3
ATCC 55025	6.5	18.5	0.5	22.1

to produce butanol. A comparison of butanol production of three *C. aceto-butylicum* strains, that is, ATCC 824, ATCC 4259, and ATCC 55025, is shown in Table 8.7. The data in the table clearly indicate that the strain ATCC 55025 *Clostridium beijerinckii* is the superior over the other two.

Butanol production from fermentation with fresh cultural media: Butanol production with fresh cultural media, for example, without any other organic solvent added, was tested with bacterium ATCC 55025. Butanol production rate is, in general, proportional to microbial cell mass presented in fermenter. Cell mass growth with ATCC 55025 in fresh culture media (without solvent addition) is shown in Figure 8.11. As one can see from the plot, this strain reaches its maximal cell density after 48 h growth. Butanol production is expressed here as for total solvent production (acetone, ethanol, and butanol), for a batch fermentation process, has been plotted along with glucose consumption in Figure 8.12.

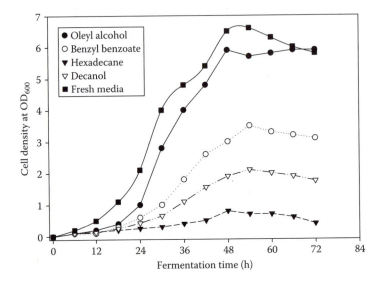

FIGURE 8.11
The effect of various solvents on microbial growth during butanol fermentation.

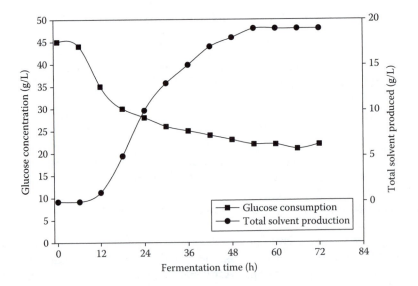

FIGURE 8.12
ABE total solvent production and glucose consumption from fresh media fermentation.

FIGURE 8.13
Photos of cell culture (ATCC 55025) saturated with decanol (flask 3), oleyl alcohol (flask 4), hexadecane (flask 5), and benzyl benzoate (flask 7).

Cell culture with solvent saturated media: The traditional growth media used for *C. acetobutylicum* (see previous section) was saturated with selected organic solvent as listed in Table 8.6. The cell growth of ATCC 55025 with oleyl alcohol, benzyl benzoate, hexadecane, and decanol saturated media has been demonstrated in Figure 8.13, and the cell mass data is plotted in Figure 8.11. The results of the tests indicated that oleyl alcohol is the most biocompatible solvent among these four and it is consistent with literature data (Evans and Wang, 1988). The solvent list presented in Table 8.6 is generated based on some previous studies at Resodyn and the information gathered from published literature (Evans and Wang, 1988; Kollerup and Daugulis, 1985; Job et al., 1989; Roffler et al., 1987).

Butanol production with solvent saturated media: The continuous solvent extraction of butanol will lead to lower concentration of inhibitory compounds but the media will also be saturated with extraction solvent (extractant). The extractant solvent saturated media is recycled back to fermentation process to fully utilize the residual glucose and other nutrients. Butanol production or total solvent production under such condition is presented in Figure 8.14, with selected four extractants: oleyl alcohol (OA), benzyl benzoate (BB), hexadecane (HA), and decanol (DA), as targeting compounds.

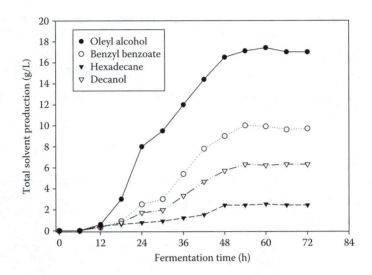

FIGURE 8.14
ABE production with extractant saturated culture media.

The culture media was prepared as stated in a previous section and it was saturated with above mentioned extractant at 37°C. Results have indicated that the total solvent production follows similar profile of dependence on organic solvent with cell mass production as shown in Figure 8.11. It is reasonable to assume that the more biocompatible solvents, which lead to higher *Clostridium* cell mass production, will also have less inhibitory effect on butanol production.

8.6.4.2 Butanol Partition

Mass balance for solvent extraction for butanol recovery can be expressed as

$$V_a C_i = V_a C_a + V_o C_o \tag{8.6}$$

where
V_a and V_o are volumes of aqueous media and organic solvent, respectively
C_i, C_a, and C_o represent butanol concentration in initial aqueous media, lean phase aqueous media, and organic solvent, respectively

The partition coefficient of extractant (butanol) is defined as the mass ratio of organic solvent phase to aqueous phase. The mathematical expression is

$$P = \frac{V_o C_o}{V_a C_a} = \frac{V_o C_o}{V_a C_i - V_o C_o} = \frac{1}{\dfrac{V_a C_i}{V_o C_o} - 1} \tag{8.7}$$

If organic solvent volume used for extraction is the same as aqueous volume, the calculation for P can be simplified as

$$P = \frac{C_o}{C_i - C_o} = \frac{C_o}{C_a} \tag{8.8}$$

After reaching equilibrium and waiting for a period of time to allow organic solvent to be separated from aqueous solution, two phases can be separated using separation funnel.

The rule of selecting organic solvent for butanol extraction is that it must have a high partition coefficient for butanol and be compatible with butanol-producing cells (biocompatibility). Another key consideration is to minimize the energy input when extractant butanol is separated from the extraction solvent. Thus, the higher the concentration of butanol that can be extracted into the solvent, the lower the energy needed to distill the butanol out.

Figure 8.15 shows the equilibrium concentrations of butanol in both organic solvent (y-axis) and aqueous residual (x-axis) for four different solvents, that is, oleyl alcohol (OA), benzyl benzoate (BB), hexadecane (HA), and decanol (DA). The data shown in Figure 8.15 indicate that up to 6.0% (v/v) butanol can exist in aqueous media and that the extraction is linear, that is, the concentration distribution ratio between aqueous media and an organic solvent is a constant (straight line). This constant is the partition coefficient of butanol between aqueous medial and the organic solvent.

The relationship between butanol partition coefficient and organic solvent to aqueous media for the four different organic solvents (DA, OA, BB, and

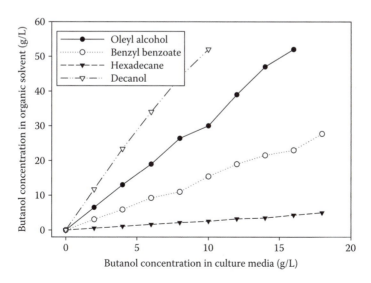

FIGURE 8.15
Butanol concentration distribution between organic solvent and fermentation media.

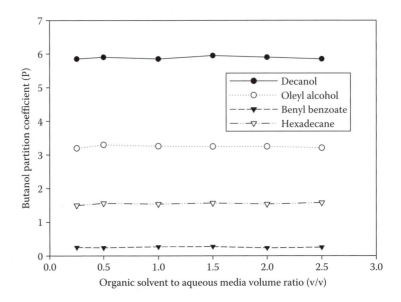

FIGURE 8.16
The relationship between butanol partition coefficient and organic solvent to aqueous media volume ratio at 25°C.

HA) is shown in Figure 8.16. As long as neither phase (organic or aqueous) is saturated with butanol, the partition coefficient will be constant, that is, independent from the two phase volume ratio.

8.6.4.3 Energy Consumption Calculation

If distillation is used for the final butanol purification process, energy consumption for distillation separation of solvent and butanol depends on the solvent selection. If the extraction solvent has a higher boiling point than butanol, the energy input is the sum of heating the whole content (mixture of solvent and butanol) to the butanol boiling point (117.6°C) and the latent heat of butanol present in the solution. If the solvent has lower boiling point than 117.6°C, then the energy consumption will be the heating of whole mixture to that solvent's boiling point plus the latent heat of that solvent present in the mixture. The energy consumption for these two scenarios can be expressed in the following equations:

For the former case, the energy consumption is

$$E = x_S C_{pS}(117.6 - T_m) + x_B C_{pB}(117.6 - T_m) + x_B \Delta H_B \qquad (8.9)$$

For the latter case, the energy consumption is

$$E = x_S C_{pS}(T_{bS} - T_m) + x_B C_{pB}(T_{bS} - T_m) + x_S \Delta H_S \qquad (8.10)$$

where x, C_p, T_b, and ΔH are molar fraction, specific heat capacity, boiling point, and vaporization heat, respectively. The subscripts S, B, and m refer to organic solvent, butanol, and mixture, respectively.

8.6.4.4 Extraction Kinetics

For each two-phase system (organic solvent and aqueous phase) kinetic and equilibrium data were collected to establish the transfer rate of extractant to organic solvent and to determine how much the organic solvent can be loaded (solubility) for extraction. The rate of butanol transfer from aqueous media to organic phase was determined by analyzing butanol concentration in organic phase at exit of the reactor with bar resonant mixing time as the variable. This kinetic behavior was used to judge the efficiency of the RSEMC. The transfer rate for the system with impeller as agitation, or mixing vehicle, was also determined with the same vessel.

Isotherm: Isotherm data for extracting butanol from aqueous media to organic solvent was collected over a phase ratio (O/A) from 2.5:1 to 1:4. Figure 8.17 shows the isotherm at the organic solvent to media volume ratio of 1:1 at 20°C. This O/A volume ratio was used for consistency of the investigations. Similar graphs were generated for impeller diving mixing with the same vessel except that resonant bar was replaced with a Lightnin (Labmaster,

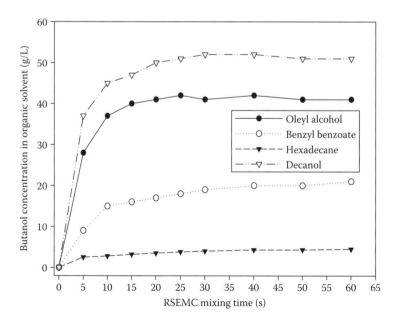

FIGURE 8.17
Butanol extraction isotherm with RSMEC. Bar frequency—41.0 Hz. Temp—20°C. O/A ratio—1:1. Starting butanol concentration—15 g/L.

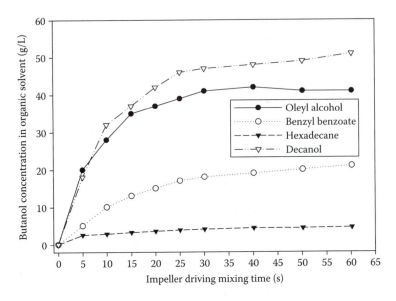

FIGURE 8.18
Butanol extraction isotherm with impeller mixing. Impeller speed—300 rpm. Temp—20°C.
O/A ratio—1:1. Starting butanol conc—15 g/L.

Lightnin, Wytheville, VA) impeller mixing device. The data are shown in
Figure 8.18. A comparison of the data shown in Figures 8.17 and 8.18 has
shown that the extraction rate of butanol to organic solvent is slower with
impeller mixing than that with RSEMC except for low butanol partition solvent such as hexadecane.

Stripping and loading kinetics: Stripping (for fermentation broth) and/or loading (for organic solvent) kinetics can be induced from the isotherm data
shown in Figures 8.17 and 8.18. Initial extraction rate can be obtained by
dividing concentration increase of butanol in organic solvent within linear
portion of isotherm curve by the corresponding mixing time. The stripping
kinetics (extraction rates) is mainly used as guidance for estimating the mixing time to ensure the extraction is close to equilibrium stage.

8.6.5 Pervaporation

Pervaporation, in its simplest form, is an energy-efficient combination of
membrane permeation and evaporation. It is considered as an attractive alternative to other separation methods for a variety of processes. For example,
with the low temperatures and pressures involved in pervaporation, it often
has cost and performance advantages for the separation of constant-boiling
azeotropes. Pervaporation is also used for the dehydration of organic solvents and the removal of organics from aqueous streams. Additionally,

pervaporation has emerged as a good choice for separation heat-sensitive products. Pervaporation involves the separation of two or more components across a membrane by differing rates of diffusion through a thin polymer and an evaporative phase change comparable to a simple flash step. A concentrate and vapor pressure gradient is used to allow one component to preferentially permeate across the membrane. A vacuum applied to the permeate side is coupled with the immediate condensation of the permeated vapors. Pervaporation is typically suited to separating a minor component of a liquid mixture, thus high selectivity through the membrane is essential. Figure 8.19 shows a typical pervaporation process. Pervaporation is a technique for the separation of mixtures of liquid by partial vaporization through a nonporous polymeric membrane or molecularly porous inorganic membrane (such as a zeolite membrane), differing from other membrane processes due to vapor phase occurring when the transported compounds permeate through the membrane (Leland, 2005). Due to different species in the feed mixture having different affinities to the membrane and different diffusion rates through the membrane, even a component at low concentration in the feed can be highly enriched in the permeate (Leland, 2005). The resulting vapor, referred to as "the permeate," is then separated. We could choose the properties of the membrane material for various separation processes. For example, if the membrane is hydrophobic, organic compounds will be preferentially permeated through the membrane. Alternatively, if the membrane is hydrophilic, water will be enriched in the permeate and the feed liquid will be dehydrated (Leland, 2005).

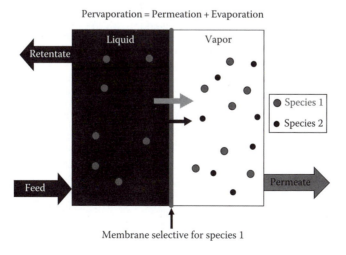

FIGURE 8.19
Schematic diagram of pervaporation process.

8.6.6 Reverse Osmosis

RO is a relatively mature technology for desalination and water purification applications. RO systems remove not only a variety of metals and ions but also certain organic, inorganic, and bacterial contaminants. RO employs a semipermeable membrane, under pressure a large proportion of water within the fermentation broth could pass through the membrane than do the solvent molecules. Therefore, biobutanol will be concentrated. In butanol purification, for example, the membrane is permeable to water molecules, rather than butanol molecules. Figure 8.20 shows a typical process flow diagram of the RO system.

RO is probably a simpler and more economical process for biobutanol recovery from the broth than many other separation methods. The dilute fermentation broth with butanol and acetone could be concentrated by RO to dewater the fermentation liquor using polyamide membrane, which exhibits butanol rejection rate as high as 85%. Optimum rejection of butanol could be obtained under the optimized conditions (operation pressure of 5.5–6.5 MPa, hydraulic recovery of 50%–70%, flux of 0.5–1.8 L/m²/min) (Garcia et al., 1984).

8.6.7 Perstraction

Several problems are associated with liquid–liquid extraction, including cell toxicity, emulsion formation, extraction solvent loss, rag layer formation

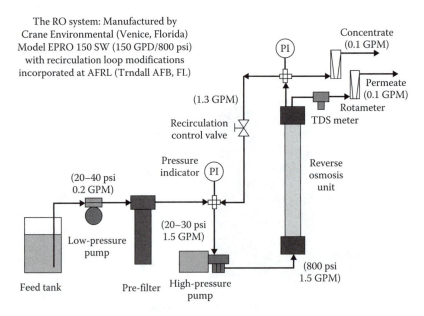

FIGURE 8.20
Process flow diagram of the RO test unit.

(i.e., cell accumulation at the interphase) (Ezeji et al., 2007). In a typical perstraction process, the extractant and the broth are separated by a membrane contactor, which provides surface area to exchange butanol between the two immiscible phases. As there is no direct contact between the two phases, extractant toxicity, phase dispersion, emulsion, and rag layer formation are drastically reduced or eliminated. In such a system, butanol would diffuse preferentially across the membrane, while other components, including fermentation intermediates (e.g., acetic and butyric acids), could be retained in the aqueous phase (Qureshi et al., 2005). The total mass transport largely depends on the diffusion rate of butanol across the membrane (Ezeji et al., 2007).

Perstraction allows continuous removal of butanol as it is produced in the reactor, thus alleviating the inhibition problem and increasing the solvent concentration up to 9.9% (Qureshi et al., 2005). Acetone–butanol–ethanol (ABE) were produced from whey permeate medium, supplemented with lactose, in a batch reactor using *C. acetobutylicum* P262, coupled with ABE removal by perstraction. It should be noted that the ratio of acids to solvents was significantly lower in the perstraction experiment compared to the control batch process suggesting that acids were converted to solvents. The perstraction experiment results are superior to the control batch fermentation where 9.34 g/L ABE was produced. It was determined that lactose at 250 g/L was a strong inhibitor to the cell growth of *C. acetobutylicum* and fermentation. A membrane with an area of 0.1130 m^2 was used as the perstraction membrane while oleyl alcohol as the perstraction solvent. Removal of ABE by perstraction was faster than their production in the reactor, and the maximum concentration of ABE in the oleyl alcohol was 9.75 g/L. It is viewed that recovery of ABE from oleyl alcohol (at this concentration) would be more economical than recovery from the fermentation broth (Qureshi and Maddox, 2005).

8.6.8 Integration Process for Biobutanol Fermentation

The major problems associated with butanol fermentation were uneconomical product recovery and the use of dilute sugar solution (usually 60 g/L for batch fermentation), which results in a large reactor and process stream volumes thereby making butanol fermentation noncompetitive when compared to butanol obtained from petrochemical sources. Simultaneous recovery and fermentation when using novel recovery techniques have solved these problems because concentrated sugar solutions (130–135 g/L) can be used in combination with economic product recovery systems (Maddox et al., 1995; Qureshi and Maddox, 1991; Qureshi and Blaschek, 1999; Qureshi and Blaschek, 2000; Qureshi et al., 2002; Qureshi et al., 1992). Product recovery techniques such as gas stripping, pervaporation, and liquid–liquid extraction have been applied to high-productivity reactors (Qureshi and

Maddox, 1991, 1995; Friedl et al., 1991) thereby making butanol fermentation and recovery much more attractive. These reactors offer the advantages of high productivity, use of concentrated sugar solutions, reduction in reactor size, and efficient recovery of the product. With these developments, only a few problems remain with the butanol fermentation. It is anticipated that with the recent developments in integrated fermentation and recovery, the biobutanol fermentation would be closer to commercialization. Using batch fermentation and distillative recovery the biobutanol price is estimated to be $0.55/kg, while using integrated technology it is estimated to be $0.12–0.37/ kg, which is competitive to petrochemically derived butanol. Integration of fermentation and product recovery makes the butanol fermentation economically more attractive.

8.7 Development Potential for Biobutanol Production

In a recent independent study on biobutanol commercialization conducted by the college of Commerce at the University of Illinois (Urbana, IL), it was determined that fermentatively produced biobutanol is competitive with petrochemically derived butanol. This is because of the use of integrated process technology in combination with *C. beijerinckii* BA101, which can produce 25–33 g/L total ABE in batch process. However, it is anticipated that petrochemical industries would reduce price of butanol in an attempt to prevent the fermentative production of biobutanol from being successful. Currently, petrochemical industries have a monopoly with respect to butanol market. We suggest that researchers should focus on the following problems.

8.7.1 Issues for Biological Process of Biobutanol Production

Although butanol fermentation is more economically competitive today than it was two decades ago, certain problems still remain to be resolved. These include the development of a superior culture that is able to produce elevated levels of butanol. Attempts have been made along these lines, but with limited success. *Clostridium beijerinckii* BA101 and *C. acetobutylicum* PJC4BK are examples of two strains that have been modified to produce elevated levels of solvents. However, the maximum level of solvents that these strains can produce is in the order of 25–33 g/L, which is still low from a product recovery point of view. Another problem is the low ABE yield. Due to the release of CO_2 and H_2, approximately 53% of the substrate is lost in the form of these gases. An improvement in yield would further improve the economics of the biobutanol production. Alternately, CO_2 could be captured and converted to acetone and butanol.

Another problem with biobutanol fermentation is the inability of these cultures to use sugars derived from economically available substrates such as corn fiber hydrolysate (Ebener et al., 2003). As with corn fiber hydrolysate, it is anticipated that sugars derived from hydrolyzed corn stalks, wheat straw, and rice husk could not be utilized without pretreatment of these substrates, which would add to the cost. In order to meet these challenges, new strains capable of utilizing the hydrolysates from agricultural biomass should be developed. Simultaneous saccharification and fermentation is another approach that should be considered.

Although the fermentation seems to be economically competitive or even superior to the petrochemical synthesis of butanol (as demonstrated by building new plants and reopening old ones), there is still potential for further improvement. Usually, most of the total costs of a biotechnological process are required for substrate delivery and processing. Currently, sugars (molasses, sugarcane, sugar beet) will be used as substrate (new plants in Brazil or the United Kingdom) or starch (stemming from corn) will serve this purpose (pilot plant in the United States). These compounds are at present comparatively cheap. First, this may change in future. Second, although there is still plenty of arable land that could be used for such crops, there is also an ongoing public discussion about the ethics of using sugar or corn for fuel preparation instead of nutrition.

8.7.2 New Technologies Being Developed

For future industrial application of the biobutanol fermentation, the search for an abundant substrate that does not compete with food offers a tremendous economic benefit. Such a substance might well be cellulose. Already in the former Soviet Union, the early biobutanol fermentation had been based on lignocellulose as a substrate (which only later became economically noncompetitive) (Zverlov et al., 2006). The cellulosome systems of *Clostridia* and other anaerobic bacteria are meanwhile well characterized (Schwarz, 2001; Leschine, 2005); substrate conversion over time, however, is still unsatisfactory for industrial purposes. This renders metabolic engineering of solventogenesis genes into such bacteria less promising. However, a combination of cellulose pretreatment and biobutanol fermentation certainly has a lot of potential that awaits exploitation. Examples of economically suitable pretreatments are the Iogen process (cellulose ethanol) as well as methods that have been developed by Green Biologics, Abingdon, Oxfordshire, United Kingdom (Green Biologics, 2008), and Green Sugar GmbH (Green Sugar, 2008). They claim to be able to convert all cellulosic components of biomass (including grass, leaves, etc.) into sugars that can be used for fermentation (Durre, 2007).

The second focus of development is on metabolic engineering (Durre, 2007). In principle, there are two different options: (1) the genes required for biobutanol formation can be cloned into a desired production organism, and (2) a biobutanol producer such as *C. acetobutylicum* can be mutated in a way that it only produces biobutanol (homobiobutanol producer) plus some CO_2 and H_2. The first possibility would be suitable for biotechnology workhorses such as *Escherichia coli*. Since this organism does not possess the ability to synthesize C4 compounds, the respective genes would have to come from *C. acetobutylicum* as well. A problem might become the expression of oxygen-sensitive essential clostridial enzymes, such as AdhE. Such an approach also does not take into account tolerance of the envisaged production strain against biobutanol. As already mentioned before, this alcohol damages membranes and certain membrane proteins. This seems to be less of a problem in *C. acetobutylicum*. There have already been a number of attempts to improve solvent tolerance in this organism (Tomas et al., 2003, 2004; Borden and Papoutsakis, 2007).

Another approach used over-expression of cyclopropane fatty acid synthase. Tolerance could indeed be improved, but biobutanol production was decreased, too. However, even without such targeted constructs, a number of other mutants have produced biobutanol in concentrations considered impossible a few years ago (Durre, 2005). The current best performer (238 mM biobutanol) is a *orf5*-negative strain, over-expressing *adhE* (Harris et al., 2001). Thus, it may be wise to start with *C. acetobutylicum* to construct a tailor-made production strain. Acetate and butyrate production could be inactivated by mutations in the phosphotransacetylase and phosphotransbutyrylase genes or the corresponding kinase genes, respectively. Lactate formation can be inhibited by inactivating the lactate dehydrogenase gene, acetone synthesis by knockout of *adc*, and acetoin production by a mutation in the 2-acetolactate synthase gene. A gene knockout procedure has recently been developed for the *Clostridia* (Chai, 2008). Finally, there is still another approach. A patent from the Fraunhofer-Institute in Stuttgart, Germany, describes biobutanol formation from butane by means of a specific monooxygenase (Fraunhofer IGB, 2008). Over-expression of the respective gene might result in a different, yet suitable recombinant production strain.

The third area for improvement is downstream processing. Previously, distillation has been the method to separate biobutanol from the fermentation broth. However, this procedure has a high requirement for energy, as biobutanol has a boiling point higher than water. Therefore, other recovery methods have been intensively tested (Ezeji et al., 2004). Alternative possibilities are adsorption, gas stripping, liquid–liquid extraction, perstraction, pervaporation, and RO. Adsorption with molecular sieves such as silicalite is attractive, as the energy requirement is lower than that for gas stripping and pervaporation (heat is needed for desorption) (Qureshi et al., 2005). Disadvantages, however, are removal of nutrients from the broth, low selectivity, and high prices of the resin. Gas stripping seems to be a simple, yet successful method. It has a low selectivity and does not completely remove solvents from the fermentation broth, but continuous use is possible. Recovery of products is achieved by condensation. The gases do not interfere with the microorganisms. A recent study showed it to be better suited than pervaporation (Ezeji et al., 2004). However, foam formation might cause problems (Ezeji et al., 2005). All other methods have a much higher selectivity, but suffer from other problems. Liquid–liquid extraction requires a solvent that is nontoxic to the bacteria. In perstraction, the same principle is used, but a membrane separates culture and extracting solvent. In general, membranes are expensive and often suffer from clogging and fouling problems. This limits also the use of pervaporation, by which products such as biobutanol selectively diffuse across a membrane and are evaporated for recovery, and RO, in which high pressure is used to separate the water via a semipermeable membrane and thus to yield a concentrated product broth. However, it is obvious that numerous recovery methods are available that allow a much more economic biobutanol purification than in the past (Durre, 2007).

8.7.3 Biobutanol Fermentation Waste and By-Products

In the traditional biobutanol fermentation plants a number of by-products and waste streams were generated, including a large wastewater stream, waste cell mass, and CO_2 and H_2 gases. However, due to recent developments in bioreactor design, process integration, and recovery technology, the size of some of these streams has been reduced significantly. As mentioned, the application of process integration allows for the use of concentrated sugar solutions thereby enabling the recycle of water and a reduction in waste streams. Recycling the wastewater stream allows for economic benefits in addition to being environmentally friendly. Use of highly productive bioreactors permits biomass reuse, thereby reducing waste biomass. However, the biomass that is generated can be used as an animal feed. Hydrogen gas can also be separated from the mixture of CO_2 and H_2 and used as an excellent fuel. CO_2 can be either recycled as a substrate (which is under investigation) or sold as a gas for other applications (Qureshi, 2005).

8.8 Commercialization and Economic Evaluation

8.8.1 Technology

At present, there is not any pilot commercial biobutanol plant operational in the world. However, as previously noted, in June 2006, DuPont announced that it would bring commercial volumes of biobutanol to the market in 2007. The biobutanol plant with a capacity of 30,000 t is currently under construction at British Sugar's bioethanol plant in Wissington (United Kingdom). Surprisingly, however, no comparable biobutanol projects have been announced in the United States. As with bioethanol, there is a need for cost reduction of the biobutanol by using low-cost, cellulosic waste materials as feedstock. In addition, with the entry of biobutanol, bioethanol, and biobutanol will compete for the same raw materials. For sustainability, the use of abundant, low-cost biomass is the answer for these biofuels to be cost competitive with petroleum-based fuels and to ultimately be sustainable.

In anticipation of this trend and Montana's surplus of wheat straw, Resodyn Corporation is developing a biobutanol technology, by using Montana's wheat straw as a demonstration of the feasibility for such a concept. Resodyn Corporation's process and technology designs are pragmatic, simple, and economical. The process being developed uses a pretreatment batch-type, oxidative hydrogen peroxide treatment under mild caustic conditions, an enzymatic hydrolysis of the lignocellulose into sugars, a semi-continuous two-stage fermentation and recovery stages to produce the refined biobutanol. In addition to these unit operations, the ultimate design will involve feedstock handling and storage, product purification, wastewater treatment, lignin combustion, product storage, and other utilities. Enzymes are required for the splitting of the lignocellulose into sugars and can be obtained at commercial scale at current time, therefore an on-site enzyme production will be deferred to a later point of time, but it is included in the refinement of the economical analysis. The process chemicals being used are environmentally benign and their use will be optimized to keep processing costs down. In addition, the cost and process analysis includes the utilization of a waste byproduct, for example, lignin will be used as a fuel to heat the boilers in the biobutanol plant. Resulting byproduct chemicals, for example, mineral salts are suitable as a fertilizer and can be directly applied to the cropland from which the straw was harvested.

8.8.2 Economics

A first-cut economic comparison of a corn starch-to-biobutanol with a wheat-straw-to-biobutanol plants has been completed. There are some major differences in processing starch (corn) to biobutanol versus wheat straw to biobutanol. In particular, wheat straw requires more handling: delivered straw bales must be washed, shredded, and finally milled to achieve the appropriate particle size. Corn requires milling to a meal. The chemical pretreatment to open the wheat straw to deliver the lignocellulose and the enzymatic hydrolysis take longer and are more energy intensive than for corn starch. However, corn is much more costly than wheat straw and one also is faced with the ever-present downside risks of using a price-volatile feedstock, corn, that is both an animal feed and human food.

For the corn-to-biobutanol process, starch is converted using two enzymes. However, in the wheat straw-to-biobutanol process, the hydrolysis of ligno-cellulose requires a blend of three, or more, enzymes. In the concept design aspects, for example, minimum of water and energy consumption and recycle are being taken into consideration, as well as the economic aspects. The process being developed will be operated in a semi-continuous manner; as a goal a continuous operation is envisioned for the later pilot operations and the commercial plants.

Both feedstock, corn and wheat straw, will produce, as a consequence of the anaerobic fermentation of the sugars, hydrogen, acetone, and ethanol besides other biomass. These by-products have value that must be collected, utilized, or sold for process economic viability. As an additional benefit, the wheat straw processing results to lignin residues that can be used as a fuel to supply processing heat, hence, helping to reduce processing costs.

A first-order capital investment calculation shows that the capital requirement for a wheat straw–based biobutanol plant is about $40 million higher than for a corn starch-based processing facility. This higher cost is primarily due to the additional pretreatment and separation steps needed to delignify the cellulose feedstock and to remove the lignin from the sugar syrup in the straw-to-biobutanol plant. The clear advantage of using wheat straw instead of corn starch as a raw material can be depicted from a comparison of the manufacturing costs that is shown in Table 8.8 (Yang and Biermann, 2007). Except for the feedstock price, operating chemicals, and depreciation of capital, all other expenses were kept at the same level for input to Table 8.8. This simplification is justified based upon the similarities of the two processes in this regard. Moreover, the fact that this economic analysis is provided as a preliminary estimate of the processing costs is used to justify this simplification. The raw material cost saving for using wheat straw instead of corn starch is about $0.53 per gal of biobutanol.

The total manufacturing cost for biobutanol from wheat straw is approximately $2.15 per gal, compared to approximately $2.56 per gal using corn starch as a feedstock. Table 8.8 also shows that obtaining credits from all the byproducts is critical to the biobutanol processing economics. Table 8.9 gives a summary of all critical line items. A preliminary profit-and-loss statement is given in Table 8.10.

The preliminary analysis shows that, given the underlying cost assumptions and a full plant utilization, the straw-to-biobutanol fermentation process can be very profitable with a net profit margin of 39% and a return of investment of about 40% in the first year. The values used for the byproduct credits were obtained from Qureshi and Blaschek (2001). A more detailed analysis for the byproduct credits is needed and should be adjusted with time and market condition

There are various strategies feasible to establish the economic value for the proposed wheat straw-to-biobutanol process. The direction taken will

TABLE 8.8

First Comparative Economical Analysis for Biobutanol from Wheat Straw and Corn Starch

	Using Wheat Straw		Using Corn Starch	
	Annual	Per Gallon	Annual	Per Gallon
Direct expenses				
Wheat straw	$26,870,000.00	$0.54	$53,584,500.00	$1.07
Alkali, hydrogene peroxide, enzymes	$1,000,000.00	$0.02	$792,759.00	$0.02
Operating labor	$1,650,000.00	$0.03	$1,650,000.00	$0.03
Supervisor, clerical	$250,000.00	$0.01	$250,000.00	$0.01
Utilities: steam, power, water	$11,000,000.00	$0.22	$11,000,000.00	$0.22
Maintenance, repair	$4,680,000.00	$0.09	$4,680,000.00	$0.09
Lab charges	$240,000.00	$0.00	$240,000.00	$0.00
Waste disposal	$1,000,000.00	$0.02	$1,000,000.00	$0.02
Total direct expenses	$46,690,000.00	$0.93	$73,197,259.00	$1.46
Indirect expenses				
Local taxes	$1,000,000.00	$0.02	$1,000,000.00	$0.02
Insurance	$5,000,000.00	$0.10	$5,000,000.00	$0.10
Depreciation of capital	$10,753,333.00	$0.22	$7,900,000.00	$0.16
Interest	$14,517,000.00	$0.29	$11,600,000.00	$0.23
Total indirect expenses	$31,270,333.00	$0.63	$25,500,000.00	$0.51
General expenses				
Administration	$160,000.00	$0.00	$160,000.00	$0.00
RD	$600,000.00	$0.01	$600,000.00	$0.01
Annual tax after profit	$19,700,000.00	$0.39	$19,700,000.00	$0.39
Federal taxes	$9,000,000.00	$0.18	$9,000,000.00	$0.18
Total general expense	$29,460,000.00	$0.59	$29,460,000.00	$0.59
Total manufacturing expenses	$107,420,333.00	$2.15	$128,157,259.00	$2.56
By-product credits				
Electricity credit lignin	−$4,500,000.00	−$0.09	$0.00	
Credit from biomass	−$79,000,000.00	−$1.58	−$79,000,000.00	−$1.58
Total by-product credits	−$83,500,000.00	−$1.67	−$79,000,000.00	−$1.58
Net production cost	$23,920,333.00	$0.48	$49,157,259.00	$0.98
Total revenue	$253,500,000.00		$249,000,000.00	
Gross profit	$146,079,667.00	$2.92	$120,842,741.00	$2.42
Gross profit margin	86%	86%	71%	71%

TABLE 8.9

Summary of Economical Analysis for Biobutanol Production
from Corn Starch and Wheat Straw

	Wheat Straw	Corn Starch
Capital costs	$161,300,000	$118,200,000.00
Feed cost	$26,870,000	$53,584,500.00
Operational cost	$80,550,333	$74,572,759.00
Biobutanol revenue @ # 3.4 per gallon	$170,000,000	$170,000,000.00
Revenue from byproducts	$79,000,000	$79,000,000
Credit lignin	$4,500,000	
Total credits	$83,500,000	$79,000,000.00
Total revenue	$253,500,000	$249,000,000.00
Total manufacturing expenses	$107,420,333	$128,157,259.00
Total cost per gallon	$2.15	$2.56
Gross profit	$146,079,667	$120,842,741.00
Gross profit per gallon	$2.92	$2.42
Gross margin	86%	71%

depend upon various factors. Of great importance will be the ultimate biobutanol fuel standard. For example, it must be decided on:

1. The mixture of biobutanol, acetone, and ethanol will be produced as fuel.
2. The solvent mixture will be fractionated into the individual solvents and sold to the fuel market at solvent quality and value. (Biobutanol sold as a chemical is $3.40 per gal as of January 2007.)

In addition, solid and gaseous biomass is generated. The solid biomass includes fiber, proteins, polysaccharides, and cell mass. The biogas byproducts hydrogen and carbon dioxide can be used within the plant infrastructure: hydrogen could be used as a fuel, while carbon dioxide could be used as inert gas in the anaerobic butanol fermentation process. A separation and compression of the individual gasses and its sale has also been considered (Qureshi and Blaschek, 2001). Other potential value-added products from wheat straw, for example, waxes, phytosterols, and policosanol and its sale can further improve the process economics. Therefore, thrust for the near-term research and development activity has to be on recovering and capturing the value of these byproducts in the market. The dynamics in the byproduct market in terms of price and volume development are critical to understand. Therefore, the dynamics of the by-product markets need to be included and their proximity to a biobutanol plant location need to be considered. In order to reduce the dependency on those markets, the type and amount of byproducts has to be reduced. Ultimately, research and development investment

TABLE 8.10

Preliminary Profit-and-Loss Statement for Biobutanol Production from Wheat Straw and Corn Starch

	With Wheat Straw as Feed	With Corn Starch as Feed
Revenue		
Total revenue	$179,000,000.00	$174,500,000
Costs of goods sold		
Fixed costs		
Labor	$2,140,000	$2,140,000
RD, Admin	$760,000	$760,000
Maintenance	$4,680,000	$4,680,000
Total fixed costs	$7,580,000.00	$7,580,000
Variable costs		
Feedstock	$26,870,000	$53,584,500
Process chemicals	$1,000,000	$792,759
Utilities	$11,000,000	$11,000,000
Depreciation	$16,000,000	$7,900,000
Waste disposal	$1,000,000	$1,000,000
Total variable costs	$55,870,000.00	$74,277,259
Total cost of goods sold	$63,450,000.00	$81,857,259
Total costs/gallon	$1.27	$1.64
Total costs/gallon - tax credit		
Gross profit before tax	$115,550,000.00	$92,642,741
Gross profit/gallon	$2.31	$1.85
Gross profit margin	68%	54%
Operating expenses		
Insurance	$5,000,000.00	$5,000,000
Total operating expenses	$5,000,000.00	$5,000,000
EBIT	$110,550,000.00	$87,642,741
EBIT/gallon	$2.21	$1.75
EBIT margin	65%	51%
Interest and taxes		
Interest income		
Interest expense	$14,517,000	$11,600,000
Taxes	$10,000,000	$10,000,000
Annual profit tax	$19,700,000	$19,700,000
Total interest and taxes	$44,217,000.00	$41,300,000
Net profit	$66,333,000.00	$46,342,741
Net profit per gallon	$1.33	$0.93
Net profit margin	39.00%	27%

has to be made to improve the overall yield of biobutanol at the expense of acetone and ethanol. Finally, the development of a wheat-straw-to-biobutanol technology has to be viewed as part of an overall biorefinery concept, which needs to include at a minimum the entire wheat plant and in addition other agricultural crops and residues.

8.9 Problems

1. Please identify butanol isomers and draw their chemical structure.

2. Explain why oxygenated fuel such as butanol burns more completely than petroleum fuel? Please use combustion chemical reaction scheme to illustrate your argument. Is this also the reason why gasoline needs to be added with oxygenated additives?

3. If wheat straw is used as feedstock for biobutanol production, how many pounds of feedstock are needed to make 1 gal of biobutanol? You can assume that wheat straw is composed (by weight) of 75% cellulose and hemicellulose, 20% lignin, and 5% ash. Assume that cellulose and hemicellulose are 100% recoverable and will be hydrolyzed to sugar, and 40% of it is fermentable to produce acetone, ethanol, and butanol in a ratio of 3:1:6 in weight percentage. Sixty percent of sugar goes to carbon dioxide during the fermentation. Use chemical composition of cellulose and hemicellulose you can find in the literature and follow the stoichiometric relationship to obtain your results.

4. Please identify the annul usage of gasoline in the United States and the amount of wheat straw produced annually. Use the butanol production amount you derived from problem 3 and calculate the amount of butanol that can be produced from wheat straw. What percentage of auto fuel can be replaced by wheat straw–derived butanol?

5. What is the main difference between ethanol fermentation and butanol fermentation (ABE process)? Please tabulate the pros and cons of these two processes. What is the most significant drawback of ABE fermentation?

6. The growth of *C. acetobutylicum* population in acidogenesis stage is a function of pH and is given by the following equation:

$$\mu = \frac{1}{x}\frac{dx}{dt} = \frac{\mu_m S}{K_s\left(1 + \dfrac{H^+}{k_1}\right) + S}$$

a. With a given experimental data (x and S versus t), describe how you would determine the constants μ_m, K_s, and k_1. Here x and S are cell mass and substrate concentration, respectively.

b. How would the double-reciprocal plot $1/\mu$ versus $1/S$ change with pH (or H^+ concentration)?

7. Heating water is used to maintain the fermentation temperature of a 100 L fermenter with a 80 L working volume. The production rate for butanol is targeted at 5.0 g/L-h. The desired fermentation temperature is 37°C and the feed (broth) stream is at room temperature 20°C. A heating coil is to be used. The heating water is available at 50°C. Assume that the fermenter is insulated very well. The heat loses will only come from CO_2 discharging from fermenter to the atmosphere. The heat capacity of the broth and the heating water are roughly equal. The heat capacity of CO_2 can be found in the literature. The ABE fermentation products ratio CO_2:butanol:acetone:ethanol is 60:25:12:3 by weight. You are asked to calculate the following items:

a. Heating water flow rate

b. Broth feeding flow rate

c. Cooling coil length if the coil has a 1.0 cm diameter and the overall heat transfer coefficient is 14 J/s m² °C

8. Butanol is extracted from the fermentation broth using an organic solvent in a countercurrent staged extraction unit. Partition coefficient of butanol between organic solvent and broth at pH 6.0 is $P = C_{org}/C_{broth} = 40$. The flow rate of broth is 10 L/min. If five equal volume extraction units are available to reduce the butanol concentration from 20 to 0.2 g/L in the broth (aqueous phase), determine the flow rate of the organic solvent (organic phase) as well as the size of each extraction unit.

9. Butanol is separated from ABE fermentation broth by adsorption on resin beads in a packed bed. The bed is 5 cm in diameter and contains 0.75 cm³ resin/cm bed. The overall mass transfer coefficient is 12/h. If butanol concentration in the feed is 20 g/L and is desired to be 0.2 g/L in the effluent in a 50-cm column, determine the feed flow rate through the column. The equilibrium relationship is $C_s^* = 20(C_L^*)^{1/2}$ and the operating line relationship is $C_s = 5C_L$, where C_s is g solute/L resin and C_L is g solute/L solution.

Acknowledgment

The authors would like to thank Xu-Hu Gu, Chun-Hui Gu, and Yin Liu, PhD candidates in Qingdao Institute of Bioenergy and Bioprocess Technology,

Chinese Academy of Sciences, Qingdao, China, for their contribution in collecting data and references.

References

Alsaker, K. V., Spitzer T. R., and Papoutsakis, E. T. (2004) Transcriptional analysis of spo0A overexpression in *Clostridium acetobutylicum* and its effect on the cell's response to butanol stress. *J. Bacteriol.*, 186, 1959–1971.

Bahl, H., Andersch, W. A., Braun, K., and Gottschalk, G. (1982) Effect of pH and butyrate concentration on the production of acetone and butanol by *Clostridium acetobutylicum* grown in continuous culture. *Eur. J. Appl. Microbiol. Biotechnol.*, 14, 17–20.

Borden, J. R. and Papoutsakis, E. T. (2007) Dynamics of genomic-library enrichment and identification of solvent tolerance genes for *Clostridium acetobutylicum. Appl. Environ. Microbiol.*, 73, 3061–3068.

Chai, (2008) https://hcai.nottingham.ac.uk/minton.pdf

Chandra, R. P., Mabee, W. E., Berlin, A., Pan, X., and Saddler, J. N. (2007) Substrate pretreatment: The key to effective enzymatic hydrolysis of lignocellulosics? *Adv. Biochem. Eng./Biotechnol.*, 108, 67–93.

Colby, G. D. and Chen, J.-S. (1992) Purification and properties of 3-hydroxybutyryl-coenzyme A dehydrogenase from *Clostridium beijerinckii* ("*Clostridium butylicum*") NRRL B593. *Appl. Environ. Microbiol.*, 58, 3297–3302.

Cornillot, E., Croux, C., and Soucaille, P. (1997) Physical and genetic map of the *Clostridium acetobutylicum* ATCC 824 chromosome. *J. Bacteriol.*, 179, 7426–7434.

DuPont, (2006) DuPont biobutanol fact sheet. http://www2.dupont.com/Biofuels/en_US/facts/ BiobutanolFactSheet.html

Durre, P. (1998) New insights and novel developments in clostridial acetone/butanol/isopropanol fermentation. *Appl. Microbiol. Biotechnol.*, 49, 639–648.

Durre, P. (2005) Formation of solvents in clostridia. In Durre, P. (Ed.) *Handbook on Clostridia*. CRC Press, Boca Raton, FL.

Durre, P. (2007) Biobutanol: An attractive biofuel. *Biotechnol. J.*, 2, 1525–1534.

Durre, P. and Bahl, H. (1996) Microbial production of acetone/butanol/isopropanol. *Biotechnology*, 6, 229–268.

Durre, P., Bohringer, M., Nakotte, S., Schaffer, S., Thormann, K., and Zickner, B. (2002) Transcriptional regulation of solventogenesis in *Clostridium acetobutylicum. J. Molec. Microbiol. Biotechnol.*, 4, 295–300.

Ebener, J., Qureshi, N., and Blaschek, H. P. (2003) Com fiber hydrolysis and fermentation to butanol using *Clostridium beijerinckii* BA101. *Proceedings of 25th Biotechnology Symposium on Fuels and Chemicals*, Denver, CO.

Eckert, G. and Schugerl, K. (1987) Continuous acetone–butanol production with direct product removal. *Appl. Microbiol. Biotechnol.*, 27, 221–228.

EEI (Environmental Energy Inc.), (2005) http://www.butanol.com/

Ennis, B. M., Gutierrez, N. A., and Maddox, I. S. (1986) The acetone–butanol–ethanol fermentation: A current assessment. *Process Biochem.*, 21, 131–147.

Evans, P. J. and Wang, H. Y. (1988) Enhancement of butanol formation by *Clostridium acetobutylicum* in the presence of decanol-oleyl alcohol mixed extractants. *Appl. Environ. Microbiol.*, 54, 1662–1667.

Ezeji, T. C., Karcher, P. M., Qureshi, N., and Blaschek, H. P. (2005) Improving performance of a gas stripping-based recovery system to remove butanol from *Clostridium beijerinckii* fermentation. *Bioprocess Biosyst. Eng.*, 27, 207–214.

Ezeji, T. C., Qureshi, N., and Blaschek, H. P. (2003) Production of acetone, butanol and ethanol by *Clostridium beijerinckii* BA101 and in situ recovery by gas stripping. *World J. Microbiol. Biotechnol.*, 19, 595–603.

Ezeji, T. C., Qureshi, N., and Blaschek, H. P. (2004) Butanol fermentation research: Upstream and downstream manipulations. *Chem. Rec.*, 4, 305–314.

Ezeji, T. C., Qureshi, N., Karcher, P., and Blaschek, H. P. (2006) Butanol production from corn. In Minteer, S. (Ed.) *Alcoholic Fuels: Fuels for Today and Tomorrow.* Taylor & Francis, New York.

Ezeji, T. C., Qureshi, N., and Blaschek, H. P. (2007) Bioproduction of butanol from biomass: From genes to bioreactors. *Curr. Opin. Biotech.*, 18, 220–227.

Fan, L. T. and Beardmore, D. H. (1981) The influence of major structural features of cellulose on rate of enzymatic hydrolysis. *Biotechnol. Bioeng.*, 23, 419–424.

Fontaine, L., Meynial-Salles, I., Girbal, L., Yang, X., Croux, C., and Soucaille, P. (2002) Molecular characterization and transcriptional analysis of adhE2, the gene encoding the NADH-dependent aldehyde/alcohol dehydrogenase responsible for butanol production in alcohologenic cultures of *Clostridium acetobutylicum* ATCC 824. *J. Bacteriol.*, 184, 821–830.

Fraunhofer IGB, (2008) http://www.igb.fhg.de/start.en.html

Friedl, A., Qureshi, N., and Maddox, I. S. (1991) Continuous acetone–butanol–ethanol (ABE) fermentation using immobilized cells of *Clostridium acetobutylicum* in a packed bed reactor and integration with product removal by pervaporation. *Biotechnol. Bioeng.*, 38, 518–527.

Gabriel, C. L. (1928) Butanol fermentation process. *Ind. Eng. Chem.*, 28, 1063–1067.

Gapes, J. R., Nimcevic, D., and Friedl, A. (1996) Long-term continuous cultivation of *Clostridium beijerinckii* in a two-stage chemostat with on-line solvent removal. *Appl. Environ. Microbiol.*, 62, 3210–3219.

Garcia, A., Iannotti, E. L., and Fischer, J. R. (1984) *Reverse Osmosis Application for Butanol–Acetone Fermentation.* ASAE Tech. Pap. (United States), Winter meeting of the American Society of Agricultural Engineers, Chicago, IL.

George, H. A. and Chen, J.-S. (1983) Acidic conditions are not obligatory for onset of butanol formation by *Clostridium beijerinckii* (synonym, *C. butylicum*). *Appl. Environ. Microbiol.*, 46, 321–327.

Green Biologics, (2008) http://www.greenbiologics.com/biofuels

Green Sugar, (2008) http://www.green-sugar.eu/products03_de.html

Groot, W. J., Van Der Lans, R. G. J. M., and Luyben, K. C. A. M. (1992) Technologies for butanol recovery integrated with fermentations. *Process Biochem.*, 27, 61–75.

Harris, L. M., Blank, L., Desai, R. P., and Welker, N. E. E. A. (2001) Fermentation characterization and flux analysis of recombinant strains of *Clostridium acetobutylicum* with an inactivated solR gene. *J. Ind. Microbiol. Biotechnol.*, 27, 322–328.

Harris, L. M., Desai, R. P., Welker, N. E., and Papoutsakis, E. T. (2000) Characterization of recombinant strains of the *Clostridium acetobutylicum* butyrate kinase inactivation mutant: Need for new phenomenological models for solventogenesis and butanol inhibition? *Biotechnol. Bioengin.*, 67, 1–11.

Holt, R. A., Stephens, G. M., and Morris, J. G. (1984) Production of solvents by *Clostridium acetobutylicum* cultures maintained at neutral pH. *Appl. Environ. Microbiol.*, 48, 1166–1170.

Hon, D. N.-S. (1994) Cellulose: A random walk along its historical path. *Cellulose*, 1, 1–25.

Huang, W.-C., Ramey, D. E., and Yang, S.-T. (2004) Continuous production of butanol by *Clostridium acetobutylicum* immobilized in a fibrous bed bioreactor. *Appl. Biochem. Biotechnol.*, 113–116: 887–898.

Husemann, M. H. W. and Papoutsakis, E. T. (1988) Solventogenesis in *Clostridium acetobutylicum* fermentations related to carboxylic acid and proton concentrations. *Biotechnol. Bioengin.*, 32, 843–852.

Ismaiel, A. A., Zhu, C.-X., Colby, G. D., and Chen, J.-S. (1993) Purification and characterization of a primary-secondary alcohol dehydrogenase from two strains of *Clostridium beijerinckii*. *J. Bacteriol.*, 175, 5097–5105.

Job, C., Schertler, C., Staudenbauer, W. L., and Blass, E. (1989) Selection of organic solvents for in situ extraction of fermentation products from *Clostridium thermohydrosuluricum* cultures. *Biotechnol. Techn.*, 3(5), 315–320.

Jones, D. T. and Woods, D. R. (1986) Acetone–butanol fermentation revisited. *Microbiol.*, 50, 484–524.

Jones, D. T. and Woods, D. R. (1989) Solvent production. In Minton, P. N. and Clarke, D. J. (Eds.) *Biotechnology Handbooks* #3, Plenum Press, New York, p. 135.

Keis, S., Sullivan, J. T., and Jones, D. T. (2001) Physical and genetic map of the *Clostridium saccharobutylicum* (formerly *Clostridium acetobutylicum*) NCP 262 chromosome. *Microbiology*, 147, 1909–1922.

Kollerup, F. (1986) Ethanol production by extractive fermentation. PhD thesis, Queen's University at Kingston, Ontario, Canada.

Kollerup, F. and Daugulis, A. J. (1985) Screening and identification of extractive fermentation solvents using a database. *Can. J. Chem. Eng.* 63, 919–927.

Leland, M. V. (2005) A review of pervaporation for product recovery from biomass fermentation processes. *J. Chem. Tech. Biot.*, 80, 603–629.

Leschine, S. (2005) Degradation of polymers: Cellulose, xylan, pectin, starch. In Dürre, P. (Ed.) *Handbook on Clostridia*, CRC Press, Boca Raton, FL, pp. 101–131.

Luukkonen, P., Maloney, T., Rantanen, J., Paulapuro, H., and Yliruusi, J. (2001) Microcrystalline cellulose-water interaction—a novel approach using thermoporosimetry. *Pharm. Res.*, 18, 1562–1569.

Maddox, I. S., Qureshi, N., and Gutierrez, N. A. (1993) Utilization of whey by clostridia and process technology. In Woods, D. R. (Ed.) *The Clostridia and Biotechnology*. Butterworth-Heinemann, Stoneham, MA.

Maddox, I. S., Qureshi, N., and Roberts-Thompson, K. (1995) Production of acetone–butanol–ethanol from concentrated substrates using *Clostridium acetobutylicum* in an integrated fermentation-product removal process. *Proc. Biochem.*, 30, 209–215.

Maddox, L. S. (1989) The acetone–butanol–ethanol fermentation: Recent progress in technology. *Biotechnol. Genet. Eng. Rev.*, 7, 190–220.

Mitchell, W. J. (1998) Physiology of carbohydrate to solvent conversion by clostridia. *Adv. Microb. Physiol.*, 39, 31–130.

Montoya, D., Arevalo, C., Gonzales, S., Aristizabal, F., and Schwarz, W. H. (2001) New solvent-producing *Clostridium* sp. strains, hydrolyzing a wide range of polysaccharides, are closely related to *Clostridium butyricum*. *J. Ind. Microbiol. Biotechnol.*, 27, 329–335.

Nair, R. V., Bennett, G. N., and Papoutsakis, E. T. (1994) Molecular characterization of an alcohol/aldehyde dehydrogenase gene of *Clostridium acetobutylicum* ATCC 824. *J. Bacteriol.*, 176, 871–885.

Palosaari, N. R. and Rogers, P. (1988) Purification and properties of the inducible coenzyme A-linked butyraldehyde dehydrogenase from *Clostridium acetobutylicum*. *J. Bacteriol.*, 170, 2971–2976.

Phillips, J. A. and Humphrey, A. E. (1985) Microbial production of energy: Liquid fuels. In Ghose, T. K. (Ed.) *Biotechnology and Bioprocess Engineering*, pp. 157–186. New Delhi: United India Press.

Qureshi, N. (2001) ABE production from corn: A recent economic evaluation. *J. Ind. Microbiol. Biotechnol.*, 27, 292–297.

Qureshi, N. (2005) Butanol production from agricultural biomass. In Shetty, K., Pometto, A., and Paliyath, G., (Eds.) *Food Biotechnology*. Taylor & Francis Group, Boca Raton, FL.

Qureshi, N. and Blaschek, H. P. (1999) Production of acetone butanol ethanol (ABE) by a hyper-producing mutant strain of *Clostridium beijerinckii* BA101 and recovery by pervaporation. *Biotechnol. Prog.*, 15, 594–602.

Qureshi, N. and Blaschek, H. P. (2000) Butanol production using hyper-butanol producing mutant strain of *Clostridium beijerinckii* BA101 and recovery by pervaporation. *Appl. Biochem. Biotechnol.*, 84, 225–235.

Qureshi, N. and Blaschek, H. P. (2001) ABE production from corn: A recent economic evaluation. *J. Ind. Microbiol. Biotechnol.*, 27, 292–297.

Qureshi, N., Hughes, S., Maddox, I., and Cotta, M. (2005) Energy-efficient recovery of butanol from model solutions and fermentation broth by adsorption. *Bioprocess Biosyst. Eng.*, 27, 215–222.

Qureshi, N. and Maddox, I. S. (1991) Integration of continuous production and recovery of solvents from whey permeate: Use of immobilized cells of *Clostridium acetobutylicum* in fluidized bed bioreactor coupled with gas stripping. *Bioproc. Eng.*, 6, 63–69.

Qureshi, N. and Maddox, I. S. (2005) Reduction in butanol inhibition by perstraction: Utilization of concentrated lactose/whey permeate by *Clostridium acetobutylicum* to enhance butanol fermentation economics. *Food Bioprod. Process.*, 83, 43–52.

Qureshi, N. and Maddox, I. S. (1995) Continuous production of acetone butanol ethanol using immobilized cells of *Clostridium acetobutylicum* and integration with product removal by liquid–liquid extraction. *J. Ferment. Bioeng.*, 80, 185–189.

Qureshi, N., Maddox, L. S., and Friedl, A. (1992) Application of continuous substrate feeding to the ABE fermentation: Relief of product inhibition using extraction, perstraction, stripping and pervaporation. *Biotechnol. Prog.*, 8, 382–390.

Qureshi, N., Meagher, M. M., Huang, J., and Hutkins, R. W. (2002) Acetone butanol ethanol (ABE) recovery by pervaporation using silicalite-silicone composite membrane from fed-batch reactor of *Clostridium acetobutylicum*. *J. Membrane Sci.*, 187, 93–102.

Roffler, S. R., Blanch, H. W., and Wike, C. R. (1987) In-situ recovery of butanol during fermentation Part 2: Fed-batch extractive fermentation. *Bioprocess. Eng.*, 2, 181–190.

Rogers, P. (2006) Organic acid and solvent production Part I: Acetic, lactic, gluconic, succinic and polyhydroxyalkanoic acids. *Prokaryotes*, 1, 511–755.

Schwarz, W. H. (2001) The cellulosome and cellulose degradation by anaerobic bacteria. *Appl. Microbiol. Biotechnol.*, 56, 634–649.

Sullivan, L. G. N. B. (2006) Proteome analysis and comparison of *Clostridium acetobutylicum* ATCC 824 and SpoOA strain variants. *J. Ind. Microbiol. Biotechnol.*, 33, 298–308.

Tomas, C. A., Beamish, J. A., and Papoutsakis, E. T. (2004) Transcriptional analysis of butanol stress and tolerance in *Clostridium acetobutylicum*. *J. Bacteriol.*, 186, 2006–2018.

Tomas, C. A., Welker, N. E., and Papoutsakis, E. T. (2003) Overexpression of groESL in *Clostridium acetobutylicum* results in increased solvent production and tolerance, prolonged metabolism, and changes in the cell's transcriptional program. *Appl. Environ. Microbiol.*, 69, 4951–4965.

Toth, J., Ismaiel, A. A., and Chen, J.-S. (1999) Purification of a coenzyme A-acylating aldehyde dehydrogenase and cloning of the structural gene from *Clostridium beijerinckii* NRRL B593. *Appl. Environ. Microbiol.*, 65, 4973–4980.

Traxler, R. W., Woods, E. M., Mayer, J., and Wilson, M. P. (1985) Extractive fermentation for the production of butanol. *Dev. Ind. Microbiol.*, 26, 519–525.

USDOE, (2003) Energy Information Administration: http://www.eia.doe.gov/

Walton, M. T. and Martin, J. L. (1979) Production of butanolacetone by fermentation. In Perlman, H. J. P. A. D., (Ed.) *Microbial Technology*, 2nd edn. Academic Press, New York.

Wang, D. I. C., Cooney, C. L., Demain, A. L., Gomez, R. F., and Sinskey, A. J. (1979) Production of acetone and butanol by fermentation. MIT Quarterly report to the U.S. Department of Energy, COG-4198-9, 141–149.

Waterson, R. M., Castellino, F. J., Hass, G. M., and Hill, R. L. (1972) Purification and characterization of crotonase from *Clostridium acetobutylicum*. *J. Biol. Chem.*, 247, 5266–5271.

Weizmann, C. (1919) Production of acetone and alcohol by bacteriological processes. United States Patent 1,315,585, September 9, 1919.

Welch, R. W., Rudolph, F. B., and Papoutsakis, E. T. (1989) Purification and characterization of the NADH-dependent butanol dehydrogenase from *Clostridium acetobutylicum* (ATCC 824). *Arch. Biochem. Biophys.*, 273, 309–318.

Westheimer, F. H. (1969) Acetoacetate decarboxylase from *Clostridium acetobutylicum*. *Methods Enzymol.*, 14, 231–241.

Wiesenborn, D. P., Rudolph, F. B., and Papoutsakis, E. T. (1988) Thiolase from *Clostridium acetobutylicum* ATCC 824 and its role in the synthesis of acids and solvents. *Appl. Environ. Microbiol.*, 54, 2717–2722.

Wiesenborn, D. P., Rudolph, F. B., and Papoutsakis, E. T. (1989) Coenzyme A transferase from *Clostridium acetobutylicum* ATCC 824 and its role in the uptake of acids. *Appl. Environ. Microbiol.*, 55, 323–329.

Wilkinson, S. R. and Young, M. (1995) Physical map of the *Clostridium beijerinckii* (formerly *Clostrdium acetobutylicum*) NCIMB 8052 chromosome. *J. Bacteriol.*, 177, 439–448.

Winzer, K., Lorenz, K., and Zickner, B. (2000) Differential regulation of two thiolase genes from *Clostridium acetobutylicum* DSM 792. *J. Molec. Microbiol. Biotechnol.*, 2, 531–541.

Yan, R.-T. and Chen, J.-S. (1990) Coenzyme A-acylating adlehyde dehydrogenase from *Clostridium beijerinckii* NRRL B592. *Appl. Environ. Microbiol.*, 56, 2591–2599.

Yang, F. X. (2008) The ResonantSonic Enhanced Mixer and Coalescer (RSEMC) as an advanced solvent extraction technology. NSF SBIR Phase II finale report. NSF contract DMII-0321499.

Yang, F. X. and Biermann, M. (2007) Butanol production with Montana wheat straw (Montana Board of Research and Commercialization Technology MBRCT), Contract No. 08-56, 2007–2009.

Zverlov, V. V., Berezina, O., Velikodvorskaya, G. A., and Schwarz, W. H. (2006) Bacterial acetone and butanol production by industrial fermentation in the Soviet Union: Use of hydrolyzed agricultural waste for biorefinery. *Appl. Microbiol. Biotechnol.*, 71, 587–597.

9

Chemical Conversion Process for Biodiesel Production

Dong-Zhi Wei, Fangxiao Yang, and Erzheng Su

CONTENTS

9.1 Introduction 338
9.2 Feedstocks for Biodiesel Production 342
    9.2.1 Biolipid Feedstocks 342
        9.2.1.1 Plant-Derived Oils..346
        9.2.1.2 Animal Fats...346
        9.2.1.3 Waste Oils and Fats...349
        9.2.1.4 Microalgal Oils 351
    9.2.2 Properties of Oils/Fats That Affect Biodiesel Production 353
    9.2.3 Acyl Receptors..354
9.3 Production of Biodiesel 356
    9.3.1 Direct Use of Vegetable Oil and Blending.................................357
    9.3.2 Microemulsification..358
    9.3.3 Pyrolysis or Thermal Cracking...359
        9.3.3.1 Pyrolysis of Vegetable Oils and Fats 359
        9.3.3.2 Pyrolysis of Vegetable Oil Soaps......................................361
        9.3.3.3 Catalytic Cracking of Vegetable Oils..............................362
        9.3.3.4 Pyrolysis Mechanisms of Vegetable Oils.......................363
    9.3.4 Esterification 365
    9.3.5 Transesterification...367
        9.3.5.1 General Aspects of Transesterification 367
        9.3.5.2 Acid-Catalyzed Transesterification 368
        9.3.5.3 Alkali-Catalyzed Transesterification 369
        9.3.5.4 Kinetics of Acid/Alkali-Catalyzed
                Transesterification... 370
        9.3.5.5 Supercritical Alcohol Transesterification....................... 372
        9.3.5.6 Lipase-Catalyzed Transesterification.............................. 374
        9.3.5.7 *In Situ* Transesterification 391

9.3.6   Effect of Different Parameters on Transesterification 393
         9.3.6.1   Molar Ratio of Alcohol to Triglyceride and Type
                   of Alcohol 393
         9.3.6.2   Catalyst Type and Concentration 394
         9.3.6.3   Moisture and Free Fatty Acid Content 395
         9.3.6.4   Purity of Reactants.. 396
         9.3.6.5   Reaction Temperature and Time.................................... 396
         9.3.6.6   Intensity of Mixing 397
         9.3.6.7   Organic Cosolvents... 397
9.4   Unit Operation Processes for Biodiesel Production............................ 398
         9.4.1   General Flowchart for Biodiesel Production Process 398
         9.4.2   Feedstock Pretreatment 398
                  9.4.2.1   Oils/Fat Feedstocks Pretreatment 398
                  9.4.2.2   Pretreatment of High Free Fatty Acid Feedstocks 404
                  9.4.2.3   Preparation of Other Feedstocks 406
         9.4.3   Transesterification Process 406
                  9.4.3.1   Transesterification with Batch Processing................... 406
                  9.4.3.2   Transesterification with Continuous Processing........ 408
                  9.4.3.3   Noncatalyzed Transesterification System.................... 409
         9.4.4   Alkyl Esters Purification... 409
                  9.4.4.1   Ester–Glycerol Separation.. 409
                  9.4.4.2   Ester Washing.. 411
                  9.4.4.3   Other Ester Treatments 412
         9.4.5   Recovery of Side Streams.. 412
                  9.4.5.1   Alcohol Recovery 412
                  9.4.5.2   Glycerol Refining... 413
                  9.4.5.3   Wastewater Treatment.. 414
         9.4.6   Basic Plant Equipment for Biodiesel Production...................... 414
         9.4.7   Integrated Biodiesel Production Processes 416
                  9.4.7.1   The BIOX Processing System.. 416
                  9.4.7.2   High Free Fatty Acid Grease Feed Biodiesel
                            Production Process 416
         9.4.8   Biodiesel Production Economic Analysis................................. 419
9.5   Problems.. 421
References.. 422

9.1 Introduction

The majority of the world's energy needs is supplied through petrochemical sources, coal, and natural gases; these sources are finite and at current usage rates will be consumed shortly (Srivastava and Prasad, 2000). Diesel fuel is largely utilized in the transport, agriculture, commercial, domestic, and industrial sectors for the generation of power/mechanical energy. The high energy demand in the industrialized world as well as in the developing countries and pollution problems caused due to the widespread use of fossil fuels make it increasingly necessary to develop the renewable energy sources of limitless duration and smaller environmental impact than the traditional one. This has stimulated recent interest in alternative sources for petroleum-based fuels. An alternative fuel must be technically feasible, economically competitive, environmentally acceptable, and readily available. One possible alternative to fossil fuel is the use of derivatives of oils of plant origin such as vegetable oils. This alternative diesel fuel can be termed as biodiesel.

The history of the use of vegetable oils in diesel engines goes back more than a 100 years. Transesterification of a vegetable oil was conducted as early as 1853 by scientists Duffy and Patrick, many years before the first diesel engine became functional. Rudolf Diesel demonstrated a diesel engine running on peanut oil at the World Fair in Paris, France in 1900 (Nitske and Wilson, 1965). This engine stood as an example of Diesel's vision because it was powered by peanut oil, though not biodiesel, since it was not transesterified. Diesel believed that the utilization of biomass fuel was the real future of his engine. As he stated in his 1912 speech saying "the use of vegetable oils for engine fuels may seem insignificant today, but such oils may become, in the course of time, as important as petroleum and the coal-tar products of the present time." In remembrance of Rudolf Diesel's first German run, August 10 has been declared International Biodiesel Day.

However, during the 1920s, diesel engine manufacturers decided to alter their engines utilizing the lower viscosity of petrodiesel (a fossil fuel), rather than vegetable oil (a biomass fuel). All petroleum industries were able to make inroads in fuel markets because their fuel was much, much cheaper to produce than the biomass alternatives, ignoring that years ahead it would bring high pollution costs. The result, for many years, was a near elimination of the biomass fuel production infrastructure.

Despite the widespread use of fossil petroleum-derived diesel fuels, interest in vegetable oils as fuels in internal combustion engines was reported in several countries during the 1920s and 1930s and later during World War II. Belgium, France, Italy, the United Kingdom, Portugal, Germany, Brazil, Argentina, Japan, and China had been reported to have tested and used vegetable oils as diesel fuels during this time (Feuge and Gros, 1949).

On August 31, 1937, Chavanne of the University of Brussels (Belgium) was granted a patent for a "Procedure for the transformation of vegetable oils for their uses as fuels" (Chavanne, 1937). This patent described the alcoholysis (often referred to as transesterification) of vegetable oils using methanol and ethanol in order to separate the fatty acids from the glycerol by replacing the glycerol with short linear alcohols. This appears to be the first account of the production of what is known as "biodiesel" today (Knothe, 2001).

More recently, in 1977, Brazilian scientist Expedito Parente produced biodiesel using transesterification with ethanol, and again filed a patent for the same process. This process is classified as biodiesel by international norms, conferring a "standardized identity and quality. No other proposed biofuel has been validated by the motor industry." Research into the use of transesterified sunflower oil, and refining it to diesel fuel standards, was initiated in South Africa in 1979. By 1983, the process for producing fuel-quality, engine-tested biodiesel was completed and published internationally (SAE Technical Paper series no. 831356, 1983). An Austrian company, Gaskoks, obtained the technology from the South African Agricultural Engineers; the company erected the first biodiesel pilot plant in November 1987, and the first industrial-scale plant in April 1989 (with a capacity of 30,000 tons of rapeseed per annum).

Throughout the 1990s, plants were opened in many European countries, including the Czech Republic, Germany, and Sweden. France launched local production of biodiesel fuel from rapeseed oil, which is mixed into regular diesel fuel at a level of 5%, and into the diesel fuel used by some captive fleets at a level of 30%. During the same period of time, nations in other parts of the world also saw local production of biodiesel starting to shoot up; by 1998, the Austrian Biofuels Institute had identified 21 countries with commercial biodiesel projects. Hundred percent biodiesel fuel is now available at many normal service stations across Europe.

In September 2005, Minnesota became the first U.S. state to mandate that all diesel fuel sold in the state contain part biodiesel, requiring a content of at least 2% biodiesel. The technical and economic aspects of the production and use of biodiesel have been extensively studied for the past two decades. Biodiesel has progressed from experimental fuels to the initial stage of commercialization. To date, biodiesel could be approaching a milestone in its development: it is being accepted as a standard item on new vehicle technology.

Biodiesel is defined as the fatty acid alkyl monoesters derived from renewable feedstocks, such as vegetable oils, animal fats, and used cooking oils. Because plants produce oils from sunlight and air, and can do so year after year on cropland, these oils are renewable. Animal fats are produced when the animal consumes plant oils and other fats, and they too are renewable. Used cooking oils are mostly made from vegetable oils, but may also contain animal fats, and are both reusable and renewable. Biodiesel is biodegradable, and does not contain any sulfur, aromatic hydrocarbons, metals, or crude oil residues. Moreover, it helps to reduce the greenhouse effect. A 1998 biodiesel lifecycle study, jointly sponsored by the U.S. Department of Energy (USDOE) and the U.S. Department of Agriculture (USDA) (NBB, 2003), concluded that biodiesel reduces net CO_2 emissions by 78% compared to petroleum diesel (Sheehan et al., 1998a). This is due to biodiesel's closed carbon cycle. The CO_2 released into the atmosphere when biodiesel is burned is recycled by growing plants, which are later processed into fuel. In addition, the emissions of blended biodiesel with petroleum diesel can also be reduced. Sheehan et al. found that the B20 (20% biodiesel/80% conventional No. 2 diesel blend) reduces net CO_2 emission by 16%. Thus biodiesel reduces greenhouse gas emissions and helps mitigate "global warming." Combustion of biodiesel can also bring in other environment benefits because it has lower particulate matter emissions compared to that of petroleum diesel. According to research at the Southwest Research Institute, compared with low-sulfur No. 2 diesel, B20 with an oxidation catalyst reduced particulate matter by 45%, total hydrocarbons by 65%, and carbon monoxide by 41% (Von Wedel, 1999). Biodiesel is the only alternative fuel to fully comply with the health effects testing requirements of the Clean Air Act (NBB, 2003). Since it is made domestically from renewable resources such as soybeans, its use decreases dependence on foreign oil and contributes to economy.

There is a great structural similarity between biodiesel and petrodiesel molecules, as highlighted in Figure 9.1, where methyl palmitate and methyl oleate are examples of typical biodiesel molecules and cetane represents a typical petrodiesel molecule.

This structural similarity allows biodiesel to replace petrodiesel in conventional engines with little or no modifications. Like petroleum diesel, biodiesel operates in compression-ignition engines, such as private vehicles, industrial fleet trucks, off-road and farm equipment, boat engines, and generators. Biodiesel maintains the payload capacity and range of diesel. Because biodiesel is oxygenated, it is a better lubricant than diesel fuel, and increases the life of engines. Diesel engines combust biodiesel more completely than petroleum diesel (National Biodiesel Board). The flash point of biodiesel in its pure form is more than 149°C (300°F) compared to about 52°C (125°F) for regular No. 2 diesel. This makes biodiesel the safest fuel to use, handle, and store. With its relatively low emission profile, it is an ideal fuel for use in sensitive environments, such as marine areas, national parks, and forests.

Biodiesel can be used in several different ways. You can use 1%–2% biodiesel as a lubricity additive, which could be especially important for ultralow sulfur diesel fuels, which may have poor lubricating properties. You can blend 20% biodiesel with 80% diesel fuel (B20) for use in most applications that use diesel fuel. You can even use it in its pure form (B100) if you take proper precautions. A number following the "B" indicates the percentage of biodiesel in a gallon of fuel, where the remainder of the gallon can be No. 1 or No. 2 diesel, kerosene, jet A, JP8, heating oil, or any other distillate fuel.

Today, B20 is the most common biodiesel blend because it balances property differences with conventional diesel, performance, emission benefits, and costs. B20 can be used in equipment designed to use diesel fuel. Equipment that can use B20 includes compression-ignition engines, fuel oil and heating oil boilers, and turbines. Higher blend levels, such as B50 or B100, require

$CH_3OCO(CH_2)_7CH=CH(CH_2)_7CH_3$	Methyl oleate	(biodiesel)
$CH_3OCO(CH_2)_{14}CH_3$	Methyl palmitate	(biodiesel)
$CH_3(CH_2)_{14}CH_3$	Cetane	(petrodiesel)

FIGURE 9.1
The chemical structure similarity of biodiesel with petrodiesel.

special handling and fuel management and may require equipment modifications such as the use of heaters or changing seals and gaskets that come in contact with the fuel to those compatible with high blends of biodiesel.

9.2 Feedstocks for Biodiesel Production

Biodiesel can be produced from any material that contains fatty acids, either bonded to other molecules or present as free molecules. Thus, various vegetable fats and oils, animal fats, waste greases, and edible oil processing wastes can be used as feedstocks for biodiesel production. The choice of feedstock is based on such variables as local availability, cost, government support, and performance as a fuel. Different countries are looking for different types of fats and oils as feedstocks for biodiesel. For example, soybean oil in the United States, rapeseed and sunflower oils in Europe, palm oil in southeast Asia (mainly Malaysia and Indonesia), and coconut oil in the Philippines are being considered (Ghadge and Raheman, 2005; Demirbas, 2006; Meher et al., 2006a,b; Sarin et al., 2007).

Fats and oils are primarily water-insoluble, hydrophobic substances in the plant and animal kingdom. The main constituent of oils and fats is triglycerides, which compose about 90%–98% of total mass (Srivastava and Prasad, 2000). In a triglyceride molecule, the weight of the glycerol is about 41 g whereas the weights of fatty acid radicals are in the range of 650–790 g. Thus, it is understood that fatty acid radicals comprise most of the reactive groups in the triglyceride molecule, and they greatly affect the characteristics of oils and fats. Therefore, the importance of comprehensive investigation on fatty acids comprising about 94%–96% (w/w) of the triglyceride molecule is clearly seen. The physical and chemical properties of biodiesel basically depend on the fatty acid distribution of the triglyceride used in the production. The common fatty acids and the corresponding fatty acid methyl esters (biodiesel) are shown in Table 9.1. The fatty acid distributions of some feedstocks commonly used in biodiesel production are shown in Table 9.2.

9.2.1 Biolipid Feedstocks

A variety of biolipids (biolipds are lipids from biological sources) can be used as feedstocks to produce biodiesel. There is no industry-wide standard or specification for feedstocks used for biodiesel production. The standard or specification required by a production facility depends on the process used for making biodiesel as well as the requirements for product yield and purity. Hence, feedstock specifications vary among producers. Biodiesel produced today is mainly made from plant-derived oils, animal fats, waste oils and fats, and microbial oils.

TABLE 9.1

Chemical Structures of Common Fatty Acids and Their Methyl Esters

Fatty Acid	Chemical Structure	Methyl Ester	Chemical Structure
Caprylic(C8)	$CH_3(CH_2)_6COOH$	Caprylate	$CH_3(CH_2)_6COOCH_3$
Capric(C10)	$CH_3(CH_2)_8COOH$	Caprote	$CH_3(CH_2)_8COOCH_3$
Lauric(C12)	$CH_3(CH_2)_{10}COOH$	Laurate	$CH_3(CH_2)_{10}COOCH_3$
Myristic(C14)	$CH_3(CH_2)_{12}COOH$	Myristate	$CH_3(CH_2)_{12}COOCH_3$
Palmitic(C16)	$CH_3(CH_2)_{14}COOH$	Palmitate	$CH_3(CH_2)_{14}COOCH_3$
Palmitoleic(C16:1)	$CH_3(CH_2)_5CH=CH(CH_2)_7COOH$	Palmitoleate	$CH_3(CH_2)_5CH=CH(CH_2)_7COOCH_3$
Stearic(C18)	$CH_3(CH_2)_{16}COOH$	Stearate	$CH_3(CH_2)_{16}COOCH_3$
Oleic(C18:1)	$CH_3(CH_2)_7CH=CH(CH_2)_7COOH$	Oleate	$CH_3(CH_2)_7CH=CH(CH_2)_7COOCH_3$
Linoleic(C18:2)	$CH_3(CH_2)_4CH=CHCH_2CH=CH (CH_2)_7COOH$	Linoleate	$CH_3(CH_2)_4CH=CHCH_2CH=CH(CH_2)_7COOCH_3$
Linolenic(C18:3)	$CH_3(CH_2)_2CH=CHCH_2CH=CHCH_2CH=CH(CH_2)_7COOH$	Linolenate	$CH_3(CH_2)_2CH=CHCH_2CH=CHCH_2$ $CH=CH(CH_2)_7COOCH_3$
Arachidic(C20:0)	$CH_3(CH_2)_{18}COOH$	Arachidate	$CH_3(CH_2)_{18}COOCH_3$
Eicosenoic(C20:1)	$CH_3(CH_2)_7CH=CH(CH_2)_9COOH$	Eicosenoate	$CH_3(CH_2)_7CH=CH(CH_2)_9COOCH_3$
Behenic(C22:0)	$CH_3(CH_2)_{20}COOH$	Behenate	$CH_3(CH_2)_{20}COOCH_3$
Eurcic(C22:1)	$CH_3(CH_2)_7CH=CH(CH_2)_{11}COOH$	Eurcate	$CH_3(CH_2)_7CH=CH(CH_2)_{11}COO CH_3$

Sources: Adapted from Canakci, M. and Sanli, H., *J. Ind. Microbiol. Biotechnol.*, 35, 5, 431, 2008; Jüri, K. et al., *Proc. Estonian Acad. Sci. Chem.*, 51, 2, 75, 2002.

TABLE 9.2

Fatty Acid Distributions of Some Biodiesel Feedstocks

Oils/Fats	Fatty Acids (%)											
	$C_{16:0}$	$C_{16:1}$	$C_{18:0}$	$C_{18:1}$	$C_{18:2}$	$C_{18:3}$	$C_{20:0}$	$C_{20:1}$	$C_{22:0}$	$C_{22:1}$	$C_{24:0}$	Others
Almond kernel	6.5	0.5	1.4	70.7	20.0							0.9
Bay laurel leaf	25.9	0.3	3.1	10.8	11.3	17.6						31
Borage	12.9	0.2	4.3	19.1	39.0	18.7	0.3	3.5		2.0		
Candlenut	5.1	0.4	2.7	18.8	46.9	25.4	0.2	0.4				
Corn	11.7		1.9	25.2	60.6	0.5	0.2					
Cottonseed	28.3		0.9	13.3	57.5							
Crambe	2.1		0.7	18.9	9.0	6.9	2.1		0.8	58.5	1.1	
Groundnut	8.5		6.0	51.6	26.0							
Hazelnut kernel	4.9	0.2	2.6	81.4	10.5							0.3
Jatropha	16.4	1.0	6.2	37.0	39.2		0.2					
Karanj	10.2		7.0	51.8	17.7	3.6	1.6	1.2			1.5	5.4
Linseed	5.0		2.0	20.0	18.0	55.0						
Olive	11.8	1.5	2.7	74.2	8.5	0.7	0.4	0.3				
Palm oil	42.6	0.3	4.4	40.5	10.1	0.2			2.5			1.1
Peanut	11.4		2.4	48.3	32.0	0.9	1.3				1.2	

Poppy seed	12.6	0.1	1.9	13.5	77.0			0.2
Rapeseed	3.5		0.9	64.4	22.3	8.2		
Rice bran	11.7–16.5		1.7–2.5	39.2–43.7	26.4–35.1		0.4–0.6	0.4–0.9
Rubber seed	10.1	0.3	8.8	24.6	38.9	17.1		
Sesame	13.0		4.0	53.0	30.0			
Soybean	11.4		4.4	20.8	53.8	9.3	0.3	
Soybean soapstock	17.2		4.4	15.7	55.6	7.1		
Sunflower	7.1		4.7	25.5	62.4		0.3	
Tung				4–13	8–15	72–88		
Wheat grain	20.6	1	1.1	16.6	56	2.9		1.8
Tallow	29.0		24.5	44.5				
Lard	28–30		12–18	4–50	7–13			1–2
Yellow grease	23.24		3.79	12.96	44.32	6.07	0.67	2.43
Brown grease	22.83	3.13	12.54	42.36	12.09	0.82		1.66
Butter	25–32	2–5	25–32	22–29	3			

Sources: Data summarized from Canakci, M. and Sanli, H., *J. Ind. Microbiol. Biot.*, 35, 5, 431, 2008; Jüri, K. et al., *Proc. Estonian Acad. Sci. Chem.*, 51, 2, 75, 2002; Casimir, C.A. et al., *J. Agric. Food Chem.*, 55, 8995, 2007.

9.2.1.1 Plant-Derived Oils

Plant-derived oils as a substitute raw material for diesel were considered a good energy supply because they are carbon dioxide neutral. This stems from the fact that green plants grow through the photosynthesis process with CO_2 as a carbon source. Therefore, the combustion of plant-derived oils will release carbon dioxide, which has previously been fixed through photosynthesis. Plant-derived fuels are renewable, inexhaustible, nontoxic, and biodegradable, with energy content similar to that of fossil diesel fuel. Obviously, obtaining fuel from vegetable oils and fats is more expensive than doing so from petroleum-based fuels at the present time. This is due, in part, to the competition between vegetable oils and fats for biodiesel production with that for the food, feed, and oleochemical industries. Genetic engineering may help lower the cost in the near future as new crops will be developed with high oil content specifically for nonfood use, especially from underutilized crops and plants. If petroleum prices continue to increase, then biodiesel from plant oils will become competitive and commercially feasible. The source of oil crops will depend on their availability and varies by regions and countries, but these crops must be cheap and easy to grow, capable of large-scale production (high yield/acre), and rich in oil percentage, and have a low cost of production (Pinto et al., 2005). Refined oils are more expensive but have a low production scale. Of all the vegetable oils available, high oleic acid containing oils are preferred because of the increased stability of their alkyl esters on storage and improved fuel properties (Casimir et al., 2007). The main plants whose oils have been considered as feedstock for biodiesel are soybean, rapeseed, palm kernel, sunflower, cottonseed, safflower, peanut/groundnut, and Jatropha oil. Others in the contention are mustard, hemp, castor oil, and so on. There is ongoing research into finding more suitable crops and improving oil yield. A list of oils that have been tried for biodiesel production is provided in Table 9.3 along with the researchers who initiated the work. Table 9.4 shows the leading biodiesel sources used by various researchers since 2000.

9.2.1.2 Animal Fats

Animal fats are rendered tissue fats that can be obtained from a variety of animals. Although mentioned frequently, they have not been studied to the same extent as vegetable oils. Some methods applicable to vegetable oils are not applicable to animal fats because of natural property differences.

Animal fats are divided into several categories, and there is sometimes overlap. This can lead to confusion when trying to identify animal-based products with specific characteristics. Generally, animal fats are products derived from meat-processing facilities and are solid at room temperature. They include tallow from beef processing, lard and choice white grease from pork processing, and poultry fat (chicken fat) from poultry processing. Other animal fats, such as cod liver oil, crocodile oil, dippel's oil, emu oil,

TABLE 9.3

Biodiesel Production from Different Plant-Derived Oils

Year	Feedstock	Acyl Acceptor	Catalyst	Reference
1999	Artichoke	Methanol/ethanol	KOH/NaOH	Encinar et al. (2002)
2002				Encinar et al. (1999)
2006	Canola	Methanol/ethanol	KOH	Kulkarni et al. (2006)
2007		Mixtures of methanol/ethanol	KOH	Kulkarni et al. (2007)
2007		Methanol	H$_2$SO$_4$	Ataya et al. (2007)
2007		Methanol	Heterogeneous base	D'Cruz et al. (2007)
2007		Methanol	Mg–Al hydrotalcite	Ilgen et al. (2007)
2004	Castor	Ethanol	Lipase	De Oliveira et al. (2004)
2005		Methanol	Lipase	Yang et al. (2005)
2007		Methanol/ethanol	Supercritical conditions/lipase	Varma and Madras (2007)
2006	Coconut	Methanol	Acidic and basic solids	Jitputti et al. (2006)
2007	Corn	Methanol	Lipase	Li et al. (2007a,b)
2006	Cottonseed	Methanol	Ionic liquids	Wu et al. (2006)
2007		Methanol	Lipase	Royon et al. (2007)
2007		Dimethyl carbonate	Lipase	Su et al. (2007)
2007		Methanol/ethanol	Solid basic	Cui et al. (2007)
2007		Methanol	KOH	Anan and Danisman (2007)
2008		Methanol/ethanol	Supercritical condition	Demirbas (2008)
2004	Jatropha	Methanol/ethanol	Lipase	Shah et al. (2004)
2007		Methanol	Micro-NaOH	Tang et al. (2007)
2008		Methanol	H$_2$SO$_4$	Berchmans and Hirata (2008)
2006	Jojoba	Methanol	Sodium methoxide	Canoira et al. (2006)
2007		Methanol	KOH	Bouaid et al. (2007)
2006	Karanj	Propan-2-ol	Lipase	Modi et al. (2006)
2007		Ethyl acetate	Lipase	Modi et al. (2007)

(continued)

TABLE 9.3 (continued)

Biodiesel Production from Different Plant-Derived Oils

Year	Feedstock	Acyl Acceptor	Catalyst	Reference
2005	Mustard	Methanol	KOH	Vicente et al. (2005)
2006	Neem	Methanol	NaOH	Nabi et al. (2006)
2006	Olive	Methanol	Lipase	Sanchez and Vasudevan (2006)
2007		Methanol/methyl acetate	Lipase	Coggon et al. (2007)
2001	Palm	Methanol	H_2SO_4	Crabbe et al. (2001)
2006		Methanol	Lipase	Mahabubur et al. (2006)
2007		Methanol	KF/Al_2O_3	Bo et al. (2007)
2008		Ethanol	KOH	Alamu et al. (2008)
2001	Rapeseed	Methanol	Supercritical condition	Saka and Kusdiana (2001)
2004		Methanol	KOH	Li et al. (2006)
2006		Methanol	Lipase	Jeong et al. (2004)
2008		Methanol	Supported CaO/MgO	Yan et al. (2008)
2005	Rice bran	Methanol	Lipase	Lai et al. (2005)
2005		Methanol	H_2SO_4	Lai et al. (2005)
2008		Methanol	Tin compounds	Einloft et al. (2008)
2008	Safflower	Methanol	KOH, NaOH, $KOCH_3$, and $NaOCH_3$	Rashid and Anwar (2008)
2004	Soybean	Methanol	Zeolite and metal	Suppes et al. (2004)
2004		Methanol	H_2SO_4	Goff et al. (2004)
2006		Methanol	Zn/I^{2-}	Li et al. (2006)
2007		Methanol	Tin(IV) complexes	Ferreira et al. (2007)
2007		Methanol	SrO	Liu et al. (2007)
2007		Ethyl acetate	Lipase	Kim et al. (2007)
2002	Sunflower	Methanol	Lipase	Belafi-Bako et al. (2002)
2007		Methanol	CaO	Granados et al. (2007)
2007		Methanol	KOH	Stamenkovic et al. (2007)
2008	Tung	Methanol	Amberlyst-15	Park et al. (2008)

TABLE 9.4

Leading Biodiesel
Sources Cited in
Scientific Articles

Source	Article Number
Soybean oil	49
Recycled oil	26
Rapeseed oil	25
Sunflower oil	21
Palm kernels oil	9
Canola oil	9
Olive oil	8
Animal fats	8
Cotton seed oil	8
Corn oil	6
Fatty acids	6
Castor oil	5
Linseed oil	4
Safflower oil	4
Brassica carinata	4
Hazelnut	4
Clays with oils	4

neatsfoot oil, shark liver oil, skunk oil, and whale oil, also can be potential feedstocks for biodiesel production. Animal fats are generally marketed directly by renderers or animal processors (http://en.wikipedia.org/wiki/Category:Animal_fats).

Many researchers have produced biodiesel from different animal fats. Nelson reported the energetic and economic feasibility associated with the production, processing, and conversion of beef tallow to biodiesel (Nelson and Schrock, 2006). Lu and coworkers investigated the enzymatic synthesis of biodiesel from lard with immobilized *Candida* sp. 99–125 as the catalyst (Lu et al., 2007). Lee and coworkers have also produced biodiesel from fractionated lard by lipase- or base-catalyzed reaction (Lee et al., 2002). Researchers at the University of Arkansas have investigated supercritical methanol (SCM) as a method of converting chicken fat into biodiesel fuel (http://www.newswise.com/articles/view/536409/). The production of fatty acid ethyl esters (FAAEs) from fish oil using ultrasonic energy and alkaline catalysts dissolved in ethanol has been evaluated by a group of investigators (Armenta et al., 2007).

9.2.1.3 Waste Oils and Fats

Most biodiesel is currently produced from high-quality food-grade vegetable oils and animal fats using methanol and a catalyst. The end cost of the biodiesel mainly depends on the price of feedstock. The high cost of the food-grade oils and fats is the major factor that prevents biodiesel's large-scale application in daily use. Unfortunately, biodiesel's economic situation has gone from bad to worse because of the increase in the vegetable oil and animal fats prices in the last years. In the mid-1990s, the cost of feedstock accounted for 60%–75% of the total cost of biodiesel fuel (Krawczyk, 1996), but today more than 85% of the costs of production are filed up in feedstock costs (Haas et al., 2006). To become an economically viable alternative fuel and to survive in the market, biodiesel must compete economically with diesel fuel. However, the raw material cost of biodiesel is already higher than the final cost of diesel fuel. Nowadays, biodiesel unit price is 1.5–3.0 times higher than that of petroleum-derived diesel fuel depending on feedstock (Zhang et al., 2003a,b; Demirbas, 2007). In order to make biodiesel an economically suitable fuel and increase its marketability, its high cost must be lowered. The low-cost and profitable biodiesel can be produced from waste oils and fats such as used frying oils and fats, greases, and soapstocks (Canakci and Gerpen, 2001; Schnepf, 2003; Zhang et al., 2003a,b). The decrease in the feedstock cost will reduce the biodiesel final price, and therefore price gaps between biodiesel and petrodiesel can be lowered to an acceptable value.

Every year many millions of tons of waste oils and fats are collected and used in a variety of ways throughout the world. These oils contain some degradation products, water, and foreign material from cooking or other processing processes. However, analyses of used vegetable oils indicate that the differences between used and unused fats are not significant and in most cases simple heating and filtration can remove most of water and solid particles and the resulting material is good enough for subsequent transesterification (Report of the committee on development of biofuel. India 2003).

Used frying oils and fats are currently collected from large food processing and service facilities where they are rendered and used primarily in animal food, while cheap, they may have some disadvantages because of the contents of high polymerization products, high free fatty acid (FFA) contents, susceptibility to oxidation, and high viscosity. Therefore, preliminary treatment such as the use of adsorbent materials to reduce the FFA content and polar contaminants may be necessary to improve the oil quality prior to transesterification to produce biodiesel catalyzed by a basic catalyst. Poor-quality oils may inactivate the basic catalysts or even enzyme catalysts (Casimir et al., 2007). Grease can also be used as feedstock for biodiesel production. The most common types are yellow grease and brown (trap) grease. The grease having content of FFAs at level of less than 15% is known as yellow grease. If the FFA level exceeds 15%, the grease is typically called as brown grease (Canakci and Sanli, 2008). Yellow grease generally refers to grease generated from used cooking oil and other fats from large-scale cooking operations such as restaurants and other types of commercial food service. Renderers filter out the solids and remove enough moisture to meet industry specifications for yellow grease. Yellow grease is currently used for producing livestock and pet foods. Prices of yellow grease are generally well below, sometimes less than half, the prices of soybean and other vegetable oils.

Most biodiesel production currently in the United States made with a feedstock other than soybean oil is manufactured from yellow grease. Trap or brown grease is grease that is collected in facilities that separate oil and grease from wastewater (such as waste treatment plants). Generally, grease traps are installed in sewage lines and grease and oil that is flushed with water down a drain floats into the top of the trap. The top of the traps can be opened allowing for the recovery of the oil and grease. Trap grease generally has limited demand because it cannot be used in the animal feeds market. Also, trap grease has very high water content, meaning it would require significant pretreatment before it would be useable as a biodiesel feedstock (Fortenbery, 2005).

Soapstock (SS), a by-product of the refining of vegetable oils and fats, is another potential biodiesel feedstock. Many researchers have reported methods for the production of fatty acid esters from SS (Stern et al., 1995).

SS consists of a heavy alkaline aqueous emulsion of lipids, containing about 50% water, with the balance made up of FFAs, phosphoacylglycerols, triacylglycerols, pigments, and other minor nonpolar components. SS is generated at a rate of about 6% of the volume of crude oil refined, which equates to an annual U.S. production of approximately one billion pounds. Its market value is approximately $0.11 per kg on a dry weight basis, that is, about one-fifth the price of crude soybean oil. Michael and coworkers have recently investigated methods for the production of fuel-grade fatty acid esters from SS and provided an overview of that work (Michael, 2005).

9.2.1.4 Microalgal Oils

Microalgal oils represent another cheap source of renewable raw materials for biodiesel production that has received little or no attention. Microalgae are sunlight-driven cell factories that convert carbon dioxide to potential biofuels, foods, feeds, and high-value bioactives (Metting, 1996; Singh et al., 2005; Spolaore et al., 2006). Microalgae can provide several different types of renewable biofuels. These include methane produced by anaerobic digestion of the algal biomass (Spolaore et al., 2006); biodiesel derived from microalgal oil (Roessler et al., 1994; Sheehan et al., 1998a,b); and photobiologically produced biohydrogen (Ghirardi et al., 2000). The idea of using microalgae as a source of fuel is not new, but it is now being taken seriously because of the escalating price of petroleum and, more significantly, the emerging concern about global warming.

Algae (singular alga) is a term that encompasses many different groups of living organisms. Algae capture light energy through photosynthesis and convert inorganic substances into simple sugars using the captured energy. Algae range from single-celled organisms to multicellular organisms, some with fairly complex differentiated form. The main branches/lines of algae are (1) Chromista—this line includes the brown algae, golden brown algae, and diatoms; (2) The Red Line—this is an early branch of marine algae; (3) Dinoflagellates; (4) The Euglenids—this independent line of single-celled organisms that include both photosynthetic and non-photosynthetic species; (5) The Green Line, is related to plants. The three most prominent lines of algae are the Chromista, the red algae, and the green algae, of which some of the most complex forms are found among the green algae (http://en.wikipedia.org/wiki/Algae). A representative of green algae is shown in Figure 9.2.

Algae are some of the most robust organisms on earth, able to grow in a wide range of conditions. They are usually found in damp places or bodies of water and thus are common in terrestrial as well as aquatic environments. Algae are made up of eukaryotic cells. These are cells with nuclei and organelles. All algae have plastids, the bodies with chlorophyll that carry out photosynthesis. But the various lines of algae can have different combinations of chlorophyll molecules; some have just Chlorophyll A, some with both

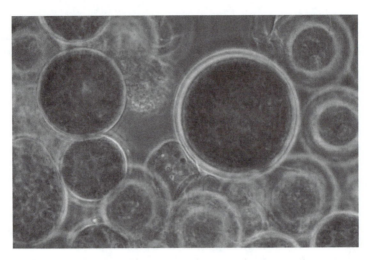

FIGURE 9.2
Light microscopic view of microalgae *Haematococcus pluvialis* (vegetative stage).

TABLE 9.5

Oil Content of Some Microalgae

Microalgae	Oil Content (%)
Botryococcus braunii	25–75
Chlorella sp.	28–32
Crypthecodinium cohnii	20
Cylindrotheca sp.	16–37
Dunaliella primolecta	23
Isochrysis sp.	25–33
Monallanthus salina	More than 20
Nannochloris sp.	20–35
Nannochloropsis sp.	31–68
Neochloris oleoabundans	35–54
Nitzschia sp.	45–47
Phaeodactylum tricornutum	20–30
Schizochytrium sp.	50–77
Tetraselmis sueica	15–23

Source: Data from Yusuf, C., *Biotechnol. Adv.*, 25, 294, 2007.

A and B, and other lines, with A and C. All algae are primarily comprised the following, in varying proportions: proteins, carbohydrates, fats, and nucleic acids. While the percentages vary with the type of algae, there are some algae types that are comprised up to 40% of their overall mass by oils (Table 9.5). It is this oil that can be extracted and converted into biodiesel.

Producing biodiesel from algae has been touted as the most efficient way to make biodiesel fuel. The highlighted advantages of deriving biodiesel from algae oil include (1) rapid growth rates, certain species of algae can be harvested daily; (2) a high per-acre yield (7–31 times greater than the next best crop-palm oil); and (3) algae biodiesel contains no sulfur (Leonard, 2007). Many researchers have demonstrated the feasibility of producing biodiesel from algae oil. Li and coworkers have reported the production of biodiesel on a large scale using the oils from microalgae, *Chlorella prototothecoides*, in bioreactors. The lipid content of the microalgae was increased up to 44%–48% of the cell dry weight. The oils were then used to produce biodiesel (98% conversion to FAAE) by a reaction catalyzed by immobilized *Candida* sp. lipase at a substrate molar ratio of 3:1 and a reaction time of 12 h. The product was said to be comparable to conventional biodiesel in physical properties (Li et al., 2007a,b). Xu and coworkers have obtained their high-quality biodiesel product from a microalgae *Chlorella prototothecoides* through the technology of transesterification. The heterotrophic *C. protothecoides* contained the crude lipid content of 55.2%. Large amount of microalgal oil was efficiently extracted from the heterotrophic cells by using *n*-hexane, and then transmuted into biodiesel through an acid-catalyzed transesterification (Xu et al., 2006). Miao and Wu also studied the production of biodiesel from heterotrophic microalgal oil. The study illustrated that *Chlorella prototothecoides* under a heterotrophic growth conditions can be used for oil production. The extracted oil from *Chlorella prototothecoides* was converted to methyl ester using an acid catalyst since the oil had high acid value (Miao and Wu, 2006).

A novel method has been developed by a group of scientists for the production of biodiesel from monosodium glutamate wastewater having COD of 10,000 mg/L. After the treatment, the COD removal has been achieved at 85% and with 10% by weight of crude lipid production. The microalgae species used for crude lipid biosynthesis was *Rhodotorula glutinis* and the lipid has been converted to methyl ester via a transesterification reaction with yield of 92.54 ($\pm$2.0%) (Xue et al., 2006).

9.2.2 Properties of Oils/Fats That Affect Biodiesel Production

Key physical and chemical characteristics that affect the ability of oils and fats to be used for biodiesel include the titre, FFA content, moisture content, calorific content, and other impurities. The rendering industry processes waste oils, greases, and animal fats into rendered products, such as edible and inedible lards, tallows, greases, and animal feed-grade fats. The rendered products are then sold to a variety of markets that use them to produce food, soap, animal feed, and other items. Standard definitions and specifications for oils and fats have been developed by trade associations, such as the American Fats and Oils Association and the National Renderers Association. The biodiesel industry refers to and uses some specifications developed for the rendering industry when describing and assessing potential biodiesel feedstocks.

Titre: Titre is the temperature at which an oil changes from a solid to a liquid. Titre is a standard specification used by the rendering industry. Titre is important since the transesterification process is basically a liquid process, and oils with high titre (i.e., a high temperature is required to change the oil to a liquid) may require heating, which increase the energy requirements and production costs for a biodiesel plant.

Free fatty acid content: FFA content is the amount of fatty acids (in weight percent) in an oil that is not connected to triglyceride molecules. The FFA content is a standard specification used by the rendering industry. During transesterification, FFAs react with alkalis and hydroxides to form soaps and water, both of which must be removed during ester purification in order to produce biodiesel. Hence, the use of oils with a high FFA can decrease biodiesel yield and thereby increase production costs. Oils with relatively high FFA contents include used or waste oils and greases that were heated during food processing and preparation. Heating oils can cause hydrolysis and oxidation, which result in fatty acids to disconnect from triglyceride molecules and result in the fatty acids becoming "free." Ambient summer temperatures can increase the FFA content of oils.

Moisture, impurities, and unsaponifiables: The moisture content, presence of impurities, and presence of other nontriglycerides in an oil should be considered when evaluating potential feedstocks for biodiesel. Referred to as MIU (moisture content, impurities, and unsaponifiables), it is defined as the amount of water, filterable solids (such as bone fragments, food particles, or other solids), and other nontriglycerides in an oil (measured in weight percent). The MIU content is a standard specification used by the rendering industry. MIUs must be removed prior to biodiesel production or during ester purification.

Calorific content: The calorific content of oil refers to the energy content of the material and is measured in Btus or calories/unit weight. The energy content of potential feedstocks affects the energy content of the resulting biodiesel. Feedstocks with higher energy content are preferred; however, there is no biodiesel specification regarding energy content.

9.2.3 Acyl Receptors

Transesterification appears to be the simplest and the most economical route to produce biodiesel in large quantity. Transesterification is the process of exchanging acyl groups between an ester and an acid (*acidolysis*), between an ester and another ester (interesterification or ester–ester interchange), or between an ester and an alcohol (*alcoholysis*) (Akoh et al., 2002). Alcoholysis is the most used method for biodiesel production. For alcoholysis, methanol is most commonly used to produce FAME. Biodiesel can be FAME or FAAE. Methanol is cheaper, more reactive, and more volatile than ethanol.

However, ethanol is preferred because it is considered more renewable than methanol and because it is obtained from agricultural products and hence is more environmentally friendly. Of course, the main purpose of alcoholysis is to reduce the viscosity of the fat and increase volatility and FAME combustion in a diesel engine without any engine modification. However, methanol may be potentially toxic. Other acyl acceptors for the alcoholysis reaction are propanol, iso-propanol (Shaw et al., 1991; Lee et al., 2004), butanol, branched-chain alcohols (Nelson et al., 1996; Modi et al., 2006; Kose et al., 2002), and octanol (Marchetti et al., 2007). There are a few articles that explored the use of acyl acceptors other than alcohols. Some researchers used alkyl acetate (such as methyl or ethyl acetate) to transesterified vegetable oil to its alkyl esters. For example, Mukherjee et al. investigated transesterification of mustard oil with various alkyl acetates by *Rhizomucor miehei* (Lipozyme) at 20°C–22°C in order to enrich very-long-chain monounsaturated fatty acids. By using ethyl acetate as alkyl source, they found that yield of ethyl esters reach up to 74% after 4 h reaction (Mukherjee and Kiewit, 1996).

A new enzymatic route for methyl esters production from soybean oil using methyl acetate as a novel acyl acceptor has been developed by Xu et al. Xu et al. found that Novozym 435 gave the highest methyl ester yield of 92%. They claim that this method is very convenient for recycling the catalyst because no glycerol was produced in this process. Triacetylglycerol was found as by-product (Xu et al., 2003). Ethyl acetate was explored as an acyl acceptor for immobilized lipase-catalyzed preparation of biodiesel from the crude oils of Jatropha, Karanj, and sunflower by Modi et al. The maximum yield of ethyl esters was 91.3%, 90%, and 92.7%, respectively (Modi et al., 2007). Kim and coworkers have proposed ethyl acetate as an alternative acyl acceptor for the production of biodiesel from soybean oil. Ethyl acetate mixed well with soybean oil, and only slightly inhibited the lipase activity by 5%. The highest biodiesel production yield, 63.3 (±0.6)%, was obtained (Kim et al., 2007).

The disadvantage for technical application of transesterification with alcohol or alkyl acetates that are reported above is that it is an equilibrium reaction as shown in Figure 9.3A. The first transesterification of fatty acids with dialkyl carbonates was reported by a research group in 1991 (Pioch et al., 1991). Lipase-catalyzed conversion of plant oil with dialkyl carbonates has been found not an equilibrium reaction, because the intermediate compound (carbonic acid monoacyl ester) decomposes immediately to carbon dioxide and alcohol (Figure 9.3C) (Warwel et al., 1999).

Su et al. (2007) also investigated the lipase-catalyzed transesterification of different vegetable oils for biodiesel production with dimethyl carbonate as the acyl acceptor, and found a very high conversion of 96.4%. Dialkyl carbonates are eco-friendly, cheap, and nontoxic. In view of these, lipase-catalyzed transesterification using dialkyl carbonates, especially dimethyl or diethyl carbonate, may provide an alternative route for biodiesel production.

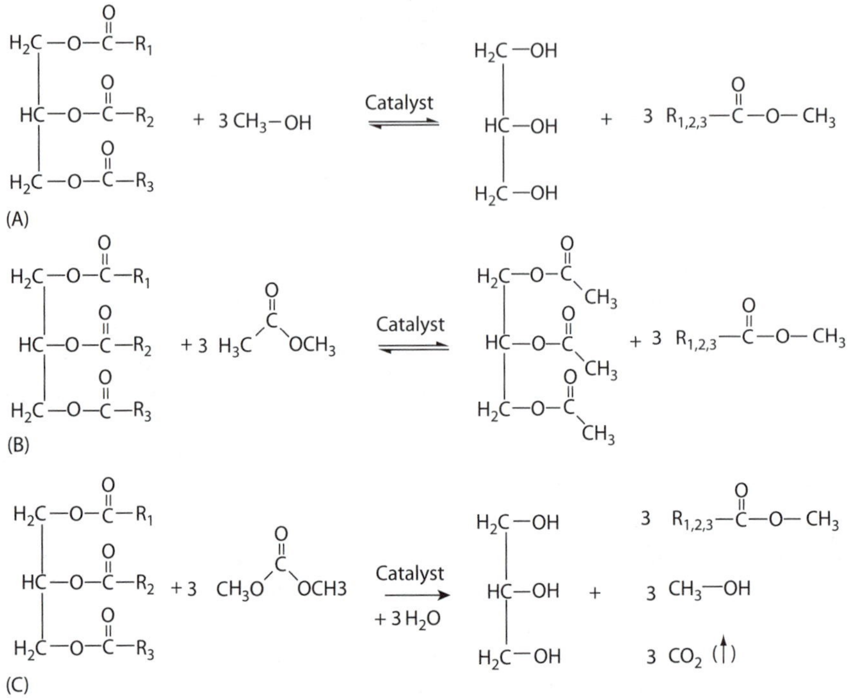

FIGURE 9.3
Transesterification of oils/fats with methanol (A), methyl acetate (B), and dimethyl carbonate (C).

9.3 Production of Biodiesel

Biodiesel is mainly produced from oils and fats. The oils contain higher levels of unsaturated fatty acids and are thus liquids at room temperature. Their direct uses as fuel were tried more than a 100 years ago. But the problems with directly using oils as fuel are mostly associated with their high viscosities, low volatilities, and polyunsaturated character. Fats, however, contain more saturated fatty acids. They are solid at room temperature and cannot be used as fuel in a diesel engine in their original form. Because of the problems, such as carbon deposits in the engine, engine durability, and lubricating oil contamination, associated with the direct use of oils and fats as internal combustion engine fuels, they must be derivatized to be compatible with existing engines. Four primary production methodologies for producing biodiesel have been studied extensively: direct use and blending, microemulsification, pyrolysis or thermal cracking, esterification and transesterification.

9.3.1 Direct Use of Vegetable Oil and Blending

Before turning oil into biodiesel, one might ask if vegetable oils can be used in diesel engines without any modification. This section has briefly summarized the history of such attempts. Beginning in 1980, there was considerable discussion regarding use of straight vegetable oil (SVO) as internal combustion engine fuel. The most comprehensive research work of using SVO as fuel has been conducted with sunflower oil in South Africa in the eighties. In 1980, Caterpillar Brazil used precombustion chamber engines with a mixture of 10% vegetable oil to maintain total power without any alterations to the engine. At that point, it was not practical to substitute 100% vegetable oil for diesel fuel, but with a blend of 20% vegetable oil and 80% petrodiesel fuel.

Short-term performance tests were conducted to evaluate the feasibility of using crude soybean oil as fuel. Crude-degummed soybean oil and soybean ethyl ester as complete substitutes for No. 2 diesel fuel in a 2.59 L, 3 cylinder 2600 series Ford diesel engine has been tested with promising results (Pryor et al., 1983). A longer-term evaluation of the engine with 100% crude soybean oil, however, was prematurely terminated. Severe injector coking led to decreases in power output and thermal efficiency. A long-term performance test was then tried with a fuel blend of 75% unrefined mechanically expelled soybean oil and 25% diesel fuel (Schlautman et al., 1986). The fuel blend was burned in a direct injection mode diesel engine for 159 h before the test was terminated because a constant power output could not be held on the engine. The test has shown the failure sign after 90 h into the combustion test due to a 670% increase in the lubricating oil viscosity. Other tests with the same type of engine were conducted with mechanically expelled-unrefined soybean oil and sunflower oil blended with No. 2 diesel fuel on a 25:75 v/v basis (Schlick et al., 1988). The power remained constant throughout 200 h of operation. An excessive carbon deposit on all combustion chamber parts precludes the use of these fuel blends.

Two most severe problems associated with the use of vegetable oils as fuels were oil deterioration and incomplete combustion (Peterson et al., 1983). Polyunsaturated fatty acids were very susceptible to polymerization and gum formation caused by oxidation during storage or by complex oxidative and thermal polymerization at the higher temperature and pressure of combustion. The gum did not combust completely, resulting in carbon deposits and lubricating oil thickening. The reason why winter rapeseed oil was used as subject is that the oil contains very high (46.7%) erucic acid content (Peterson et al., 1983). The rate of gum formation of winter rapeseed oil was five times slower than that of high linoleic (75%–85%) oil. The viscosities of 50/50 and 70/30 blends of winter rapeseed oil and diesel and whole winter rape oil were however much higher (6–18 times) than No. 2 diesel.

In summary, the advantages of SVO as engine fuel are (1) portability, (2) heat content (80% of diesel fuel), (3) availability, and (4) renewability. The disadvantages are (1) higher viscosity, (2) lower volatility, and (3) the reactivity of unsaturated hydrocarbon chains (Pryde, 1983). Problems appear only after the engine has been operating on vegetable oils for longer periods of time, especially with direct-injection engines. The problems include (1) coking and trumpet formation on the injectors to such an extent that fuel atomization does not occur properly or is even prevented as a result of plugged orifices, (2) carbon deposits, (3) oil ring sticking, and (4) thickening and gelling of the lubricating oil as a result of contamination by the vegetable oils (Ma and Hanna, 1999).

9.3.2 Microemulsification

In order to address the problems raised with SVO as fuel, other nonchemical modification approaches to use oil as fuel have been explored. One of these methods is to prepare oil in a microemulsion form. Microemulsions are the colloidal equilibrium dispersion of optically isotropic fluid microstructures with dimensions in the range of 1–150 nm. These dispersions are formed spontaneously from two immiscible liquids and one or more ionic or nonionic amphiphiles or called surfactants (Schwab et al., 1987; Ma, 1998). Microemulsification is aimed to solve the problem of high viscosity of pure vegetable oils. The formulation reduces the viscosity of vegetable oils with solvents such as methanol, ethanol, and 1-butanol. They can improve spray characteristics by explosive vaporization of the low boiling constituents in the micelles (Pryde, 1984). Short-term performances of formulated fuel with both ionic and nonionic microemulsions of aqueous ethanol in soybean oil were nearly as good as that of No. 2 diesel, in spite of the lower cetane number and energy content (Goering et al., 1982). One group has prepared an emulsion of 53% (vol) alkali-refined and winterized sunflower oil, 13.3% (vol) 190-proof ethanol, and 33.4% (vol) 1-butanol (Ziejewski et al., 1984). This nonionic emulsion had a viscosity of 6.31 cSt at 40°C, a cetane number of 25, and an ash content of less than 0.01%. Lower viscosities and better spray patterns were observed with an increase of 1-butanol. In a 200 h laboratory engine endurance test, no significant deteriorations in performance were observed, but irregular injector needle sticking, heavy carbon deposits, incomplete combustion, and an increase of lubricating oil viscosity were reported.

Another nonionic emulsion called Shipp fuel containing 50% No. 2 diesel fuel, 25% degummed and alkali-refined soybean oil, 5% 190-proof ethanol, and 20% 1-butanol was evaluated in the 200 h engine manufacturers' association (EMA) screening test (Goering and Fry, 1984). The fuel passed the 200 h EMA test, but carbon and lacquer deposits on the injector tips, intake valves, and tops of the cylinder liners were major problems. The Shipp nonionic fuel performed better than a 25% blend of sunflower oil in diesel oil. The engine performances were the same for a microemulsion of 53% sunflower oil and the 25% blend of sunflower oil in diesel (Ziejewski et al., 1983).

A microemulsion prepared by blending soybean oil, methanol, 2-octanol, and cetane improver in the ratio of 52.7:13.3:33.3:1.0 also passed the 200 h EMA test (Goering, 1984).

Schwab and coworkers used the ternary phase equilibrium diagram and the plot of viscosity versus solvent fraction to determine the emulsified fuel formulations (Schwab et al., 1987). All microemulsions with butanol, hexanol, and octanol met the maximum viscosity requirement for No. 2 diesel. The 2-octanol was an effective amphiphile in the micellar solubilization of methanol in triolein and soybean oil. Methanol was often used due to its economic advantage over ethanol.

9.3.3 Pyrolysis or Thermal Cracking

9.3.3.1 Pyrolysis of Vegetable Oils and Fats

Pyrolysis is the chemical decomposition of organic materials by means of heat or by heat with the aid of a catalyst in the absence of air or oxygen (Sonntag, 1979; Srivastava and Prasad, 2000). It is a series of thermally driven chemical reactions that decompose organic compounds (Brown, 2003) into shorter carbon chain compounds. It involves heating in the absence of air or oxygen and cleavage of chemical bonds to yield small molecules (Weisz et al., 1979), which results in the production of alkanes, alkenes, alkadienes, carboxylic acids, aromatics, and small amounts of gaseous products. The properties of liquid fractions of the thermally decomposed vegetable oil are likely to approach diesel fuels. The pyrolyzate had lower viscosity, flash point, and pour point than diesel fuel and equivalent calorific values. The cetane number of the pyrolyzate was lower. The pyrolysed vegetable oils contain acceptable amounts of sulfur, water, and sediment and give acceptable copper corrosion values but unacceptable ash and carbon residue. The pyrolyzed material can be vegetable oils, animal fats, natural fatty acids, and methyl esters of fatty acids. Depending on the operating conditions, the pyrolysis process can be divided into three subclasses: conventional pyrolysis, fast pyrolysis, and flash pyrolysis.

The pyrolysis of fats has been investigated for more than 100 years, especially in those areas of the world that lack deposit of petroleum (Sonntag, 1979). The first pyrolysis of vegetable oil was conducted in an attempt to synthesize petroleum from vegetable oil. Pyrolysis is conceptually simple to perform. The difference between transesterification and pyrolysis of triglycerides is shown in Figure 9.4.

As one can see, transesterification (approach (i) in the Figure 9.4) leads to fatty acid esters (biodiesel) and pyrolysis (approach (ii) in the figure) leads to a complex mixture, part of which can be used as engine fuel. A simple experimental setup for pyrolysis is depicted in Figure 9.5 (Demirbas, 2003).

The main components of pyrolysis of vegetable oils are alkanes and alkenes, which accounted for approximately 60% of the total feeder weight. Carboxylic acids accounted for another 9.6%–16.1%. It is believed that as

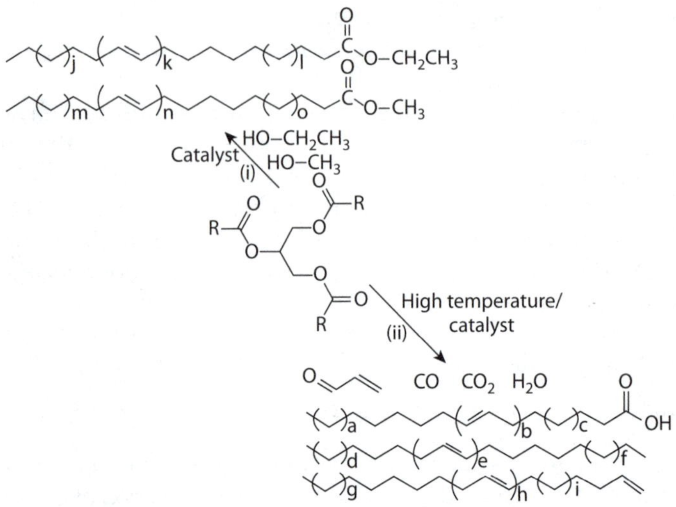

FIGURE 9.4

Illustration of the difference between transesterification and pyrolysis of triglycerides.

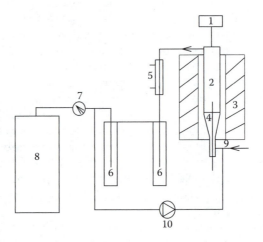

FIGURE 9.5

Simplified experimental setting for vegetable oil pyrolysis. (1) Vegetable oil feeder, (2) pyrolysis chamber, (3) electric furnace, (4) packing material, (5) condenser, (6) cold trap, (7) gas volume measure, (8) tight gas tank, (9) sweeping gas stream, and (10) peristaltic pump (Redrawn from Demirbas, A., *Energ. Convers. Manage.*, 44, 2093, 2003.).

the reaction progresses, the residue becomes less reactive and forms stable chemical structures, and consequently, the activation energy increases as the decomposition level of the vegetable oil residue increases (Demirbas, 2002a).

There are several parameters that should be controlled and can be optimized during pyrolysis such as temperature, residence times, and catalyst content.

Three vegetable oils (soybean, palm, and castor oils) were pyrolyzed to obtain light fuel products at 503–673 K with $(Al_2O_3)_X(SnO)_Y(ZnO)_Z$ or Na_2CO_3 as catalysts (Lima et al., 2004). The results show that soybean, palm, and castor oils present a similar behavior depending on the pyrolysis temperature range. Palm oil reacts in a lower temperature range with a higher yield in the heavy fraction (Lima et al., 2004). A short pyrolysis time (less than 10 s) leads to a high amount of alkanes, alkenes, and aldehydes instead of carboxylic acids. On the other hand, higher temperature and long pyrolysis times do not favor a pyrolysis of this material. In this case, a process like desorption becomes more likely than the pyrolytic process. The liquid products can be improved by deoxygenation in order to obtain an enriched hydrocarbon diesel-like fuel (Fortes and Baugh, 1999; Lima et al., 2004). The reaction conditions of pyrolysis have significant affect on the outcome. Increasing Na_2CO_3 (as adsorbent) content and temperature increases the formation of liquid hydrocarbon and gas products and decreases the formation of aqueous-phase, acid-phase, and coke-residual oil. The highest $C_5–C_{11}$ yield (36.4%) was obtained using 10% Na_2CO_3 and a packed column of 180 mm at 693 K. The use of a packed column increased the residence times of the primer pyrolysis products in the reactor and packed column by the fractionating of the products, which caused the additional catalytic and thermal reactions in the reaction system and increased the content of liquid hydrocarbons in the gasoline boiling range (Dandik and Aksoy, 1998).

The process equipment for pyrolysis is expensive for modest throughputs. In addition, while the products are chemically similar to petroleum-derived gasoline and diesel fuel, the removal of oxygen during the thermal processing also removes any environmental benefits of using an oxygenated fuel. It produced some low value materials and, sometimes, more gasoline than diesel fuel.

9.3.3.2 Pyrolysis of Vegetable Oil Soaps

The soaps obtained from vegetable oils can be pyrolyzed into hydrocarbon-rich products (Demirbas, 2002a,b). The saponification and pyrolysis of sodium soap of vegetable oil proceed as follows:

Saponification:

$$\text{Vegetable oils or fats} + NaOH \rightarrow RCOONa + Glycerin \tag{9.1}$$

Pyrolysis of Na soaps:

$$2RCOONa \rightarrow R–R + Na_2CO_3 + CO \tag{9.2}$$

The vegetable oil soaps can be pyrolyzed according to Equation 9.2, with higher yields at the higher temperatures. Table 9.6 shows the yields of pyrolysis products from different vegetable oil sodium soaps at different temperatures.

TABLE 9.6

Pyrolysis Yields from Vegetable Oil Soaps (Percent by Weight)

Temperature (K)	Sunflower Oil	Corn Oil	Cottonseed Oil	Soybean Oil
400	2.8	2.3	3.1	2.9
450	8.4	8.6	8.5	8.8
500	29.0	28.5	31.3	32.6
520	45.4	46.2	48.0	49.2
550	62.4	65.5	67.2	68.0
570	84.6	84.0	83.9	85.1
590	92.7	93.0	93.5	93.4
610	97.5	97.1	97.5	97.8

Source: Data from Demirbas, A., *Energ. Convers. Manage.*, 44, 2093, 2003.

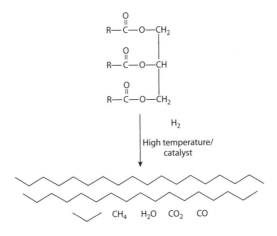

FIGURE 9.6
Reaction scheme of triglyceride hydrocracking.

In 1947, a large-scale thermal cracking of tung oil calcium soap was reported (Chand and Wan, 1947). Tung oil was saponified with lime and then thermally cracked to yield a crude oil, which was refined to produce diesel fuel and small amounts of gasoline and kerosene. In that process, 68 kg of the tung oil soap produced 50 L of crude oil.

Pyrolyses of the vegetable oil soaps were performed in the same apparatus designed for vegetable oil (see Figure 9.6). The main element of this device was a tubular reactor of height 95.1 mm, ID 17.0 mm, and OD 19.0 mm inserted vertically into an electrically heated tubular furnace.

9.3.3.3 Catalytic Cracking of Vegetable Oils

The thermal degradation of aliphatic long-chain compounds is known as "cracking." Higher-molecular-weight molecules generally convert into lower-molecular-weight molecules by the cracking process (Figure 9.6). Large alkane molecules are converted into smaller alkanes and hydrogen gas in the cracking process. Smaller hydrocarbons can also be obtained by a "hydrocracking" process. The hydrocracking is carried out in the presence of a catalyst and hydrogen, at high pressure and at much lower temperatures (525–725 K).

Higher boiling point petroleum fractions (typically gas oil) are obtained from silica–alumina-catalyzed cracking process at temperature of 725–825 K and pressure of 1–5 atm. The catalytic cracking not only increases the yield of gasoline by breaking large molecules into smaller ones but also improves the quality of the gasoline.

Palm oil stearin and copra oil were subjected to conversion over different catalysts like silica–alumina and zeolite (Pioch et al., 1993). It was found that the conversion of palm and copra oil was 84 and 74 wt %, respectively. The silica–alumina catalyst was highly selective for obtaining aliphatic hydrocarbons, mainly in the kerosene boiling point range (Katikaneni et al., 1995). The organic liquid products obtained with a silica–alumina catalyst contained between 4 and 31 wt % aliphatic hydrocarbons and 14 and 58 wt % aromatic hydrocarbons. The conversion was high and ranged between 81 and 99 wt %. Silica–alumina catalysts are suitable for converting vegetable oils into aliphatic hydrocarbons. Zinc chloride catalyst, as a Lewis acid, contributed to hydrogen transfer reactions and the formation of hydrocarbons in the liquid phase.

Palm oil has been cracked at atmospheric pressure at a reaction temperature of 723 K to produce biofuel in a fixed-bed microreactor. The reaction was carried out over microporous HZSM-5 zeolite, mesoporous MCM-41, and composite micromesoporous zeolite as catalysts. The products obtained were gas, organic liquid, water, and coke. The organic liquid product was composed of hydrocarbons corresponding to the gasoline, kerosene, and diesel boiling point ranges. The maximum conversion of palm oil, 99 wt %, and gasoline yield of 48 wt % was obtained with composite micromesoporous zeolite (Sang et al., 2003).

9.3.3.4 Pyrolysis Mechanisms of Vegetable Oils

Pyrolysis involves thermal degradation of vegetable oils or fats by heat in the absence of oxygen, which results in the production of charcoal (solid), bio-oil (liquid), and fuel gaseous products. The pyrolysis of biomass has been studied with the ultimate objective of recovering a biofuel with a medium-low heating value (Maschio et al., 1992; Barth, 1999; Bridgwater et al., 1999). Soybean, rapeseed, sunflower, and palm oils are the most studied for the preparation of bio-oil.

A mechanism for catalytic decarboxylation of vegetable oils is presented in Figure 9.7 (Schwab et al., 1988). Vegetable oils contain mainly palmitic, stearic, oleic, and linoleic acids. These fatty acids underwent various reactions, resulting in the formation of different types of hydrocarbons.

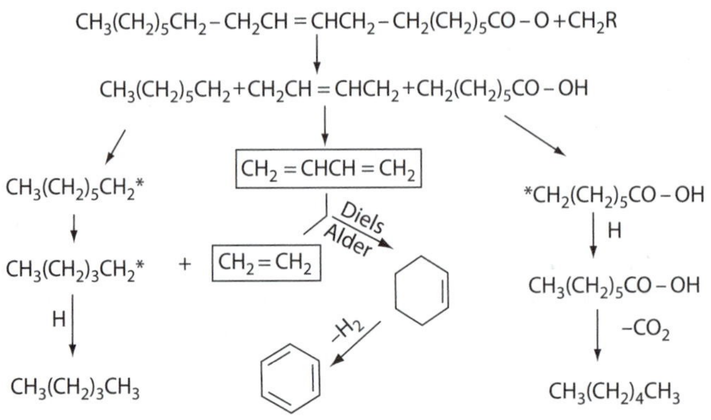

$$CH_3(CH_2)_5CH_2 - CH_2CH = CHCH_2 - CH_2(CH_2)_5CO - O + CH_2R$$

$$CH_3(CH_2)_5CH_2 + CH_2CH = CHCH_2 + CH_2(CH_2)_5CO - OH$$

$$CH_3(CH_2)_5CH_2*$$

$$\boxed{CH_2 = CHCH = CH_2}$$

$$*CH_2(CH_2)_5CO - OH$$

$$CH_3(CH_2)_3CH_2* \quad + \quad \boxed{CH_2 = CH_2}$$

Diels
Alder

$$CH_3(CH_2)_5CO - OH$$

H

H

$$-CO_2$$

$$CH_3(CH_2)_3CH_3$$

$$CH_3(CH_2)_4CH_3$$

FIGURE 9.7
The mechanism of thermal decomposition of triglycerides.

The variety of reaction paths and intermediates makes it difficult to describe the reaction mechanism. In addition, the multiplicity of possible reactions of mixed triglycerides makes pyrolysis reactions more complicated (Zhenyi et al., 2004). Generally, thermal decomposition of triglycerides proceeds through either a free-radical or carbonium ion mechanism (Srivastava and Prasad, 2000). Vegetable oil is converted into lower molecular products by two simultaneous reactions: cracking and condensation. The heavy hydrocarbons produced from primary and secondary deoxygenation and cracking yield light olefins and light paraffins, water, carbon dioxide, and carbon monoxide. Hydrocarbon formation can be identified as deoxygenation, cracking, and aromatization with H-transfer. Deoxygenation can take place via decarboxylation and dehydration (Chang and Silvestri, 1977).

The distribution of pyrolysis products depends on the dynamics and kinetic control of different reactions. The maximum gasoline fraction can be obtained under appropriate reaction conditions. Thermodynamic calculation shows that the initial decomposition of vegetable oils occurs with the breaking of C–O bonds at lower temperatures, and fatty acids are the main product. The pyrolysis temperature should be higher than 675 K; at this temperature, the maximum diesel yield with high oxygen content can be obtained (Zhenyi, 2004). The effect of temperature, the use of catalysts, and the characterization of the products have been investigated (Srivastava and Prasad, 2000). In pyrolysis, the high-molecular-weight materials are heated to high temperatures, so their macromolecular structures are broken down into smaller molecules and a wide range of hydrocarbons are formed. These pyrolytic products can be divided into a gas fraction, a liquid fraction comprising paraffins, olefins, and naphthenes, and solid residue (Demirbas, 1991a,b). The cracking process yields a highly unstable low-grade fuel oil that can be acid-corrosive, tarry, and discolored along with a characteristically foul odor (Demirbas, 2004).

It was proposed that thermal and catalytic cracking of triglyceride molecules occurs at the external surface of the catalysts to produce small molecular size components, comprised mainly of heavy liquid hydrocarbons and oxygenates (Leng et al., 1999). In general, it is assumed that the reactions occur predominantly within the internal pore structure of a zeolite catalyst.

The catalyst acidity and pore size affect the formation of aromatic and aliphatic hydrocarbons. Hydrogen transfer reactions, essential for hydrocarbon formation, are known to increase with catalyst acidity. The high acid density of $ZnCl_2$ catalysts contribute greatly to high amounts of hydrocarbons in the liquid product.

9.3.4 Esterification

In the case of using waste oils or fats as feedstock for biodiesel production, FFAs may pose a problem. A FFA is one that has already separated from the glycerol molecule. This is usually the result of the oil breaking down after many cycles of use. FFA creates four major problems for one-stage (transesterification only process) biodiesel production: (1) more catalyst will need to be used leading to higher cost; (2) soap (fatty acid salt) is formed, making washing the finished product more difficult; (3) water is formed, which will retard the main reaction (transesterification, described in the following text); (4) the FFAs are not converted into fuel, reducing the yield. When the oil has less than 2.5% FFA, the problems listed previously are negligible by using the single step (transesterification) production mode.

There are several methods to treat high-FFA waste oils or fats in small-scale systems. The easiest is to mix the high-FFA ones with low-FFA ones. This will work for an occasional high-FFA batch. Other options require intentionally making soap or esterification (two-stage process, esterification–transesterification). These options are (1) add catalyst and water to change FFA to soap, and remove the soap (see Equation 9.1); (2) add acid and a large percentage of alcohol to covert FFA to usable product; (3) add acid, heat, and a smaller percentage of alcohol to covert FFA to a usable product. Adding catalyst and water to high-FFA oil is the easiest solution, but it also has a significant disadvantage of low yield. For example, if feedstock has 10% of FFA, 100 gal of waste oil will lose more than 10 gal product. Adding acid and large quantities of methanol to the oil is the most common method among small-scale producers. The disadvantage to this method besides time is the cost of the methanol. Adding acid with high heat (90°C) and smaller quantities of methanol is not widely used.

The commonly used catalyst during esterification is sulfuric acid (H_2SO_4) (Ramadhas et al., 2005; Veljkovic et al., 2006). Figure 9.8 shows the mechanism of acid-catalyzed esterification of fatty acids. The initial step is protonation of the acid to give an oxonium ion (1), which can undergo an exchange reaction with an alcohol to give the intermediate (2), and this in turn can lose

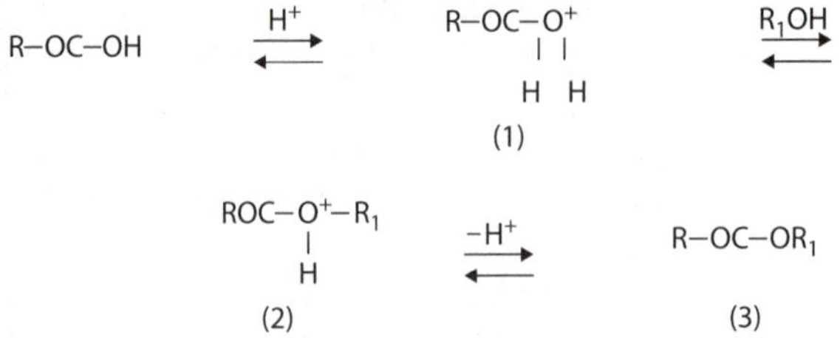

R–OC–OH $\xrightleftharpoons{H^+}$ R–OC–O⁺ $\xrightleftharpoons{R_1OH}$
H H

(1)

ROC–O⁺–R₁ $\xrightleftharpoons{-H^+}$ R–OC–OR₁
H

(2)   (3)

FIGURE 9.8
Mechanism of acid-catalyzed esterification of fatty acids.

a proton to become an ester (3). Each step in the process is reversible, but in the presence of a large excess of the alcohol, the equilibrium position of the reaction is favored to esterification completion direction.

During sulfuric acid-catalyzed esterification, water is produced along with ester from the reaction of FFA with alcohol, which inhibits the transesterification of glycerides (Canakci, 2007). Therefore, the conversion reported is low (82%) and the alcohol required for the reaction is high (200% excess of ethanol) (Wang et al., 2007). Wang and coworkers have then tried a new catalyst $Fe_2(SO_4)_3$ (ferric sulfate) as an alternative to sulfuric acid and have reported much better conversion (97.02%) (Wang et al., 2007). Since ferric sulfate is insoluble in oil, it can be centrifuged from the liquid after acid esterification process and reused for the next batch.

Diazomethane (CH_2N_2) was also investigated to catalyze the esterification of FFA (Schelenk and Gellerman, 1960). The CH_2N_2 is generally prepared in ethereal solution by the action of an alkali (a 30% solution of KOH) on a nitrosamide, for example, N-methyl-N-nitroso-p-toluene-sulfonamide or nitroso-methyl-urea. CH_2N_2 reacts rapidly with FFA to give methyl esters but does not affect transesterification of other lipids. The reaction is not instantaneous, however, as has sometimes been assumed, unless a little methanol is present as a catalyst (Schelenk and Gellerman, 1960). Carboxylic acids {R–C(=O)–O– H} can be converted into methyl esters {R–C(=O)–O–CH₃} by the action of CH_2N_2:

$$R-C(=O)-O-H + CH_2N_2 \rightarrow R-C(=O)-O-CH_3 + N_2 \qquad (9.3)$$

Notice that the CH_2N_2 appears to insert itself between the O and the H of the O–H bond (Equation 9.3). The high reactivity of diazomethane arises from the fact that it possesses an exceedingly reactive leaving group, the nitrogen molecule (N_2). A nucleophilic substitution reaction on the protonated diazomethane molecule transfers a methyl group to the oxygen atom of the carboxylic acid, while liberating a very stable product (N_2 gas). This process is very favorable energetically, owing to the great stability of N_2.

9.3.5 Transesterification

9.3.5.1 General Aspects of Transesterification

Chemically, transesterification means taking a triglyceride molecule or a complex fatty acid, neutralizing the FFAs, removing the glycerin, and creating an alkyl ester. When the original ester is reacted with an alcohol, the transesterification process is called alcoholysis. Transesterification consists of a number of consecutive, reversible reactions: the triglyceride is converted stepwise to diglyceride, monoglyceride, and finally glycerol. The overall chemistry of transesterification can be represented as in Figure 9.3A. The overall reaction consists of three consecutive and reversible reactions as follows: (1) the formation of diglycerides (Figure 9.9A); (2) the formation of monoglycerides (Figure 9.9B); (3) the formation of glycerol (Figure 9.9C).

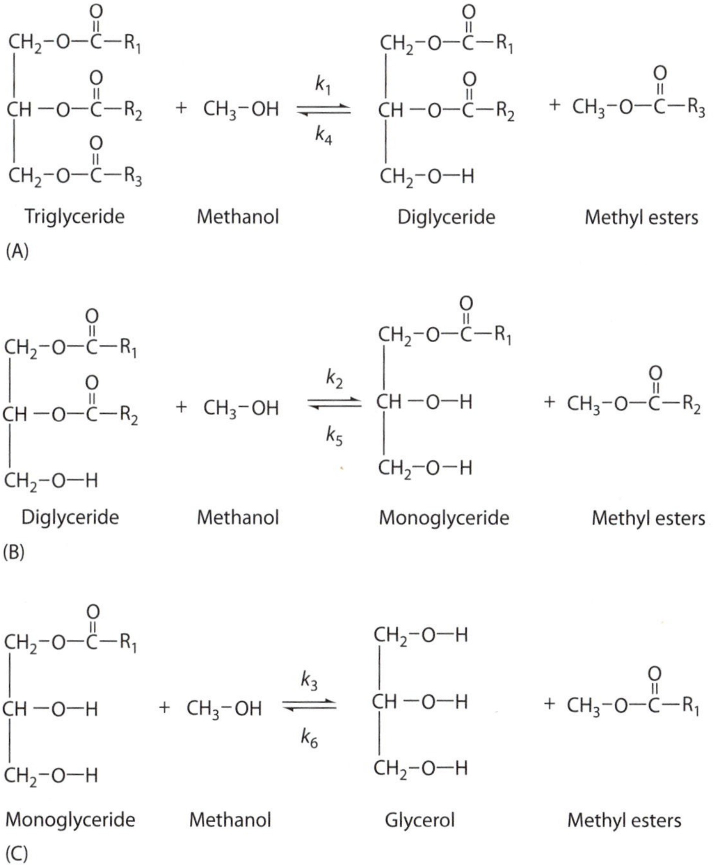

FIGURE 9.9
Sequential transesterification of triglycerol.

Alcohols used are primary or secondary monohydric aliphatic alcohols having 1–8 carbon atoms (Sprules and Price, 1950). Among the alcohols that can be used in the transesterification reaction are methanol, ethanol, propanol, butanol, and amyl alcohol. Methanol and ethanol are used most frequently. Ethanol is a preferred alcohol in the transesterification process compared to methanol because it is derived from agricultural products and is renewable and biologically less objectionable in the environment; however, methanol is used because of its low cost and its physical and chemical advantages (polar and shortest chain alcohol). The stoichiometric reaction requires 1 mol of a triglyceride and 3 mol of the alcohol. However, a larger amount of alcohol was used to shift the reaction equilibrium to the right side and produce more alkyl esters, the proposed product.

A catalyst is usually used to improve the transesterification reaction rate and yield. The catalysts can be homogeneous ones such as alkalis, acids, and supercritical fluids and heterogeneous ones such as lipases. The alkalis include NaOH, KOH, carbonates, and corresponding sodium and potassium alkoxides such as sodium methoxide, sodium ethoxide, sodium propoxide, and sodium butoxide. Sulfuric acid, sulfonic acids, and hydrochloric acid are usually used as acid catalysts. Alkali-catalyzed transesterification is much faster than acid-catalyzed transesterification and is most often used commercially.

9.3.5.2 Acid-Catalyzed Transesterification

Acids used for transesterification include sulfuric, phosphoric, hydrochloric, and organic sulfonic acids (Freedman et al., 1986; Stern and Hillion, 1990). These catalysts give very high yields in alkyl esters, but the reactions are slow, requiring, typically, temperatures above 100°C and more than 3 h to reach complete conversion (Freedman et al., 1984). Pryde et al. showed that the methanolysis of soybean oil, in the presence of 1 mol % of H_2SO_4, with an alcohol/oil molar ratio of 30:1 at 65°C, takes 50 h to reach complete conversion of the vegetable oil (>99%), while the butanolysis (at 117°C) and ethanolysis (at 78°C), using the same quantities of catalyst and alcohol, take 3 and 18 h, respectively (Freedman et al., 1986). Although the acid-catalyzed reaction requires a longer reaction time and a higher temperature, acid catalysis is more efficient when the amount of FFAs in the oil exceeds 1% (Freedman et al., 1984; Liu, 1994; Canakci and Gerpen, 1999). An economic analysis study has shown that the acid-catalyzed procedure, being a one-step process, is more economical than the alkali-catalyzed process, which requires an extra step to convert FFAs to methyl esters, thus avoiding soap formation (Zhang et al., 2003a,b).

The mechanism of the acid-catalyzed transesterification of vegetable oils is shown in Figure 9.10, for a monoglyceride. However, it can be extended to di- and triglycerides (Stoffel et al., 1959). The protonation of the carbonyl group of the ester leads to the carbocation II, which, after a nucleophilic

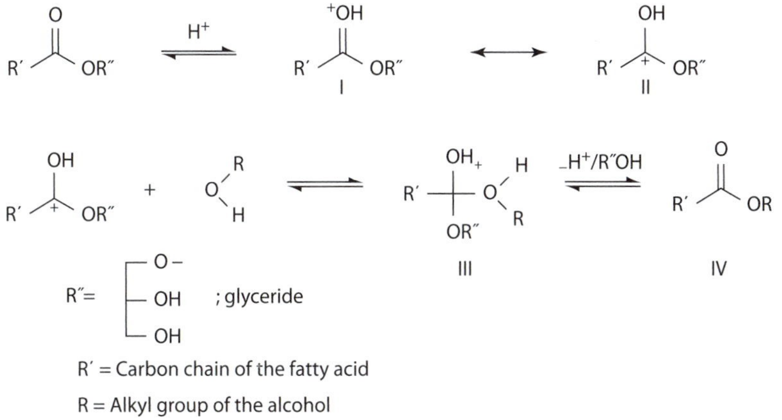

FIGURE 9.10
Mechanism of the acid-catalyzed transesterification of vegetable oils.

attack of the alcohol, produces the tetrahedral intermediate III, which eliminates glycerol to form the new ester IV, and to regenerate the catalyst H^+. According to this mechanism, carboxylic acids can be formed by the reaction of carbocation II with water present in the reaction mixture. This suggests that an acid-catalyzed transesterification should be carried out in the absence of water, in order to avoid the competitive formation of carboxylic acids, which reduce the yields of alkyl esters (Ulf et al., 1998).

9.3.5.3 Alkali-Catalyzed Transesterification

For oil samples with FFA below 2.0%, alkali-catalyzed transesterification is preferred over the acid-catalyzed transesterification as the former is reported to proceed about 4000 times faster than the latter (Fukuda et al., 2001). Due to this reason, together with the fact that the alkaline catalysts are less corrosives than acidic compounds, industrial processes usually favor base catalysts, such as alkaline metal alkoxides (Freedman et al., 1984, 1986; Schwab et al., 1987) and hydroxides (Aksoy et al., 1990; Wimmer, 1994) as well as sodium or potassium carbonates (Graille et al., 1985; Filip et al., 1992).

Alkaline metal alkoxides (as CH_3ONa for the methanolysis) are the most active catalysts, since they give very high yields (>98%) in short reaction times (30 min) even if they are applied at low molar concentrations (0.5 mol %). However, they require the absence of water, which makes them inappropriate for typical industrial processes (Freedman et al., 1984).

Alkaline metal hydroxides (KOH and NaOH) are cheaper than metal alkoxides, but less active. Nevertheless, they are a good alternative since they can give the same high conversions of vegetable oils just by increasing the catalyst concentration to 1 or 2 mol %. However, even if a water-free alcohol/oil mixture is used, some water is produced in the system by the reaction of

$$K_2CO_3 + ROH \quad \rightleftharpoons \quad ROK + KHCO_3$$

R = alkyl group of the alcohol

FIGURE 9.11
Reaction of potassium carbonate with the alcohol.

the hydroxide with the alcohol. The presence of water gives rise to hydrolysis of some of the produced ester, with consequent soap formation. This undesirable saponification reaction reduces the ester yields and makes the recovery of the glycerol considerably difficult due to the formation of emulsions (Freedman et al., 1984).

Potassium carbonate, used in a concentration of 2 or 3 mol % gives high yields of fatty acid alkyl esters and reduces the soap formation (Filip et al., 1992). This can be explained by the formation of bicarbonate instead of water (Figure 9.11), which does not hydrolyze the esters.

The use of homogeneous catalyst such as sodium hydroxide and potassium hydroxide has been successful at industrial level for the production of biodiesel. However, the biodiesel and glycerol produced have to be purified to remove the basic catalyst and need its separation by washing with hot distilled water twice or thrice. Thus, heterogeneous catalyst has also been tried by researchers to overcome this drawback of time consumption and colossal consumption of water. The heterogeneous catalyst can be separated from the final product by filtration, which checks time consumption and prevents the consumption of large volume of water. The filtered solid then can be reused. The application of a heterogeneous catalyst, CaO has been tested by Granados et al. (2007) for its feasibility. The experiments confirmed that CaO could be used as a catalyst for the transesterification reaction without significant deactivation up to eight runs with significant amount of CaO.

The mechanism of the base-catalyzed transesterification of vegetable oils is shown in Figure 9.12. The first step is the reaction of the base (B) with the alcohol (ROH), producing an alkoxide and the protonated catalyst. The nucleophilic attack of the alkoxide at the carbonyl group of the triglyceride generates a tetrahedral intermediate (Taft et al., 1950; Guthrie, 1991), from which the alkyl ester and the corresponding anion of the diglyceride are formed. The latter deprotonates the catalyst, thus regenerating the active species, which is now able to react with a second molecule of the alcohol, starting another catalytic cycle. Diglycerides and monoglycerides are converted by the same mechanism to a mixture of alkyl esters and glycerol.

9.3.5.4 Kinetics of Acid/Alkali-Catalyzed Transesterification

The kinetics of transesterification of vegetable oils has been studied extensively for the past two decades. An understanding of the kinetics of transesterification is indispensable for determining the optimum reaction conditions

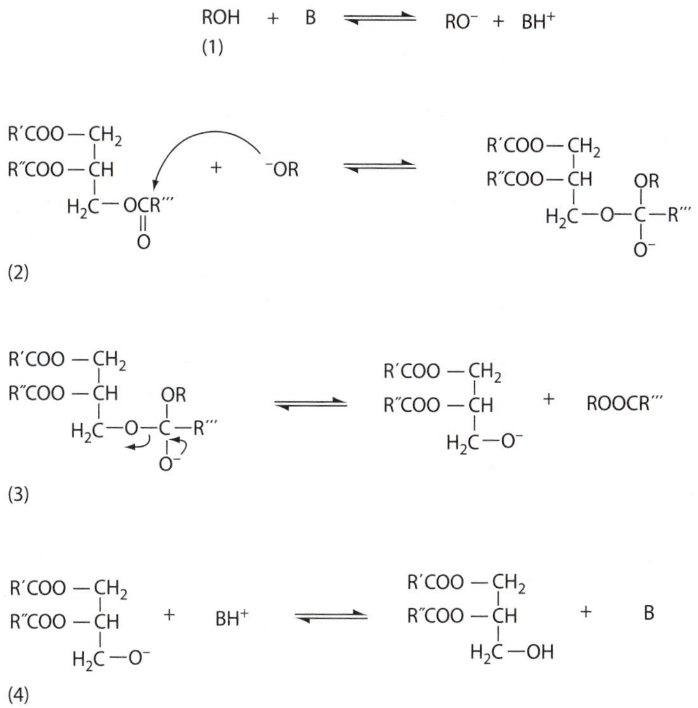

$$ROH + B \rightleftharpoons RO^- + BH^+$$
(1)

(2)

(3)

(4)

FIGURE 9.12
Mechanism of base-catalyzed transesterification of vegetable oil.

for maximum conversion into ester. Moreover, transesterification reaction variables such as molar ratio and types of catalysts and alcohols have an important effect on the subsequent workup procedure, thereby affecting yield and purity of the final product esters.

In 1986, Freedman et al. studied the transesterification kinetics of soybean oil (Freedman, 1986). The S-shaped curves of the effects of time and temperature on ester formation for a 30:1 ratio of butanol and soybean oil, 1% H_2SO_4, and 77°C –117°C at 10°C intervals indicated that the reaction began at a slow rate, proceeded at a faster rate, and then slowed again as the reaction neared completion. With acid or alkali catalysis, the forward reaction followed pseudo-first-order kinetics for butanol:soybean oil of 30:1. However, with alkali catalysis the forward reaction followed consecutive, second-order kinetics for butanol:soybean oil of 6:1. The reaction of methanol with soybean oil at 6:1 molar ratio with 0.5% sodium methoxide at 20°C–60°C was a combination of second-order consecutive and fourth-order shunt reactions. The reaction rate constants for the alkali-catalyzed reaction were much higher than those for the acid-catalyzed reactions. Rate constants increased with an increase in the amount of catalyst used. The activation energies ranged from 8 to 20 kcal/mol. *Ea* for the shunt reaction triglyceride-glycerol was 20 kcal/mol.

In 1997, Noureddini and Zhu conducted a kinetic study using soybean oil with methanol. They used a MLAB computer program to solve the differential equations for the TG, DG, and MG (Noureddini and Zhu, 1987). They found that including the shunt mechanism used by Freedman to describe the transesterification kinetics of soybean oil was not necessary. In 2000, Darnoko et al. investigated the kinetics of palm oil with methanol using a 6:1 mol ratio of alcohol to TG. They found that the best kinetic model to describe their data was a pseudo-second-order kinetic for the first stages of the transesterification reaction followed by a first-order or zero-order model (Darnoko and Cheryan, 2000). The pseudo-second-order was obtained because of methanol excess.

Boocock et al. (1996) emphasized the need for an alternate reaction mechanism, since it was highly unlikely for the three methoxide molecules to attack the triglyceride molecule in concert. Their study for reaction kinetics in both acid- and base-catalyzed reactions revealed that methanolysis was 15 times slower at 40°C than butanolysis at 30°C; also there was a lag of 4 min before the appearance of esters in the methanol system. The reason for this is that whereas butanol and soybean oil are completely miscible and exist in one phase, two distinct phases exist in the methanol–soybean oil system, and since the catalyst is in the methanol phase the reaction is mass transfer limited. Besides explaining the deviation from second-order kinetics for methanolysis, Boocock et al. (1998) also found that the addition of a cosolvent such as tetrahydrofuran (THF) could speed up the methanolysis by creating one phase reaction mixture. Further, the proximity of the boiling points of methanol and THF made it easy to distil and recycle the unreacted methanol and THF from the products. Although the initial reaction rate for methanolysis is very high, the rate decreases drastically. This was attributed to the separation of glycerol, in which the catalyst is preferentially soluble. In a later study, the authors suggested that, this could be prevented by using high alcohol/oil molar ratios of up to 27:1, above which the reaction rate decreased due to the dilution effect (Boocock et al., 1998).

9.3.5.5 Supercritical Alcohol Transesterification

In general, methyl and ethyl alcohols are used in supercritical alcohol transesterification. In the conventional transesterification of fats and oils for biodiesel production, FFAs and water always produce negative effects since the presence of FFAs and water causes soap formation, consumes the catalyst, and reduces catalyst effectiveness, all of which results in a low conversion (Komers et al., 2001). The transesterification reaction may be carried out using either basic or acidic catalysts, but these processes require relatively time-consuming and complicated separation of the product and the catalyst, which results in high production costs and energy consumption. To overcome these problems, Saka and Kusdiana (2001) and Demirbas (2002a,b, 2003) have proposed that biodiesel fuels may be prepared from vegetable oil via noncatalytic transesterification with SCM. A novel process of biodiesel fuel production has been developed by a noncatalytic SCM method. SCM is believed to solve the problems associated with the two-phase nature of normal methanol/oil mixtures by forming a single phase as a result of the lower value of the dielectric constant of methanol in the supercritical state. As a result, the reaction was found to be complete in a very short time. Compared with the catalytic processes under barometric pressure, the SCM process is noncatalytic, involves a much simpler purification of products, has a lower reaction time, is more environmentally friendly, and requires lower energy use. However, the reaction requires temperatures of 525–675 K and pressures of 35–60 MPa (Kusdiana and Saka, 2001; Demirbas, 2003).

Noncatalytic SCM transesterification is performed in a stainless steel cylindrical reactor (autoclave) at 520 K (Demirbas, 2002a,b). In a typical run, the autoclave is charged with a given amount of vegetable oil and liquid methanol with changed molar ratios. After each run, the gas is vented and the autoclave is poured into a collecting vessel. The rest of the contents are removed from the autoclave by washing with methanol. Table 9.7 shows the comparison between the catalytic methanol method and the SCM method for biodiesel from vegetable oils by transesterification. The SCM process is non-catalytic, involves simpler purification, has a lower reaction time, and is less energy intensive. Therefore, the SCM method would be more effective and efficient than the common commercial process (Kusdiana and Saka, 2004). The parameters affecting methyl ester formation are reaction temperature, pressure, molar ratio, water content, and FFA content. It is evident that at sub-critical states of alcohol, the reaction rate is so low and gradually increased as either pressure or temperature rises. It was observed that increasing the reaction temperature, especially to supercritical conditions, had a favorable

TABLE 9.7

Comparisons between Catalytic Methanol Method and Supercritical Methanol (SCM) Method for Transesterification

	Catalytic Methanol Method	**Supercritical Methanol Method**
Methylating agent	Methanol	Methanol
Catalyst	Alkali (NaOH or KOH)	None
Reaction temperature (K)	303–338	523–573
Reaction pressure (MPa)	0.1	10–25
Reaction time (min)	60–360	7–15
Methyl ester yield	96	98
Removal for purification	Methanol, catalyst, glycerol, soaps	Methanol
Free fatty acids	Saponified products	Methyl esters, water
Smelling from exhaust	Soap smell	Sweet smelling

influence on the yield of ester conversion. The yield of alkyl ester increased when the molar ratio of oil to alcohol was increased (Demirbas, 2002a,b). In the supercritical alcohol transesterification method, the yield of conversion rises from 50% to 95% for the first 10 min.

SCM transesterification also can be carried out in an autoclave in the presence of 1%–5% NaOH, CaO, and MgO as catalyst at 520 K. In the catalytic SCM transesterification method, the yield of conversion rises to 60%–90% for the first minute (Demirbas, 2008).

9.3.5.6 Lipase-Catalyzed Transesterification

Although chemical transesterification process gives high conversion levels of triglycerides to their corresponding alkyl esters in short time, the reaction has several drawbacks: it is energy intensive, recovery of glycerol is difficult, the acidic or alkaline catalyst has to be removed from the product, alkaline wastewater requires treatment, and FFAs and water interfere with reaction. Recently, lipase-catalyzed transesterification for biodiesel production has received much attention since it has many advantages over chemical methods such as mild reaction conditions, specificity, reuse, and enzymes or whole cells can be immobilized, can be genetically engineered to improve their efficiency, are more thermostable, and are considered natural, and the reactions they catalyze are considered "green" reactions, and separation of product will be easier (Casimir et al., 2007).

Lipases (triacylglycerol acylhydrolase, E.C. 3.1.1.3) are enzymes widely distributed among animals, plants, and microorganisms that catalyze the reversible hydrolysis of glycerol ester bond and, therefore, also the synthesis of glycerol esters. In nature, lipases were typically used for hydrolysis. Under certain circumstances, lipases can also catalyze a number of transesterification reactions. Lipases can be used in low-water environment as an excellent tool for the transformation of commercial triglycerides, and/or their derivates, to synthesize a growing range of products of potential industrial interest (Pirozzi, 2003). Lipases have been successfully used as catalysts for the synthesis of esters, both in small-scale work and on an industrial scale.

Lipases are broadly classified as intracellular and extracellular. They are also classified on the basis of the sources from which they are obtained. Lipases can be produced in high yields from microorganisms such as bacteria and fungi. In practice, microbial lipases are commonly used by the industry. The selection of a lipase for lipid modification is based on the nature of modification sought, for instance, position-specific modification of triacylglycerol, fatty acid–specific modification, modification by hydrolysis, and modification by synthesis (direct synthesis and transesterification). The literature survey showed the use of lipases from some of the following sources. Microbial lipases are derived from *Aspergillus niger, Bacillus thermoleovorans, Candida cylindracea, Candida rugosa, Chromobacterium viscosum, Geotrichum candidum, Fusarium heterosporum, Fusarium oxysporum, Humicola lanuginose, Mucor miehei, Oospora lactis, Penicillium cyclopium, Penicillium roqueforti, Pseudomonas aeruginosa, Pseudomonascepacia, Pseudomonas fluorescens, Pseudomonas putida, Rhizopus arrhizus, Rhizopus boreas, Rhizopus thermosus, Rhizopus usamii, Rhizopus stolonifer, Rhizopus fusiformis, Rhizopus circinans, Rhizopus delemar, Rhizopus chinensis, Rhizopus japonicus NR400, Rhizopus microsporus, Rhizomucor miehei, Rhizopus nigricans, Rhizopus niveus, Rhizopus oryzae, Rhizopus rhizopodiformis, Rhizopus stolonifer NRRL 1478, Rhodotorula rubra,* and *Staphylococcus hyicus,* to name a few (Sellappan and Akoh, 2005). Animal sources are from pancreatic lipases, and plant lipases are from papaya latex, oat seed lipase, and castor seed lipase (Sellappan and Akoh, 2005).

There are at least 35 lipases available commercially, but only a few can be obtained in industrial quantities. Some of the most promising commercially available microbial lipases are from *Candida antarctica, Candida rugosa* (ex. *Candida cylindracea*), *Rhizomucor miehei, Pseudomonas fluorescens, Aspergillus niger,* and *Chromobacterium viscosum*; in addition, a number of proteolytic enzymes, such as papain and pancreatin, catalyze the hydrolysis and synthesis of ester bonds (Wu et al., 1996).

Currently, there are extensive reports about lipase-mediated alcoholysis for biodiesel production. Enzymatic production is possible using both extracellular and intracellular lipases. In both the cases, the lipase is immobilized and used, which eliminates downstream operations like separation and recycling. Hence in all the works reported in the literature either immobilized (extracellular) lipases or immobilized whole cells (intracellular lipases) are used for catalysis. Both the processes are reported to be highly efficient compared to using free enzymes.

9.3.5.6.1 Immobilized Extracellular Lipase-Catalyzed Transesterification

In terms of enzymatic approaches for biodiesel production, most researches are focused on immobilized extracellular lipases since it is usually more stable than free lipases and can be used repeatedly without complex separation. The lipases that were found to be capable of catalyzing methanolysis are obtained from microorganisms like *Mucor miehei, Rhizopus oryzae, Candida antarctica,* and *Pseudomonas cepacia* (Iso et al., 2001; Orcaire et al., 2006; Du et al., 2007).

Extracellular lipase used for biodiesel production can be immobilized with different carriers, such as porous kaolinite, fiber cloth, hydrotalcite, macroporous resins, and silica gel. Iso et al. investigated the production of biodiesel fuel from triglycerides and alcohol using different immobilized lipases and found that the immobilized *Pseudomonas fluorescens* lipase showed the highest activity in this reaction. The immobilization of *Pseudomonas fluorescens* lipase was carried out using porous kaolinite particle as a carrier (Iso et al., 2001). Yagiz et al. studied the use of hydrotalcite and different types of zeolite as immobilization material for immobilizing free enzyme of Lipozyme TL IM. The results showed that hydrotalcite was more efficient than zeolites. Free Lipozyme TL IM enzyme immobilized on hydrotalcite gave a yield of

92.8% in the transesterification of waste cooking oil and methanol for biodiesel production (Yagiz et al., 2007). Orcaire et al. (2006) also reported that the encapsulation of lipases in silica aerogels could be used for biodiesel production, but the substrate diffusion inside the aerogels should be taken into consideration.

It has been reported that the immobilization carriers might influence the acyl immigration, which had some influence on the biodiesel yield accordingly when 1,3-regiospecificity enzymes are used for biodiesel production (Du et al., 2005). Lipozyme TL IM is one of the typical lipases with 1,3-regiospecificity, so theoretically, biodiesel yield is only about 66%. While during the process of Lipozyme TL IM-mediated methanolysis for biodiesel production, a biodiesel yield of over 90% could be obtained. Further study showed that silica gel acting as the immobilized carrier contributed significantly to the promotion of acyl migration in the transesterification processes (Du et al., 2005).

Different acyl acceptors have been studied for lipase-catalyzed biodiesel production. The most commonly used acyl acceptors were short-chain alcohols. Several primary short-chain alcohols like methanol, ethanol, propanol, and butanol, as well as secondary alcohols like iso-propanol and 2-butanol, have been attempted as acyl acceptors for lipase-catalyzed biodiesel production (Mittelbach, 1990; Nelson et al., 1996; Chen and Wu, 2003). It was found that different lipases have different preference over primary and secondary alcohols (Nelson et al., 1996).

Mittelbach reported transesterification of sunflower oil with primary alcohols like methanol, ethanol, and butanol using *M. miehei* and *C. antarctica* (Novozym 435) in the presence and absence of the solvent, petroleum ether (Mittelbach, 1990). Yields obtained for ethanol and butanol were high even without the solvent but methanol was found to produce only traces of methyl esters without the solvent. Nelson et al. conducted batch experiments and found that *C. antarctica* was suitable for secondary alcohols (80% conversion) like iso-propanol and 2-butanol and *M. miehei* was efficient for primary short-chain alcohols (95% conversion) like methanol, ethanol, propanol, and butanol in the presence of hexane as a solvent. However, in the absence of the solvent, methanol was the least efficient with a methyl ester yield of 19.4% (Nelson et al., 1996). The low yield was attributed to the inhibitory effects caused by methanol on the immobilized enzyme. This was again confirmed by Abigor et al. (2000) who reported the conversion of palm kernel oil using methanol and ethanol as 15% and 72%, respectively. Noureddini et al. used methanol and ethanol for the tranesterification of soybean oil in the presence of immobilized enzyme obtained from *Pseudomonas flourescens* and reported conversions 67% and 65% for methanol and ethanol, respectively (Noureddini et al., 2001). But conversions as high as 97% is possible, which was demonstrated by Linko et al. (1998) using 2-ethyl-1-hexanol for the transesterification of rapeseed oil. Similarly, usage of other alcohols as alternate acyl acceptors instead of methanol have been experimented and conversions as high as 90% were constantly obtained. Iso et al. (2001) reported 90% conversion of vegetable oil using *P. fluorescens* enzyme with butanol as the acyl acceptor. The reaction was carried out in a solvent-free medium under an optimum condition of 0.3% water and 60°C. Propane-2-ol was used as an acyl acceptor by Modi et al. (2006) for the transesterification of Jatropha, Karanj, and sunflower oil achieving a maximum conversion of 92.8%, 91.7%, and 93.4%, respectively. With propane-2-ol, the reusability of lipase was maintained over 12 cycles while it dropped to zero after seven cycles when methanol was used. Recently, acyl acceptors other than alcohols were experimented to improve the efficiency of transesterification process.

Pretreatment of the immobilized lipase was believed to minimize the deactivation of the enzyme, and Samukawa et al. (2000) illustrated this by preincubation of the enzyme in methyl oleate for 0.5 h and subsequently in soybean oil for 12 h. Inhibitory effects were considerably reduced and high conversions were obtained. Chen and Wu (2003) suggested the usage of *tert*-butanol and 2-butanol for regenerating the activity of deactivated enzymes. They conducted experiments by completely deactivating Novozyme 435 with methanol and regenerating by washing the enzyme with *tert*-butanol and 2-butanol. The activity of the enzyme increased by 10 times compared to the untreated enzyme and the completely deactivated enzyme was restored to 56% and 75% of its original value when washed with 2-butanol and *tert*-butanol, respectively. The same process was also patented by Chen and Wu.

9.3.5.6.2 *Intracellular Lipase-Catalyzed Transesterification*

Utilizing intracellular lipase (whole cell immobilization) instead of conventional immobilized lipase as the catalyst for biodiesel production is a potential way to reduce the cost of biocatalyst since it can avoid the complex procedures of isolation, purification, and immobilization of traditional immobilized lipase preparation. A comparison of the immobilization process of extracellular and intracellular lipases is shown in Figure 9.13. It can be clearly seen that considerable reduction in cost can be achieved with intracellular lipase.

R. oryzae, as a sort of whole cell catalyst producing intercellular lipase, has been studied extensively for biodiesel production. Ban et al. (2001) utilized immobilized whole cell *R. oryzae* for the transesterification of vegetable oils and investigated the culture conditions, cell pretreatment effects, and effect of water content on the production process. To enhance the methanolysis activity of the immobilized cells, several substrate-related compounds were added to the medium, of which olive oil and oleic acid were found to be effective. It was reported that, with stepwise addition of methanol and with 15% water content, a high conversion of 90% was obtained, which was comparative with the extracellular process. In order to stabilize *R. oryzae* cells, cross-linking treatment with 0.1% glutaraldehyhde was further examined

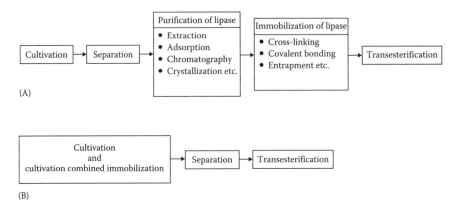

FIGURE 9.13
Comparison of lipase production processes for transesterification with extracellular (A) and intracellular lipases (B).

by Ban et al. (2002). It was reported that without glutaraldehyde treatment, in the stepwise methanol addition process, conversion level dropped to 50% after sixth batch cycle whereas with glutaraldehyde treatment conversion can be maintained at 72%–83% even after six batch cycles.

Fukuda and Kondo patented the process of producing biodiesel utilizing whole cell biocatalyst by pretreating the cells with lower alcohols. They claimed a 350–600 times increase in the reaction rate using cells treated with lower alcohols compared to untreated cells (Fukuda and Kondo, 2003).

Hama et al. studied the production of biodiesel using a packed-bed reactor utilizing *R. oryzae* whole cell biocatalyst by plant oil methanolysis. Compared with methanolysis reaction in a shaken bottle, the packed-bed reactor enhanced repeated batch methanolysis by protecting immobilized cells from physical damage and excess amounts of methanol (Hama et al., 2006). Lipase-producing *R. oryzae* cells were immobilized within 6×6×3 mm cubical polyurethane foam biomass support particles (BSPs) during batch cultivation in a 20-l air-lift bioreactor. The technique using BSPs, developed by Atkinson et al., is the most available immobilization method. It has several advantages over other methods in terms of industrial application: (1) no chemical additives are required, (2) there is no need for preproduction of cells, (3) aseptic handling of particles is unnecessary, (4) there is a large mass transfer rate of substrate and production within BSPs, (5) the particles are reusable, (6) the particles are durable against mechanical shear, (7) bioreactor scale-up is easy, and (8) costs are low compared to other methods (Atkinson et al., 1979).

During the process of using whole cell catalyst for biodiesel production, mass transfer resistance should be taken into consideration. Cell surface display might be one of the useful ways to reduce the mass transfer resistance. Ueda et al. (2002) succeeded in using cell surface-engineered yeast displaying *R. oryzae* lipase and Matsumoto et al. (2002) constructed yeast whole cell biocatalyst overproducing lipase with a pro-sequence from *R. oryzae* IFO4697. Some other ways such as proper pretreatment of whole cells also might be an effective way for the reduction of mass transfer resistance.

Instead of using intracellular lipase of cultivated whole cell for biodiesel production, Kalscheuer et al. reported biodiesel (referred to as microdiesel) production during the cultivation of whole cell. Biosynthesis of biodiesel-adequate FAEEs was established in metabolically engineered *Escherichia coli* (Kalscheuer et al., 2006). The process was achieved by heterogonous expression in *E. coli* of the *Zymomonas mobilis* pyruvate decarboxylase and alcohol dehydrogenase and the unspecific acyltransferase from *Acinetobacter baylyi* strain ADP1. By this approach, ethanol formation was combined with subsequent esterification of the ethanol with the acyl moieties of coenzyme A thioesters of fatty acids if the cells were cultivated under aerobic conditions in the presence of glucose and oleic acid. This novel approach might pave the way for industrial production of biodiesel equivalents from renewable resources by employing engineered microorganisms, enabling a broader use of biodiesel-like fuels in the future. Table 9.8 summarizes the significant works reported in the literature with respect to enzymatic production of biodiesel (Ranganathan et al., 2008).

9.3.5.6.3 Lipase-Catalyzed Transesterification in Different Reaction Mediums

Lipase-catalyzed transesterification for biodiesel production has been exploring in different reaction medium. These reaction systems can be classified into solvent-free system, hydrophobic organic solvent system, hydrophilic organic solvent system, cosolvent mixture system, water-containing system, and ionic liquid system (Kaieda et al., 1999; Du et al., 2008; Su and Wei, 2008).

Solvent-free system: Solvent-free system was proposed firstly for lipase-catalyzed methanolysis for biodiesel production (Shimada et al., 1999). In this system, methanol has very low solubility in oils, and too much methanol existing as drops in reaction medium will have some negative effects on lipase activity. Therefore, stepwise methanol addition was recommended to reduce the negative effects during producing biodiesel through lipase catalysis in solvent-free system (Shimada et al., 1999). Through three-step methanol addition, a very high conversion (ex. 98.4%) of the oil to its corresponding alkyl esters could be achieved by immobilized lipase. Following this study, there are many other reports related to biodiesel production using stepwise addition of methanol during lipase-catalyzed methanolysis for biodiesel production (Watanabe et al., 2000; Shimada et al., 2002; Xu et al., 2004).

In the process of biodiesel production, glycerol will be produced as a byproduct. Glycerol is very hydrophilic and insoluble in the oils, so it is easily

TABLE 9.8

Comparison of Various Works on Enzymatic Production of Biodiesel

S.No.	Authors/Year	Oil/Enzyme	Acyl-Acceptor	Conversion (%)	Technique Employed	Cost of Production
1.	Watanabe et al. (2000)[a]	Vegetable oil, Novozyme 435	Methanol	90–93	Stepwise addition of methanol	Moderate
2.	Samukawa et al. (2000)[a]	Soybean oil, Novozyme 435	Methanol	97	Stepwise addition methanol and pre-incubation of enzyme in methyl oleate and soybean oil	High
3.	Ban et al. (2001)[b]	Vegetable oil, R. Oryzae	Methanol	90	Stepwise addition of methanol and application of glutaraldehyde for stability of enzyme	Low
4.	Iso et al. (2001)[a]	Triolein, P. flourescens	Butanol	90	Butanol was used as acyl acceptor and no solvent was used	Moderate
5.	Shimada et al. (2002)[a]	Waste cooking oil, Novozyme 435	Methanol	90	Stepwise addition of methanol	Low
6.	Bako et al. (2002)[a]	Sunflower oil, Novozyme 435	Methanol	97	Stepwise addition of methanol and removal of glycerol dialysis	High

No.	Reference	Oil / Lipase	Acyl acceptor	Yield	Remarks	
7.	Du et al. (2004)[a]	Soybean oil, Novozyme 435	Methyl acetate	92	A novel acyl acceptor, methyl acetate, which had no inhibitory effects, was used	High
8.	Xu et al. (2004)[a]	Soybean oil, Novozyme 435	Methanol	93	Stepwise addition of methanol and removal of glycerol using the solvent, *iso*-propanol	High
9.	Li et al. (2006)[a]	Rapeseed oil, Novozyme 435 and Lipozyme TL IM	Methanol	95	Combined use of Lipozyme TL IM and Novozyme 435 along with *tert*-butanol as solvent	High
10.	Royon et al. (2007)[a]	Cotton seed oil, Novozyme 435	Methanol	97	*tert*-Butanol was used as a solvent	High
11.	Modi et al. (2007)[a]	Jatropha oil, Novozyme 435	Elhyl acetate	91.3	Ethyl acetate having no inhibitory effects was used	High
12.	Hama et al. (2007)[b]	Soybean oil, *R. Oryzae*	Methanol	90	Stepwise addition of methanol in a packed-bed reactor	Low

[a] Extra cellular lipase.
[b] Intracellular lipase.

adsorbed onto the surface of the immobilized lipase, leading to negative effect on lipase activity and operational stability (Soumanou and Bornscheuer, 2003). Several methods have also been proposed to eliminate the negative effect caused by glycerol: addition of silica gel into the reaction system to absorb the glycerol (Stevenson et al., 1994) or washing the lipase with some organic solvents periodically to remove glycerol (Dossat et al., 1999; Soumanou and Bornscheuer, 2003; Su et al., 2007). Actually, these methods can only be used to reduce the negative effect caused by glycerol to some extent, but it is a little bit complicated especially for large-scale continuous production of biodiesel. Obviously, how to solve the negative effects caused by methanol and glycerol are the key points for immobilized lipase-catalyzed biodiesel production in solvent-free system.

Hydrophobic organic solvent system: In order to solve the problems caused by methanol and glycerol in solvent-free system, some researchers had carried out the research of lipase-catalyzed methanolysis for biodiesel production in organic solvent mediums. The enzyme activity in organic media is often correlated with the solvent hydrophobicity (log P) with the highest activities found at high log P (>2) values (Laane et al., 1987; Valivety et al., 1992; Van et al., 1995; Ducret et al., 1998). Further, the stability of the enzyme is usually higher in the more hydrophobic solvents. Based on these knowledges, the hydrophobic organic solvents such as n-hexane and petroleum ether have been tried as reaction medium for biodiesel production (Soumanou and Bornscheuer, 2003; Ghamguia et al., 2004; Lara and Park, 2004). However, with such relatively hydrophobic organic solvents as the reaction medium for lipase-catalyzed biodiesel production, methanol and glycerol still have poor solubility in the system. So, the negative effects caused by methanol and glycerol cannot be eliminated, and lipase still exhibits poor operational stability in such reaction media (Dossat et al., 1999). Dossat et al. studied lipozyme-catalyzed transesterification of sunflower oil with butanol for biodiesel production in n-hexane in a continuous packed-bed reactor. By-product glycerol, which is insoluble in n-hexane, was found to remain in the reactor and adsorbed onto the enzymatic support, leading to a drastic decrease in enzymatic activity, and intermittent rinsing of the catalyst bed with a solution of tertiary alcohol is needed. It seems that conventional hydrophobic organic solvents such as hexane and petroleum ether are not suitable as the reaction media for immobilized lipase-catalyzed biodiesel production since the negative effects caused by methanol and glycerol still exist in such systems.

Hydrophilic organic solvent system: From the above introduction, we could deduce that a solvent, which could dissolve the hydrophobic oil, hydrophilic methanol and glycerol, and in which lipase could show higher stability, might be an appropriate reaction medium. *tert*-Butanol, a relatively hydrophilic organic solvent, has been developed as a novel reaction medium for lipase-catalyzed methanolysis of biodiesel production (Li et al., 2006; Wang et al., 2006). In this novel medium, both methanol and glycerol are soluble, so the negative effects caused by them on lipase performance can be eliminated totally. With combined use of Lipozyme TL IM and Novozym 435 in *tert*-butanol system, the highest biodiesel yield of 95% could be obtained, and there was no obvious loss in lipase activity even after being reused for 200 cycles (Li et al., 2006). This technology has been applied to the industrialization of biodiesel production with a capacity of 20,000 tons/year in China. Royon et al. also reported lipase-catalyzed methanolysis of cottonseed oil for biodiesel production with *tert*-butanol as the reaction medium. A methanolysis yield of 97% was observed, and experiments with the continuous reactor over 500 h did not show any appreciable decrease in ester yields (Royon et al., 2007).

According to the nonaqueous enzymology principle, an enzyme could maintain high catalytic activity in organic solvents, which are rather hydrophobic (usually log P >3), and in those organic solvents with log P <2, an enzyme showed much lower catalytic activity (Soumanou and Bornscheuer, 2003; Lara and Park, 2004). Therefore, generally solvents with a log P value below 2 have been considered unsuitable as media for enzyme catalysis (Ducret et al., 1998). However, the log P value of *tert*-butanol is just about 0.80, but the lipase still can express rather high activity and good stability. This result shows that the *tert*-butanol can dissolve methanol and glycerol and thus overcome their negative effects on lipase to a high degree. Du et al. (2007) have also explored the related mechanism, and log P environment was proposed to consider the effect of whole environment on lipase activity instead of just considering the effect of organic solvent itself as traditional ways did.

Introducing relatively hydrophilic solvents such as *tert*-butanol into biodiesel production system could solve the negative effects caused by methanol and glycerol existing in conventional reaction mediums. The operational life of the lipase could be prolonged significantly, and this technology showed great prospect. However, the freezing point of *tert*-butanol is 25.2°C, so it is solid at the room temperature and this must increase the operational complexity. Furthermore, the low-temperature lipase cannot be used as the biocatalyst in the *tert*-butanol system. For example, a psychrophilic lipase from *Pseudomonas fluorescens* was found in our lab, which showed the optimal transesterification activity for biodiesel production at 20°C (Luo et al., 2006).

Cosolvent mixture system: To overcome the disadvantages of *tert*-butanol system, a solvent engineering strategy was applied to the lipase-catalyzed methanolysis of triacylglycerols for biodiesel production by Su and Wei (doi:10.1016/j.molcatb.2008.03.001). The effect of different pure organic solvents and cosolvent mixtures on the lipase-catalyzed methanolysis was compared. The substrate conversions in the cosolvent mixtures were all higher than those of the corresponding pure organic solvents. Further study showed that addition of cosolvent decreased the values of $|\log P_{\text{interface}} - \log P_{\text{substrate}}|$ and thus led to a faster reaction. The more the values of $|\log P_{\text{interface}} - \log P_{\text{substrate}}|$ decreased, the faster the reaction proceeded and the

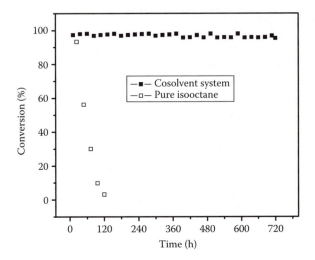

FIGURE 9.14
Operational stability of Novozym 435 during batch reaction.

higher the conversion attained. Different cosolvent ratio was further investigated. The cosolvent mixture of 25% *t*-pentanol:75% isooctane (v/v) was
optimal, with which both the negative effects caused by excessive methanol
and by-product glycerol could be eliminated. The lipase could even tolerate
a very high methanol/oil ratio (9:1), and there was no obvious loss in lipase
activity even after being repeatedly used for 60 cycles (720 h, Figure 9.14)
with this cosolvent mixture as reaction medium. Other lipases and lipase
combinations can also catalyze methanolysis in this cosolvent mixture.
Furthermore, other vegetable oils were also explored for biodiesel production in this cosolvent mixture, and it had been found that this cosolvent
mixture media has extensive applicability.

From the results of Su and Wei's work, it can be concluded that the use of
the cosolvent mixture in the enzymatic biodiesel production has the following advantages: (1) both the negative effects caused by excessive methanol
and by-product glycerol can be eliminated completely; (2) high reaction rates
and conversion can be obtained; (3) no catalyst regeneration steps are needed
for lipase reuse; (4) the operational stability of the catalyst is high. In a word,
the cosolvent mixture is a very prospective media for lipase-catalyzed methanolysis of triacylglycerols in biodiesel production.

Water-containing system: Naturally, lipase catalyzes the reversible hydrolysis of triacylglycerols in the aqueous system. However, it also can catalyze
the transesterification reaction in the nonaqueous system. Lipase-catalyzed
biodiesel production makes use of the latter characteristics. Therefore, when
using lipase as the biocatalyst to produce biodiesel, the reaction medium

should hardly contain water. Waste oils and fats are cheap feedstocks for biodiesel production and are used as the substrate inevitably contains a certain amount of water, it is necessary to find lipases that efficiently catalyze methanolysis even in the presence of water.

Studies on methyl ester synthesis in aqueous medium have been reported with lipases from *Geotrichum candidum* (Okumura et al., 1979) and *Candida deformans* (Boutur et al., 1994). However, the yields of the reaction were 62% and 58%, respectively, with the substrate oleic acid and triolein.

Lipase from *Rhizopus oryzae* efficiently catalyzed the methanolysis of soybean oil in the presence of 4–30 wt % water in the starting materials; however, the lipase was nearly inactive in the absence of water (Kaieda et al., 1999). The methyl ester (ME) content in the reaction mixture reached 80–90 wt % by stepwise additions of methanol to the reaction mixture. The kinetics of the reaction appears to be in accordance with the successive reaction mechanism. That is, the oil is first hydrolyzed to FFAs and partial glycerides, and the fatty acids produced are then esterified with methanol.

A number of factors affecting the methanolysis of vegetable oils in aqueous medium by *Cryptococcus* spp. S-2 lipase were investigated by Kamini et al. (2001). The crude lipase from the yeast efficiently catalyzed the methanolysis of vegetable oils (oil/methanol molar ratio of 1:1) in the presence of 40 wt % water. The methyl ester content was high with rice bran oil at 30°C for 96 h and further optimization studies were carried out with varying amount of enzyme, water, or methanol. The enzyme was not inactivated by shaking in a mixture containing 4 Meq of methanol against the oil and 100 wt % water by weight of the substrate and the methyl ester contents increased with increasing molar equivalents of methanol and water contents from 60 to 100 wt %. Thus, the reaction was conducted in a single step to avoid the stepwise addition of methanol and the methyl ester contents reached 80.2 wt % at 120 h.

A biphasic oil–aqueous system for biodiesel production by enzymatic catalysis was studied (Oliveira and Rosa, 2006). The transesterification of sunflower oil with methanol was catalyzed by free or immobilized lipases from *Rhizomucor miehei* and *Humicola insolens*. The effects of protein amount, temperature, pH, and molar ratio of methanol to sunflower oil on the enzymatic reaction using free lipase were evaluated; the best results were obtained with *H. insolens*, at pH 5, 40°C, and 36.8 mg of protein. By using this lipase immobilized at a 6:1 methanol/oil molar ratio and a 2:1 volumetric oil/water phase ratio, an ester content of 96.1% and a conversion of 91.2% were achieved.

Ionic liquid system: Ionic liquids, known as organic salts and entirely composed of ions, are emerging as desirable substitutes for volatile, toxic, and flammable organic solvents. Recently, production of biodiesel in ionic liquids through immobilized *C. antarctica* lipase-catalyzed methanolysis of soybean oil was demonstrated by Ha et al. (2007). Among the 23 tested ionic liquids, the highest FAMEs production after 12 h at 50°C was achieved in [Emim][TfO]. The production yield of 80% was eight times higher compared to the conventional solvent-free system. It was around 15% higher than the FAMEs production system using *tert*-butanol as an additive. The optimum substrate molar ratio of methanol to soybean oil for FAMEs production in [Emim][TfO] was found to be 4:1. Their results of high production yield in ionic liquids show that ionic liquids are potential reaction media for biodiesel production. However, more studies are needed, such as the recovery of ionic liquids.

9.3.5.6.4 Kinetic Mechanism of Lipase-Catalyzed Biodiesel Production

The application of lipase in the production of biodiesel from vegetable oils has been thoroughly addressed in the literature, but these studies were purely experimental. Furthermore, all the kinetic studies found in the literature investigated the esterification of FFAs rather than the transesterification of vegetable oil (Janssen et al., 1996, 1999; Marty et al., 1992). However, the main industrial interest is on the production of biodiesel from the oils or fat, not the FFAs. Therefore, it was desired to develop a reliable kinetic model for transesterification reaction. An earlier attempt to model vegetable oil transesterification was done by Al-Zuhair, indicating that the reaction took place in two consecutive steps (Al-Zuhair, 2005). In the first step, triglycerides were hydrolyzed to produce FFAs, and in the second step, the FFAs produced were esterified to produce fatty acids methyl esters. However, in the recent work of Al-Zuhair et al., it was shown that it was more accurate to assume that transesterification takes place by direct alcoholysis of the triglycerides (Al-Zuhair et al., 2007). The main difference between esterification and transesterification is that in the first O–H bonds are broken, whereas in the second ester bonds are broken. In addition, the by-product of esterification is water whereas it is glycerol in transesterification.

In the study of Al-Zuhair et al. (2007) experimental determination of the separate effects of palm oil and methanol concentrations on the rate of lipase-catalyzed transesterification were used to understand the reaction behavior and propose suitable mechanism steps and to test the generated kinetic model. The reaction took place in n-hexane and the lipase from *Mucor miehei* was used. At a constant methanol concentration of 300 mol m^{-3}, it was found that, initially as the palm oil concentration increased, the initial reaction rate increased. However, the initial rate dropped sharply at substrate concentrations larger than 1250 mol m^{-3}. Similar behavior was observed for methanol concentration effect, where at a constant substrate concentration of 1000 mol m^{-3}, the initial rate of reaction dropped at methanol concentrations larger than 3000 mol m^{-3}. Ping-Pong Bi Bi mechanism with inhibition by both reactants was adopted as it best explains the experimental findings.

The proposed mechanism of transesterification was based on the enzymatic hydrolysis mechanism (Bailey and Ollis, 1986). Acidic or basic functional groups found at specific locations in the active sites. The enzyme catalyzes the reaction by donating or accepting protons during the course of the reaction. For example, conjugate acids of amines are proton donors, whereas amines and carboxylate ions are proton acceptors. By transferring protons to and from these groups to the substrate, an enzyme carries out acid and base-catalyzed reactions within the active sites. The active sites of the lipase have been studied by chemical and X-ray techniques (Panalotov and Verger, 2000). Two functional groups that are part of the active sites have been identified as being particularly important to the catalytic process. One is hydroxyl group that acts as a nucleophile, and the other is the nitrogen atom of an amine group, which accepts a proton and then gives it back during the reaction.

The mechanism of esterase-catalyzed alcoholysis is shown in Figure 9.15: (a) nucleophilic addition to form enzyme–substrate complex, where the nucleophile is the oxygen in the O–H group on the enzyme; (b) proton transferred from the conjugate acid of the amine to the alkyl oxygen atom of the substrate, and a glycerol moiety is formed. If a triacylglyceride was the initial substrate, then a diacylglyceride would form, whereas if diacylglyceride was the substrate, then monoacylglyceride would form and so on; (c) the oxygen atom from a methanol molecule is added to the carbon atom of the C=O of the acyl enzyme intermediate to form acylated enzyme–alcohol complex; (d) the enzyme oxygen atom of the complex is eliminated and a proton is transferred from the conjugate acid of the amine, resulting in fatty acid methyl ester, that is, biodiesel.

These steps represent a typical Ping-Pong Bi Bi mechanism, which agrees with most of the earlier kinetic studies on lipase-catalyzed esterifications (Janssen et al., 1996, 1999; Marty et al., 1992; Rizzi et al., 1992). However, in the esterification, the first product was assumed to be the FAME, then followed by glycerol moiety (in these studies, water was the second product, as FFA was the substrate used). On the other hand, it has been shown by proposed mechanism of transesterification that the first product should be the glycerol moiety, followed by the FAME as a final product. The mechanism steps explained above were represented by Equations (9.4)–(9.9):

$$E + S \underset{k_{-1}}{\overset{k_1}{\longleftrightarrow}} E.S \tag{9.4}$$

$$E.S \longleftrightarrow E.Ac.G \tag{9.5}$$

$$E.Ac.G \underset{k_{-2}}{\overset{k_2}{\longleftrightarrow}} E.Ac + G \tag{9.6}$$

$$E.Ac + A \underset{k_{-3}}{\overset{k_3}{\longleftrightarrow}} E.Ac.A \tag{9.7}$$

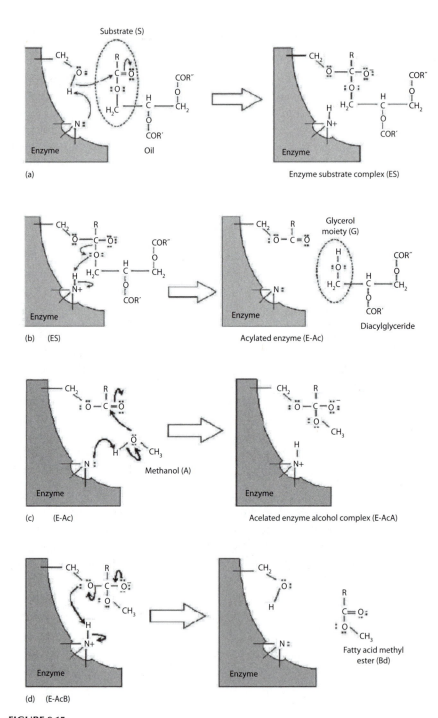

FIGURE 9.15
The mechanism of enzymatic production of FAME from triacylglycerides.

$$E.Ac.A \longleftrightarrow E.Bd \tag{9.8}$$

$$E.Bd \underset{k_{-4}}{\overset{k_4}{\longleftrightarrow}} E+Bd \tag{9.9}$$

where

k_1 and k_{-1}, k_2 and k_{-2}, and k_3 and k_{-3} are the rate constants for the reversible formation of enzyme–substrate complex, E.S, acylated enzyme–glycerol moiety complex, E.Ac.G, acylated enzyme–alcohol complex, E.Ac.A, respectively

k_4 and k_{-4} are the rate constants for the product formation and enzyme regeneration, respectively

To account for the inhibition by alcohol, competitive inhibition was assumed when an alcohol molecule reacts with the enzyme directly to produce a dead-end enzyme–alcohol complex (E.A). And to account for the inhibition by the substrate, competitive inhibition was also assumed when a substrate molecule reacts with the acylated enzyme to produce another dead-end complex, namely, acylated enzyme–substrate complex (E-Ac.S). The two competitive inhibition reactions mentioned above were presented by Equations 9.10 and 9.11, respectively:

$$E+A \underset{k_{-5}}{\overset{k_5}{\longleftrightarrow}} E.A \tag{9.10}$$

$$E.Ac+S \underset{k_{-6}}{\overset{k_6}{\longleftrightarrow}} E.Ac.S \tag{9.11}$$

where k_5 and k_{-5} and k_6 and k_{-6} are the rate constants for the reversible formation of dead-end complexes, enzyme–alcohol, E.S, and acylated enzyme–substrate, respectively. With the above mechanism and assumptions, the reaction rate presented in the following equation (Equation 9.12) was derived:

$$v = \frac{V_{max}}{1+\left(K_{IA}/[A]\right)\left[1+\left([S]/K_S\right)\right]+\left(K_{IS}/[S]\right)\left[1+\left([A]/K_A\right)\right]} \tag{9.12}$$

where

v is the initial reaction rate
V_{max} is the maximum reaction rate
K_S and K_A are the dissociation constants for the substrate (S) and the alcohol (A), respectively
K_{IS} and K_{IA} are the inhibition constants for the substrate and the alcohol, respectively

Equation 9.9 describes the rate of transesterification reaction of triglycerides using lipase, which was derived from Equations 9.1 through 9.8. The equation is applicable for oil, alcohol, and enzyme concentrations in the region of complete homogeneity and could be used to predict the optimal operating conditions. The kinetic model obtained (Equation 9.9) is nonmonotonic, which has important implications on optimal design. Under specific condition, the nonmonotonic kinetics might give rise to complex bifurcation behavior when applied in a continuous process; the simplest of it is multiplicity of the steady states. This bifurcation occurs when a small smooth change made to the parameter values (e.g., substrate concentrations) of a system causes a sudden qualitative or topological change in the system's long-term dynamical behavior. The development of reaction kinetic model, such as Equation 9.9, would sturdily assist in predicting the dynamic behavior of the system and avoid any unexpected responses. Al-Zuhair et al. carried out extensive study on kinetic mechanism of lipase-catalyzed transesterification (Al-Zuhair, 2005, 2006; Al-Zuhair et al., 2007). So the aforementioned discussion is mainly cited from their works.

9.3.5.6.5 *Future Prospects for Lipase-Catalyzed Biodiesel Production*

The potential for using lipases as biocatalysts for biodiesel are enormous. Advantages of using lipase include ease of product recovery; low energy and temperature requirements; ease of enzyme recovery; mild reaction conditions of pH, temperature, and pressure; regeneration and reuse of the enzyme several times; use of reactors for continuous production; thermal stability at a relatively low temperature of operation; operational stability of the enzyme; flexibility of accepting various substrates and alcohols; and reaction in solvent and solvent-free systems; lipases allow reactions in systems that contain acceptable levels of water and can esterify FFAs present in the crude, waste oils, and used frying oils as well as deodorizer distillates; immobilization allows reuse, confers stability, and allows higher enzyme loading for faster reaction; enzymes are natural, and recombinant lipases with enhanced or altered activities can be mass-produced after over-expression in relevant microorganisms, therefore making the overall process economically viable.

Generally speaking, all kinds of oils or fats can be adopted as the feedstocks for biodiesel production. But from an industrial perspective, the cost of refined vegetable oils could be prohibitive especially for long-term development of biodiesel. Compared to chemical approaches, enzymatic ways have many advantages, and one of them is that enzymatic ways can be used for tinpot oils' transformation. Based on the premise of enzymatic ways being economically competitive to chemical process, this advantage will make enzymatic ways quite promising in the future. So the high cost of lipase production is the main hurdle for commercialization of the lipase-catalyzed process.

Several attempts have been made to develop cost-effective systems. To avoid serious degradation of lipase activity in the presence of a high concentration of methanol, a novel operation with stepwise addition of methanol has been developed (Watanabe et al., 2000; Shimada et al., 2002; Xu et al.,

2004). The use of intracellular lipase as a whole cell biocatalyst is also an effective way to lower the lipase production cost, since complex purification is not necessary. Immobilization of whole cells and pretreatment of extracellular lipases to be more tolerant to short-chain alcohols such as methanol can also be ways to reduce the cost of the enzyme process.

Another useful approach to reducing the production cost is to use solvent-tolerant lipases. Several microorganisms that produce solvent-tolerant lipases, either 1, 3-specific from *Fusarium* sp. (Shimada et al., 1993) or nonspecific from *Pseudomonas* and *Bacillus* sp. (Sugihara et al., 1991; Ogino et al., 1999), have been reported. These lipases are stable in most water-immiscible solvents, but their stability generally decreases somewhat in water-miscible solvents such as methanol and ethanol. However, activity of the lipase from *Fusarium heterosporum* was found to slightly increase in the presence of a low concentration of methanol (Shimada et al., 1993).

Further enhancement of lipase production may be achieved by genetic engineering. High levels of expression of lipases from several microorganisms have been successfully achieved using *Saccharomyces cerevisiae* as the host (Bertolini et al., 1995; Nagao et al., 1996). Among these, the lipase cDNA from *F. heterosporum* increased lipase productivity threefold over that of the original strain (Nagao et al., 1996).

In the light of these findings, combining a whole cell biocatalyst with the use of a recombinant microorganism and stepwise addition of alcohol can be expected to considerably decrease the cost of lipase production. Such a novel system is promising for the industrial-scale enzymatic production of biodiesel.

9.3.5.7 In Situ Transesterification

Perhaps the largest impediment to wider adoption of biodiesel is its cost. When produced from refined oils, feedstock cost contributes more than 70% to the cost of the product. Consideration of the current route from oilseeds to biodiesel caused many researchers to inquire whether isolation of the oil from the seed, and its refining, was necessary. In contrast, transesterification reagents such as alcohol might be able to access acylglycerides resident in oilseeds and achieve their transesterification directly, *in situ*. Such a route to biodiesel eliminates the expense associated with solvent extraction and oil cleanup, and simplify the steps in biodiesel production. This could result in a decrease in the cost of the product. *In situ* transesterification differs from the conventional reaction in that the oil-bearing material contacts alcohol directly instead of reacting with purified oil and alcohol. That is, extraction and transesterification proceed within the same process, the alcohol acting both as an extraction solvent and an esterification reagent.

In situ transesterification was first performed by transesterification of sunflower seed oil with acidified methanol, and it turned out that this process produce fatty acid methyl esters in yields significantly greater than those obtained from conventional reaction with pre-extracted seed oil. Yield improvement of over 20% was achieved and could be related to the moisture content of the seed. The presence of moisture in the oilseed reduces the yield of methyl esters (Harrington and D'Arcy-Evans, 1985).

Özgül and Türkay (1993) investigated *in situ* esterifications of high-acidity rice bran oil with methanol and ethanol using sulfuric acid as catalyst. In the esterification with methanol, all FFAs dissolved in methanol were interest-erified within 15 min, and it was possible to obtain nearly pure methyl esters. The amount of methyl esters obtained from rice bran was dependent on the FFA content of the rice bran oil. It was not possible to obtain the pure esters in the esterification with ethanol, because the solubilities of oil component in ethanol were much higher than those in methanol.

In situ alcoholysis of soybean oil with various monohydroxy alcohols in the presence of acid catalyst was investigated by Kildirin et al. The result showed that ethyl, propyl, and butyl esters of soybean fatty acid could be obtained directly in high yields (80%–85%) by *in situ* alcoholysis just in a 3 h reaction. Because methanol is a poor solvent for soybean oil, the amount of oil dissolved in methanol and converted to methyl esters was low after *in situ* alcoholysis (41%). The FFA contents of the esterified products were higher than that of the crude soybean oil in seed (2–6 times), suggesting that a hydrolysis reaction occurred during *in situ* alcoholysis (Kildiran et al., 1996).

Özgül-Yücel et al. also investigated the effect of specific factors on *in situ* methanolic esterification of rice bran oil using sulfuric acid catalyst. The result showed that the ester content of the extract increased about 67% when the FFA contain of the oil was increased from 16.6% to 84.5%. Increasing the reaction time beyond 30 min did not affect yields, but increasing the temperature from 20°C to 65°C elevated the fatty acid methyl esters yield by about 30%. Increasing the amount of acid catalyst above 5 mL did not enhance yield and resulted in bran darkening and an increase in bran stickiness that interfered with filtration operation due to bran cell wall disruption by the acid. Moisture content of rice bran and amount of methanol used had a little effect on the esters content (Özgül-Yücel and Türkay, 2002).

By using ethanol as solvent, Özgül-Yücel et al. found that ethyl esters formation led to an increase in the amount of fatty acid dissolved in ethanol through dissolution and esterification, and the solubility of neutral oil component was reduced. Water content of ethanol was also affected on extraction and *in situ* esterification. When the water content of ethanol was decreased, the amount of total oil and neutral oil in alcohol increased considerably even though the amounts of fatty acid dissolved were nearly the same for both extraction and *in situ* esterification. The ethyl ester content of crude esters obtained with 99.1% ethanol was higher than that obtained with 96% ethanol (Özgül-Yücel and Türkay, 2003).

Some of *in situ* transesterification were also conducted with enzyme as catalyst. Rüsch Gen Klaas and Warwe proposed *in situ* transesterification using a commercial immobilized enzyme (Novozym 435), deposited at the bottom of a soxhlet apparatus, to transesterify plant oil extracted from the oilseed reservoir using either dimethyl or diethyl carbonate (Rüsch Gen Klaas and Warwe, 2001). Torres et al. investigated one pot triacylglycerides extraction and fatty acid methyl esters formation using fungal resting cells and oilseeds (rapeseed) at moderate temperature. By using methanol and isooctane as solvent at 50°C, the formation of fatty acid methyl esters already was apparent after 24 h of reaction although the yield was only 25%; however, the triglycerides yield was 62% and the FFA was 8%. After 120 h, the yield of fatty acid methyl esters rose to 65% (Torres et al., 2003). Su et al. adopted methyl acetate and ethyl acetate as extraction solvents and transesterification reagents at the same time for *in situ* transesterification of *Pistacia chinensis Bunge* seed and *Jatropha curcas L* seed with the Novozym 435 as catalyst. Fatty acid methyl esters and ethyl esters were respectively obtained with higher yields than those achieved by conventional two-step extraction/transesterification. The improvement ranged from 5.3% to 22%. The highest *Pistacia chinensis Bunge* and *Jatropha curcas L* methyl/ethyl esters could attain 92.8%, 89.5%, 86.1%, and 87.2%, respectively, under the optimized conditions (Su et al., 2007).

Several researchers also proposed a simultaneous supercritical fluid extraction and transesterification. But much of them only focused on analytical determination of fatty acids content. For example, Snyder et al. proposed this method for analysis of several seed oils such as canola, soybean, and sunflower. They found that fatty acid composition of such oils was higher than that found by conventional method (extraction and transesterification in separate way) (Snyder et al., 1997). Turner et al. used this method for produced cis-vaccenic acids methyl esters from Milkweed seeds. Novozym 435 was chosen as a catalyst because in an analytical context only this lipase has been used for the production of esters in dynamic supercritical fluid extraction (Turner et al., 2002).

9.3.6 Effect of Different Parameters on Transesterification

9.3.6.1 Molar Ratio of Alcohol to Triglyceride and Type of Alcohol

One of the most important variables affecting the transesterification is the molar ratio of alcohol to triglyceride. The stoichiometric ratio for transesterification requires 3 mol of alcohol and 1 mol of glyceride to yield 3 mol of fatty acid ester and 1 mol of glycerol. However, transesterification is an equilibrium reaction in which a large excess of alcohol is required to drive the reaction to the right (Meher et al., 2006a,b). The excess of alcohol depends on the catalyst that is used in the reaction. Commonly for alkaline-catalyzed reactions, a 100% excess of alcohol is used, 6 mol of alcohol per mole of triglyceride (Freedman et al., 1986); but for the acid-catalyzed reaction, 30 mol of alcohol are used per mole of triglyceride. For an alkali-catalyzed reaction with 6 mol of alcohol and 1 mol of triglyceride at 60°C, the expected conversion of triglyceride to methyl esters is 98% after 60 min.

Higher molar ratios result in greater ester conversion in a shorter time. In the ethanolysis of peanut oil, a 6:1 molar ratio liberated significantly more glycerol than did a 3:1 molar ratio. Rapeseed oil was methanolyzed using 1% NaOH or KOH. They found that the molar ratio of 6:1 of methanol to oil gave the best conversion. When a large amount of FFAs was present in the oil, a molar ratio as high as 15:1 was needed under acid catalysis. Freedman et al. (1984) studied the effect of molar ratio (from 1:1 to 6:1) on ester conversion with vegetable oils. Soybean, sunflower, peanut, and cotton seed oils behaved similarly and achieved highest conversions (93%–98%) at a 6:1 molar ratio. Tanaka et al., in his novel two-step transesterification of oils and fats such as tallow, coconut oil, and palm oil, used 6:1–30:1 molar ratios with alkali catalysis to achieve a conversion of 99.5%.

A very high ratio of alcohol to oil should be avoided, because if this ratio is too large it interferes with the separation of the glycerol by increasing the solubility (Meher et al., 2006a,b). If the glycerol remains in solution, it can help driving the equilibrium back to the reactants, decreasing the yield of alkyl esters (Meher et al., 2006a,b).

The type of alcohol used is also important, because when using short-chain alcohols such as MeOH the oil is not miscible with the alcohol at ambient temperatures. A previous study showed that longer chain alcohols required longer reaction times if the same temperature is used (Canakci and Gerpen, 1999). The addition of every methylene (CH_2) group doubles the reaction time (Canakci and Gerpen, 1999). Canakci's group has conducted reactions using soybean oil with MeOH and H_2SO_4 as the catalyst; their mole ratio of MeOH to soybean oil was 6:1. They showed that longer chain alcohols provided higher reaction rates when compared to methyl esters (Canakci and Gerpen, 1999). The reason of obtaining higher conversion might be because the boiling point of the long-chain alcohols is larger than MeOH allowing the reaction temperatures to be higher at ambient pressures (Canakci and Gerpen, 1999).

9.3.6.2 Catalyst Type and Concentration

Catalysts used for the transesterification of triglycerides are classified as alkali, acid, and enzyme, among which alkali catalysts like sodium hydroxide, sodium methoxide, potassium hydroxide, and potassium methoxide are more effective (Ma and Hanna, 1999). If the oil has high FFA content and more water, acid-catalyzed transesterification is suitable. The acids could be sulfuric acid, phosphoric acid, hydrochloric acid, or organic sulfonic acid.

Methanolysis of beef tallow was studied with catalysts NaOH and $NaOCH_3$. Comparing the two catalysts, NaOH was significantly better than $NaOCH_3$ (Ma et al., 1998). The catalysts NaOH and $NaOCH_3$ reached their maximum activity at 0.3% and 0.5% w/w of the beef tallow, respectively. As a catalyst in the process of alkaline methanolysis, mostly sodium hydroxide or potassium hydroxide have been used, both in concentration from 0.4% to 2% w/w of oil. Refined and crude oils with 1% either sodium hydroxide or potassium hydroxide catalyst resulted successful conversion. Methanolysis of soybean oil with the catalyst 1% potassium hydroxide has given the best yields and viscosities of the esters (Tomasevic and Marinkovic, 2003).

Acid-catalyzed transesterification was studied with waste vegetable oil. The reaction was conducted at four different catalyst concentrations, 0.5, 1.0, 1.5, and 2.25 M HCl in presence of 100% excess alcohol and the result was compared with 2.25 M H_2SO_4 and the decrease in viscosity was observed. H_2SO_4 has superior catalytic activity in the range of 1.5–2.25 M concentration (Mohamad and Ali, 2002).

Enzymatic catalysts like lipases are also able to effectively catalyze the transesterification of triglycerides in either aqueous or nonaqueous systems, which can overcome the problems caused by chemical catalysts (Fukuda et al., 2001). In particular, the by-product glycerol can be easily removed without any complex process, and also that FFAs contained in waste oils and fats can be completely converted to alkyl esters.

9.3.6.3 Moisture and Free Fatty Acid Content

The FFA and moisture content are very important parameters determining the viability of the transesterification reaction (Meher et al., 2006a,b). When using a base catalyst, a value of 3% or lower of FFA is needed in order to carry the reaction to completion (Meher et al., 2006a,b). The higher the acidity of the oil, the smaller is the conversion efficiency. The FFA and moisture contribute to a saponification reaction (Ma et al., 1998).

Ma et al. studied the transesterification of beef tallow catalyzed by NaOH in presence of FFAs and water (Ma et al., 1998). Without adding FFA and water, the apparent yield of beef tallow methyl esters (BTME) was highest. When 0.6% of FFA was added, the apparent yield of BTME reached the lowest, less than 5%, with any level of water added. The products were solid at room temperature, similar to the original beef tallow. When 0.9% of water was added, without addition of FFA, the apparent yield was about 17%.

Kusdiana and Saka (2004) observed that water could pose a greater negative effect than presence of FFAs and hence the feedstock should be water free. Romano (1982) and Canakci and Gerpen (1999) insisted that even a small amount of water (0.1%) in the transesterification reaction would decrease the ester conversion from vegetable oil. Demirbas too reported a decrease in yield of the alkyl ester due to the presence of water and FFA as they cause soap formation, consume catalyst, and reduce the effectiveness of catalyst (Demirbas, 2006). Ellis et al. (2008) found that even a small amount of water in the feedstock or from esterification reaction producing water from FFA might cause reduction in conversion of fatty acid methyl ester and formation of soap instead.

Most of the biodiesel is currently made from edible oils. However, there are large amounts of low cost oils and fats that could be converted to biodiesel. The problems with processing these low cost oils and fats are that they often contain large amounts of FFAs. Commonly for running reactions of these oils and fats, a two-step process is needed (Ramadhas et al., 2005). The first step is the esterification followed by the alkaline transesterification. In the esterification step, the FFA reacts with an alcohol by an acid catalysis to be converted to alkyl esters. Then the transesterification reaction is completed using alkaline catalysts (Meher et al., 2006a,b; Ramadhas et al., 2005).

9.3.6.4 Purity of Reactants

Impurities present in the oil also affect conversion levels. Under the same conditions, 67%–84% conversion into esters using crude vegetable oils can be obtained, compared with 94%–97% when using refined oils (Freedman et al., 1984). The FFAs in the original oils interfere with the catalyst. However, under conditions of high temperature and pressure, this problem can be overcome.

It was observed that crude oils were equally good compared to refined oils for the production of biodiesel. However, the oils should be properly filtered. Oil quality is very important in this regard. The oil settled at the bottom during storage may give lesser biodiesel recovery because of the accumulation of impurities like wax.

9.3.6.5 Reaction Temperature and Time

Transesterification can occur at different temperatures, depending on the oil used. In methanolysis of castor oil to methyl ricinoleate, the reaction proceeded most satisfactorily at 20°C–35°C with a molar ratio of 6:1–12:1 and 0.005%–0.35% (by weight of oil) of NaOH catalyst (Smith, 1949). For the transesterification of refined soybean oil with methanol (6:1) using 1% NaOH, three different temperatures were used (Freedman et al., 1984). After 0.1 h, ester yields were 94%, 87%, and 64% for 60°C, 45°C, and 32°C, respectively. After 1 h, ester formation was identical for the 60°C and 45°C runs and only slightly lower for the 32°C run. Generally, the transesterification reaction is conducted close to the boiling point of methanol (60°C–70°C) at atmospheric pressure. The maximum yield of esters occurs at temperatures ranging from 60°C to 80°C at a molar ratio (alcohol to oil) of 6:1. Further increase in temperature is reported to have a negative effect on the conversion.

The conversion rate increases with reaction time. Freedman et al. transesterified peanut, cottonseed, sunflower, and soybean oils under the condition of methanol to oil ratio of 6:1, 0.5% sodium methoxide catalyst and 60°C. An approximate yield of 80% was observed after 1 min for soybean and sunflower oils. After 1 h, the conversions were almost the same for all four oils (93%–98%) (Freedman et al., 1984). Ma et al. studied the effect of reaction time on transesterification of beef tallow with methanol. The reaction was very slow during the first minute due to the mixing and dispersion of methanol into beef tallow. From 1 to 5 min, the reaction proceeded very fast. The apparent yield of BTME surged from 1 to 38. The production of beef tallow slowed down and reached the maximum value at about 15 min. The di- and monoglycerides increased at the beginning and then decreased. At the end, the amount of monoglycerides was higher than that of diglycerides (Ma et al., 1998).

9.3.6.6 Intensity of Mixing

The mixing of the alcohol and triglyceride is an important variable that affects the yield of alkyl esters. The reaction rate is initially controlled by mass transfer and does not follow expected homogeneous kinetics (Meher et al., 2006a,b). For alkali-catalyzed reactions, the catalyst is originally dissolved in the alcohol. The reaction involves two phases because the triglyceride and the alcohol are not miscible at ambient temperature (Ma and Hanna, 1999; Meher et al., 2006a,b). Mixing is very important to disperse the alcohol into the triglyceride in order to initiate the reaction (Ma and Hanna, 1999). Once the alcohol is dispersed into the triglyceride, the reaction takes place in the triglyceride phase. Ma et al. studied the effect of mixing in the transesterification of beef tallow; without mixing no reaction was observed. Mixing is used to increase the contact between the oil and the alcohol, to enhance mass transfer, and to facilitate the initiation of the reaction (Ramadhas et al., 2005; Meher et al., 2006a,b).

9.3.6.7 Organic Cosolvents

The methanol/oil system is inherently immiscible, so the transesterification reaction is mass transfer limited. Intensive mixing does not result in a satisfactory increase in the reaction rate due to the unavoidable separation of a glycerol phase, which results in the removal of the catalyst. In addition, high temperature and pressure conditions also do not fair well because of the problems associated with emulsion formation, increase in saponification, and moreover the hazards associated with high temperature and pressure conditions. Boocock et al. (1998) discovered that the use of a cosolvent, which dissolves both the polar and nonpolar phases, facilitates the creation of one-phase. In order to conduct the reaction in a single phase, cosolvents like THF, 1,4-dioxane, and diethyl ether were tested. At the 6:1 methanol/oil molar ratio, the addition of 1.25 volume of THF per volume of methanol produces an oil-dominant one phase system in which methanolysis speeds up dramatically and occurs as fast as butanolysis. In particular, THF is chosen because its boiling point of 67.8°C is only 2°C higher than that of methanol. Therefore at the end of the reaction the unreacted methanol and THF can be codistilled and recycled (Boocock et al., 1996).

Using THF, transesterification of soybean oil was carried out with methanol at different concentrations of sodium hydroxide. The ester contents after 1 min for 1.1%, 1.3%, 1.4%, and 2.0% sodium hydroxide were 82.5%, 85%, 87%, and 96.2%, respectively. Results indicated that the hydroxide concentration could be increased up to 1.3 wt %, resulting in 95% methyl ester after 15 min. Similarly for transesterification of coconut oil using THF/MeOH volume ratio 0.87 with 1% NaOH catalyst, the conversion was 99 % in 1 min (Boocock et al., 1998).

9.4 Unit Operation Processes for Biodiesel Production

9.4.1 General Flowchart for Biodiesel Production Process

The transesterification process for biodiesel production consists of four principal steps. (1) Pretreatment of the feedstocks to remove components that will be detrimental to subsequent processing steps. (2) Transesterification, where the pretreated oils or fat are reacted with alcohol to form the raw alkyl esters and glycerol. There are two basic steps: the reaction process followed by separation of the alkyl esters and glycerol streams. In most technologies, these two steps are undertaken twice to push the transesterification closer to completion by reducing the concentration of glycerol in the second stage. The reaction is also pushed closer to completion by using an excess of alcohol. Processes are generally designed to a high level of conversion, and alkyl esters purity (>98%), as lower conversion rates result in increased levels of mono- and diglycerides, causing processing problems with emulsion formation and low-temperature hazing problems with the biodiesel itself as these compounds have higher melting points (and viscosity) than the alkyl esters (John, 2003). (3) Alkyl esters purification, which removes the excess methanol, catalyst, and glycerol carried from the transesterification process. Methanol removed is recycled to the transesterification process. (4) Glycerol purification, removing methanol for recycling to the transesterification process. Further impurities, such as catalyst, are carried along with the glycerol and may be removed to produce a higher grade of glycerol if economics dictate.

Implementing the above-described processing steps results in the process flow chart shown in Figure 9.16.

9.4.2 Feedstock Pretreatment

9.4.2.1 Oils/Fat Feedstocks Pretreatment

It has been mentioned above that biodiesel can be produced from a wide variety of feedstocks including vegetable oil, animal fats, restaurant waste oils,

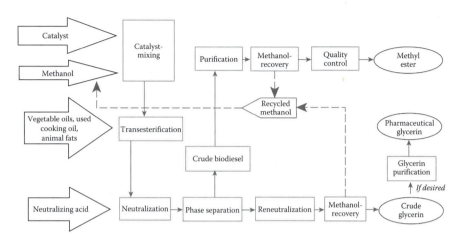

FIGURE 9.16
General flow chart of biodiesel production process.

and trap grease. Another way to categorize these feedstocks is to identify those coming from oilseeds and those that originate from rendering plants.

9.4.2.1.1 Production of Oil from Oilseeds

Soybean oil is the largest source of vegetable oil in the United States and rapeseed oil is the largest source in Europe. Most of these oils are used as food products but they require considerable processing before they are considered edible. Fully refined oils are also known as RBD oils where RBD strands for refined, bleached, and deodorized. Crude soybean oil typically contains 1.5%–2.5% phosphatides, 1.6% unsapoifiable matter including 0.15%–0.21% tocopherol (Vitamin E), and 0.3%–0.7% FFAs. These compounds contribute color, odor, and strong flavors to the oil that are usually undesirable for cooking. Some of the compounds also contribute to poor storage stability.

It has been shown that high-quality biodiesel can be produced even from crude soybean oil (Van Gerpen and Dvorak, 2002). However, use of partially or fully refined oil simplifies the biodiesel production process so some producers have chosen to use degummed oils or even RBD oils. The following discussion describes current practice for processing soybean oil. Similar processes are used for other vegetable oils.

Oil is removed from soybeans by two techniques: mechanical crushing and solvent extraction. Mechanical crushing is preferred for smaller plants because it requires a smaller investment. Typically plants that process less than 100,000 kg/day use mechanical crushing and plants that process more than 300,000 kg/day use solvent extraction (Williams, 1997).

Before the oil can be extracted, the seeds must be prepared. This involves removing stems, leaves, stones, sands, dirt, and weed seeds. The stems and stones can be screened out because they are usually larger than the seeds.

Weed seeds, sand, and dirt are usually smaller and can be removed by finer screens. Magnets can remove iron particles. After cleaning, the seeds are often dehulled. The hulls are abrasive and contain very little oil (less than 1%). Removing the hulls reduces the wear on the screw press. It can also increase the extraction of oil because the hulls absorb oil. On oilseeds with thick shells, dehulling typically involves cracking the shells and then separating the shells by screening or aspiration. With soybeans, dehulling generally includes removal of residual fines.

Oilseeds that are to be processed by solvent extraction are usually flaked to increase the exposure of the oil to the solvent. Hexane in a percolation extractor is the most common technology used today. This type of extractor drips the solvent down on the flaked soybeans so it can dissolve the oil in a manner similar to a coffee percolator. The oil-laden solvent, or miscella, is then filtered and removed from the extractor. To remove the solvent, the miscella is heated to vaporize the solvent. This gets the miscella to about 5% solvent (Williams, 1997). Then the remaining solvent is stripped by injecting steam into the oil. The steam and hexane vapors are condensed and, since the two fluids are insoluble, they can be separated in a settling tank.

The alternative to solvent extraction is mechanical extraction or crushing. The beans are generally preheated to destroy enzymes that cause problems with using the meal as animal feed. One popular way of cooking the bean is to use an expanding extruder. This device compresses the beans to very high pressure so that the temperature rises to 150°C –160°C, which is high enough to deactivate the enzymes. The extrusion also frees the oil and produces a frothy liquid that quickly solidifies as the water flashes to vapor and reduces the moisture level from 10%–14% to 6%–7%. After cooking or extrusion, the oil seeds are usually fed to a screw press where they are compressed to high pressure. Drainage slots are provided for the oil to leave the press. Moisture control is important to maintain proper temperature. If the moisture level is too high, then beans will be soft and the oil will contain excessive solids. If the moisture level is too low, then the oilseeds will overheat. Ideally, the moisture should be 2%–3% for full pressing and 4%–6% for prepressing (Williams, 1997). If the beans have not been extruded, then prepressing may be used to break the cell walls and release the oil from the seed. It can be processed with both mechanical and solvent extraction.

After extrusion from the seed, crude soybean oil contains impurities that can be classified into two categories: oil insoluble and oil soluble (Hui, 1996). The oil-insoluble impurities consist of seed fragments and meal fines, free water, and long-chain hydrocarbons or waxes that cause cloudiness when the oil is refrigerated. Most of this material can be removed by filtration. The oil-soluble materials include FFAs, phosphatides, gummy or mucilaginous substances, color bodies, protein, tocopherols, sterols, hydrocarbons, ketones, and aldehydes. Some of these compounds, particularly the ketones and aldehydes, can cause the oil to have an unpleasant taste and smell. Other compounds are actually desirable, such as the tocopherols, which act as antioxidants. The sterols are relatively inert and can remain in the oil. The food-grade oil typically contains 0.5%–1.5% nonoil compounds that are known collectively as unsaponifiable matter.

Processing of crude soybean oil to food-grade oil generally consists of three steps: refining (degumming and caustic refining), bleaching, and deodorization. As mentioned above, these oils are known as RBD oils (refined, bleached, and deodorized).

9.4.2.1.2 Refining

The refining step is designed to remove the phospholipids and FFAs from the crude oil. Crude soybean oil typically contains 500–700 ppm of phosphorus, which corresponds to 1.5%–2.1% phospholipids (Hui, 1996). The main reason to remove these phospholipids is that some of the compounds, particularly the calcium and magnesium salts of phosphatidic and lysophatidic acids, are strong emulsifiers. If these compounds are still present during the later alkali neutralization step, they will inhibit the separation of the soaps and lower the yield of neutral oil. Phospholipids will also react with water to form insoluble sediments that are not desirable when cooking with the oil.

Refining may consist of two steps. The first step is degumming. In this step, crude soybean oil is mixed with 1%–3% of water and the mixture is agitated mechanically for 30–60 min at 70°C. This hydrates the phospholipids and gums and these hydrates are insoluble in the oil. They can be separated by settling, filtering, or centrifuging. The phosphorus content can be lowered to 12–170 ppm by this procedure. The by-product of water degumming has value as a feedstock for lecithin production.

A portion of the phosphatides is not hydratable by contact with water alone. The addition of citric or phosphoric acid will hydrate those remaining. Citric acid is preferred if the by-product is to be used for lecithin. The acid hydration is accomplished by adding 0.05–0.2 wt % of concentrated (85%) phosphoric acid to crude oil, which has been heated to 60°C –85°C. The residence time varies from a few seconds to 1–2 min depending on the type and quality of the oil. The extent of gum removal by these techniques is related to the strength of the acid treatment.

The second phase of refining is neutralization, or caustic refining. This process is to remove the FFAs present in the crude oil. An alkali solution, usually sodium hydroxide, is added that reacts with the FFAs to produce soaps. The soaps are insoluble in the oil and easily separated by water washing. The alkali solution also neutralizes any acid remaining from the degumming stage. The alkali will react with the triglycerides in the oil also, so the neutralization parameters (type of alkali, solution strength, temperature, agitation, and time) must be optimized to minimize the yield loss. There can be additional losses from emulsification and suspension of oil droplets in the soap solution. A by-product of caustic refining is a mixture of soap, water, and oil known as SS. This has been considered as a feedstock for biodiesel as its cost is low, but its high water content and converting the soaps to methyl esters are significant obstacles to cost-effective utilization.

9.4.2.1.3 Bleaching

The primary purpose of bleaching is to remove the color pigments from the oil. It also helps to remove remaining soap, trace metals, phosphatides, and sulfur compounds. Hydroperoxides, the initial products of oxidation, are broken down during the bleaching process and some of the final oxidation products, the carbonyl compounds, are removed. Bleaching involves mixing bleaching clays, comprising naturally occurring bentonite and montmorillonite clays, with the oil and agitating for 10–30 min. The oil is usually heated to 90°C–120°C and the process occurs under a slight vacuum to exclude oxygen. A consequence of the vacuum and high temperature is that residual water from the neutralization and water washing is removed. This is not always desirable because the effectiveness of the clays is usually enhanced by the presence of a small amount of water. The clays are usually activated before use with a mineral acid such as sulfuric acid. This removes some of the minerals from the clay and produces a larger volume of micropores and smaller clay particles. The bleaching clays are mixed with the oil, either directly or by premixing with a small amount of oil to produce a slurry and then adding the slurry to the oil. After the adsorption processes have reached equilibrium, the clay is filtered out by self-cleaning filters. The peroxide value of the oil is very low after bleaching because the hydroperoxides have been eliminated but the FFA level may be slightly higher. This may be due to some of the sulfuric acid leaching out of the clay or to hydrolysis of the oil.

9.4.2.1.4 Deodorization

Deodorization is the final processing stage before the oil is ready for use in food products. The deodorization removes the trace components that give the oil an unpleasant taste and odor. Deodorization is essentially a distillation process that occurs at high temperature (200°C–260°C) and low pressure (2–7 torr, 2.5–9.2 mbar). The odor-causing compounds are more volatile than the oil and will be removed by extended heating at high temperature and low pressure.

The first step of deodorization is deaeration. This step removes all the dissolved oxygen so that the oil does not oxidize at the high temperatures used in deodorization. In batch deodorizers, the deaeration occurs naturally while the oil is heated from the typical inlet temperature of 40°C–80°C to the deodorization temperature range of 200°C–260°C. If adequate vacuum is available, the oil should be deaerated by the time the oil reaches 100°C–120°C. Preheating is sometimes identified as the second step of deodorization. In continuous flow deodorization processes, a portion of the heat required for heating the oil comes from cooling the deodorized oil. As much heat as possible is transferred from the final product to incoming oil. After preheating, the oil is heated to its final temperature by steam or another heating fluid.

After the oil has been held at high temperature for sufficient time (2–5 h in a batch deodorizer), it is cooled to 120°C where it is common practice to add 50–100 mg/kg of citric acid. The citric acid is added to chelate trace metals in the oil so that they can be removed by filtration. Chelation is a chemical process where a compound, known as a chelating agent, binds to a metal compound and makes it unavailable for further reaction. Then the oil is cooled further to its storage or bottling temperature. One final stage of filtration, known as polish filtration, occurs immediately before the oil leaves the plant.

Biodiesel can be produced from oil at any stage in the various processes. The compounds removed during refining, bleaching, and deodorization are mostly removed by water washing or collected in the glycerol phase. Biodiesel produced from crude soybean oil retains somewhat more color than biodiesel from refined oil but phosphorus and FFA levels seem to be equivalent (Van Gerpen and Dvorak, 2002).

9.4.2.1.5 Rendering

Rendering is the process whereby by-products of the food industry, including animal fat, bone, offal, hides, feathers, and blood are recycled into usable products. In the case of rendering operations that may be attached to slaughtering facilities, the products may be edible fats and proteins that are suitable for human consumption. These edible fats may be refined, bleached, and deodorized in a manner similar to that described above for vegetable oils. This provides a product that is suitable for food, cosmetics, and other high-value uses.

Larger volumes of material are processed in plants that produce so-called inedible products, although the products are suitable for use in animal feeds. These inedible rendering plants may process fat and bone trimmings, meat scraps, restaurant grease, fallen animals, and other sources of fat and protein. This material is then fed to a cooker that heats the material to 121°C–135°C. The high temperature encourages the removal of water and separates a portion of the fat from the solids. The fat and solids are fed to a drain pan where the fat is drained off and the solids are fed to a screw press where the remaining fat can be squeezed out. The dry solids are ground to a fine powder and sold as meat and bone meal. The fat is centrifuged or filtered to remove particulates. Since the fat may still contain some water, it is common to reheat the fat with steam injection to encourage separation of the water. After heating, the dry fat will rise to the top, the water will settle to the bottom and an interphase layer will appear that consists of emulsified water and oil and possibly protein and minerals. This material is separated and returned to the cooker for reprocessing.

Inedible fats with FFA levels below 15% and moisture, insolubles, and unsaponifiables (MIU) less than 2% are generally sold as yellow grease. Fats with higher FFA levels are sold as brown grease. Although their products are referred to as inedible, most rendered material is reintroduced to the food chain as animal feed. Meat and bone meal are used to add protein and minerals to feed and grease is used to increase calories or energy content. Renderers are coming under increasing scrutiny because of concerns about disease transmission. Most renderers will no longer accept sheep carcasses in the United States because of concerns about scrapes. The same concern also exists with deer and elk carcasses, due to chronic wasting disease, a relative of bovine spongiform encephalopathy (BSE-mad cow disease). Products from ruminants (cattle) are not fed to other cattle. This requirement has been reinforced by recent events in England where BSE has spread to large herds of cattle through use of infected animal products in the feed. The risks of disease transmission are thought to be minimal in the United States and use of animal products across species is generally considered safe. However, concerns about BSE may cause future legislation further restricting the use of animal by-products in animal feed and if this occurs, biodiesel may become the most significant market for animal fat.

9.4.2.2 Pretreatment of High Free Fatty Acid Feedstocks

Many low-cost feedstocks are available for biodiesel production. Unfortunately, most of these feedstocks contain large amounts of FFAs. These FFAs will react with alkali catalysts to produce soaps that inhibit the reaction. The following ranges of FFA are commonly found in biodiesel feedstocks: refined vegetable oils <0.05%; crude vegetable oil 0.3%–0.7%; restaurant waste grease 2%–7%; animal fat 5%–30%; trap grease 40%–100%. Generally, when the FFA level is less than 1%, and certainly if it is less than 0.5%, the FFAs can be ignored.

Soaps may allow emulsification that causes the separation of the glycerol and ester phases to be less sharp. Soap formation also produces water that can hydrolyze the triglycerides and contribute to the formation of more soap. Further, catalyst that has been converted to soap is no longer available to accelerate the reaction. When FFA levels are above 1%, it is possible to add extra alkali catalyst. This allows a portion of the catalyst to be devoted to neutralizing the FFAs by forming soap, while still leaving enough to act as the reaction catalyst.

This approach to neutralizing the FFAs will sometimes work with FFA levels as high as 5%–6%. The actual limit can depend on whether other types of emulsifiers are present. It is especially important to make sure that the feedstock contains no water. Two to three percent FFA may be the limit if traces of water are present. For feedstocks with higher amounts of FFA, the addition of extra catalyst may create more problems than it solves. The large amount of soap created can gel. It can also prevent the separation of the glycerol from the ester. Moreover, this technique converts the FFAs to a waste product when they could be converted to biodiesel (Van Gerpen et al., 2004).

When working with feedstocks that contain 5%–30% FFA or even higher, it is important to convert the FFAs to biodiesel or the process yield will be low.

There are at least four techniques for converting the FFAs to biodiesel (Van Gerpen et al., 2004):

1. Enzymatic methods: these methods require lipases, which seem to be less affected by water. At the present time, no one is using these methods on a commercial scale.

2. Glycerolysis: this technique involves adding glycerol to the feedstock and heating it to high temperature (200°C), usually with a catalyst such as zinc chloride. The glycerol reacts with the FFAs to form mono- and diglycerides. This technique produces a low-FFA feed that can be processed using traditional alkali-catalyzed techniques. The drawback of glycerolysis is the high temperature and that the reaction is relatively slow. An advantage is that no methanol is added during the pretreatment so that water formed by the reaction immediately vaporizes and can be vented from the mixture.

3. Acid catalysis: this technique uses a strong acid such as sulfuric acid to catalyze the esterification of the FFAs and the transesterification of the triglycerides. The reaction does not produce soaps because no alkali metals are present. The esterification reaction of the FFAs to alcohol esters is relatively fast, proceeding substantially to completion in 1 h at 60°C. However, the transesterification of the triglycerides is very slow, taking several days to complete. Heating to 130°C can greatly accelerate the reaction but reaction times will still be 30–45 min. Another problem with acid catalysis is that the water production during the esterification reaction stays in the reaction mixture and ultimately stops the reaction, usually well before reaching completion.

4. Acid catalysis followed by alkali catalysis, this approach solves the reaction rate problem by using each technique to accomplish the process for which it is best suited. Since acid catalysis is relatively fast for converting the FFAs to methyl esters, it is used as a pretreatment for the high-FFA feedstocks. Then, when the FFA level has been reduced to 0.5%, or lower, an alkali catalyst is added to convert the triglycerides to methyl esters. This process can convert high FFA feedstocks quickly and effectively. Water formation is still a problem during the pretreatment phase. One approach is to simply add so much excess methanol during the pretreatment that the water produced is diluted to the level where it does not limit the reaction. Molar ratios of alcohol to FFA as high as 40:1 may be needed. The disadvantage of this approach is that more energy is required to recover the excess methanol. Another approach would be to let the acid-catalyzed esterification proceed as far as it will go until it is stopped by water formation. Then, boil off the alcohol and water. If the FFA level is still too high, then additional methanol and, if

necessary, acid catalyst can be added to continue the reaction. This process can be continued for multiple steps and will potentially use less methanol than the previous approach. Again, the disadvantage is the large amount of energy required by the distillation process. A less energy-intensive approach is to let the acid-catalyzed reaction mixture settle. After a few hours, a methanol–water mixture will rise to the top and can be removed. Then, additional methanol and acid can be added to continue the reaction. It is also possible to use fluids such as glycerol and ethylene glycol to wash the water from the mixture.

9.4.2.3 Preparation of Other Feedstocks

Catalysts may either be base, acid, or enzyme materials. The most commonly used catalyst materials for converting triglycerides to biodiesel are sodium hydroxide, potassium hydroxide, and sodium methoxide. The base catalysts are highly hygroscopic and they form chemical water when dissolved in the alcohol reactant. They also absorb water from the air during storage. If too much water has been adsorbed, the catalyst will perform poorly and the biodiesel may not meet the total glycerin standard. So some care must be taken to ensure the base catalysts do not absorb water in storage. The catalyst is prepared by dissolving in the methanol using a simple mixing process.

The most commonly used primary alcohol used in biodiesel production is methanol, although other alcohols, such as ethanol, iso-propanol, and butanol, can be used. A key quality factor for the primary alcohol is the water content. Water interferes with transesterification reactions and can result in poor yields and high levels of soap, FFAs, and triglycerides in the final fuel. Unfortunately, all the lower alcohols are hygroscopic and are capable of absorbing water from the air. So some care must be taken to avoid the alcohol to absorb water.

9.4.3 Transesterification Process

9.4.3.1 Transesterification with Batch Processing

The simplest method for producing alcohol esters is to use a batch, stirred tank reactor. Alcohol-to-triglyceride ratios from 4:1 to 20:1 have been reported, with a 6:1 ratio most common. The reactor may be sealed or equipped with a reflux condenser. The operating temperature is usually about 65°C, although temperatures from 25°C to 85°C have been reported (Knothe et al., 1997; Ma and Hanna, 1999; Lang et al., 2001; Demirbas, 2002a,b; Bala, 2005; Wang et al., 2007). The most commonly used catalyst is sodium hydroxide, though potassium hydroxide also used. Typical catalyst loadings range from 0.3% to about 1.5%. Transesterification completion rates of 85%–95% have been reported.

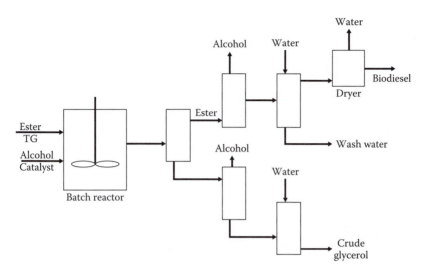

FIGURE 9.17
Batch reaction system.

Thorough mixing is necessary at the beginning of the reaction to bring the oil, catalyst, and alcohol into intimate contact. Toward the end of the reaction, less mixing can help increase the extent of reaction by allowing the inhibitory product, glycerol, to phase separate from the ester-oil phase.

Some groups use a two-step reaction, with glycerol removal between steps, to increase the final reaction extent to 95%. Higher temperatures and higher alcohol to oil ratios also can enhance the percent completion. Typical reaction times range from 20 min to more than 1 h.

Figure 9.17 shows a process flow diagram for a typical batch system (Van Gerpen et al., 2004). The oil is first charged to the system, followed by the catalyst and methanol. The system is agitated during the reaction time. Then agitation is stopped. In some processes, the reaction mixture is allowed to settle in the reactor to give an initial separation of the esters and glycerol. In other processes, the reaction mixture is pumped into a settling vessel, or is separated using a centrifuge. The alcohol is removed from both the glycerol and ester stream using an evaporator or a flash unit. The esters are neutralized, washed gently using warm, slightly acid water to remove residual methanol and salts, and then dried. The finished biodiesel is then transferred to storage. The glycerol stream is neutralized and washed with soft water. The glycerol is then sent to the glycerol refining section.

For high FFA feedstocks, the system is slightly modified with the addition of an acid esterification vessel and storage for the acid catalyst. The feedstock is sometimes dried (down to 0.4% water) and filtered before loading the acid esterification tank. The sulfuric acid and methanol mixture is added and the system is agitated. Similar temperatures to transesterification are used and sometimes the system is pressurized or a cosolvent is added. Glycerol is not produced. If a two-step acid treatment is used, the stirring is suspended until the methanol phase separates and is removed. Fresh methanol and sulfuric acid is added and the stirring resumes. Once the conversion of the fatty acids to methyl esters has reached equilibrium, the methanol/water/acid mixture is removed by settling or with a centrifuge. The remaining mixture is neutralized or sent straight into transesterification where it will be neutralized using excess base catalysts. Any remaining FFAs will be converted into soaps in the transesterification stage. The transesterification batch-stage processes as described above.

9.4.3.2 Transesterification with Continuous Processing

A popular variation of the batch process is the use of continuous stirred tank reactors (CSTRs) in series. The CSTRs can be varied in volume to allow for a longer residence time in CSTR 1 to achieve a greater extent of reaction. After the initial product glycerol is decanted, the reaction in CSTR 2 is rather rapid, with 98% completion not uncommon.

An essential element in the design of a CSTR is sufficient mixing input to ensure that the composition throughout the reactor is essentially constant. This has the effect of increasing the dispersion of the glycerol product in the ester phase. The result is that the time required for phase separation is extended. There are several processes that use intense mixing, either from pumps or motionless mixers, to initiate the esterification reaction. Instead of allowing time for the reaction in an agitated tank, the reactor is tubular. The reaction mixture moves through this type of reactor in a continuous plug, with little mixing in the axial direction. This type of reactor, called a plug-flow reactor (PFR), behaves as if it were a series of small CSTRs chained together. The result is a continuous system that requires rather short residence times, as low as 6–10 min, for near completion of the reaction. The PFRs can be staged, as shown, to allow decanting of glycerol. Often this type of reactor is operated at an elevated temperature and pressure to increase reaction rate. A PFR system is shown in Figure 9.18 (Van Gerpen et al., 2004).

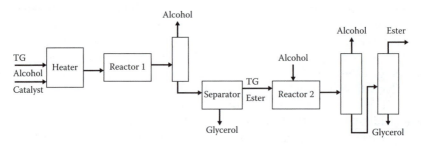

FIGURE 9.18
Plug-flow reaction system.

9.4.3.3 Noncatalyzed Transesterification System

Saka and Kusdiana (2001) and Demirbas (2002a,b) were the first to propose that biodiesel fuels may be prepared from vegetable oils via noncatalytic transesterification with SCM (Saka and Kusdiana, 2001; Demirbas, 2002a,b). A novel process of biodiesel fuel production has been developed by a noncatalytic SCM method. SCM is believed to solve the problems associated with the two-phase nature of normal methanol/oil mixtures by forming a single phase as a result of the lower value of the dielectric constant of methanol in the supercritical state. As a result, the reaction was found to be complete in a very short time (Han et al., 2005). In contrast to catalytic processes under barometric pressure, the SCM process is noncatalytic, involves much simpler purification of products, has a lower reaction time, is more environmentally friendly, and requires lower energy use. However, the reaction requires temperatures of 525–675 K and pressures of 35–60 MPa (Saka and Kusdiana, 2001; Demirbas, 2002a,b).

Supercritical transesterification is carried out in a high-pressure reactor (autoclave). In a typical run, the autoclave is charged with a given amount of vegetable oil and liquid methanol with changed molar ratios. The autoclave is supplied with heat from an external heater, and power is adjusted to give an approximate heating time of 15 min. The temperature of the reaction vessel can be measured with an iron-constantan thermocouple and controlled at ± 5 K for 30 min. Transesterification occurs during the heating period. After each run, the gas is vented, and the autoclave is poured into a collecting vessel. The remaining contents are removed from the autoclave by washing with methanol.

An intriguing example of this process has been demonstrated in Japan, where oils in a very large excess of methanol have been subjected to very high temperatures and pressures for a short period of time. The result is a very fast (3–5 min) reaction to form esters and glycerol. The reaction must be quenched very rapidly so that the products do not decompose.

9.4.4 Alkyl Esters Purification

The objective of this section is to describe in more detail the steps in processing the alkyl ester phase resulting from the transesterification. This section will discuss the recovery of the esters from the reaction mixture, and, then, the refining needed to meet the requirements of ASTM D 6751-2. The topics include ester/glycerol separation, ester washing, ester drying, other ester treatments (Van Gerpen et al., 2004).

9.4.4.1 Ester–Glycerol Separation

The ester/glycerol separation is typically the first step of product recovery in most biodiesel processes. The separation process is based on the facts that fatty acid alcohol esters and glycerol are sparingly mutually soluble, and that there is a significant difference in density between the ester and glycerol phases.

The ester washing step is used to neutralize any residual catalyst, to remove any soaps formed during the esterification reaction, and to remove residual free glycerol and methanol.

Ester drying is required to meet the stringent limits on the amount of water present in the final biodiesel product. In addition, there may be other treatments used to reduce color bodies in the fuel, remove sulfur, and/or phosphorus from the fuel, or to remove glycerides.

Fatty acid alcohol esters have a density of about 0.88 gm/cc, while the glycerol phase has a density of the order of 1.05 gm/cc, or more. The glycerol density depends on the amount of methanol, water, and catalyst in the glycerol. This density difference is sufficient for the use of simple gravity separation techniques for the two phases. However, the rate of separation is affected by several factors. Most biodiesel processes use relatively intense mixing, at least at the beginning of the reaction, to incorporate the sparingly soluble alcohol into the oil phase. If this mixing continues for the entire reaction, the glycerol can be dispersed in very fine droplets throughout the mixture. This dispersion requires from 1 h to several hours to allow the droplets to coalesce into a distinct glycerol phase. For this reason, mixing is generally slowed as the reaction begins to progress, to reduce the time required for phase separation.

The more nearly neutral the pH, the quicker the glycerol phase will coalesce. This is one reason to minimize the total catalyst use. In some batch systems, the reaction mixture is neutralized at the beginning of the glycerol/ester phase separation step.

The presence of significant quantities of mono-, di-, and triglycerides in the final mixture can lead to the formation of an emulsion layer at the ester–glycerol interface. At best, this layer represents a net loss of product, unless it is recovered and separated. At worst, the ester phase will not meet the biodiesel specification and will have to be re-run. If problems with mono, di, and triglycerides occur, you should reevaluate the entire reaction to see where improvements can be made to improve process yields in the preceding steps.

The esterification process is run with an excess of alcohol to ensure complete reaction and to attain higher reaction rates. The residual alcohol is distributed between the ester and glycerol phases. The alcohol can act as a dispersant for the ester into the glycerol phase and for the glycerol into the ester phase. The result can be a need for additional processing of the products to meet specifications. Other people claim that methanol aids in phase separation, which is one reason that product is generally phase separated before methanol recovery.

There are three categories of equipment used to separate the ester and glycerol phases (Decanter Systems, Centrifuge System, Hydrocyclone). Decanter systems rely solely on the density difference and residence time to achieve the separation. For relatively small throughput, or batch processes, the 1–8 h required for complete separation of the phases may be acceptable. Many of the continuous plants use a centrifuge for the phase separation. The centrifuge creates an artificial, high gravity field by spinning at very high speeds. The separation can be completed rapidly and effectively. The disadvantage of the centrifuge is its initial cost, and the need for considerable and careful maintenance. An intriguing, relatively new device that has been considered for use in biodiesel plants is the hydrocyclone. A liquid–liquid hydrocyclone uses an inverted conical shape and the incompressibility of the liquids to accelerate the liquid entering the cyclone. The effect is similar to a centrifuge, with the heavier material being forced toward the wall and downward, and the lighter material forced to the center and upward. The result is a density-based separation. Although they are now used for oil–water separations, hydrocyclones are at the experimental stage for application in biodiesel production.

9.4.4.2 Ester Washing

The primary purpose of the ester washing step is the removal of any soaps formed during the transesterification reaction. In addition, the water provides a medium for addition of acid to neutralize the remaining catalyst and a means to remove the product salts. The residual methanol should be removed before the wash step. This prevents the addition of methanol to the wastewater effluent. However, some processes remove the methanol with the wash water and remove it from the wash water.

The use of warm water (120°F–140°F) prevents precipitation of saturated fatty acid esters and retards the formation of emulsions with the use of a gentle washing action. Softened water eliminates calcium and magnesium contamination and neutralizes remaining base catalysts. Similarly, removal of iron and copper ions eliminates a source of catalysts that decrease fuel stability. Gentle washing prevents the formation of emulsions and results a rapid and complete phase separation. The phase separation between esters and water is typically very clean and complete. However, the equilibrium solubility of water in esters is higher than the specified water content for B100. Therefore, after the washing step there will be more than the equilibrium amount of water present.

Vacuum driers can either be batch or continuous devices for removing water. The system is operated at a highly reduced pressure, which allows the water to evaporate at much lower temperature. A variation that also allows for rather high heating and evaporation rates is the falling film evaporator. This device operates at reduced pressure. As the esters pour down, the inside wall of the evaporator the direct contact with the heated wall evaporates the water rapidly. Care should be taken with high-temperature evaporators to avoid darkening of the fuel, which is a sign that the polyunsaturated methyl esters are polymerizing. Because the total water burden in the esters is low, molecular sieves, silica gels, etc., can also be used to remove the water. An advantage of these systems is that they are passive. However, a disadvantage is that these units must be periodically regenerated.

9.4.4.3 Other Ester Treatments

There are absorbents on the market that selectively absorb hydrophilic materials such as glycerol and mono- and diglycerides (i.e., Magnesol from the Dallas Group). This treatment, followed by an appropriate filter, has been shown to be effective in lowering glycerides and total glycerol levels.

Some vegetable oils and many yellow greases and brown greases leave an objectionable color in the biodiesel. There is no color specification in ASTM D 6751, but an activated carbon bed is an effective way to remove excessive color. The fats and oil industry literature has other bleaching technologies that may also be explored for biodiesel producers.

The European specification for sulfur content is much tighter than the U.S. requirements. As a result, a number of producers in Europe are resorting to the use of vacuum distillation for the removal of sulfur compounds from the biodiesel product. Vacuum distillation has the added benefit of deodorization and the removal of other minor contaminants, which may provide a benefit to those firms that use highly degraded feedstocks such as trap grease.

Filtering is an essential part of all biodiesel production. While feedstocks entering the plant should be filtered to at least 100 μm, biodiesel leaving the plant should be filtered to at least 5 μm to ensure no contaminants are carried with the fuel that could damage the engine. It has been suggested that the fuel could be cooled before filtering to capture some of the saturated esters as they crystallize and thereby lower the cloud point of the fuel. The crystallized esters could be melted by heating and used within the plant as boiler fuel.

9.4.5 Recovery of Side Streams

There are three nonester side streams that must be treated as a part of the overall biodiesel process. These streams are (1) the excess alcohol that is recycled within the process; (2) the glycerol coproduct, and (3) the wastewater stream from the process (Van Gerpen et al., 2004).

9.4.5.1 Alcohol Recovery

There are several physical parameters that are important to the recovery and recycle of alcohol. Methanol's relatively low boiling point, 64.7°C, means that it is fairly volatile and can largely be removed from the oil, ester, and aqueous streams by flash evaporation and recondensation.

Methanol is fully miscible with water and with glycerol. However, it has a low solubility in fats and oils. Methanol is more soluble in esters, but it is not fully miscible. The solubility in glycerol and water means that methanol will prefer these phases when there is a two-phase system present. The low solubility in fats and oils is the reason for the solubility-limited phase of the overall transesterification reaction.

When the two phases present are esters and glycerol, the methanol will distribute between the phases. At 90:10% wt/wt ester and glycerol, the methanol distributes approximately 60:40 wt % between the phases. This fact is important, since the reaction is complete at 90:10 wt %. If the methanol is allowed to remain in the system during phase separation, the methanol acts as a phase stabilizer, retarding the rate of gravity separation. It is advantageous to remove the methanol before phase separation.

Methanol can be recovered using distillation, either conventional or vacuum, or partially recovered in a single-stage flash. An alternative to distillation is a falling-film evaporator. Residual methanol in the ester phase can be removed in the water wash step in ester post-processing. Product esters are typically washed with warm (140°F), softened water to remove soaps and residual methanol.

9.4.5.2 Glycerol Refining

The recovered glycerol from the transesterification reaction contains residual alcohol, catalyst residue, carry-over fat/oil, and some esters. The glycerol from rendered feedstocks may also contain phosphatides, sulfur compounds, proteins, aldehydes and ketones, and insolubles (dirt, minerals, bone, or fibers).

9.4.5.2.1 Chemical Refining

There are several factors that are important in the chemical refining of glycerol. First, the catalyst tends to concentrate in the glycerol phase where it must be neutralized. The neutralization step leads to the precipitation of salts. Also, the soaps produced in the esterification must be removed by coagulation and precipitation with aluminum sulfate or ferric chloride. The removal may be supplemented by centrifuge separation. The control of the pH is very important because low pH leads to dehydration of the glycerol and high pH leads to polymerization of the glycerol. The glycerol may then be bleached using activated carbon or clay.

9.4.5.2.2 Physical Refining

The first step in physical refining is to remove fatty, insoluble, or precipitated solids by filtration and/or centrifugation. This removal may require pH adjustment. Then the water is removed by evaporation. All physical processing is typically conducted at 150°F–200°F, where glycerol is less viscous, but still stable.

9.4.5.2.3 Glycerol Purification

The final purification of glycerol is completed using vacuum distillation with steam injection, followed by activated carbon bleaching. The advantages of this approach are that this is a well-established technology. The primary disadvantage is that the process is capital and energy intensive. Ion exchange purification of glycerol is an attractive alternative to vacuum distillation for smaller capacity plants. The ion exchange system uses cation, anion, and mixed-bed exchangers to remove catalyst and other impurities. The glycerol is first diluted with soft water to a 15%–35% glycerol-in-water solution. The ion exchange is followed by vacuum distillation or flash drying for water removal, often to an 85% partially refined glycerol. The advantage of this process is the fact that all purification takes place in the resin vessels so the system is suited to smaller capacity operations. The disadvantages are that the system is subject to fouling by fatty acids, oils, and soaps. The system also requires regeneration of the beds producing large quantities of wastewater. Regeneration requires parallel systems to operate and regenerate simultaneously.

9.4.5.3 Wastewater Treatment

Ester washing produces about 1 gal of water per gallon of ester per wash. All process water must be softened to eliminate calcium and magnesium salts and treated to remove iron and copper ions. The ester wash water will have a fairly high BOD from the residual fat/oil, ester, and glycerol. The glycerol ion exchange systems can produce large quantities of low salt waters as a result of the regeneration process. In addition, water softening, ion exchange, and cooling water blow down will contribute a moderate dissolved salts burden.

The aggregate process wastewaters should meet local municipal waste treatment plant disposal requirements, if methanol is fully recovered in the plant and not present in the wastewater. In many areas, internal treatment and recycle of the process water may lead to cost savings and easier permitting of the process facility.

9.4.6 Basic Plant Equipment for Biodiesel Production

The basic plant equipments used in biodiesel production are reactors, pumps, settling tanks, centrifuges, distillation columns, and storage tanks (Demirbas, 2008).

The reactor is the only place in the process where chemical conversion occurs. Reactors can be grouped into two broad categories: batch reactors and continuous reactors. In the batch reactor, the reactants are fed into the reactor at the determined amount. The reactor is then closed, and the desired reaction conditions are set. The chemical composition within the reactor changes with time. The construction materials are an important consideration for the reactor and storage tanks. For the base-catalyzed transesterification reaction, stainless steel is the preferred material for the reactor. Key reactor variables that dictate conversion and selectivity are temperature, pressure, reaction time (residence time), and degree of mixing. In general, increasing the reaction temperature increases the reaction rate and, hence, the conversion for a given reaction time. Increasing the temperature in the transesterification reaction does impact the operating pressure. Two reactors within the continuous reactor category are CSTRs and PFRs. For CSTRs, the reactants are fed into a well-mixed reactor. The composition of the product stream is identical to the composition within the reactor. Hold-up time in a CSTR is given by a residence time distribution. For PFRs, the reactants are fed into one side of the reactor. The chemical composition changes as the material moves in plug flow through the reactor.

The pumps play the key role in moving chemicals through the manufacturing plant. The most common type of pump in the chemical industry is a centrifugal pump. The primary components of a centrifugal pump are (1) a shaft, (2) a coupling attaching the shaft to a motor, (3) bearings to support the shaft, (4) a seal around the shaft to prevent leakage, (5) an impeller, and (6) a volute, which converts the kinetic energy imparted by the impeller into the feet or head. The gear pumps are generally used in biodiesel plants. There are a number of different types of positive displacement pumps including gear pumps (external and internal) and lobe pumps. External gear pumps generally have two gears with an equal number of teeth located on the outside of the gears, whereas internal gear pumps have one larger gear with internal teeth and a smaller gear with external teeth.

The separation of biodiesel and glycerin can be achieved using a settling tank. While a settling tank may be cheaper, a centrifuge can be used to increase the rate of separation relative to a settling tank. Centrifuges are most typically used to separate solids and liquids, but they can also be used to separate immiscible liquids of different densities. In a centrifuge, separation is accomplished by exposing the mixture to a centrifugal force. The denser phase will be preferentially separated to the outer surface of the centrifuge. The choice of appropriate centrifuge type and size is predicated on the degree of separation needed in a specific system.

An important separation device for miscible fluids with similar boiling points (e.g., methanol and water) is the distillation column. Separation in a distillation column is predicated on the difference in volatilities (boiling points) between chemicals in a liquid mixture. In a distillation column, the concentrations of the more volatile species are enriched above the feed point and the less-volatile species are enriched below the feed point.

9.4.7 Integrated Biodiesel Production Processes

As mentioned above, many commercial process have been developed in the past few years at the wave of crude oil price rising. Two of these processes have been described in the following sections.

9.4.7.1 The BIOX Processing System

The BIOX processing system is a new Canadian process developed originally by Professor David Boocock at the University of Toronto, which has attracted considerable attention. Dr. Boocock has transformed the production process through the selection of inert cosolvents that generate an oil-rich one-phase system. This reaction is over 99% complete in seconds at ambient temperatures, compared to previous processes that required several hours. BIOX is a technology development company that is a joint venture of the University of Toronto Innovations Foundation and Madison Ventures Ltd. Its process uses base-catalyzed transesterification of fatty acids to produce methyl esters. It is a continuous process and is not feedstock specific. The unique feature of the BIOX process is that it uses inert reclaimable cosolvents in a single-pass reaction taking only seconds at ambient temperature and pressure. The developers are aiming to produce biodiesel that is cost competitive with petrodiesel. The BIOX process handles not only grain-based feedstocks but also waste cooking greases and animal fats (Van Gerpen et al., 2004).

The BIOX process uses a cosolvent, THF, to solubilize the methanol. Cosolvent options are designed to overcome slow reaction times caused by the extremely low solubility of the alcohol in the triglyceride phase. The result is a fast reaction, of the order of 5–10 min, and no catalyst residues in either the ester or the glycerol phase.

9.4.7.2 High Free Fatty Acid Grease Feed Biodiesel Production Process

U.S. government has also strongly supported biofuel research and development through its various agencies such as USDOE, USDA, Environment Protection Agency (USEPA), and National Science Foundation (NSF) to fund the research. One of the programs those agencies have supported is called Small Business Innovation and Research (SBIR). Resodyn Corporation has obtained generous supports from USDA and EPA in the past several years to develop processes to utilize waste grease as feedstock for biodiesel and other chemical production. In particular, Resodyn Corporation developed an innovative (patented) process to make biodiesel form waste greases and rendered biomaterials. The successful development of this technology was conducted in a three-stage process as described above (Yang, 2007). The first stage was funded by the USDA and Kenosha Beef International, (KBI), of Kenosha, WI. The second stage was funded by KBI, the USDA, and Resodyn Corporation. The completed production process is illustrated in Figure 9.19. The process

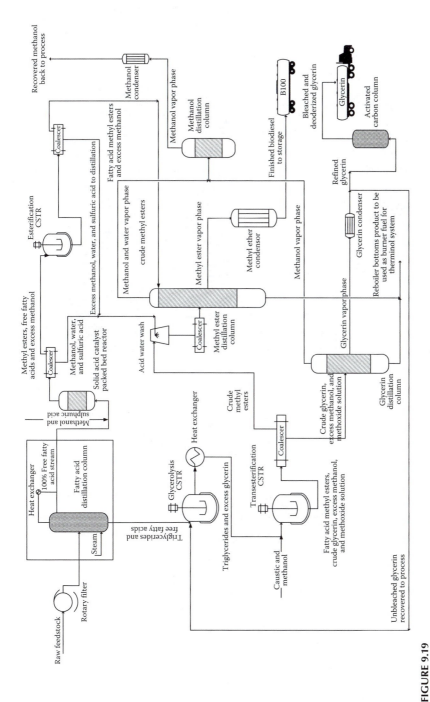

FIGURE 9.19
Resodyn Corporation process for biodiesel production with waste.

was developed based on a throughput of 10 gal/h pilot plant as shown in Figure 9.20, which was built and operated in the Resodyn Corporation facilities in Butte, Montana, United States. The starting material and the products from the pilot plant are shown in Figure 9.21. The third stage was funded by the sale of a 10 million gal/year plant to a client who was backed by a

FIGURE 9.20
Resodyn Corporation biodiesel pilot plant in Butte, Montana.

FIGURE 9.21
Starting feedstock (used cooking grease) and finished products, biodiesel and glycerin produced from pilot plant.

FIGURE 9.22
Ten million gallons biodiesel plant under construction.

group of investors that purchased the technology from KBI and Resodyn Corporation. The new owners of this technology are a publicly traded company, Nova Biofuels (See NVBF for a listing of the stock). Nova is currently designing and/or building six biodiesel plants nationwide that range in size from 10 to 40 million gal/year. One of the facility under construction in the state of Iowa in United States is shown in Figure 9.22.

9.4.8 Biodiesel Production Economic Analysis

As discussed earlier, biodiesel's overall cost consists of raw material (production and processing), catalyst, biodiesel processing (energy, consumables, and labor), transportation (raw materials and final products), and local and national taxes. However, the major factor that contributes the cost of biodiesel production is the feedstock, which is about 80% of the total operating cost (Demirbas, 2007). As one can see from data shown in Figure 9.23, the cost of feedstock has a very significant effect on biodiesel production cost. The plot has also shown that the overall production depends on the process efficiency (raw material utilization).

Hass and associates have established a generic process model for estimating biodiesel costing (Haas et al., 2006). The model examines the cost of biodiesel production on the basis of degummed vegetable oil using a continuous transesterification process. The model was based on a process plant with a 37, 854, 118 L (10×10^6 gal) capacity. The model also excluded some economic factors, such as internal rate of return, economic life, corporate tax rate, salvage value, debt fraction, construction interest rate and long-term interest rate, working capital, environmental control equipment, marketing and distribution expenses, the cost of capital, and the existence of governmental credits

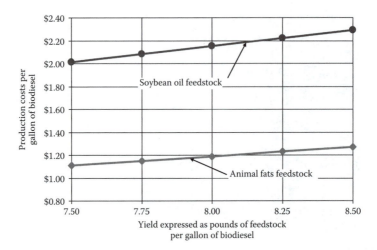

FIGURE 9.23
Illustration of dependence of biodiesel production cost on feedstock.

or subsidies, from these calculations. The model estimated a final biodiesel production cost of $0.53/L ($2.00 gal) and illustrated that the raw material cost constitutes the greatest component of the overall cost. The degummed soybean oil contributed 88% of the overall production cost. The Haas model also illustrated that some gains can be achieved by the sale of crude glycerol from the process. However, it should be noted that this is very negligible and will not have much effect, as the amount of glycerol increases in the market as biodiesel production increases.

Nas and Berktay too concluded that feedstock cost is the major component contributing to the cost of biodiesel production (Nas and Berktay, 2007). In a comparison of feedstock, namely, soybean oil and yellow grease, it was estimated that yellow grease is quite less expensive than soybean oil. On the contrary, the supply of yellow grease is limited and has other applications and hence it cannot be used in large-scale production. An 8 year comprehensive study (from 2004–2005 to 2012–2013), on the price rise of petroleum diesel, yellow grease, and soybean oil is reported. Cost of petroleum has been estimated to rise from 0.67 to 0.75 (i.e., 11.94% increase). Biodiesel fuel obtained from yellow grease is estimated to rise from 1.41 to 1.55 (i.e., 9.93% increase). Biodiesel fuel from soybean oil will rise from 2.54 to 2.80 (i.e., 10.24% increase). Author is of the opinion that biodiesel will not be produced at a cost comparable with that of petrodiesel unless the soybean oil price declines.

The cost of biodiesel after blending with petrodiesel will reduce as the cost of biodiesel becomes less significant in blended form. At present, biodiesel can be blended with 80% petrodiesel (B20) without any engine modification. B100 (biodiesel) cost $3.76 per gal in United States in June 2006, whereas B20 (20% biodiesel and 80% petrodiesel) cost only $2.98 per gal of biodiesel.

Even with higher production costs, biodiesel production has begun to expand exponentially as more governments, concerned with the growing volatility and increasing prices in the international oil market, discover the benefits of biodiesel as a petroleum fuel substitute. The implementation of the Kyoto protocol has also been an incentive for many countries to move toward alternative fuels such as biodiesel.

9.5 Problems

1. Find the heat of combustion of biodiesel (represented as oleic acid methyl ester), petroleum diesel (represented as cetane), and vegetable oil (oleic acid triglycerides).

2. Why do diesel engines combust biodiesel more completely than petroleum diesel? Use a combustion chemical reaction scheme to illustrate your argument.

3. If yellow grease is being used as feedstock for biodiesel production, how many pounds of feedstock are needed to make 1 gal of biodiesel? You can assume that yellow grease consists (by weight) of 15% FFA, 5% water, 0.5% insoluble solid, and the rest is triglycerides. Use oleic acid for FA and oleic acid methyl ester for biodiesel.

4. Identify the methods for FFA content determination in oil/grease. Titration is one of the methods you can use. Outline the procedures for titration method and point out the limitation of this method.

5. Identify the methods for oil/grease water content determination. Karl-Fischer titration is one method for such a purpose. Describe the principle of this method and outline the procedures for the measurement.

6. Explain why acid-catalyzed transesterification is much slower than alkali-catalyzed transesterification.

7. The most often used kinetic modeling equation in enzymatic reactions is called the Michaelis–Menten equation. Under what conditions can Equation 9.12 in this chapter be simplified into the Michaelis–Menten equation?

8. If a feedstock consists (by weight) of 15% FFA, 5% water, 0.5% insoluble solid, and the rest triglycerides, do the following mass and heat balance. In the final product mixture, there will be water in the stream. Assume the conversion is 100%, based on 100 pounds of feedstock, how much water will be there in the final mixture? If the water has to be removed either by distillation or by adsorption, give

estimation to show which method is more economic. You need some preliminary data on adsorbent capacity and cost on regeneration, etc., if you use adsorption method. You need heating cost data, etc., to evaluate distillation operation.

9. In a glycerolysis process, the feed is typically heated to 220°C and the FFA is converted to mono-, di-, and triglycerides. If a feedstock consists (by weight) of 15% FFA, 5% water, 0.5% insoluble solid, and the rest triglycerides, how much pure glycerin does it need to convert all FFA to mono-, di-, and triglycerides in a molar ratio of 30%, 30%, and 40% based on 100 pounds of feedstock? If the feedstock needs to be heated to 200°C from room temperature at 25°C, how much heat does it need?

10. The color from crude glycerin stream in biodiesel production is typically removed by feeding the stream to an activated carbon-packed column. Assume that the activated carbon particle is a rigid and nonporous sphere and the color-causing particles in crude glycerol stream are also rigid spheres. Calculate how many grams of activated carbon is needed for removing the color of feed stream comprising 60% (w/w) glycerol, 35% water, 3% FAME, and 2% insoluble solid particle.

References

Abigor, R., Uadia, P., Foglia, T., Haas, M., Jones, K., Okpefa, E., Obibuzor, J., Bafor, M. Lipase-catalysed production of biodiesel fuel from some Nigerian lauric oils. *Biochem Soc T*. 2000, 28, 979–981.

Akoh, C. C., Sellappan, S., Fomuso, L. B., Yankah, V. V. Enzymatic synthesis of structured lipds. In: *Lipid Biotechnology*, Kuo, T. M., Gardner, H. W., Eds., Marcel Dekker: New York, 2002, 433–478.

Aksoy, H. A., Becerik, I., Karaosmanoglu, F., Yamaz, H. C., Civelekoglu, H. Utilization prospects of Turkish raisin seed oil as an alternative engine fuel. *Fuel*. 1990, 69(5), 600–603.

Alamu, O. J., Waheed, M. A., Jekayinfa, S. O. Effect of ethanol-palm kernel oil ratio on alkali-catalyzed biodiesel yield. *Fuel*. 2008, 87(8–9), 1529–1533.

Al-Zuhair S. Production of biodiesel by lipase-catalysed transesterification of vegetable oils: A kinetics study. *Biotechnol Prog*. 2005, 21(5), 1442–1448.

Al-Zuhair, S. The effect of substrate concentrations on the production of biodiesel by lipase-catalysed transesterification of vegetable oils. *J Chem Technol Biot*. 2006, 81, 299–305.

Al-Zuhair, S., Ling, F. W., Jun, L. S. Proposed kinetic mechanism of the production of biodiesel from palm oil using lipase. *Process. Biochem*. 2007, 42, 951–960.

Anan, N., Danisman, A. Alkali catalyzed transesterification of cottonseed oil by microwave irradiation. *Fuel*. 2007, 86(17–18), 2639–2644.

Armenta, R. E., Vinatoru, M., Burja, A. M., Kralovec, J. A., Barrow, C. J. Transesterification of fish oil to produce fatty acid ethyl esters using ultrasonic energy. *JAOCS*. 2007, 84(11), 1045–1052.

Ataya, F., Dube, M. A., Ternan, M. Acid-catalyzed transesterification of canola oil to biodiesel under single- and two-phase reaction conditions. *Energ Fuel*. 2007, 21(4), 2450–2459.

Atkinson, B., Black, G. M., Lewis, P. J. S., Pinches, A. Biological particles of given size, shape, and density for use in biological reactors. *Biotechnol Bioeng*. 1979, 21, 193–200.

Bailey, J. E., Ollis, D.F. *Biochemical Engineering Fundamentals*, 2nd edn., McGraw Hill: New York, 1986.

Bala, B. K. Studies on biodiesels from transformation of vegetable oils for diesel engines. *Energy Edu Sci Technol*. 2005, 15, 1–43.

Ban, K., Hama, S., Nishizuka, K., Kaieda, M., Matsumoto, T., Kondo, A., Noda, H., Fukuda, H. Repeated use of whole cell biocatalysts immobilized within biomass support particles for biodiesel fuel production. *J Mol Catal B: Enzym*. 2002, 17, 157–165.

Ban, K., Kaieda, M., Matsumoto, T., Kondo, A., Fukuda, H. Whole-cell biocatalyst for Biodiesel fuel production utilizing *Rhizopus oryzae* cells immobilized within biomass support particles. *Biochem Eng J*. 2001, 8, 39–43.

Barth, T. Similarities and differences in hydrous pyrolysis and source rocks. *Org Geochem*. 1999, 30, 1495–1507.

Belafi-Bako, K., Kovacs, F., Gubicza, L., Hancsok, J. Enzymatic biodiesel production from sunflower oil by *candida antarctica* lipase in a solvent-free system. *Biocatal Biotransfor*. 2002, 20(6), 437–439.

Berchmans, H. J., Hirata, S. Biodiesel production from crude *Jatropha curcas* L. seed oil with a high content of free fatty acids. *Bioresource Technol*. 2008, 99(6), 1716–1721.

Bertolini, M. C., Schrag, J. D., Cygler, M., Ziomek, E., Thomas, D. Y., Vet-net, T. Expression and characterization of *Geotrichum candidum* lipase I gene. *Eur J Biochem*. 1995, 228, 863–869.

Bo, X., Xiao, G. M., Cui, L. F., Wei, R. P., Gao, L. J. Transesterification of palm oil with methanol to biodiesel over a KF/Al_2O_3 heterogeneous base catalyst. *Energ Fuel*. 2007, 21(6), 3109–3112.

Boocock, D. G. B., Konar, S. K., Mao, V., Sidi, H. Fast one-phase oil-rich processes for the preparation of vegetable oil methyl esters. *Biomass Bioenerg*. 1996, 11, 43–50.

Boocock, D. G. B., Konar, S. K., Mao, V., Lee, C., Bulligan, S. Fast formation of high-purity methyl esters from vegetable oils. *JAOCS*. 1998, 75, 1167–1172.

Bouaid, A., Bajo, L., Martinez, M., Aracil, J. Optimization of biodiesel production from jojoba oil. *Process Saf Eviron*. 2007, 85(B5), 378–382.

Boutur, O., Dubreucq, E., Galzy, P. Methyl esters production, in aqueous medium, by an enzymatic extract from Candida deformans (Zach) Langeron and Guerra. *Biotechnol Lett*. 1994, 16, 1179–1182.

Bridgwater, A. V., Meier, D., Radlein, D. An overview of fast pyrolysis of biomass. *Org Geochem*. 1999, 30, 1479–1493.

Brown, R. *Biorenewable Resources: Engineering New Product From Agriculture*, Iowa State Press, Ames, IA, 2003.

Canakci, K., Gerpen, J. V. Biodiesel production via acid catalysis. *T ASAE*. 1999, 42, 1203–1210.

Canakci, M. The potential of restaurant waste lipids as biodiesel feedstocks. *Bioresource Technol.* 2007, 98, 183–190.

Canakci, M., Gerpen, J. V. Biodiesel production from oils and fats with high free fatty acids. *T ASAE.* 2001, 44, 1429–1436.

Canakci, M., Sanli, H. Biodiesel production from various feedstocks and their effects on the fuel properties. *J Ind Microbiol Biot.* 2008, 35(5), 431–441.

Canoira, L., Alcantara, R., Garcia-Martinez, J., Carrasco, J. Biodiesel from Jojoba oil-wax: Transesterification with methanol and properties as a fuel. *Biomass Bioenerg.* 2006, 30(1), 76–81.

Casimir, C. A., Chang, S. W., Lee, G. C., Shaw, J. F. Enzymatic approach to biodiesel production. *J Agr Food Chem.* 2007, 55, 8995–9005.

Chand, C. C., Wan, S. W. China's motor fuels from tung oil. *Ind Eng Chem.* 1947, 39, 1543–1548.

Chang, C. D., Silvestri, A. J. The conversion of methanol and other compounds to hydrocarbons over zeolite catalysts. *J Catal.* 1977, 47, 249–259.

Chavanne, G. Procedure for the transformation of vegetable oils for their uses as fuels. Belgian Patent, 1937, 422, 877.

Chen, J. W., Wu, W. T. Regeneration of immobilized *Candida antarctica* lipase for trans-esterification. *J Biosci Bioeng.* 2003, 95, 466–469.

Coggon, R., Vasudevan, P. T., Sanchez, F. Enzymatic transesterification of olive oil and its precursors. *Biocatal Biotransfor.* 2007, 25(2–4), 135–143.

Crabbe, E., Nolasco-Hipolito, C., Kobayashi, G., Sonomoto, K., Ishizaki, A. Biodiesel production from crude palm oil and evaluation of butanol extraction and fuel properties. *Process Biochem.* 2001, 37(1), 65–71.

Cui, L. F., Xiao, G. M., Xu, B., Teng, G. Y. Transesterification of cottonseed oil to biodiesel by using heterogeneous solid basic catalysts. *Energ Fuel.* 2007, 21(6), 3740–3743.

D'Cruz, A., Kulkarni, M. G., Meher, L. C., Dalai, A. K. Synthesis of biodiesel from canola oil using heterogeneous base catalyst. *JAOCS.* 2007, 84(10), 937–943.

Dandik, L., Aksoy, H. A. Pyrolysis of used sunflower oil in the presence of sodium carbonate by using fractionating pyrolysis reactor. *Fuel Process Technol.* 1998, 57, 81–92.

Darnoko, D., Cheryan, M. Kinetics of palm oil transesterification in a batch reactor. *JAOCS.* 2000, 77, 563–567.

De Oliveira, D., Di Luccio, M., Faccio, C., Dalla Rosa, C., Bender, J. P., Lipke, N., Menoncin, S., Amroginski, C., De Oliveira, J. V. Optimization of enzymatic production of biodiesel from castor oil in organic solvent medium. *Appl Biochem Biotech.* 2004, 113, 771–780.

Demirbas, A. Analysis of beech wood fatty acids by supercritical acetone extraction. *Wood Sci Technol.* 1991a, 25, 365–370.

Demirbas, A. Fatty and resin acids recovered from spruce wood by supercritical acetone extraction. *Holzforschung.* 1991b, 45, 337–339.

Demirbas, A. Combustion characteristics of different biomass fuels. *Progress Energy Combus. Sci.* 2004, 30, 219–230.

Demirbas, A. Biodiesel production via non-catalytic SCF method and biodiesel fuel characteristics. *Energ Convers Manage.* 2006, 47, 2271–2282.

Demirbas, A. Studies on cottonseed oil biodiesel prepared in non-catalytic SCF conditions. *Bioresource technol.* 2008, 99(5), 1125–1130.

Demirbas, A. Importance of biodiesel as transportation fuel. *Energ Policy*. 2007, 35, 4661–4670.

Demirbas, A. Biodiesel fuels from vegetable oils via catalytic and non-catalytic supercritical alcohol transesterifications and other methods: A survey. *Energ Convers Manage*. 2003, 44, 2093–2109.

Demirbas, A. Biodiesel from vegetable oils via transesterification in supercritical methanol. *Energ Convers Manage*. 2002a, 43, 2349–56.

Demirbas, A. Diesel fuel from vegetable oil via transesterification and soap pyrolysis. *Energ Source*. 2002b, 24, 835–841.

Dossat, V., Combes, D., Marty, A. Continuous enzymatic transesterification of high oleic sunflower oil in a packed bed reactor: Influence of the glycerol production. *Enzyme Microb Tech*. 1999, 25, 194–200.

Du, W., Liu, D. H., Li, L. L., Dai, L. M. Mechanism exploration during lipase-mediated methanolysis of renewable oils for biodiesel production in a tert-butanol system. *Biotechnol Progr*. 2007, 23, 1087–1090.

Du, W., Li, W., Sun, T., Chen, X., Liu, D. H. Perspectives for biotechnological production of biodiesel and impacts. *Appl Microbiol Biot*. 2008, 79(3), 331–337.

Du, W., Xu, Y. Y., Zeng, J., Liu, D. H. Novozym 435-catalysed transesterification of crude soya bean oils for biodiesel production in a solvent-free medium. *Biotechnol. Appl. Biochem*. 2004, 40(Pt 2), 187–190.

Du, W., Xu, Y. Y., Liu, D. H., Li, Z. B. Study on acyl migration in immobilized lipozyme TL-catalyzed transesterification of soybean oil for biodiesel production. *J Mol Catal B: Enzym*. 2005, 37, 68–71.

Ducret, A., Trani, M., Lortie, R. Lipase-catalyzed enantioselective esterification of ibuprofen in organic solvents under controlled activity. *Enzyme Microb Tech*. 1998, 22, 212–216.

Einloft, S., Magalhaes, T. O., Donato, A., Dullius, J., Ligabue, R. Biodiesel from rice bran oil: Transesterification by tin compounds. *Energ Fuel*. 2008, 22(1), 671–674.

Ellis, N., Guan, F., Chen, T., Poon, C. Monitoring biodiesel production (transesterification) using in-situ viscometer. *Chem Eng J*. 2008, 138, 200–206.

Encinar, J. M., Encinar, J. F., González, E. S. et al. Biodiesel fuels from vegetable oils: Transesterification of Cynara cardunculus L. oils with ethanol. *Energ Fuel*. 2002, 16(2), 443–450.

Encinar, J. M., Encinar, J. F. González, E. S., Ramiro, M. J. Preparation and properties of biodiesel from Cynara cardunculus L. oil. *Ind Eng Chem Res*. 1999, 38(8), 2927–2931.

Ferreira, D. A. C., Meneghetti, M. R., Meneghetti, S. M. P., Wolf, C. R. Methanolysis of soybean oil in the presence of tin (IV) complexes. *Appl Catal A-Gen*. 2007, 317(1), 58–61.

Feuge, A. J., Gros, A. T. Modification of peanut oil with ethanol. *JAOCS*. 1949, 26, 97–102.

Filip, V., Zajic, V., Smidrkal, J. *Rev Fr Corps Gras*. 1992, 39, 91.

Fortenbery, T. R. *Biodiesel Feasibility Study: An Evaluation of Biodiesel Feasibility in Wisconsin*. Agricultural & Applied Economics, University of Wisconsin, Madison, WI, Staff Paper Series, 2005.

Fortes, I. C. P., Baugh, P. J. Study of analytical on-line pyrolysis of oils from macauba fruit (Acrocomia sclerocarpa M) via GC/MS. *J Brazil Chem Soc*. 1999, 10, 469–477.

Freedman, B., Pryde, E. H., Mounts, T. L. Variables affecting the yields of fatty esters from transesterified vegetable oils. *JAOCS*. 1984, 61, 1638–1643.

Freedman, B., Butterfield, R. O., Pryde, E. H. Transesterification kinetics of soybean oil. *JAOCS*. 1986, 63, 1375–1380.

Fukuda, H., Kondo, A., Noda, H. Biodiesel fuel production by transesterification of oils. *J Biosci Bioeng*. 2001, 92(5), 405–416.

Fukuda, H., Kondo, A., 2003. Patent No.: US 6524839 B1.

Ghadge, S. V., Raheman, H. Biodiesel production from mahua (Madhuca indica) oil having high free fatty acids. *Biomass Bioenerg*. 2005, 28, 601–605.

Ghamguia, H., Karra-Chaabouni, M., Gargouri, Y. 1-Butyl oleate synthesis by immobilized lipase from *Rhizopus oryzae*: A comparative study between n-hexane and solvent-free system. *Enzyme Microb Tech*. 2004, 35(4), 355–363.

Ghirardi, M. L., Zhang, J. P., Lee, J. W., Flynn, T., Seibert, M., Greenbaum, E. et al. Microalgae: A green source of renewable H2. *Trends Biotechnol*. 2000, 18, 506–511.

Goering, C. E., Camppion, R. N., Schwab, A. W., Pryde, E. H., 1982. In: Vegetable oil fuels. *Proceedings of the International Conference on Plant and Vegetable Oils as Fuels*, Fargo, North Dakota. American Society of Agricultural Engineers, St. Joseph, MI, 4, 279–286.

Goering, C. E., Fry, B. Engine durability screening test of a diesel oil/soy oil/alcohol microemulsion fuel. *JAOCS*. 1984, 61, 1627–1632.

Goering, C. E. Final report for project on effect of nonpetroleum fuels on durability of direct-injection diesel engines under contract 59-2171-1-6-057-0, 1984, USDA, ARS, Peoria, IL.

Goff, M. J., Bauer, N. S., Lopes, S., Sutterlin, W. R., Suppers, G. J. Acid-catalyzed alcoholysis of soybean oil. *JAOCS*. 2004, 81(4), 415–420.

Graille, J., Lozano, P., Pioch, D., Geneste, P. *Oléagineux*. 1985, 40, 271–276.

Granados, M. L., Poves, M. D. Z., Alonso, D. M., Mariscal, R., Galisteo, F. C., Tost, R. M. et al. Biodiesel from sunflower oil by using activated calcium oxide. *Appl Catal B-Environ*. 2007, 73(3–4), 317–326.

Guthrie, J. P. Concerted mechanism for alcoholysis of esters: An examination of the requirements. *J Am Chem Soc*. 1991, 113(10), 3941–3949.

Ha, S. H., Lanb, M. N., Lee, S. H., Hwang, S. M., Koo, Y. M. Lipase catalyzed biodiesel production from soybean oil in ionic liquids. *Enzyme Microb Technol*. 2007, 41, 480–483.

Haas, M. J., McAloon, A. J., Yee, W. C., Foglia, T. A. A process model to estimate biodiesel production costs. *Bioresource Technol*. 2006, 97(4), 671–678.

Hama, S., Yamaji, H., Fukumizu, T., Tamalampudi, S., Kondo, A., Miura, K., Fukuda, H. Lipase localization in *Rhizopus oryzae* cells immobilized within biomass support particles for use as whole cell biocatalysts in bioidiesel fuel production. *J Biosci Bioeng*. 2006, 101, 328–333.

Hama, S., Yamaji, H., Fukumizu, T., Numata, T., Tamalampudi, S., Kondo, A., Noda, H., Fukuda, H. Biodiesel-fuel production in a packed-bed reactor using lipase-producing *Rhizopus oryzae* cells immobilized within biomass support particles. *Biochem. Eng. J*. 2007, 34, 273–278.

Han, H., Cao, W., Zhang, J. Preparation of biodiesel from soybean oil using supercritical methanol and CO_2 as co-solvent. *Process Biochem*. 2005, 40, 3148–3151.

Harrington, K. J., D'Arcy-Evans, C. A. Comparison of conventional and in situ methods of transesterification of seed oil from a series of sunflower cultivar. *JAOCS.* 1985, 62(6), 1009–1013.

Hui, Y. H., Ed., *Bailey's Industrial Oil and Fat Products, Vol. 2. Edible Oil and Fat Products: Oils and Oilseeds,* Chapter 11 Soybean oil, 5th edn., Wiley-Interscience: New York, 1996.

Ilgen, O., Dincer, I., Yildiz, M., Alptekin, E., Boz, N., Canakci, M., Akin, A. N. Investigation of biodiesel production from canola oil using Mg–Al hydrotalcite catalysts. *Turk J Chem.* 2007, 31(5), 509–514.

Iso, M., Chen, B., Eguchi, M., Kudo, T., Shrestha, S. Production of biodiesel fuel from triglycerides and alcohol using immobilized lipase. *J Mol Catal B: Enzym.* 2001, 16, 53–58.

Janssen, A. E. M., Sjursnes, B. J., Vakurov, A. V., Halling, P. J. Kinetics of lipase catalyzed esterification in organic media: Correct model and solvent effects on parameters. *Enzyme Microb Tech.* 1999, 24, 463–470.

Janssen, A. E. M., Vaidya, A. M., Halling, P. J. Substrate specificity and kinetics of *Candida rugosa* lipase in organic media. *Enzyme Microb Tech.* 1996, 18, 340–346.

Jeong, G. T., Park, D. H., Kang, C. H., Lee, W. T., Sunwoo, C. S., Yoon, C. H., Choi, B. C., Kim, H. S., Kim, S. W., Lee, U. T. Production of biodiesel fuel by transesterification of rapeseed oil. *Appl Biochem Biotechnol.* 2004, 113, 747–758.

Jitputti, J., Kitiyanan, B., Rangsunvigit, P., Bunyakiat, K., Attanatho, L., Jenvanitpanjakul, P. Transesterification of crude palm kernel oil and crude coconut oil by different solid catalysts. *Chem Eng J.* 2006, 116(1), 61–66.

John, D. Costs of biodiesel production. Prepared for Energy Efficiency and Conservation Authority. May 2003.

Jüri, K., Heino, R., Jüri, Kr. Advances in biodiesel fuel research. *Proc. Estonian Acad Sci Chem.* 2002, 51(2), 75–117.

Kaieda, M., Samukawa, T., Matsumoto, T. et al. Biodiesel fuel production from plant oil catalyzed by *Rhizopus oryzae* lipase in a water-containing system without an organic solvent. *J Biosci Bioeng.* 1999, 88(6), 627–631.

Kalscheuer, R., Stölting, T., Steinbüchel, A. Microdiesel: *Escherichia coli* engineered for fuel production. *Microbiology.* 2006, 152, 2529–2536.

Kamini, N. R., Iefuji, H. Lipase catalyzed methanolysis of vegetable oils in aqueous medium by *Cryptococcus* spp. S-2. *Process Biochem.* 2001, 37, 405–410.

Katikaneni, S. P. R., Adjaye, J. D., Bakhshi, N. N. Catalytic conversion of canola oil to fuels and chemicals over various cracking catalysts. *Can J Chem Eng.* 1995, 73, 484–497.

Kildiran, G., Yücel, S. Ö., Türkay, S. In-situ alcoholysis of soybean oil. *JAOCS.* 1996, 73(2), 225–228.

Kim, S. J., Jung, S. M., Park, Y. C., Park, K. Lipase catalyzed transesterification of soybean oil using ethyl acetate, an alternative acyl acceptor. *Biotechnol Bioproc Eng.* 2007, 12(4), 441–445.

Knothe, G., Dunn, R. O., Bagby, M. O. Biodiesel: The use of vegetable oils and their derivatives as alternative diesel fuels. *Am Chem Soc Symp Ser.* 1997, 666, 172–208.

Knothe, G. Historical perspectives on vegetable oil-based diesel fuels. *Inform.* 2001, 12(11), 1103–1107.

Komers, K., Machek, J., Stloukal, R. Biodiesel from rapeseed oil and KOH 2. Composition of solution of KOH in methanol as reaction partner of oil. *Eur J Lipid Sci Tech.* 2001, 103, 359–362.

Kose, O., Tuter, M., Aksoy, H. A. Immobilized *Candida antarctica* lipase-catalyzed alcoholysis of cottonseed oil in a solvent-free medium. *Bioresource Technol.* 2002, 83, 125–129.

Krawczyk, T. Biodiesel-alternative fuel make inroads but hurdles remain. *Inform.* 1996, 7, 801–829.

Kulkarni, M. G., Dalai, A. K., Bakhshi, N. N. Utilization of green seed canola oil for biodiesel production. *J Chem Technol Biotechnol.* 2006, 81(12), 1886–1893.

Kulkarni, M. G., Dalai, A. K., Bakhshi, N. N. Transesterification of canola oil in mixed methanol/ethanol system and use of esters as lubricity additive. *Bioresource Technol.* 2007, 98(10), 2027–2033.

Kusdiana, D., Saka, S. Kinetics of transesterification in rapeseed oil to biodiesel fuels as treated in supercritical methanol. *Fuel.* 2001, 80, 693–698.

Kusdiana, D., Saka, S. Effects of water on biodiesel fuel production by supercritical methanol treatment. *Bioresource Technol.* 2004, 91, 289–295.

Laane, C., Boeren, S., Vos, K., Veeger, C. Rules for optimization of biocatalysis in organic solvents. *Biotechnol Bioeng.* 1987, 30, 81–87.

Lai, C. C., Zullaikah, S., Vali, S. R., Ju, Y. H. Lipase-catalyzed production of biodiesel from rice bran oil. *J Chem Technol Biotechnol.* 2005, 80(3), 331–337.

Lang, X., Dalai, A. K., Bakhshi, N. N., Reaney, M. J., Hertz, P. B. Preparation and characterization of biodiesels from various bio-oils. *Bioresource Technol.* 2001, 80, 53–63.

Lara, P. V., Park, E. Y. Potential application of waste activated bleaching earth on the production of fatty acid alkyl esters using *Candida cylindracea* lipase in organic solvent system. *Enzyme Microb Tech.* 2004, 34, 270–277.

Lee, K. T., Foglia, T. A., Chang, K. S. Production of alkyl ester as biodiesel from fractionated lard and restaurant grease. *JAOCS.* 2002, 79(2), 191–195.

Lee, G. C.,Wang, D. L., Ho, Y. F., Shaw, J. F. Lipase-catalyzed alcoholysis of triglycerides for short-chain monoglyceride production. *JAOCS.* 2004, 81, 533–536.

Leng, T. Y., Mohamed, A. R., Bhatia, S. Catalytic conversion of palm oil to fuels and chemicals. *Can J Chem Eng.* 1999, 77,156–162.

Leonard, W. Biodiesel from algae oil. Research report, July 2007.

Li, H. T., Xie, W. L. Transesterification of soybean oil to biodiesel with Zn/I-2 catalyst. *Catal Lett.* 2006, 107(1–2), 25–30.

Li, L. L., Du, W., Liu, D. H., Wang, L., Li, Z. B. Lipase-catalyzed transesterification of rapeseed oils for biodiesel production with a novel organic solvent as the reaction medium. *J Mol Catal B: Enzym.* 2006, 43, 58–62.

Li, N. W., Wu, H., Zong, M. H., Lou, W. Y. Immobilization of lipase from Penicillium expansum and its application to transesterification of corn oil. *Chinese J Catal.* 2007a, 28(4), 333–338.

Li, X., Xu, H., Wu, Q. Large-scale biodiesel production from microalga Chlorella protothecoids through heterotrophic cultivation in bioreactors. *Biotechnol Bioeng.* 2007b, 98(4), 764–771.

Li, Z. B. Lipase-catalyzed transesterification of rapeseed oils for biodiesel production with a novel organic solvent as the reaction medium. *J Mol Catal B: Enzym.* 2006, 43, 58–62.

Lima, D. G., Soares, V. C. D., Ribeiro, E. B., Carvalho, D. A., Cardoso, E. C. V., Rassi, F. C., Mundim, K. C., Rubim, J. C., Suarez, P. A. Z. Diesel-like fuel obtained by pyrolysis of vegetable oils. *J Anal Appl Pyrol*. 2004, 71, 987–996.

Linko, Y. Y., Liimsii, M., Wu, X., Uosukainen, W., Sappiilii, J., Linko, P. Biodegradable products by lipase biocatalysis. *J Biotechnol*. 1998, 66, 41–50.

Liu, K. S. Preparation of fatty acid methyl esters for gas chromatographic analysis of lipids in biological material. *JAOCS*. 1994, 71, 1179–1187.

Liu, X. J., He, H. Y., Wang, Y. J., Zhu, S. L. Transesterification of soybean oil to biodiesel using SrO as a solid base catalyst. *Catal Commun*. 2007, 8(7), 1107–1111.

Lu, J. K., Nie, K. L., Xie, F., Wang, F., Tan, T. W. Enzymatic synthesis of fatty acid methyl esters from lard with immobilized Candida sp 99–125. *Process Biochem*. 2007, 42(9), 1367–1370.

Luo, Y., Zheng, Y. T., Jiang, Z. B., Ma, Y. S., Wei, D. Z. A novel psychrophilic lipase from pseudomonas fluorescens with unique property in chiral resolution and biodiesel production via transesterification. *Appl Microb Biotechnol*. 2006, 73(2), 349–355.

Ma, F., Hanna, M. A. Biodiesel production: A review. *Bioresource Technol*. 1999, 70, 1–15.

Ma, F., Clements, L. D., Hanna, M. A. The effect of catalyst, free fatty acids, and water on transesterification of beef tallow. *T ASAE*. 1998, 41(5), 1261–1264.

Ma, F. Biodiesel fuel: The transesterification of beef tallow, PhD Dissertation, University of Nebraska-Lincoln. P1, 1998.

Mahabubur, M., Talukder, R., Puah, S. M., Wu, J. C., Won, C. J., Chow, Y. Lipase-catalyzed methanolysis of palm oil in presence and absence of organic solvent for production of biodiesel. *Biocatal Biotransfor*. 2006, 24(4), 257–262.

Marchetti, J. M., Miguel, V. U., Errazu, A. F. Possible methods for biodiesel production. *Renew Sust Energ Rev*. 2007, 11, 1300–1311.

Marty, A., Chulalaksananukul, W., Willemot, R. M., Condoret, J. S. Kinetics of lipase-catalyzed esterification in supercritical CO_2. *Biotechnol Bioeng*. 1992, 39, 273–280.

Maschio, G., Koufopanos, C., Lucchesi, A. Pyrolysis, a promising route for biomass utilization. *Bioresource Technol*. 1992, 42, 219–231.

Matsumoto, T., Takahashi, S., Uedab, M., Tanaka, A., Fukuda, H., Kondo, A. Preparation of high activity yeast whole cell biocatalysts by optimization of intracellular production of recombinant *Rhizopus oryzae* lipase. *J Mol Catal B: Enzym*. 2002, 17, 143–149.

Nabi, M. N., Akhter, M. S., Zaglul Shahadat, M. M. Improvement of engine emissions with conventional diesel fuel and diesel-biodiesel blends. *Bioresource Technol*. 2006, 97(3), 372–378.

Meher, L. C., Dharmagadda, V. S. S., Naik, S. N. Optimization of alkali-catalyzed transesterification of Pongamia pinnata oil for production of biodiesel. *Bioresource Technol*. 2006a, 97, 1392–1397.

Meher, L., Sagar, D., Naik, S. Technical aspects of biodiesel production by transesterification—a review. *Renew Sust Energ Rev*. 2006b, 10, 248–268.

Metting, F. B. Biodiversity and application of microalgae. *J Ind Microbiol*. 1996, 17, 477–489.

Miao, X., Wu, Q. Biodiesel production from heterotrophic microalgal oil. *Bioresource Technol*. 2006, 97(6), 841–846.

Michael, J. H. Improving the economics of biodiesel production through the use of low value lipids as feedstocks: vegetable oil soapstock. *Fuel Process Technol.* 2005, 86(10), 25, 1087–1096.

Mittelbach, M. Lipase catalyzed alcoholysis of sunflower oil. *JAOCS.* 1990, 67, 168–170.

Modi, M. K., Reddy, J. R. C., Rao, B. V. S. K., Prasad, R. B. N. Lipase-mediated transformation of vegetable oils into biodiesel using propan-2-ol as acyl acceptor. *Biotechnol Lett.* 2006, 28, 637–640.

Modi, M. K., Reddy, J. R. C., Rao, B. V. S. K., Prasad, R. B. N. Lipase-mediated conversion of vegetable oils into biodiesel using ethyl acetate as acyl acceptor. *Bioresource Technol.* 2007, 98(6), 1260–1264.

Mohamad, I. A. W., Ali, O. A. Experimental evaluation of the transesterification of waste palm oil into biodiesel. *Bioresource Technol.* 2002, 85, 253–256.

Mukherjee, K. D., Kiewit, I. Enrichment of very-long-chain monounsaturated fatty acids by lipase-catalysed hydrolysis and transesterification. *Appl Microbial Biotechnol.* 1996, 44, 557–562.

Nagao, T., Shimada, Y., Sugihara, A., and Tominaga, Y. Expression of lipase cDNA from *Fusarium heterosporum* by *Saccharomyces cerevisiae*: High-level production and purification. *J Ferment Bioeng.* 1996, 81, 488–492.

Nas, B., Berktay, A. Energy potential of biodiesel generated from waste cooking oil: an environmental approach. *Energy Sources, Part B: Econ Plan Policy.* 2007, 2, 63–71.

NBB (National Biodiesel Board). (2003). Biodiesel fuel and U.S. agriculture. Available from http://www.biodiesel.org/resources/reportsdatabase/reports/gen/20030707_gen-379.pdf.

Nelson, L. A., Foglia, T. A., Marmer, W. N. Lipase-catalyzed production of biodiesel. *JAOCS.* 1996, 73, 1191–1195.

Nelson, R. G., Schrock, M. D. Energetic and economic feasibility associated with the production, processing, and conversion of beef tallow to a substitute diesel fuel. *Biomass Bioenerg.* 2006, 30(6), 584–591.

Nitske, W. R., Wilson, C. M. *Rudolf Diesel: Pioneer of the Age of Power*, University of Oklahoma Press, Norman, OK, 1965, p. 139.

Noureddini, H., Gao, X., Phikana, R. S. Immobilized *Pseudomonas cepacia* lipase for biodiesel fuel production from soyabeen oil. *Bioresour Technol.* 2001, 96, 767–777.

Noureddini, H., Zhu, D. Kinetics of transesterification of soybean oil. *JAOCS.* 1987, 74, 1457–1463.

Ogino, H., Watanabe, F., Yamada, M., Nakagawa, S., Hirose, T., Noguchi, A., Yasuda, M., Ishikawa, H. Purification and characterization of organic solvent-stable protease from organic solvent-tolerant *Pseudomanas aeruginosa* PST-01. *J Biosci Bioeng.* 1999, 87, 61–68.

Okumura, S., Iwai, M., Tsujisaka, Y. Synthesis of various kinds of esters by four microbial lipases. *Biochem Biophys Acta.* 1979, 575, 156–165.

Oliveira, A. C., Rosa, M. F. Enzymatic transesterification of sunflower oil in an aqueous-oil biphasic system. *JAOCS.* 2006, 83, 21–25.

Orcaire, O., Buisson, P., Pierre, A. C. Application of silica aerogel encapsulated lipases in the synthesis of biodiesel by transesterification reactions. *J Mol Catal B: Enzym.* 2006, 42, 106–113.

Özgül, S., Türkay, S. In situ esterification of rice bran oil with methanol and ethanol. *JAOCS*. 1993, 70(2), 145–147.

Özgül-Yücel, S., Türkay, S. Variables affecting the yields of methyl esters derived from in situ esterification of rice bran oil. *JAOCS*. 2002, 79(6), 611–614.

Özgül-Yücel, S., Türkey, S. F. A monoalkyl ester from rice bran oil by in situ esterification. *JAOCS*. 2003, 80(1), 81–84.

Panalotov, I., Verger, R. *Physical Chemistry of Biological Interfaces*. Marcel Dekker Inc.: New York, 2000.

Park, J. Y., Kim, D. K., Wang, Z. M., Lu, P. M., Park, S. C., Lee, J. S. Production and characterization of biodiesel from tung oil. *Appl Biochem Biotechnol*. 2008, 148(1–3), 109–117.

Peterson, C. L., Auld, D. L., Korus, R. A. Winter rape oil fuel for diesel engines: Recovery and utilization. *JAOCS*. 1983, 60, 1579–1587.

Pinto, A. C., Guarieiro, L. L. N., Rezende, M. J. C., Ribeiro, N. M., Torres, E. A., Lopes, W. A., Periera, P. A. de P., de Andrade, J. B. Biodiesel: An overview. *J Braz Chem Soc*. 2005, 16, 1313–1330.

Pioch, D., Lozano, P., Graille, J. First lipase-catalyzed synthesis of fatty carbonate esters. *Biotechnol Lett*. 1991, 13, 633–635.

Pioch, D., Lozano, P., Rasoanantoandro, M. C., Graille, J., Geneste, P., Guida, A. Biofuels from catalytic cracking of tropical vegetable oils. *Oleagineux*. 1993, 48, 289–291.

Pirozzi, D. Improvement of lipase stability in the presence of commercial triglycerides. *Eur J Lipid Sci Technol*. 2003, 105(11), 608–613.

Pryde, E. H. Vegetable oil as diesel fuel: Overview. *JAOCS*. 1983, 60, 1557–1558.

Pryde, E. H. Vegetable oils as fuel alternatives—symposium overview. *JAOCS*. 1984, 61, 1609–1610.

Pryor, R. W., Hanna, M. A., Schinstock, J. L., Bashford, L. L. Soybean oil fuel in a small diesel engine. *T ASAE*. 1983, 26(2), 333–342.

Ramadhas, A. S., Jayaraj, S., Muraleedharan, C. Biodiesel production from FFA rubber seed oil. *Fuel*. 2005, 84, 335–340.

Ranganathan, S. V., Narasimhan, S. L., Muthukumar, K. An overview of enzymatic production of biodiesel. *Bioresource Technol*. 2008, 99(10), 3975–3981.

Rashid, U., Anwar, F. Production of biodiesel through base-catalyzed transesterification of safflower oil using an optimized protocol. *Energ Fuel*. 2008, 22(2), 1306–1312.

Report of the committee on development of bio-fuel. India, 2003.

Rizzi, M., Stylos, P., Riek, A., Reuss, M. A kinetic study of immobilized lipase catalysing the synthesis of isoamyl acetate by transesterification in *n*hexane. *Enzyme Microb Tech*. 1992, 14, 709–714.

Roessler, P. G., Brown, L. M., Dunahay, T. G., Heacox, D. A., Jarvis, E. E., Schneider, J. C. et al. Genetic-engineering approaches for enhanced production of biodiesel fuel from microalgae. *ACS Sym Ser*. 1994, 566, 255–270.

Romano, S. Vegetable oils—a new alternative. In: *Vegetable oils fuels-Proceedings of the International Conference on Plant and Vegetable Oils as Fuels*. ASAE Publication 4–82. Fargo, ND, 1982, 101–116.

Royon, D., Daz, M., Ellenrieder, G., Locatelli, S. Enzymatic production of biodiesel from cotton seed oil using *t*-butanol as a solvent. *Bioresource Technol*. 2007, 98, 648–653.

Rüsch, Gen., Klaas, M., Warwe, S. L. Reactive extraction of oilseeds with dialkyl carbonates. *Eur J Lipid Sci Tech.* 2001, 103, 810–814.

SAE Technical Paper series no. 831356. SAE international off highway meeting, Milwaukee, WI, 1983.

Saka, S., Kusdiana, D. Biodiesel fuel from rapeseed oil as prepared in supercritical methanol. *Fuel.* 2001, 80, 225–231.

Samukawa, T., Kaieda, M., Matsumoto, T., Ban, K., Kondo, A., Shimada, Y., Noda, H., Fukuda, H. Pretreatment of immobilized *candida antarctica* lipase for biodiesel fuel production from plant oil. *J Bioresour Bioeng.* 2000, 90, 180–183.

Sanchez, F., Vasudevan, P. T. Enzyme catalyzed production of biodiesel from olive oil. *Appl Biochem Biotechnol.* 2006, 135(1), 1–14.

Sang, O. Y., Twaiq, F., Zakaria, R., Mohamed, A., Bhatia, S. Biofuel production from catalytic cracking of palm oil. *Energ Source.* 2003, 25, 859–869.

Sarin, R., Sharma, M., Sinharay, S., Malhotra, R. K. Jatropha-palm biodiesel blends: An optimum mix for Asia. *Fuel.* 2007, 86, 1365–1371.

Schelenk, H., Gellerman, J. L. Esterification of fatty acids with diazomethane on a small scale. *Anal Chem.* 1960, 32, 1412–1414.

Schlautman, N. J., Schinstock, J. L., Hanna, M. A. Unrefined expelled soybean oil performance in a diesel engine. *T ASAE.* 1986, 29(1), 70–73.

Schlick, M. L., Hanna, M. A., Schinstock, J. L. Soybean and sunflower oil performance in a diesel engine. *T ASAE.* 1988, 31(5), 1345–1349.

Schnepf, R. Biodiesel Fuel and U.S. Agriculture, 2003. http://www.biodiesel.org/resources/reportsdatabase/reports/gen/20030707_gen-379.pdf.

Schwab, A. W., Bagby, M. O., Freedman, B. Preparation and properties of diesel fuels from vegetable oils. *Fuel.* 1987, 66, 1372–1378.

Schwab, A. W., Dykstra, G. J., Selke, E., Sorenson, S. C., Pryde, E. H. Diesel fuel from thermal decomposition of soybean oil. *JAOCS.* 1988, 65, 1781–1786.

Sellappan, S., Akoh, C. C. Applications of lipases in modification of food lipids. In: *Handbook of Industrial Catalysis*, Hou, C. T., Ed., Taylor and Francis, Boca Raton, FL, 2005, 9–1–9–39.

Shah, S., Sharma, S., Gupta, M. N. Biodiesel preparation by lipase-catalyzed transesterification of Jatropha oil. *Energ Fuel.* 2004, 18(1), 154–159.

Shaw, J. F., Wang, D. L., Wang, Y. J. Lipase-catalyzed ethanolysis and isopropanolysis of triglycerides with long-chain fatty acids. *Enzyme Microb Tech.* 1991, 13, 544–546.

Sheehan, J., Camobreco, V., Duffield, J., Graboski, M., Shapouri, H. Life-cycle inventory of biodiesel and petroleum diesel for use in an urban bus. Report: NREL/SR-580-24089, http://www.ott.doe.gov/biofuels/lifecycle pdf.html (1998a).

Sheehan, J., Dunahay, T., Benemann, J., Roessler, P. A look back at the U.S. department of energy's aquatic species program-biodiesel from algae. National Renewable Energy Laboratory, Golden, CO 1998b, 580–24190.

Shimada, Y., Watanabe, Y., Samukawa, T., Sugihara, A., Noda, H., Fukuda, H., Tominaga, Y. Conversion of vegetable oil to biodiesel using immobilized *Candida antarctica* lipase. *JAOCS.* 1999, 76, 789–793.

Shimada, Y., Watanabe, Y., Sugihara, A., Tominaga, Y. Enzymatic alcoholysis for biodiesel fuel production and application of the reaction to oil processing. *J Mol Catal B: Enzym.* 2002, 17(3–5), 133–142.

Shimada, Y., Koga, C., Sugihara, A., Nagao, T., Takada, N., Tsunasawa, S., Tominaga, Y. Purification and characterization of a novel solvent-tolerant lipase from *Fusarium heterosporum. J Ferment Bioeng*. 1993, 75, 349–352.

Singh, S., Kate, B. N., Banerjee, U. C. Bioactive compounds from cyanobacteria and microalgae: An overview. *Crit Rev Biotechnol*. 2005, 25, 73–95.

Smith, M. K. Process of producing esters. US Patent 2, 1949, 444–486.

Snyder, J. M., King, J. W., Jackson M. A. Analytical supercritical fluid extraction with lipase catalysis: Conversion of different lipids to methyl esters and effect of moisture. *JAOCS*. 1997, 74(5), 585–588.

Sonntag, N. O. V. Structure and composition of fats and oils. In: *Bailey's Industrial Oil and Fat Products*, Swern, D., Ed., Vol. 1. 4th edn., John Wiley and Sons: New York, 1979. p. 1, p. 99.

Soumanou, M. M., Bornscheuer, U. T. Improvement in lipase-catalyzed synthesis of fatty acid methyl esters from sunflower oil. *Enzyme Microb Tech*. 2003, 33, 97–103.

Spolaore, P., Joannis-Cassan, C., Duran, E., Isambert, A. Commercial applications of microalgae. *J Biosci Bioeng*. 2006, 101, 87–96.

Sprules, F. J., Price, D. Production of fatty esters. US Patent 2, 1950, 42, 366–394.

Srivastava, A., Prasad, R. Triglycerides-based diesel fuels. *Renew Sust Energ Rev*. 2000, 4, 111–133.

Stamenkovic, O. S., Lazic, M. L., Todorovi, Z. B., Veljkovic, V. B., Skala, D. U. The effect of agitation intensity on alkali-catalyzed inethanolysis of sunflower oil. *Bioresource Technol*. 2007, 98(14), 2688–2699.

Stern, R., Hillion, G., Rouxel, J. J. US Patent 5,424,466, 1995.

Stern, R., Hillion, G. Purification of esters. *Eur. Pat. Appl*. EP 356,317 (Cl. C07C67/56), 1990.

Stevenson, D. E., Stanley, R. A., Furneaux, R. H. Near-quantitative production of fatty acid alkyl esters by lipase-catalyzed alcoholysis of fats & oils with adsorption of glycerol by silica gel. *Enzyme Microb Tech*. 1994, 16(6), 478–484.

Stoffel, W., Chu, F., Ahrens, E. H. J. Analysis of long-chain fatty acids by gas–liquid chromatography. *Anal Chem*. 1959, 31, 307.

Su, E. Z., Zhang, M. H., Zhang, J. G., Gao, J. F., Wei, D. Z. Lipase-catalyzed irreversible transesterification of vegetable oils for fatty acid methyl esters production with dimethyl carbonate as the acyl acceptor. *Biochem Eng J*. 2007, 36(2), 167–173.

Su, E. Z. and Wei, D. Z. Improvement in lipase-catalyzed methanolysis of triacylglycerols for biodiesel production using a solvent engineering method. *J Mol Catal B: Enzym*. Available online March 12, 2008.

Su, E. Z., Xu, W. Q., Gao, K. L., Zheng, Y., Wei, D. Z. Lipase-catalyzed in situ reactive extraction of oilseeds with short-chained alkyl acetates for fatty acid esters production. *J Mol Catal B: Enzym*. 2007, 48, 28–32.

Sugihara, A., Tani, T., Tominaga, Y. Purification and characterization of a novel thermostable lipase from Bacillus sp. *J Biochem*. 1991, 109, 211–216.

Suppes, G. J., Dasari, M. A., Doskocil, E. J., Mankidy, P. J., Goff, M. J. Transesterification of soybean oil with zeolite and metal catalysts. *Appl Catal A-Gen*. 2004, 257(2), 213–223.

Taft, R. W. Jr., Newman, M. S., Verhoek, F. H. The kinetics of the base-catalysed methanolysis of ortho, meta and para substituted λ-menthyl benzoates. *J Am Chem Soc*. 1950, 72, 4511–4519.

Tang, Z. Y., Wang, L. Y., Yang, J. C. Transesterification of the crude *Jatropha curcas* L. oil catalyzed by micro-NaOH in supercritical and subcritical methanol. *Eur J lipid Sci Tech.* 2007, 109(6), 585–590.

Tomasevic, A. V., Marinkovic, S. S. Methanolysis of used frying oils. *Fuel Process Technol.* 2003, 81, 1–6.

Torres, C. F., Lin, B., Moeljadi, M., Hill, C. G. Jr. Lipase catalyzed synthesis of designer acyl-glycerols rich in residues of eicosapentaenoic, docosahexaenoic, conjugated linoleic, and/or stearic acids. *Eur J Lipid Sci Technol.* 2003, 105(10), 614–623.

Turner, C., Mckeon, T. The use of immobilized *Candida antarctica* lipase for simultaneous supercritical fluid extraction and in situ methanolysis of cis-vaccenic acid in milkweed seeds. *JAOCS.* 2002, 79(5), 473–478.

Ueda, M., Takahashi, S., Washida, M., Shiraga, S., Tanaka, A. Expression of *rhizopus oryzae* lipase gene in saccharomyces cerevisiae. *J Mol Catal B: Enzym.* 2002, 17, 113–124.

Ulf, S., Ricardo, S., Rogério, M. V. Transesterification of vegetable oils: A review. *J Brazil Chem Soc.* 1998, 9(1), 199–210.

Valivety, R. H., Halling, P. J., Macrae, A. Reaction rate with suspended lipase catalyst shows similar dependence on water activity in different organic solvents. *Biochem. Biophys Acta.* 1992, 1118, 218–222.

Van, T. J., Stevens, R., Veldhuizen, W., and Jongejan, J. D. Do organic solvents affect the catalytic properties of lipase? Intrinsic kinetic parameters of lipases in ester hydrolysis and formation in various organic solvents. *Biotechnol. Bioeng.* 1995, 47, 71–81.

Van Gerpen, J. H., Dvorak, B. The effect of phosphorus level on the total glycerol and reaction yield of biodiesel, presented at Bioenergy 2002, The 10th Biennieal Bioenergy Conference, Boise, ID, Sept. 2002, 22–26.

Van Gerpen, J. H., Shanks, B., Pruszko, R., Clements, D., Knothe, G. Biodiesel production technology. Report from Iowa State University for the National Renewable Energy Laboratory, NREL/SR-510-36244, July 2004.

Varma, M. N., Madras, G. Synthesis of biodiesel from castor oil and linseed oil in supercritical fluids. *Ind Eng Chem Res.* 2007, 46(1), 1–6.

Veljkovic, V. B., Lakicevic, S. H., Stamenkovic, O. S., Todorovic, Z. B., Lazic, M. L. Biodiesel production from tobacco (*Nicotiana tabacum* L.) seed oil with a high content of free fatty acids. *Fuel.* 2006, 85, 2671–2675.

Vicente, G., Martinez, M., Aracil, J. Optimization of *Brassica carinata* oil methanolysis for biodiesel production. *JAOCS.* 2005, 82(12), 899–904.

Von Wedel, R. Technical handbook for marine biodiesel. In: *Recreational Boats'*, National Biodiesel Board, http://www.biodiesel.org (1999).

Wang, L., Du, W., Liu, D. H., Li, L. L., Dai, N. M. Lipase-catalyzed biodiesel production from soybean oil deodorizer distillate with absorbent present in tert-butanol system. *J Mol Catal B: Enzym.* 2006, 43, 29–32.

Wang, Y., Ou, S., Liu, P., Zhang, Z. Preparation of biodiesel from waste cooking oil via two-step catalyzed process. *Energy Convers Manage.* 2007, 48, 184–188.

Warwel, S., Rüsch, G. M., Klaas, G. R. Lipase catalyzed conversions with diethyl and dimethyl carbonate in oleochemistry. *Proceedings of the Sixth Symposium on Renewable Resources for the Chemical Industry Bonn*, Germany 1999, 93–105.

Watanabe, Y., Shimada, Y., Sugihara, A., Noda, H., Fukuda, H., Tominaga, Y. Continuous production of biodiesel fuel from vegetable oil using immobilized *Candida antarctica* lipase. *JAOCS.* 2000, 77, 355–360.

Weisz, P. B., Haag, W. O., Rodeweld, P. G. Catalytic production of high-grade fuel (gasoline) from biomass compounds by shapedelective catalysis. *Science*. 1979, 206, 57–58.

Williams, M. A. Extraction of lipids from natural sources. In: *Lipid Technologies and Applications*, Gunstone, F. D. and Padley, F. B. Eds., Chapter 5, Marcel Dekker: New York, 1997.

Wimmer, T. PCT. Ind. Appl. WO 9309,212 (Cl C11C3/04), 1993; *Chem Abstr*. 120: P10719b (1994).

Wu, Q., Chen, H., Han, M. H., Wang, J. F., Wang, D. Z., Jin, Y. Transesterification of cottonseed oil to biodiesel catalyzed by highly active ionic liquids. *Chinese J Catal*. 2006, 27(4), 294–296.

Wu, X.Y., Jääskeläinen, S., Linko,Y. An investigation of crude lipases for hydrolysis, esterification, and transesterification. *Enzym Microb Technol*. 1996, 19, 226–231.

Xu, H., Miao, X. L., Wu, Q. Y. High quality biodiesel production from a microalga Chlorella prototothecoides by heterotrophic growth in fermenters. *J Biotech*. 2006, 126(4), 499–507.

Xu, Y., Du, W., Liu, D., Zeng, J. A novel enzymatic route for biodiesel production from renewable oils in a solvent-free medium. *Biotechnol Lett*. 2003, 25, 1239–1241.

Xu, Y. Y., Du, W., Zeng, J., Liu, D. H. Conversion of soybean oil to biodiesel fuel using lipozyme TL IM in a solvent-free medium. *Biocatal Biotransfor*. 2004, 22, 45–48.

Xue, F., Zhang, X., Luo, H., Tan, T. A new method for preparing raw material for biodiesel production. *Process Biochem*. 2006, 41, 1699–1702.

Yagiz, F., Kazan, D., Akin, A. N. Biodiesel production from waste oils by using lipase immobilized on hydrotalcite and zeolites. *Chem Eng J*. 2007, 134, 262–267.

Yan, S. L., Lu, H. F., Liang, B. Supported CaO catalysts used in the transesterification of rapeseed oil for the purpose of biodiesel production. *Energ Fuel*. 2008, 22(1), 646–651.

Yang, F. X. Resodyn Corporation. 2007. Innovative Enzymatic Reactor for Production of Alternative Fuels. SBIR Phase II final report to U.S. Department of Agriculture (USDA). A featured story about Resodyn Corporation's waste oil to biodiesel technology can be viewed at USDA website: http://www.csrees.usda.gov/newsroom/partners/fats_to_fuel.html.

Yang, J. S., Jeon, G. J., Hur, B. K., Yang, J. W. Enzymatic methanolysis of castor oil for the synthesis of methyl ricinoleate in a solvent-free medium. *J Microb Biotechnol*. 2005, 15(6), 1183–1188.

Yusuf, C. 2007. Biodiesel from microalgae. *Biotechnol Adv*. 25, 294–306.

Zhang, Y., Dube, M. A., McLean, D. D., Kates, M. Biodiesel production from waste cooking oil: 1. Process design and technological assessment. *Bioresource Technol*. 2003a, 89, 1–16.

Zhang, Y., Dube, M. A., McLean, D. D., Kates, M. Biodiesel production from waste cooking oil: 2. Economic assessment and sensitivity analysis. *Bioresource Technol*. 2003b, 90, 229–240.

Zhenyi, C., Xing, J., Shuyuan, L., Li, L. Thermodynamics calculation of the pyrolysis of vegetable oils. *Energ Source*. 2004, 26, 849–856.

Ziejewski, M., Kaufman, K. R., Schwab, A. W., Pryde, E. H. Diesel engine evaluation of a nonionic sunflower oil-aqueous ethanol microemulsion. *JAOCS*. 1984, 61, 1620–1626.

Ziejewski, M. Z., Kaufman, K. R., Pratt, G. L. 1983. In: *Vegetable oil as diesel fuel*, USDA, Argic, Rev. Man., ARM-NC. 28, 06–11.

10

Thermochemical Conversion of Biomass to Power and Fuels

Hasan Jameel, Deepak R. Keshwani, Seth F. Carter,
and Trevor H. Treasure

CONTENTS

10.1 Introduction 438
10.2 Biomass Characterization 439
    10.2.1 Moisture Content 439
    10.2.2 Proximate Analysis 441
    10.2.3 Ultimate Analysis (Elemental Composition) 442
        10.2.3.1 Heating Value 442
    10.2.4 Bulk Density 444
    10.2.5 Alkali Metal Content 445
10.3 Biomass Storage 446
10.4 Overview of Conversion Technologies 447
10.5 Combustion 448
    10.5.1 Fundamentals of Combustion 448
    10.5.2 Combustion Equipment 451
    10.5.3 Uses of Combustion 457
    10.5.4 Boiler Emissions 459
10.6 Gasification 461
    10.6.1 Fundamentals of Gasification 462
    10.6.2 Gasification Technologies 464
    10.6.3 Gas Cleanup 471
    10.6.4 Syngas Applications 473
        10.6.4.1 Heat and Power Applications 473
        10.6.4.2 Transportation Fuels from Syngas 477
10.7 Pyrolysis 483
10.8 Conclusions 486
10.9 Problems 487
References 488

10.1 Introduction

A wide range of biomass can be converted to energy using the thermochemical conversion process. The biomass can either be from wastes
that are traditionally discarded and have no apparent value or they may be
dedicated energy crops grown specifically for the production of bioenergy.
The wastes can even be a pollutant, and the major barrier to their use is
the cost of collection and transportation. Another important source of biomass is agricultural residues, which constitute the part of the crop that is
discarded after the useful products have been extracted from the harvest.
The most important sources are corn stover, rice and wheat straw, bagasse
(sugar cane residue), and even grapevine prunings. Yard and municipal
wastes can also be considered a source of biomass that can be converted
into energy. Yard waste is made up of grass clippings, leaves and tree trimmings, while municipal waste is mostly made up of waste paper, plastics, food waste, and other miscellaneous nonflammable material. Other
sources of waste can also include food processing waste and animal waste
from livestock facilities.

Dedicated energy crops are plants grown specifically for applications
other than food. Some crops can be harvested annually, such as, switchgrass
and sugar beets, while rapidly growing trees, like hybrid poplar, willow, and
sycamore, may be harvested on a 3–10 year rotation. The cycle of planting,
harvesting, and regrowing assures sustainability. Of the different biomass
sources mentioned above, agricultural residues and dedicated energy crops
are the two most important, with respect to their feasibility for producing
bioenergy on a large scale.

All the above biomass can easily be converted to energy using thermochemical conversion processes, as compared to chemical conversion, where the
source of the biomass has a very large effect on conversion efficiencies. In
both of the above conversion processes of biomass to energy, the use and
production of carbon dioxide is cyclical. The carbon dioxide produced during the production and use of the energy fuel is used for the production of
new biomass.

The thermochemical conversion of biomass to energy can be carried out
using a number of different processes. The process chosen will be determined by the end-user requirements and the source and nature of the
biomass. The end-user requirements can be grouped into power and heat
generation and transportation fuel. The production of a chemical feedstock
can also be an important consideration. The thermochemical processes that
will be discussed in this chapter include the following:

- Combustion for the production of steam and power
- Gasification for the production of a synthetic gas that can be used for
  power/heat generation, transportation fuel, and chemicals

- Pyrolysis for the production of a liquid that can be used for transportation fuel or a source for chemical feedstock

10.2 Biomass Characterization

The use of any biomass for conversion to energy will be affected by its characteristics, and this information will not only determine the conversion process but the overall economics, as well. These characteristics vary between the different sources of biomass and need to be understood in detail before any thermochemical conversion process can be considered. The properties that are of the greatest importance include the following:

1. Moisture content
2. Proximate analysis
   a. Ash content
   b. Volatile matter content
   c. Fixed carbon
3. Ultimate analysis (elemental composition)
4. Heating value
5. Bulk density
6. Alkali metal content

10.2.1 Moisture Content

Moisture content is the amount of water in the biomass expressed as a percentage of the material's weight. The complication in the use of moisture is that it can be expressed on a wet basis (MC_w), dry basis (MC_d), and on a dry ash-free basis (MC_{daf}). Since moisture has a very significant effect on the overall conversion process, the basis on which the moisture has been reported should always be properly mentioned, and moisture content on a wet basis is the most commonly used basis. Increasing the moisture content of biomass from 0% to 40% can decrease the heating value in MJ/kg by about 66%.

Moisture content of the biomass will have a significant impact on the conversion process. Biochemical conversion processes can utilize a high moisture content biomass, while thermochemical conversion processes require a low moisture content biomass, or there will be a negative impact on the overall energy balance. However some moisture is required in the gasification process for the production of hydrogen and the content of hydrogen will increase with moisture content. For gasification, decreasing the moisture content beyond 30% gives only marginal improvements in overall efficiency.

It should also be realized that the moisture content of the biomass is determined by the weather conditions during harvesting. Intrinsic moisture is the moisture content without the influence of weather, and extrinsic moisture is the moisture content that exists during harvesting. For most practical applications, it is the extrinsic moisture content that is of importance.

The moisture content of different biomass fuel sources is shown in Table 10.1 on a wet basis. The moisture content can vary from less than 10% for many of the agricultural wastes, like husks and straws, and up to 60% for bagasse. Wood, which is an important source of large quantities of biomass, has a moisture content of about 50%.

TABLE 10.1

Moisture Content (MC$_w$) of Different Biomass Fuel Sources

Biomass	MC$_w$ (%)
Bagasse	40–60
Barley straw	30
Coal (bituminous)	11
Coal (lignite)	34
Charcoal	1–10
Cotton stalks	10–20
Corn (cobs)	10–20
Corn (stover)	20–30
Palm-oil residues	15–60
Pine (loblolly)	30–60
Poplar (hybrid)	30–60
Switchgrass	30–70
Wheat straw	20
Wood chips	10–60
Wood pellets	10–20
Sawdust	15–60
Sweet sorghum	20–70

Sources: Atchison, J.E. and Hettenhaus, J.R., Innovative methods for corn stover collecting, handling, storing and transporting, Subcontractor report NREL/SR-510-33893, National Renewable Energy Laboratory, Golden, CO, 2004; Klass, D., *Biomass for Renewable Energy, Fuels and Chemicals*, Academic Press, San Diego, CA, 1998; McKendry, P., *Bioresource Technol.*, 83, 37, 2002a; Quaak, P. et al., *Energy from Biomass: A Review of Combustion and Gasification Technologies*, World Bank Technical Paper No. 422, The World Bank, Washington, DC, 1999.

10.2.2 Proximate Analysis

A proximate analysis is the evaluation of the yield of various products obtained upon heating under controlled conditions and is important in determining the performance of any thermochemical conversion process. The proximate analysis is used to determine the volatile matter, the fixed carbon content, and the amount of ash. The biomass is heated to 400°C–500°C in an inert atmosphere, and under these conditions, it decomposes into volatile matter and solid char. The amount of volatile matter determines how easily the biomass can be gasified, which will affect the design of both boilers and gasifiers. The portion that remains after the determination of the volatile matter consists of fixed carbon and ash. This residue is usually combusted in the presence of oxygen to determine the fixed carbon content and the ash content.

The proximate analysis of different biomass is shown in Table 10.2 and can also be found in many other sources. The volatile matter in biomass varies between 70% and 80%, as compared to coal, which has volatile matter content between 20% and 35%. As a result, the behavior of biomass in a gasifier

TABLE 10.2

Proximate Analysis (% wt, db) of Different Biomass Fuel Sources

Biomass	Volatiles	Ash	Fixed Carbon
Bagasse (sugarcane)	74	11	15
Barley straw	46	6	18
Coal (bituminous)	35	9	45
Coal (lignite)	29	6	31
Cotton stalk	71	7	20
Corn grain	87	1	12
Corn stover	75	6	19
Douglas fir	73	1	26
Pine (needles)	72	2	26
Plywood	82	2	16
Poplar (hybrid)	82	1	16
Redwood	80	0.4	20
Rice straw	69	13	17
Switchgrass	81	4	15
Wheat straw	59	4	21

Sources: Data from Brown, R., *Renewable Resources: Engineering New Products form Agriculture*, Blackwell Publishing, Ames, IA, 2003; Guar, S. and Reed, T., *Thermal Data for Natural and Synthetic Fuels*, Marcel Decker, New York, 1998; McKendry, P., *Bioresource Technol.*, 83, 37, 2002a; McKendry, P., *Bioresource Technol.*, 83, 47, 2002b; McKendry, P., *Bioresource Technol.*, 83, 55, 2002c; Reed, T.B., *Encyclopedia of Biomass Thermal Conversion: The Principles and Technology of Pyrolysis, Gasification & Combustion*, 3rd edn., Biomass Energy Foundation Press, Golden, CO, 2002.

is significantly different from that of coal. Biomass is easier to gasify and results in a higher amount of volatile gases that are formed upon heating. The ash content of biomass is also important because it results in the production of a waste stream that will need to be disposed. The chemical composition of the ash will affect its behavior especially at high temperatures and this will impact the removal method and the disposal cost. If the ash is in a molten state, it will be difficult to remove and may plug some of the reactor components that will increase the operating cost. The ash content in wood is usually less than 1%, while it can be very high in many agricultural residues. In agricultural residues, the ash content can vary between 5% and 10% and in rice husks it can be as high as 20%. This high ash content can be a challenge for the thermochemical conversion of agricultural residues.

10.2.3 Ultimate Analysis (Elemental Composition)

An ultimate analysis involves the determination of the elemental composition of the ash-free organic fraction of biomass, and the major components are carbon, oxygen, and hydrogen, with a small amount of nitrogen and sulfur. The elemental composition can be determined by the cellulose, hemicelluloses, lignin, and extractive content of the biomass. It is typically assumed that cellulose is $CH_{1.7}O_{0.83}$, hemicelluloses are $CH_{1.6}O_{0.8}$, and lignin is $CH_{1.1}O_{0.35}$. Realizing that the exact composition of wood will vary with the amount of cellulose, hemicellulose, and lignin, the formula of $CH_{1.4}O_{0.66}$ is used in many cases as a typical example. The ultimate analysis of select biomass is shown in Table 10.3. It should be noted that the oxygen content of biomass varies from 30% to 40% and does not contribute to the fuel value of biomass.

10.2.3.1 Heating Value

Heating value represents the net enthalpy released upon reacting a particular fuel with oxygen under isothermal conditions and is expressed as kilojoules per kilograms (MJ/kg) or BTU/lb. It is a measure of the energy that is chemically available in the fuel per unit mass. Heat of combustion is usually determined by a direct calorimeter measurement of the heat evolved. After combustion of the fuel, the combustion products are cooled to the initial temperature. The heat absorbed by the cooling medium is measured to determine the higher or gross heating value. This is referred to as the high heating value (HHV). One complication in expressing the heating value is the state of water at the end of the combustion process. If the water vapor formed during the reaction condenses at the end of the process, the latent heat of condensation contributes to the heat value and is measured in the HHV. This is the case when heat values are measured using a bomb calorimeter, where all of the water vapor condenses. If the water vapor formed during the reaction exists as a vapor after cooling, as is the case of the boiler, it is expressed as the low heating value (LHV):

TABLE 10.3

Ultimate Analysis (% wt, db) of Different Biomass Fuel Sources

Biomass	C	H	O	N	S	Ash
Bagasse (sugarcane)	44.8	5.3	39.6	0.38	0.01	9.8
Barley straw	45.7	6.1	38.3	0.4	0.1	6
Coal (bituminous)	73.1	5.5	8.7	1.4	1.7	9
Coal (lignite)	56.4	4.2	18.4	0.9	1.3	10.4
Cotton stalk	43.6	5.8	43.9	—	—	6.7
Corn grain	44.0	6.1	47.2	1.2	0.14	1.3
Corn stover	43.7	5.6	43.3	0.61	0.01	6.3
Douglas fir	50.6	6.2	43.0	0.06	0.02	0.01
Pine (bark)	52.3	5.8	38.8	0.2	—	2.9
Poplar (hybrid)	48.5	5.9	43.7	0.47	0.01	1.4
Redwood	53.5	5.9	40.3	0.1	—	0.2
Rice straw	41.8	4.6	36.6	0.7	0.08	15.9
Switchgrass	47.5	5.8	42.4	0.74	0.08	3.5
Wheat straw	43.2	5.0	39.4	0.61	0.11	11.4

Sources: Data from Brown, R., *Renewable Resources: Engineering New Products form Agriculture*, Blackwell Publishing, Ames, IA, 2003; Guar, S. and Reed, T., *Thermal Data for Natural and Synthetic Fuels*, Marcel Decker, New York, 1998; Hiler, E.A. and Stout, B.A., *Biomass Energy: A Monograph*, 1st edn., Texas A&M University Press, College Station, TX, 1985; McKendry, P., *Bioresource Technol.*, 83, 37, 2002a; McKendry, P., *Bioresource Technol.*, 83, 47, 2002b; McKendry, P., *Bioresource Technol.*, 83, 55, 2002c; Reed, T.B., *Encyclopedia of Biomass Thermal Conversion: The Principles and Technology of Pyrolysis, Gasification & Combustion*, 3rd edn., Biomass Energy Foundation Press, Golden, CO, 2002.

1. LHV = HHV − MC *2260 − [H]*20,300 (kJ/kg)
2. LHV = HHV − MC *1040 − [H]*9360 (BTU/lb)
   a. MC = moisture content as a fraction
   b. H = hydrogen content as a fraction

Since the latent heat of the water vapor formed cannot be recovered in most thermochemical processes, the LHV should be used to measure the amount of energy that will be available.

The HHV of fuels is related to the ultimate analysis (chemical composition) of a fuel and was first shown by DuLong in the nineteenth century as

- HHV (in kJ/kg) = 33,742C + 143,905[H-(O/8)] + 9,396S, where all the elements are expressed as a mass fraction.

Many improvements have been made to the above relationship, and one that has a good correlation for biomass is shown below (Channiwala, 1992):

- HHV (in kJ/kg) = 349.1C + 1178.3H–103.4O–21.1A + 100.5S–015.1N, where C is the weight fraction of carbon, H of hydrogen, O of oxygen, A of ash, S of sulfur, and N of nitrogen.

Another comprehensive source for the heat values and chemical composition of biomass is the Phyllis database by ECN Biomass. This database contains information on 2340 different biomass sources. The heat values of some common biomass fuels are shown in Table 10.4.

10.2.4 Bulk Density

Bulk density is the weight of a known volume of biomass and can be reported on a dry or wet basis. The bulk density can also be expressed on an as-produced basis or an as subsequently processed basis. This important

TABLE 10.4

Heating Value (Dry) of Different Biomass Fuel Sources

Biomass	Higher Heating Value (MJ/kg)
Bagasse (sugarcane)	17.33
Charcoal	31.10
Coal (bituminous-Pittsburgh)	33.90
Coal (bituminous-Illinois)	28.30
Coal (lignite)	24.92
Cotton stalk	22.43
Corn grain	17.20
Corn stover	17.65
Douglas fir	21.05
Paper	17.60
Pine (loblolly)	20.30
Plywood	15.77
Poplar (hybrid)	19.38
Redwood	20.72
Rice straw	16.28
Sawdust	19.97
Sawdust (pellets)	20.48
Switchgrass	18.64
Wheat straw	17.51

Sources: Data from Brown, R., *Renewable Resources: Engineering New Products form Agriculture*, Blackwell Publishing, Ames, IA, 2003; Guar, S. and Reed, T., *Thermal Data for Natural and Synthetic Fuels*, Marcel Decker, New York, 1998; Hiler, E.A. and Stout, B.A., *Biomass Energy: A Monograph*, 1st edn., Texas A&M University Press, College Station, TX, 1985; Reed, T.B., *Encyclopedia of Biomass Thermal Conversion: The Principles and Technology of Pyrolysis, Gasification & Combustion*, 3rd edn., Biomass Energy Foundation Press, Golden, CO, 2002.

TABLE 10.5

Typical Bulk Densities (Wet Basis) of Different Biomass Fuel Sources

Biomass	Bulk Density (kg/m³)
Agricultural residues	50–200
Bagasse (sugarcane-chopped)	50–75
Cereal grain straws	15–200
Coal	600–900
Hardwood chips	280–480
Poplar chips (hybrid)	150
Rice hulls	130
Softwood chips	200–340
Switchgrass (chopped)	108
Switchgrass (baled)	130–150
Wood pellets	600–700

Sources: Data from Brown, R., *Renewable Resources: Engineering New Products form Agriculture*, Blackwell Publishing, Ames, IA, 2003; Quaak, P. et al., *Energy from Biomass: A Review of Combustion and Gasification Technologies*, World Bank Technical Paper No. 422, The World Bank, Washington, DC, 1999; Scurlock, J., Bioenergy feedstock characteristics, Oak Ridge National Laboratory, Bioenergy Feedstock Development Programs, Available from http://bioenergy.ornl.gov/papers/misc/biochar_factsheet.html, 2008.

characteristic of biomass effects transportation costs, size of fuel storage, and handling equipment. Bulk density determines the radius from which the biomass must be acquired, and generally, wood or biomass must be produced within a 50–100 mile radius. It is important to transport biomass in the highest bulk density form to minimize transportation cost. Logs have a higher bulk density than chips while agricultural residues will need to be baled to achieve higher bulk density for transportation. The bulk densities of some selected biomass are shown in Table 10.5.

If bulk density information is combined with the heating value, the volumetric energy content of biomass can be calculated. This represents the enthalpy content per unit volume and is useful in determining the ultimate potential of various biomass sources for the production of energy. Gasoline and diesel have the highest volumetric energy content, while agricultural residues have the lowest volumetric energy content. As a result, gasoline can be shipped over large distances, but agricultural residues will probably have to be utilized within a 50 mile radius. If the biomass is converted to ethanol, the energy density is increased. This allows for greater shipping distances.

10.2.5 Alkali Metal Content

All types of biomass contain some alkali metals, such as Na, K, Mg, P, and Ca, and the alkali content varies depending upon the biomass and the harvesting location. These alkali metals can react with the silica and chlorine in the biomass and lead to fouling and corrosion of the combustion or gasification equipment. Annual biomass crops contain higher quantities of alkali metals than old growth biomass, like trees. The silica content of agricultural residues can be very high, and in some cases, silica contamination from the soil can increase the overall residual silica content. As a result, boilers that use agricultural residues experience more problems with slagging in the grates and fouling of the tubes.

10.3 Biomass Storage

The type of wood waste storage will be largely determined by the following:

- Form and moisture content of the residues
- Frequency and reliability of year-round deliveries to the mill and production of residues
- Availability of land
- Climatic conditions
- Need for air drying
- Volume of wood waste fuel involved
- System of waste handling and treatment adopted

Biomass can be stored in outdoor storage piles on prepared concrete or gravel pads to aid drainage and reduce the entrainment of contraries and is the least expensive means of maintaining stock. This form of storage is generally suited for stocks with 20–30 days of capacity for green forest residues, bark, moist wood slabs, or chips. Biomass can also be stored in covered storage systems to safeguard against loss and damage due to wind and rain, and it is normally provided for materials that are readily wind-borne or freely absorb moisture, such as dry sawdust, planer shavings, and sanderdust.

The moisture in residues may be reduced either by mechanical pressing, air-drying, the use of hot air dryers, or a combination of all three. Green whole chips and mixed waste, when stored in outside piles for several months, may lose up to 10%–25% of their moisture content by way of drying by the wind, the sun, and spontaneous internal heating due to bacteriological action on the materials in the interior part of the pile. Biomass for thermochemical conversion is usually dried to approximately 30% moisture using various fuel drying methods, such as a rotary drum dryer, flash- and cascade-type dryers employing waste stack gases, direct combustion of residues, or using steam or hot water as heating sources. Fuel drying will undoubtedly lead to better combustion efficiency and improved boiler utilization. Nonetheless, the use of fuel dryers in medium-sized installations is questionable, as the heat energy gained would be offset by the energy that is necessary to dry the fuel. In addition, one must take into consideration the high capital and operating costs involved with biomass drying.

10.4 Overview of Conversion Technologies

Biomass can be converted to energy via thermochemical routes using three main process options: combustion, gasification, and pyrolysis. The choice will depend on the economics of biomass availability and the end product desired. The end product can be either power/heat, transportation fuel, or chemical feedstock. The various options are shown in Figure 10.1.

Combustion is the thermal conversion of biomass using an excess of oxidant (air), where the components present in the biomass are converted to their respective oxidized form. It involves the rapid oxidation of fuel to obtain energy in the form of heat. Since the major components of biomass are carbon, hydrogen, and oxygen, the main products from the combustion of biomass are carbon dioxide and water. Combustion occurs at temperatures up to 2000°C, depending on the moisture content of the fuel, and produces hot gases at temperatures between 800°C and 1000°C, which can be used for process heat or for the production of steam.

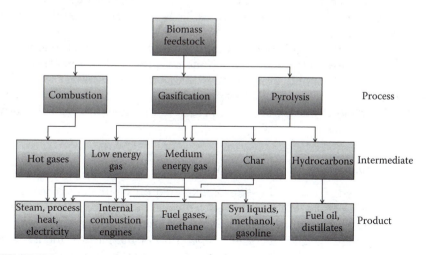

FIGURE 10.1
Options for the thermochemical conversion of biomass to fuels, power, and chemicals.

Gasification uses high temperatures under reducing conditions for the production of a combustible gas mixture. Only partial oxidation takes place under gasification conditions at temperatures that range from 800°C to 900°C. The combustible gas produced can be burned in a boiler or used in a gas engine or turbine for the production of electricity. The synthetic gas can also be used for the production of transportation fuels and chemicals.

Pyrolysis is the conversion of the biomass to a liquid, gaseous, and solid fraction by heating in the absence of air. The main goal of pyrolysis is to produce a liquid of bio-oil as the primary product. The pyrolysis oil can be used in engines, turbines, and also as a feedstock to a petroleum refinery. If it is used as a feedstock to a petroleum refinery, the pyrolysis oil will need to be upgraded to decrease its oxygen content to make it a suitable substitute for petroleum. The char and the syngas produced during pyrolysis can also be used as a source of energy.

10.5 Combustion

Combustion is usually carried out in a boiler, furnace, or stove. The fuel is burned directly in the boiler furnace to produce heat. Industrial biomass combustion facilities can burn many types of biomass fuel, including wood, agricultural residues, wood pulping liquor, municipal solid waste (MSW), and refuse-derived fuel. The objective of combustion is to release all of the chemical energy stored in a fuel while minimizing losses due to incomplete combustion. Proper combustion requires high temperatures for ignition, sufficient turbulence to mix all of the components with the oxidant, and time to complete all of the oxidation reactions. These requirements are called the three Ts of combustion.

Since combustion requires an ignition of the biomass to sustain all of the reactions, biomass with low moisture content is preferred. The moisture content of the biomass should be less than 50%, or it may be necessary to pre-dry the biomass. If the moisture content of the biomass is high, it may be preferable to use a biochemical conversion process rather than a thermochemical route.

10.5.1 Fundamentals of Combustion

The combustion of a solid, fuel-like, biomass requires four steps. These steps include heating and drying, pyrolysis, flame combustion, and char combustion. During heating and drying, heat is transferred to the solid particle phase to drive the water out. The temperature of the particle increases to about 100°C and remains there until all of the moisture has been removed. No chemical reaction occurs during heating and drying. The time required for drying is dependent on the particle size and the ignition temperature of the biomass.

After drying is completed, the temperature of the particle increases, and pyrolysis reactions begin to occur. Pyrolysis is the chemical decomposition of the biomass in the absence of oxygen. The different biomass components decompose at different temperatures. Hemicellulose breaks down first at temperatures between 225°C and 325°C. Then, cellulose breaks down at temperatures of 300°C–400°C. Lignin is the most difficult to decompose and requires temperatures up to 500°C. Decomposition rates for various biomass components have been measured using a thermogravimetric analyzer (TGA), and typical results are shown in Figure 10.2 (based on data from Shafizadeh, 1982). These results show that the xylan starts to decompose at 200°C, and it can be completely degraded as the reaction proceeds. Cellulose requires a higher temperature, and 90% of it can be degraded at temperatures up to 500°C. The decomposition of lignin varies depending upon the type of lignin. Lignin is also known to leave a high amount of residue behind. The amount of residue left is important for the design of boilers and gasifiers, because it represents the amount of char produced for the next phase of combustion.

The major products of pyrolysis are hydrogen, carbon monoxide, carbon dioxide, methane and other light hydrocarbons, and various high molecular weight hydrocarbons. The high molecular weight hydrocarbons contribute to the formation of tar if they are not combusted later in the process. Pyrolysis is rapid when compared to the overall process and requires less than one second for small particles to minutes for wood chips. Oxygen is excluded from the pyrolysis zone, and a large gaseous flow is generated by the pyrolysis reactions. Upon completion of pyrolysis, two products are formed: char (porous carbonaceous residue) and volatile gases. The proportion of volatiles

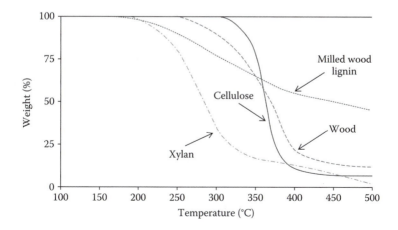

FIGURE 10.2
Degradation of biomass components with increasing temperature in a TGA. (Based on Shafizadeh, F., *J. Anal. Appl. Pyrol.*, 3, 283, 1982.)

to char is a complex interaction of temperature, heating rate, particle size, and the effect of catalysts. Yield is highly dependent on the pyrolysis temperature (the heat treatment temperature), the nature of the substrate, and the presence of incombustible materials.

Pyrolysis reactions occur at various temperatures resulting in the formation of a complex mixture of products. Cellulose is pyrolyzed via both depolymerization and dehydration. In dehydration, anhydrocellulose is formed, which is later degraded to char, tar, CO, CO_2, and water. In depolymerization, a levoglucosan intermediate is formed, which then decomposes into char and combustible volatiles. Dehydration occurs mostly at low temperatures, while depolymerization lead to the formation of products with high molecular weights that can either form tar-like compounds or degrade further to form various low molecular weight hydrocarbons. Degradation products from lignin include char (50%–65%), tar (10%), organic acids (formic, acetic, propionic, etc.), phenols (phenol, cresol, guaiacol, etc.), and cathecols. The product distribution from the pyrolysis of holocellulose and lignin is highly dependent on temperature, residence time, and heating rate.

At the end of pyrolysis, the volatile gases and the char formed continue to react independently. In flame combustion, the volatile gases react around the solid fuel, and a bright yellow flame appears on the surface. Sufficient oxygen must be present to complete the combustion process. During this phase, the three Ts of combustion are critical. Complete combustion results in the formation of mostly carbon dioxide and water. Incomplete combustion can result in the presence of carbon monoxide, polycyclic hydrocarbons, soot (long chains of carbon), and other organic compounds in the exhaust flue gas. The solid char that is left consists of mostly carbon and continues to react to form carbon monoxide and carbon dioxide. Burning the char results in a small blue flame made up of glowing char pieces. The gas–solid reaction is controlled by the mass transfer of oxygen to the char surface and is a very rapid process. Char combustion takes place either on the surface of the char (shrinking core) or inside of large pores. The gases that are formed during char combustion move away from the char particle and combine with the gases formed during volatile combustion. The carbon monoxide formed can be further oxidized to form carbon dioxide.

The various steps of combustion are shown graphically in Figure 10.3, and some examples of the reactions are shown as follows:

Drying:	no reaction
Pyrolysis:	$C_xH_yO_z \Rightarrow H_2 + CH_4 + C_2H_4 + CO + CO_2 + C$
Flame combustion:	$H_2 + 1/2O_2 \Rightarrow H_2O$
	$CH_4 + O_2 \Rightarrow CO + 2H_2O$
	$C_2H_4 + 2O_2 \Rightarrow 2CO + 3H_2O$
Char combustion:	$C + O_2 \Rightarrow CO_2$
	$C + 1/2O_2 \Rightarrow CO$
Oxidation:	$CO + 1/2O_2 \Rightarrow CO_2$

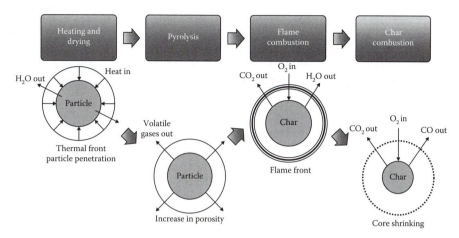

FIGURE 10.3
Steps in the conversion of biomass in a boiler.

It should be noted that drying and pyrolysis use energy, and flame and char combustion are exothermic. A summation of the energy requirement for the four steps shown represents the net energy available from the combustion process in an ideal system.

10.5.2 Combustion Equipment

Combustors convert chemical energy of fuels into high-temperature exhaust gases. These gases can be used for space heating, drying, or power generation. However, most commercial units are designed to produce steam and are called boilers. Boilers can produce low-pressure steam for process heat or high-pressure steam for power generation. A boiler is made up of a furnace, where the fuel is burned to produce hot flue gas, and a heat transfer section, where steam is produced by absorbing heat from the hot flue gas into water. In advanced boilers, the heat transfer section can be subdivided into different sections that absorb heat at different temperatures within the boiler.

Boiler efficiency is an important item of consideration for a study of the overall economics related to the combustion of biomass. This efficiency is a combination of the combustion efficiency and the efficiency with which the generated heat is transferred to water for the production of steam. The overall efficiency is defined by the following:

$$n_{boiler} = \frac{\text{Energy available for steam}}{\text{Energy supplied with the fuel}}$$

The above equation implies that all of the energy supplied with the fuel is not available for the production of steam. Various losses may be due to the following:

- Sensible heat lost to the exiting flue gas
- Sensible heat lost to the ash leaving the boiler
- Evaporation of the water from the wet biomass
- Incomplete combustion of both the char and the pyrolysis gases
- Radiation loses

A decrease in impact of any of the above factors will improve the overall boiler efficiency. Examples include the following:

- A decrease in the exiting flue gas temperature with additional heat transfer surface area
- Improving air flow (turbulence) to improve combustion
- Proper insulation of the boiler
- Maintaining a low inlet air flow (minimizing excess air) to decrease the flue gas flow rate

As a result of the many factors involved, boiler efficiencies can vary from 50% to 85%. Figure 10.4 shows how heat is utilized in a boiler and the different losses that contribute to overall boiler efficiencies. Large-scale commercial boilers also have higher efficiencies than small boilers.

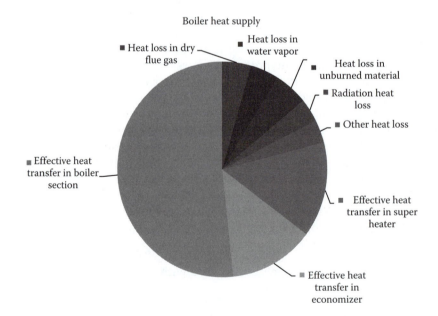

FIGURE 10.4
Utilization and heat losses in a boiler.

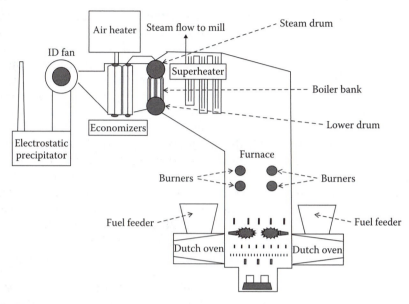

FIGURE 10.5
Schematic of a Dutch oven boiler.

Combustion systems can be classified as fixed bed or fluidized bed. Fixed-bed systems are the most commonly used method for burning biomass. Pile burners, also known as Dutch ovens, grate burners, and suspension fired furnaces are all types of fixed-bed systems. Fluidized-bed systems have recently been used for the burning of biomass and provide for improved combustion when compared to fixed-bed systems.

Pile burners consist of a lower combustion chamber, and biomass fuel burns on a grate in the lower chamber, releasing volatile gases. The gases generated are burned in the secondary combustion chamber. A schematic of a Dutch oven is shown in Figure 10.5, which shows two refractory lined ovens connected to a conventional boiler. The biomass is introduced into the Dutch oven through an opening on the top and is burned in piles. A big advantage of this boiler is that it is capable of handling high-moisture fuels and fuels mixed with dirt. However, these boilers have low efficiencies due to the high demand for excess air, and operators must shut down pile burners periodically to remove ash.

The most commonly used biomass boiler uses either a fixed or a traveling grate system. In a stationary or traveling grate combustor, an automatic feeder distributes the fuel onto a grate, where the fuel burns. Combustion air (underfire air) enters from below the grate, and secondary air (overfire air) for the combustion of the volatiles is supplied above the grates. In the stationary grate design, ashes fall into a pit for collection. In contrast, a traveling grate system has a moving grate that drops the ash into a hopper.

Most stationary grates are water cooled and formed from cast iron blocks. Fuel is distributed using a mechanical fuel distributor or air injected fuel spouts over the grates. The smaller particles burn in suspension, while the larger particles fall onto the grate and are burned there. The ash remains on the grate and must be removed by raking. Stationary grate boilers can burn fuels with moisture contents up to 55%. The major disadvantage of stationary grate boilers is the need to shut down the boiler feed for the manual removal of ash.

A significant improvement to the stationary grates, introduced in the 1940s, was the use of moving grates. These moving grates allow for the continuous removal of ash. The moving grate system consists of cast iron grates, which are attached to a chain and driven by a sprocket drive system. The biomass feed system for the stationary grate and the traveling grate systems are very similar. A vibrating grate system is also used, which allows for the intermittent and automatic removal of ash. These grates can either be water or air cooled. The main advantage of the vibrating grate system is the lower maintenance and repair costs due to a lower number of moving parts.

Suspension burners were introduced in the 1920s to distribute the fuel as a spray into the rising flue gas stream. The particles dry as they fall through the air and start to pyrolyze before landing on the grates. Suspension burners have relatively high combustion efficiencies due to the improved contact between the particles and the air. The particle size of the biomass needs to be reduced mechanically for the use of suspension burners.

A typical biomass boiler is shown in Figures 10.6 and 10.7 with most major components. The grates form the furnace floor and provide a surface on

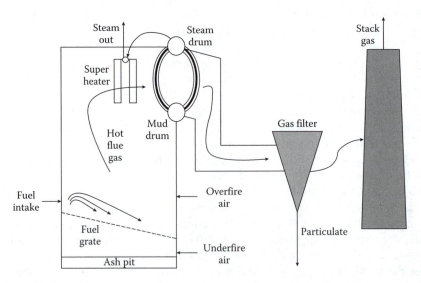

FIGURE 10.6
Simplified schematic of a biomass boiler.

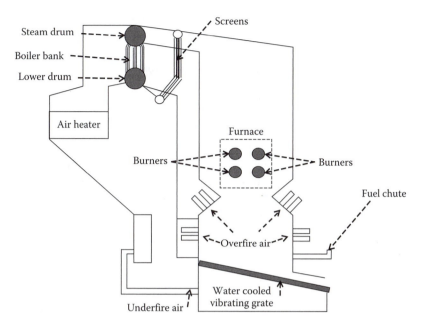

FIGURE 10.7
Schematic of a biomass boiler.

which the larger particles burn. They may be stationary, vibrating, or traveling and can either be air or water cooled. Grate blocks have venturi-type air
holes to admit undergrate air to the fuel. The fuel distributors spread fuel
as evenly as possible over the grate surface. They can be mechanical distributors made of rotating paddle wheel or wind-swept spouts using high-
pressure air. The furnace section provides a volume in which the fuel can
be burned completely and also provides for the absorption of sufficient heat
to cool the flue gas to a temperature so that the fly ash will not foul convec-
tive surfaces. The walls of most modern boilers are of membrane wall construction. The air systems are categorized as underfire (undergrate) air and
overfire air. The undergrate air is anywhere from 25% to 75% of the total
air required and is used to dry the fuel, promote the release of the volatiles,
and combust the devolatalized char resting on the grate. The overfire air
is anywhere from 25% and 75% of the total air required and provides air
for the combustion of the volatiles and mixing. In superheaters, steam from
the boiler is superheated so that more work can be extracted from a turbine
without exceeding moisture content of greater than 10%–15% within the turbine discharge. The superheater tubes are bare cylindrical tubes (1.75–2.75″
diameter) that hang in the path of the flue gas flow to exchange heat with the
steam stream. The boiler section consists of tubes, drums, and shells. These
are part of the steam–water circulation system, which are all in contact with
the hot gases. Water and steam flow inside of the tubes, and hot gases flow

over the outside surfaces. The boiler section is constructed of tubes, headers, and drums, which are joined such that water flow is provided to generate steam and cool the flue gas flowing past the tubes. An integral part of the boiler is the steam drum, which is used to separate the saturated steam from the steam–water mixture leaving the boiler heat transfer surfaces. Other auxiliary equipment include air heaters, which provide hot air to dry and burn the fuel, dust collectors, and ash handling equipment, which rakes the ash and conveys it off of the grate.

Fluidized-bed combustors can also be used to burn biomass fuel in a hot bed of granular material, such as sand. Injection of air into the bed creates turbulence, resembling a boiling liquid. Combustion takes place in the bed with high heat transfer to the furnace and low combustion temperatures. This design increases heat transfer and allows for operating temperatures below 972°C, reducing nitrogen oxide (NO_x) emissions, and it allows for improved fuel flexibility, which enables the combustion of high fouling and low-energy fuels.

The principles of fluidization are shown in Figure 10.8. The fluidization medium consists of an inert material, like sand, and air is injected into the bottom through a distributor plate. The air flows through the particle bed, and at the critical, or minimum, fluidization velocity, the upward drag forces will equal the gravitational forces of the particles, and the particles will become suspended in the fluid medium. As the velocity is increased, the height of the fluidized bed increases, and eventually, some of the bed particles will leave through the top, which creates a circulating fluidized bed.

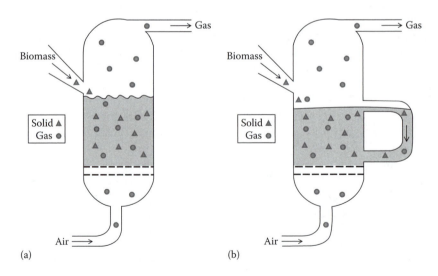

FIGURE 10.8
Principles of fluidization: (a) bubbling fluidized-bed boiler and (b) circulating fluidized-bed boiler.

Many of the new biomass boilers being built today are circulating fluidized boilers. Their advantages include the following:

- The ability to burn low-grade fuels, due to high thermal inertia and high turbulence of the fluidized bed
- High combustion efficiencies, due to the turbulent mixing of the fluidized bed and the long residence time of the fuel in the furnace
- Low SO_2 emissions, which are made possible by the reaction of limestone to sulfur in the fuel at relatively low temperatures (850°C–900°C)
- Low CO and C_xH_y emissions, due to turbulence, long residence time, and mixing in the cyclone
- Good cycling and load-following capability, due to the heat transfer being approximately proportional to the load

10.5.3 Uses of Combustion

The chemical energy in biomass is converted to heat, and eventually, it is transferred to process steam during combustion. The steam produced can then be used as a source of mechanical power to drive a shaft, or it can be used to produce power as electricity. A schematic for the conversion of biomass to power is shown in Figure 10.9. Biomass is burned in either a fixed- or fluidized-bed boiler. The steam produced is superheated and fed into a turbine generator, which converts energy from the steam into power. The

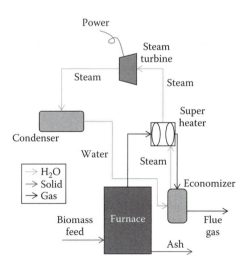

FIGURE 10.9
Schematic for the conversion of biomass to power.

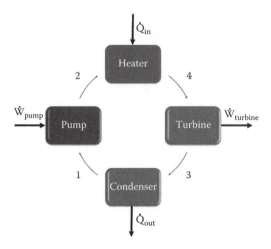

FIGURE 10.10
Components in an ideal Rankine cycle.

exhaust stream from the condenser is fed into a heat exchanger to be condensed and then pressurized to the appropriate boiler feed pressure with a pump. The working fluid (water) undergoes several phase changes as it travels through a water cycle and returns to its original state.

This ideal cycle is referred to as the Rankine cycle and is shown in Figure 10.10. The processes that make up this cycle include the following:

- 1-2: Reversible adiabatic pumping process in the pump
- 2-3: Constant-pressure transfer of heat in the boiler with super-heating
- 3-4: Reversible adiabatic expansion in the turbine
- 4-1: Constant-pressure transfer of heat in the condenser

The efficiency of the Rankine cycle determines the amount of power that is produced for a given amount of energy input and is defined as

$$n_{\text{Rankine}} = \frac{W_{\text{turbine}} - W_{\text{pump}}}{Q_{\text{in}}}$$

$$\simeq 1 - \frac{T_1}{T_2}$$

$T_1 = T_{\text{heat rejected}}$

$T_2 = T_{\text{heat absorbed}}$

The above equations are helpful in understanding the impact of different variables on the Rankine cycle efficiency. Any factor that increases the temperature at which the heat is absorbed in a boiler or allows for the rejection of heat at the lowest temperature will increase the efficiency. Newer, larger boilers with tubes made of advanced composite materials can operate at higher pressures and will have higher Rankine efficiencies. It should be noted that the Rankine efficiency is different than the boiler efficiency discussed earlier. Rankine efficiencies vary between 30% and 45%, while boiler efficiencies, representing the amount of heat recovered as steam, run between 75% and 85%. The overall efficiency for the conversion of biomass energy to power combines these two efficiencies and is quite low. Turbine efficiency also needs to be taken into consideration. Overall conversion efficiencies for a typical power plant range from 20% to 40%, with higher efficiencies being measured in systems over 100 MW. In addition, modern boilers will use reheat cycle and feed water heaters to increase the overall cycle efficiencies. Small boilers with efficiencies of less than 25% waste a significant portion of the energy in the biomass.

10.5.4 Boiler Emissions

Biomass boilers must meet the stringent environmental requirements imposed by governments in most countries. Emissions from a boiler can either be solid, thermal, or air, and the different sources are shown in Figure 10.11 for various components. Air emissions are of the greatest concern and are mainly by-products of the combustion process. Air emissions can be as follows:

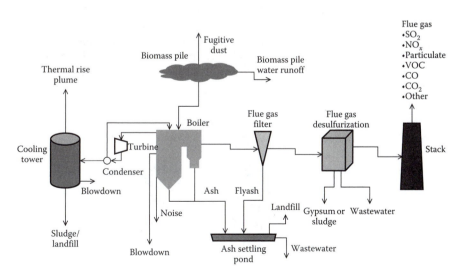

FIGURE 10.11
Sources of emissions from a biomass boiler.

- Particulates
- CO
- NO_x
- SO_x
- Volatile organic compounds

A majority of the emissions from biomass boilers are gas-borne particulates. The particulates are composed of ash, sand, and unburned char from the furnace. If there are significant metal contaminants in the biomass, some inorganic salts are also present. The amount of particulate present is a function of the combustion efficiency and aerodynamic factors. Particulate removal can be achieved with electrostatic precipitators, bag filters, or scrubbers. Electrostatic precipitators consist of vertical plates (collection electrodes) with parallel wires (discharge electrodes) positioned between the plates. It is typical for the wires to be negatively charged between 40 and 50 kV (Walkil, 1985). Gas ions form around the negatively charged wire and migrate toward the collector plates. As the ions collide with the particulate matter in the flue gas, the particles become negatively charged and move to the collector plates. The plates are rapped to remove the dust particles. Scrubbers are devices that bring flue gases into contact with a fluid that is sprayed in the gas stream shown in Figure 10.12. The dust particles become entrained in the fluid medium and are removed by drip trays. Bag filters made up of a woven textile, through which the flue gas passes, are also used to remove particulates. The collected dust is removed periodically by shaking the bags.

NO_x is formed in combustion systems at temperatures higher than 1400°C. Biomass combustion systems operate at temperatures ranging from 900°C to 1200°C, and as a result, NO_x emissions are typically low when compared to fossil fuel boilers. Nitrogen in the fuel can be converted to NO_x based on the excess air flow, heat release rate, and fuel moisture content. NO_x control technologies can be grouped into two broad categories: combustion

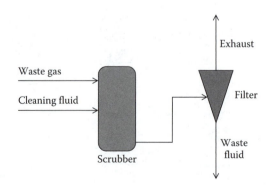

FIGURE 10.12
Schematic of wet scrubber for the removal of particulates in a biomass boiler.

modifications and post-combustion processes. Combustion modifications manage the mixing of fuel and air, thereby reducing temperature and initial turbulence, which minimizes NO_x formation in the boiler. Post-combustion control relies on various chemical processes to convert the NO_x formed in the boiler to inert nitrogen. Selective catalytic reduction (SCR), which uses base metals, zeolite, and precious metals, is being considered for NO_x control.

Wood and bark contain 0%–0.1% elemental sulfur. About 10%–30% of the sulfur in the wood can be converted to SO_2. Due to the low sulfur content in most types of biomass, sulfur emissions from most wood- and bark-fired boilers are low. However, if the biomass source contains any significant amount of sulfur, wet scrubbers (lime, dual alkali, MgO) and dry scrubbers $(Ca(OH)_2)$ may be used to decrease sulfur emissions.

10.6 Gasification

Gasification is a process that converts solid biomass to a gaseous fuel through partial oxidation. Once the fuel is in the gaseous state, undesirable substances, such as tar, sulfur compounds, and particulates, may be removed from the gas. A clean, transportable, and gaseous energy source is the result. Gasification is usually carried out at high temperatures (between 750°C and 900°C) and was initially used in the early 1800s to convert coal into a gas that could be used for illumination. Since it was difficult to transport natural gas over large distances until the mid-1900s, coal was gasified for many applications. During the energy crisis of the 1970s, interest in coal gasification grew as a source of clean energy for small-scale industrial power generation. Conversion of coal to gas was also seen as a route to decrease the dependence on petroleum from unreliable sources. Interest in biomass gasification is new and is driven by the use of energy sources that have a lower carbon footprint and do not contribute to overall carbon emissions. In addition, the use of biomass for gasification presents a potential for using wastes and residues, improved management of agricultural and forest land, and the development of a green industry.

Biomass should be easier to gasify with its high amount of volatile matter when compared to coal. The volatile matter in biomass is between 70% and 90% by weight, while in coal, it is between 30% and 40%. This results in the production of higher volumes of gas and lower amounts of char upon heating. However, biomass suffers from disadvantages, such as a lower energy density, a higher transportation cost, and a more complex gas cleanup. The supply of sufficient quantities of biomass to a large-scale gasifier could also be an issue if the biomass is dispersed over a large area.

It is important to realize that gasification is not combustion. In gasification, the conversion process produces more valuable and useful products from biomass. The gas stream exiting from gasification produces a synthetic gas, or syngas, that contains CO, H_2, H_2O, CO_2, N_2, NH_3, CH_4, and small quantities of higher hydrocarbons, H_2S, and HCl. During combustion, the gases produced consist of CO_2, H_2O, SO_2, NO, NO_2, and HCl. The combustion gases are fully oxidized, while the syngas is only partially oxidized. In gasification, about 1.5–1.8 kg of air per kg of biomass is supplied, while in combustion, the amount of air supplied is between 6 and 7 kg of air per kg of biomass. The solid by-products leaving gasification and combustion also differ. In combustion, the bottom ash consists of mineral matter, some uncombusted carbon, and nonhazardous waste. In low-pressure gasification, the bottom char is made up of unreacted carbon and mineral matter that can be a source of activated carbon. In high-pressure gasification, the slag leaving the gasifier is a glass-like inorganic material that can be disposed of as a nonhazardous waste. Combustion is an exothermic process that generates a significant amount of heat, but gasification is endothermic and requires energy for its operation. This energy can be provided externally, or it can be supplied internally by burning some of the biomass to heat the gasifier.

The use of biomass in a gasifier instead of boiler has the following advantages:

- Use of an integrated gasification combined cycle (IGCC) results in higher efficiencies
- Potential for producing chemicals
- Potential for use in fuel cells
- Lower emissions of sulfur, NO_x, and particulates
- Elimination of dioxins and furans

10.6.1 Fundamentals of Gasification

The gasification of a solid biomass requires four steps. These steps include heating and drying, pyrolysis, gas–solid reactions, and gas phase reactions. The different steps are shown schematically in Figure 10.13. Drying and pyrolysis are similar to that of combustion and have already been discussed in the combustion steps. After drying is completed, the temperature of the particle increases, and pyrolysis reactions begin to occur. Pyrolysis is the chemical decomposition of biomass in the absence of oxygen. It should be noted that both drying and pyrolysis are endothermic and require energy, which will need to be provided either directly or indirectly. If this heat is provided externally, then no oxygen is required, but if the heat is to be provided internally, a small amount of air (not more than 25% of the stoichiometric amount for complete combustion) is required for partial oxidation. During pyrolysis, the volatile matter in the biomass corresponds to the pyrolysis yield, while the carbon and ash content estimates the char yield. The gas formed during pyrolysis is composed of H_2, CH_4, CO, many hydrocarbon gases, and various oils and tars.

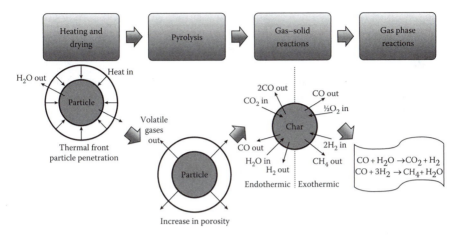

FIGURE 10.13
Steps in the conversion of biomass in a gasifier.

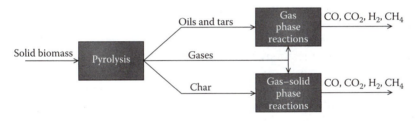

FIGURE 10.14
Gas–solid and gas-phase reactions of the pyrolysis products.

The gases formed during pyrolysis can react with the char and participate in gas–solid reactions, or they can continue to react among themselves in the gas phase. This is shown schematically in Figure 10.14. The gas–solid reactions convert solid carbon into gaseous CO, CO_2, H_2, and CH_4 and are shown by the following:

1. Carbon–oxygen reaction (combustion)
   - $C + 1/2 O_2 \leftrightarrow CO$ $\Delta H = -110\,MJ/kmol$
   - $C + O_2 \leftrightarrow CO_2$ $\Delta H = -394\,MJ/kmol$

2. Hydrogenation reaction
   - $C + 2H_2 \leftrightarrow CH_4$ $\Delta H = -75\,MJ/kmol$

3. Boudouard reaction
   - $C + CO_2 \leftrightarrow 2CO$ $\Delta H = 172\,MJ/kmol$

4. Carbon–water reaction
   - $C + H_2O \leftrightarrow H_2 + CO$ $\Delta H = 131\,MJ/kmol$

The extent of the above reactions will determine the carbon conversion in a gasifier, and the temperatures in a gasifier are hot enough that these conversions usually high. All of the above reactions are exothermic, except for the Boudouard reaction.

As the gases formed during pyrolysis react, and the gas–solid reactions continue to occur, the final mix of products is determined. The gas phase reactions can be summarized as follows:

1. Water-gas shift (WGS) reaction
   - $CO + H_2O \leftrightarrow H_2 + CO_2 \ \Delta H = -41\,MJ/kmol$
2. Methanation
   - $CO + 3H_2 \leftrightarrow CH_4 + H_2O \ \Delta H = -206\,MJ/kmol$

The final gas composition will be determined by the amount of oxygen and steam added to the system, as well as the residence time and temperature within the gasifier. Methanation is favored by low temperatures and high pressures, while the WGS reaction is favored by high temperatures.

Gasification processes should be designed so that the heat released balances with the heat required by the endothermic reactions. In addition, the thermodynamic and equilibrium characteristics determine conditions under which desired products may be maximized. Equilibrium gas compositions can be calculated based on thermodynamics, mass and energy balances, temperatures, pressures, and the biomass composition. However, these calculations can only be completed using sophisticated computer models that take these factors into account. The reliability of these models is only as good as the basic thermodynamic data used. The calculation of the final gas characteristics is further complicated by the heterogeneous nature of most gasifiers.

Most gasification models predict the fraction of CO_2, CO, H_2, CH_4, and H_2O in the syngas, and the variables used can include the following:

- Fuel composition and amount
- Amount and type of oxidant
- Water input as steam or with the fuel
- Heat loss or heat required in the gasifier
- Gasifier temperature and residence time

10.6.2 Gasification Technologies

Gasification technologies can be classified by either the gasification agent used or the type of reactor. The gasification agents most commonly used are air, oxygen, and steam. The various methods for gasification, based on the gasification agent, are shown in Figure 10.15 (Rezaiyan and Cheremisinoff,

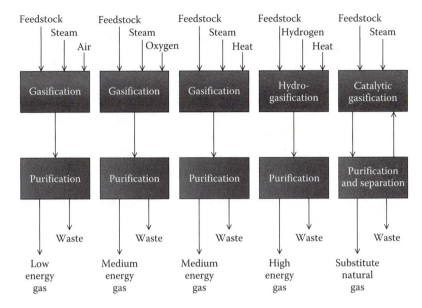

FIGURE 10.15
Various gasification technologies based on gasification agent. (Data taken from Rezaiyan, J. and Cheremisinoff, N., *Gasification Technologies: A Primer for Engineers and Scientists*, CRC Press/ Taylor & Francis, Boca Raton, FL, 2005.)

2005). The synthetic gas produced and the different methods used are shown in Table 10.6. The use of air results in lower operating costs, and the gas produced has a lower heat value. The use of oxygen results in a synthetic gas with a higher heat value that can be used for higher-value applications, such as the production of chemicals and fuel cells. Advanced gasification technologies may involve the use of hydrogen to produce syngas with higher ratios of hydrogen to carbon along with higher amounts of methane. The gas produced from hydrogen-enhanced gasification also has significantly higher heating values and is suitable for hydrogen production. Catalytic gasification results in the formation of CO and H_2, and eventually, conversion of the syngas to methane at low temperatures will occur.

Gasifiers can also be classified based on the flow pattern between the gas and the biomass, the method of contact between the fuel and gas, and the heating mode used. As a result, the major type of gasifiers can be classified as follows:

1. Fixed bed: stationary reaction zone, typically supported by a grate, and is usually fed from the top
   a. Updraft
      - Heated by the combustion of the char

TABLE 10.6

Synthetic Gas Composition (in %) from Different Commercial Gasifiers

	BFB Range	CFB Range	BCL/FERCO	MTCI	Fixed Bed	Shell
Feedstock	Various	Various	Wood	Pulp	MSW	Coal
H_2	5–26	7–20	14.9	43.3	23.4	24
CO	13–27	9–22	46.5	9.22	39.1	67
CO_2	12–40	11–16	14.6	28.1	24.4	4
H_2O	<18	10–14	Dry	5.57	dry	3
CH_4	3–11	<9	17.8	4.73	5.47	0.02
C2+	<3	<4	6.2	9.03	4.93	0
Tars	<0.11	<1	—	Scrubbed	—	0
H_2S	~0	~0	—	0.08	0.05	1
O_2	<0.02	0	0	0	—	0
NH_3	0	0	0	0	—	0.04
N_2	13–56	46–52	0	0	—	1
H_2/CO ratio	0.2–1.6	0.6–1.0	0.3	4.6	0.6	0.36
Heat value MMBTU/ft3	120–389	120–224	538	500	237	284

Source: Data from Ciferno, J. and Marano, J., *Benchmarking Biomass Gasification Technologies for Fuels, Chemicals and Hydrogen Production*, National Energy Technology Laboratory, U.S. Department of Energy, Pittsburgh, PA, 2002.

  b. Downdraft
   • Heated by the combustion of the volatiles
 2. Fluidized bed: moving bed of inert material, typically sand
  a. Bubbling fluidized bed
   • Heated by the partial combustion of volatiles and char
  b. Circulating fluidized bed
   • Heated by the partial combustion of volatiles and char
 3. Heating mode
  a. Direct heated
   • Conducts the pyrolysis and gasification in a single vessel
  b. Indirect heated
   • Utilizes a bed of hot particles, which are separated and heated, to drive the endothermic reactions

The updraft, or countercurrent, gasifier has the simplest design and can be constructed of carbon steel shells with a grate at the bottom. A lock hopper at the top is used to feed the biomass, and an air manifold is used to feed the air through the grates at the bottom. The biomass is introduced at the top of the gasifier and flows downward through the gasifier as the material is consumed by being converted to char and volatile gases. The solid material is supported on the grate. Air is taken in through the grate opening at the bottom and leaves with the synthetic gas at the top. The biomass moves through drying, pyrolysis, reduction, and combustion. This process is shown in Figure 10.16. Fresh biomass (typically 40%–50% moisture content) is dried with the hot product gases. At the same time, the product gases are cooled. The drying process takes place at a temperature of about 160°C. The major polymers in the biomass (cellulose, hemicellulose, and lignin) are broken down during pyrolysis into an approximate weight percent composition of 15% CO, 18% CO_2, 6% CH_4, 11% H_2O, and 30% tars using a dry biomass basis. A solid, highly reactive, char residue is produced during this process. The char residue typically consists of about 20% of the total carbon fed in the gasifier. Heat for pyrolysis and drying is provided by the upward flow of gas and by radiation from the combustion zone. At higher temperatures, various amounts of the char produced in the pyrolysis zone will react with H_2O and CO_2 to form H_2, CO, and CO_2 through several endothermic processes. This usually occurs within a temperature range of 500°C–1100°C. With the addition of oxygen or air, part of the char reacts with O_2 through an intermediate

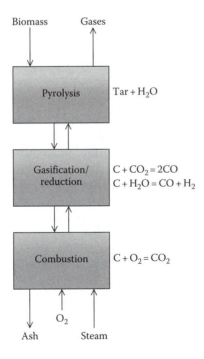

FIGURE 10.16
Zones in an updraft gasifier.

stage, which forms CO, to generate heat, which is an exothermic process, at temperatures above 1000°C.

Some of the features of an updraft gasifier include high amounts of tar and pyrolysis products, because the gas does not pass the combustion or reduction zone, low amounts of ash and dust in the gas, because the char is held at the bottom of the gasifier, and the gas temperature is low (200°C). The advantages of updraft gasification are that the gasifiers are simple and have low capital costs, the gasifiers are able to handle biomass with a high moisture and high inorganic content (e.g., municipal solid waste), and it is a proven technology. The primary disadvantage of updraft gasification is that the syngas contains 10–20 weight percent tar, which requires extensive syngas cleanup before engine, turbine, or fuel synthesis applications. Updraft gasifiers are typically used if the syngas is to be burned in a combustion device where the high tar content does not pose a significant problem.

The other type of fixed-bed gasifier is referred to as a downdraft gasifier. The design is more complex than the updraft unit, because it is constructed of two concentric shells. The biomass is contained within the inner shell, and the outer shell is used to remove the syngas from the unit. The biomass is fed into the top of the unit using a lockhop feeder, and the gas leaves at the bottom and is drawn out of the gasifier through the outer shell. The zones of the updraft and downdraft gasifiers are the same, but the order of these zones is different and is shown in Figure 10.17. After drying, the biomass is pyrolyzed. The drying and pyrolysis zones are heated through radiation from the combustion zone. Pyrolysis gases also pass through the combustion zone, where some of these gases are burned. The amount that is burned depends on the gasifier design, the biomass feed stock, and the air intake location. The remaining char and combustion products flow to the reduction zone. The reduction zone is where CO and H_2 are formed, and other gas-phase reactions also occur. The gas contains a low amount of tar, because it is vented through the combustion zone, where some of the long-chained molecules are broken down. However, the gas contains high amounts of ash and dust. This is due to the ash being in such close contact with the exiting gases. The fuel size requirement is more stringent than updraft gasifiers (uniformly sized between 4 and 10 mm), and the moisture content of biomass should be less than 25%. Gas exit temperatures are also significantly higher than that from an updraft gasifier.

Some advantages of downdraft gasification are that 99.9% of the tar formed is consumed and requires minimal or no tar cleanup and that minerals from the biomass remain with the char/ash, which reduces the need for a cyclone. It is also a proven, simple, and low-cost process, where the gas can be used in engines because of the low tar content. Some disadvantages of downdraft gasification are that it requires the feed to be dried to a low moisture content and that syngas exiting the reactor is at a high temperature (700°C), requiring a secondary heat recovery system and flue gas cleanup. Also, the carbon conversion in some downdraft designs can be less than 95%.

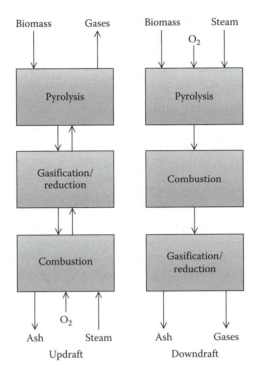

FIGURE 10.17
Comparison of the zones in an updraft and downdraft gasifier.

As gasifier sizes increase, significant issues arise with respect to the need for proper contact between the fuel particles and the heat source. Fluidized-bed systems were developed to provide uniform temperatures and efficient contact between gases and solids and to minimize the formation of hot spots within the gasifier. The fluidized bed contains either inert material, sand, or reactive materials, limestone or catalysts, and the materials are kept in suspension by a rising column of gas, which can be air, oxygen, or steam. The principles of fluidization are shown in Figure 10.8. The fuel is fed into either a suspended or a circulating fluidized bed. This is where the fuel particles mix very quickly, providing a high heat transfer rate and rapid pyrolysis. Compared to fixed-bed gasifiers, the temperatures are lower (fluidized bed: 750°C–900°C, fixed bed: 1000°C–1200°C). Most recently built large-scale gasifiers have been of the fluidized-bed design.

Bubbling fluidized-bed gasifiers contain fine inert particles of sand or alumina. A schematic of a bubbling fluidized-bed reactor is shown in Figure 10.18. The fluidized particles break up the biomass fed into the bed to ensure proper heat transfer. As gas is forced through the inert particles, a point is reached when the frictional forces between the particles and the gas counter balance the weight of the fluid. A disadvantage of bubbling fluidized-bed gasification is that the large bubbles may result in some gas bypassing through the bed. The advantages of bubbling fluidized-bed gasification include the following:

- Yields a uniform product gas
- Exhibits uniform temperature distribution throughout the reactor
- Able to accept a wide range of biomass particle sizes, including fines
- Provides high rates of heat transfer between inert material, fuel, and gas
- High conversion possible with low amounts of tar and unconverted carbon

In a circulating fluidized-bed gasifier, high gas velocities result in an entrainment of some particles, which escape from the top of the gasifier vessel. The entrained particles are separated in a cyclone and returned to the reactor. The advantages of circulating fluidized-bed gasification are as follows:

- Suitable for rapid reactions
- High heat transfer rates possible due to high heat capacity of bed material
- High conversion rates possible with low amounts of tar and unconverted carbon

The disadvantages of circulating fluidized-bed gasification include the following:

- Temperature gradients occur in direction of solid flow
- Size of fuel particles determines minimum transport velocity and high velocities may result in equipment erosion
- Less efficient heat exchange than bubbling fluidized bed

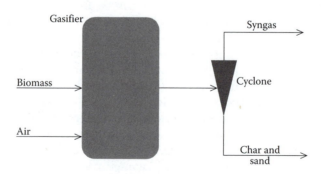

FIGURE 10.18
Schematic of bubbling fluidized-bed gasifier.

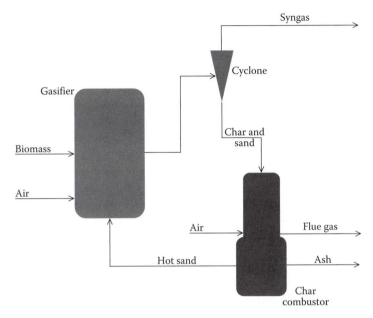

FIGURE 10.19
Schematic of a circulating biomass gasifier with two fluidized-bed reactors.

A popular concept for many new biomass gasifiers is the use of two fluid-bed reactors as shown in Figure 10.19. In this design, pyrolysis occurs in the first circulating fluidized reactor, which contains sand. The syngas leaves the gasifier with some of the sand and the char formed during pyrolysis. The char and the sand are separated from the gas in a cyclone and are sent to a combustor. The combustion reactor may be either a circulating or bubbling fluidized-bed reactor and utilizes excess air to aid the combustion of the char. The heat released during the combustion of the char is used to provide energy for the pyrolysis reactions occurring in the first reactor. In this case, the hot sand is recycled back to the pyrolysis reactor to transfer the heat from the combustion reactor. The temperature in the combustion reactor is about 950°C, while that in the pyrolysis reactor is around 800°C. The major advantage of this configuration is that the syngas has a much higher BTU content due to low levels of nitrogen and carbon dioxide since the pyrolysis and combustion are separated. With lower capital and operating costs, the resulting syngas can be converted to other chemicals and fuels through catalytic reactions.

10.6.3 Gas Cleanup

Before the syngas can be used in most applications, it must be cleaned of contaminants. Typical contaminants in syngas and their potential problems are

TABLE 10.7

Typical Contaminants in Syngas

Contaminant	Example	Problems
Tars	Oxygenated aromatics	Deposits on pipes, clogging of equipment, hinders removal of particulates
Particulates	Uncombusted biomass, ash, char, sand, fluidized-bed material	Accumulation on engine parts, equipment wear
Nitrogen compounds	Ammonia, hydrogen cyanide, NO_x	Environmental emissions
Chlorine compounds	Hydrogen chloride	Environmental emissions, corrosion, catalyst deactivation
Alkali metals	Sodium and potassium compounds	Catalyst deactivation, corrosion of filters, and turbine blades
Sulfur compounds	Hydrogen sulfide	Corrosion of turbine blades

shown in Table 10.7 (Bridgwater, 1995). Tars cause the most issues in the utilization of syngas from a biomass gasifier. Tar refers to a range of organic constituents that are produced by the partial reaction of the biomass. Tar has a molecular weight greater than benzene (MW = 78), and its actual composition is complex and dependent on the gasifier operating conditions. Biomass volatilizes as it thermally decomposes. The volatilized material can undergo further decomposition to permanent gases (CO, CO_2, H_2, light hydrocarbons), or it can undergo dehydration, condensation, and polymerization reactions to form tar. In the hot gas stream, tar exists as vaporized material or as an aerosol.

Tars can be removed by cooling the syngas so that they condense. A condenser with wide pipes can be cooled on the outside, which causes the tar to condense on the inside and is then drained from the bottom. Tar can also be removed by cooling the gas by spraying water into the gas stream with a venturi scrubber, which causes the tar to condense on the water droplets. Some of the dust, HCl, alkali metals, and sulfur oxides are also removed. The wastewater stream produced by this method of tar removal needs to be treated before disposal. Another method for tar removal that has great potential is to crack them in a bed of catalyst, such as dolomite or nickel, at high temperatures (600°C–1000°C). The heavy tars are cracked into light combustible gases. The biggest advantage of this method is that no waste streams are produced that will need to be treated before disposal. Another technique for decreasing the amount of tar in the syngas is to improve gasification such that lower amounts of tar will be formed. As mentioned earlier, fixed-bed downdraft gasifiers produce the lowest amount of tar.

Dust and particulates also need to be removed from the syngas before it can be used for most applications. Cyclones, barrier filters, and electrostatic precipitators can be used to remove particulates. Cyclones are used to remove large particles (>20 μm) and are usually the first stage in any gas conditioning process due to their simplicity. Barrier filters, like bags or rigid membranes made up of ceramic or metals, can remove small particles (<2.5 μm). The operating temperature of the filters depends on the material of construction, with ceramics being the most resistant to heat. Electrostatic precipitators operating at temperatures up to 700°C are also used to remove particulates. If barrier filters or electrostatic precipitators are used, they should be positioned after tar removal to minimize clogging as shown in Figure 10.20.

10.6.4 Syngas Applications

After the syngas has been conditioned to remove any contaminants, it can be used for power or for the production of chemicals and fuels as shown in Figure 10.21.

10.6.4.1 Heat and Power Applications

If the syngas is to be used to produce power, it can be used in a spark ignition gas engine as shown in Figure 10.22, or it can be burned in a gas turbine as shown in Figure 10.23. The syngas can also be used to produce heat by means of combustion in a boiler. The advantages of using syngas, rather than burning wood, may be related to lower emissions and an improved control of combustion.

The use of syngas in a gas engine has been demonstrated since World War II and has a low degree of technical difficulty. It can also be used on a small scale. However, the use of syngas in this application has an overall efficiency of about 30%. Efficiency is defined as

$$n_{\text{Gas engine}} = \frac{\text{Power produced}}{\text{Energy in fuel}}$$

Conversion efficiencies for typical power plants range from 20% to 40%, with the higher efficiencies being measured in systems producing over 100 MW.

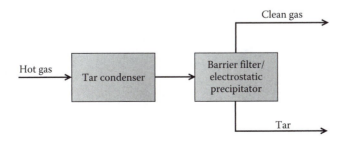

FIGURE 10.20
Gas cleanup system after gasification.

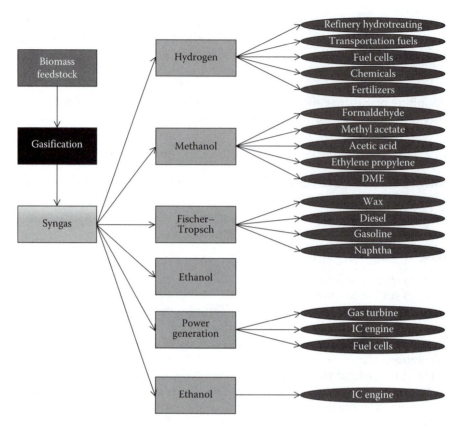

FIGURE 10.21
Options for use of syngas.

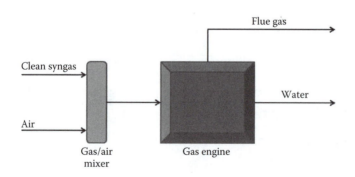

FIGURE 10.22
Use of syngas in a gas engine.

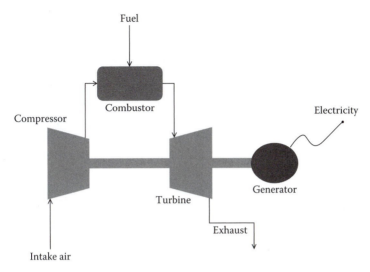

FIGURE 10.23
Use of syngas in a gas turbine.

As a result, there are benefits for overall efficiency in using biomass gasification for small-scale applications.

In order to increase the overall efficiency of power conversion, combined cycle power plants are used. In this concept, both a gas and a steam turbine are used, thereby achieving efficiencies in the range of 50%–60%. An IGCC is shown in Figure 10.24. The syngas from the biomass gasifier is cleaned to remove particulates, tar, chlorides, sulfur, and metals before being combusted in a gas (combustion) turbine. Combustion of the syngas in the combustor causes a large expansion in the gas volume, which is used to drive the turbine with nozzles aimed at the turbine blades. The turbine entry temperature in a gas turbine (Brayton cycle) is considerably higher than the peak steam temperature. Depending upon the compression ratio of the gas turbine, the turbine exhaust temperature may be high enough to permit efficient generation of steam using the "waste heat" from the gas turbine. Typical exit temperatures from gas turbines are between 400°C and 600°C. The hot exit gases are then sent to a heat exchanger that is used to produce steam. This system that produces steam from hot gases is called a heat recovery steam generator (HRSG) and is made up of evaporator, superheater, and economizer sections. The steam from the HRSG unit is used to drive a steam turbine to generate more power.

The main advantage of the IGCC plant is that electricity is produced by both the gas and the steam turbine as shown in Figure 10.24. The gas turbine operates on the Brayton cycle, while the steam turbine, which is described previously, operates on the Rankine cycle. The ideal Brayton cycle is shown in Figure 10.25 and is made up of the following processes:

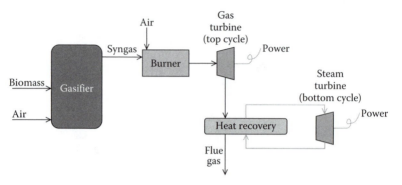

FIGURE 10.24
Use of syngas in an integrated gasification and combined cycle.

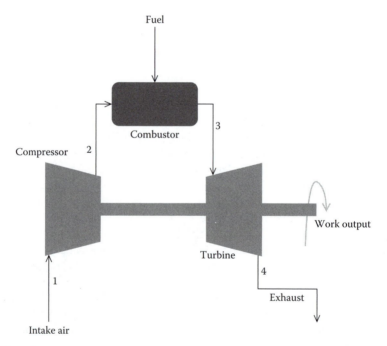

FIGURE 10.25
Elements of the Brayton cycle.

- 1-2: Fresh air is compressed to a high pressure
- 2-3: Air–fuel mixture is combusted and results in a large increase in temperature and volume at constant pressure
- 3-4: The heated air expands adiabatically through the turbine
- 4-1: The heat is rejected to the atmosphere at constant pressure

The efficiency of the Brayton cycle is defined in a similar manner to that of the Rankine cycle and is as follows:

$$n_{Brayton} = \frac{W_{turbine} - W_{compressor}}{Q_{in}}$$

$$\cong 1 - \frac{T_1}{T_2}$$

$T_1 = T_{heat\ rejected}$

$T_2 = T_{heat\ absorbed}$

The efficiency of the combined cycle can be described as follows:

$$n_{cc} = \frac{W_{gas\ turbine} + W_{steam\ turbine}}{Q_{in}}$$

$$n_{cc} = n_{Brayton} + N_{Rankine} - n_{Brayton} n_{Rankine}$$

$$n_{cc} = 1 - \frac{T_1}{T_2}$$

It should be noted that a combined cycle would not be necessary if one turbine could absorb and reject heat at temperatures similar to that from the combined cycle. These equations give insight into why combined cycles are so successful. Suppose that the gas turbine cycle has an efficiency of 40%, which is a representative value for current Brayton cycle gas turbines, and the Rankine cycle has an efficiency of 30%. The combined cycle efficiency would be 58%, which is a very large increase over either of the two simple cycles.

10.6.4.2 Transportation Fuels from Syngas

Various fuels and chemicals can be produced from the syngas that is produced through gasification. Figure 10.26 shows the pathways for the production of various chemicals. Some of the pathways shown are in the experimental stages, while others have already been demonstrated on a commercial scale. Syngas is difficult to transport and typically must be used on-site. As a result, if syngas is to be used as a fuel for other high value needs, it will need to be converted into other fuels that can be transported easily or be compatible with the energy delivery systems in existence. There is significant interest in converting the syngas into a transportation fuel, since 25% of the energy used in the United States is used for this purpose. Transportation

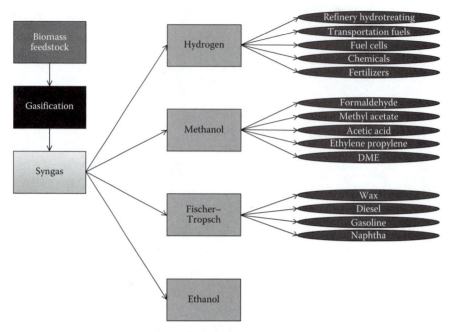

FIGURE 10.26
Production of fuel and chemicals from syngas.

fuel is a much higher value energy product than electric power. This is due to the fact that power can be produced from other energy sources, such as nuclear, coal, hydroelectric, and wind. In addition, much of the petroleum used is imported, and the use of locally grown biomass for the production of transportation fuels has a very high priority.

10.6.4.2.1 Fischer–Tropsch Process

Synthetic fuels, such as gasoline and diesel, can be produced from syngas via the Fischer–Tropsch (F–T) process. The F–T liquid that is produced can be used as a petroleum substitute. This involves catalytic reactions of H_2 and CO to form hydrocarbon chains of various lengths (CH_4, C_2H_6, C_3H_8, etc.). A simplified diagram of the F–T process is shown in Figure 10.27. The F–T synthesis reaction can be written as

$$(2n+1)H_2 + nCO \longrightarrow C_nH_{(2n+2)} + nH_2O$$

In the above reaction, n is the average chain length of the hydrocarbons that are formed. It should be emphasized that the F–T reaction produces a range of hydrocarbon chain lengths and depends upon the catalyst and the reaction conditions that are used. The most common catalysts used for this reaction are iron and cobalt. The first step in the above reaction is the formation of $-CH_2-$ groups, which are linked to one another to form a long-chained

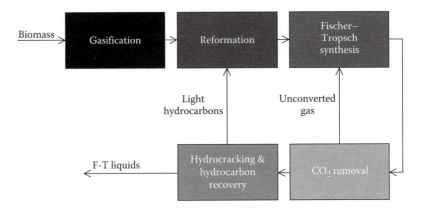

FIGURE 10.27
Simplified schematic of the Fischer–Tropsch process.

hydrocarbon. As a result, there needs to be 2 mol of hydrogen for each mole of CO. In practical applications, this H_2:CO ratio varies from between 1.7:1 and 3:1 depending upon the catalyst used and the product distribution that is desired (Rauch, 2001). The F–T reaction is very exothermic, and the reactor is designed to provide sufficient cooling to minimize unwanted reactions. The high-temperature (300°C–350°C) process uses iron as a catalyst, while the low-temperature process (200°C–240°C) uses either iron or cobalt as a catalyst. The reactors used for F–T reactions include multi-tubular fixed-bed systems cooled by water and fluidized beds with either a bubbling or circulating medium.

Before syngas can be used for the F–T reaction, it needs to be conditioned to provide the proper H_2 to CO ratio. Syngas has a higher amount of CO than is desired for F–T reactions, and the ratio is adjusted using the following WGS reaction:

$$CO + H_2O = H_2 + CO_2 \ \Delta H = -41\,MJ/kmol$$

The F–T process was developed in 1920 in Germany, and presently, only a few commercial applications exist. Royal Dutch Shell in Malaysia uses natural gas as a feedstock to produce low-sulfur diesel fuels, and South African Synthetic Oils (SASOL) uses coal and natural gas to produce a variety of synthetic petroleum products. Although the F–T process is an established technology, its implementation is hindered by high capital costs, high operation and maintenance costs, and the need for large amounts of biomass at a low cost.

10.6.4.2.2 Ethanol Production from Syngas

Ethanol can be produced using syngas with a modified F–T catalyst. A molybdenum sulfide base (MoS_2) has been suggested along with alkali metal salts (e.g., potassium carbonate) and cobalt sulfide to shift the reaction from hydrocarbons to alcohols. In addition, high pressures are suggested to ensure the production of alcohols. The reaction for the production of alcohols is as follows:

$$2nH_2 + nCO \longrightarrow C_nH_{(2n+1)}OH + (n-1)H_2O$$

The ratio of H_2:CO for the above reaction is 2.0, but the MoS_2 catalyst can also generate H_2 in concurrence with the WGS reaction. As a result, the optimum H_2:CO ratio for the production of alcohols is about 1. A wide range of reactions occur during mixed alcohol synthesis and are shown as follows:

- $CO + H_2O \leftrightarrow H_2 + CO_2$
- $CO + 2H_2 \rightarrow CH_3OH$
- $CH_3OH + H_2 \rightarrow CH_4 + H_2O$
- $CO + 2H_2 + CH_3OH \rightarrow C + H_2O$
- $CH_3OH + H_2 \rightarrow C_2H_6 + H_2O$
- $CO + 2H_2 + CH_3OH \rightarrow C_3H_7OH + H_2O$
- $C_3H_7OH + H_2 \rightarrow C_3H_8 + H_2O$
- $CO + 2H_2 + C_3H_7OH \rightarrow C_4H_9OH + H_2O$
- $C_4H_9OH + H_2 \rightarrow C_4H_{10} + H_2O$
- $CO + 2H_2 + C_4H_9OH \rightarrow C_5H_{11}OH + H_2O$
- $C_5H_{11}OH + H_2 \rightarrow C_5H_{12} + H_2O$

The above reactions show that the resulting product contains a range of alcohols and hydrocarbons. The selectivity of the conversion of carbon monoxide and hydrogen to alcohols versus hydrocarbons is very critical for the production of alcohols. With most catalysts, the amount of methanol produced is between 30% and 40%, followed by ethanol at 40%–50%, and propanol at about 10%, by weight. The other alcohols are usually less than a combined 5%–10%, by weight. The amount of CO that is converted to alcohols is between 20% and 40% in one pass. The ethanol weight percent of the alcohol product can be increased to about 70% by recycling the methanol, and this causes the methanol weight percent to decrease to around 5%. The mixed alcohols produced are then purified by distillation, and this methanol that is recovered during the distillation process can be recycled to the synthesis reactor.

   The mixture exiting from the reactor is cooled to separate the liquid alcohols from the gases. The liquid alcohols are sent to product purification, while the residual syngas, mainly consisting of CO, CO_2, H_2, and CH_4, can be used for power generation using a gas turbine. In general, integration between alcohol production and power generation is critical for the overall economics of converting syngas to ethanol. It is also possible to recycle the unconverted syngas back to the reactor to increase the overall conversion to alcohol.

In recent developments, it has also been shown that ethanol can also be produced from syngas using a bioreactor. The syngas from the gasifier is cooled to about 100°C and introduced into a fermenter, which contains the bacteria, or *Clostridium ljungdahlii*, and nutrients necessary for cell growth. The reaction in the fermenter is as follows:

- $6CO + 3H_2O \rightarrow C_2H_5OH + CO$
- $6H_2 + 2CO_2 \rightarrow C_2H_5OH + 3H_2O$

It has been demonstrated that these reactions can be conducted within a few minutes at ambient temperatures and pressures. A schematic of this process is being developed by BRI energy and is shown in Figure 10.28. Since ethanol is toxic to the bacteria, the ethanol concentration in the reactor is kept below 3% by continuously removing product from the reactor and using a molecular sieve to separate the ethanol from the bacterial cells. The bacterial cells are then recycled back through the system. BRI is constructing a full-scale demonstration unit in Labelle, Florida for the production of about 7 million gal of ethanol from wood and agricultural residues, which begins in 2010.

10.6.4.2.3 Hydrogen Production and Use in Fuel Cells

The production of hydrogen from biomass, via gasification, has a very high level of interest due to its potential use in fuels cells. A fuel cell is an electrochemical device that combines hydrogen and oxygen to produce electricity, with water and heat as its by-product. Since the conversion of the fuel to energy takes place via an electrochemical process, instead of combustion,

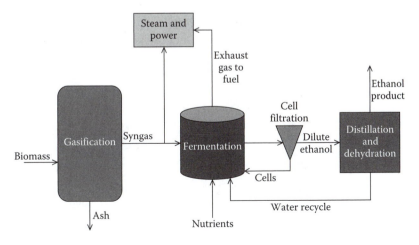

FIGURE 10.28
Simplified process for the production of ethanol using fermentation.

the process is clean, quiet, and highly efficient. If efficiency is defined as the ratio of the amount of useful energy to the total energy input, fuel cells have theoretical maximum efficiencies between 80% and 85%. In practice, fuel cell efficiencies are between 35% and 60%. It is expected that the overall tank to wheel efficiency for a car that uses hydrogen-based fuel cells will be higher than that for a gasoline- or diesel-based automobile.

A fuel cell schematic is shown in Figure 10.29. Hydrogen fuel is fed into the anode of the fuel cell, while oxygen, or air, enters the fuel cell through the cathode. At the anode, the hydrogen atom splits into a proton and an electron. The proton passes through the electrolyte, while the electrons create a separate current that can be utilized before they return to the cathode to be reunited with the hydrogen and oxygen and form a molecule of water.

Hydrogen is currently produced by steam reforming hydrocarbons over a nickel catalyst at about 800°C.

- $CH_4 + H_2O \leftrightarrow CO + H_2$

Gasification processes produce a syngas already containing hydrogen, and it can be further enriched in H_2 by the WGS reaction:

- $CO + H_2O \leftrightarrow H_2 + CO_2$

The syngas from the gasifier is reacted in two WGS reactors. The CO weight concentration from the first reactor is about 2%, and the CO weight concentration from the second reactor is about 5000 ppm. Pressure swing adsorption is used to remove the rest of the CO before the gas is vented to the atmosphere.

10.6.4.2.4 Methanol Production

Methanol can be used as a fuel, or it can be used in the production of dimethyl ether (DME), which can be used in diesel engines with minor modifications made to the engine. DME is the biofuel of choice in Europe due to the

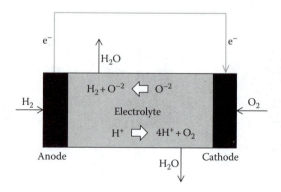

FIGURE 10.29
Schematic of a fuel cell.

large number of diesel cars compared to the United States. In addition, DME has significantly lower emissions when compared to diesel fuel due to the small chain length and low sulfur content.

Methanol is produced by reacting CO, H_2, and steam over a copper-zinc oxide catalyst at 260°C and 100 psi. The reactions are summarized as follows:

- $CO + H_2O \leftrightarrow H_2 + CO_2$ WGS
- $3H_2 + CO_2 \rightarrow CH_3OH + H_2O$ Hydrogenation

- $2H_2 + CO \rightarrow CH_3OH$ Total

Dimethyl ether is produced from methanol via the following reaction:

- $2CH_3OH \rightarrow CH_3OCH_3 + H_2O$

In DME synthesis, the syngas obtained after cleanup is reacted in packed-bed reactors in the presence of a catalyst, such as aluminum oxide or aluminum silicate, at temperatures between 200°C and 300°C. The product containing both DME and methanol is purified, and the methanol is recycled back to the reactor.

10.7 Pyrolysis

Pyrolysis is the conversion of a biomass into liquid (biofuel), solid, and gaseous fractions by heating the biomass in the absence of air. The main products of pyrolysis are gases, bio-oil, and char. The gases and the bio-oil are from the volatile fraction of biomass, while the char is mostly the fixed carbon component. The volatiles leaving the pyrolysis reactor are cooled to form the bio-oil and the gaseous fraction. The yields of these three fractions can vary significantly depending upon the pyrolysis conditions.

Heat is transferred from an external heat source in order to increase the temperature inside of the biomass particles. This initiates the primary pyrolysis reactions and results in the release of volatiles and the formation of char. Flow of hot volatiles toward the cooler solids results in heat transfer between the hot volatiles and the cooler, unpyrolyzed fuel. Condensation of some volatiles in the cooler parts of the fuel can occur, along with secondary reactions, which can form tar. Further, thermal decomposition, reforming, WGS reactions, radical recombination, and dehydrations occur in the gas phase and produce the final pyrolysis gas.

Pyrolysis can either be conducted very slowly or rapidly, which defines the type of pyrolysis. Conventional, or slow, pyrolysis has been used for thousands of years for the production of charcoal. Biomass is heated to about 500°C for between 5 and 30 min, and the main product from slow pyrolysis is char or charcoal. In fast pyrolysis, the biomass is heated rapidly in the absence of oxygen (residence time is between 0.5 and 5 s). Biomass decomposes to generate vapors, aerosols, and some char, and the vapors have a very low residence time (less than 2 s) in the reactor, which minimizes the reactions between the volatiles and the char. After cooling and condensing the vapors and aerosol particles, a dark brown liquid, or bio-oil, is formed. The heating value of the bio-oil is about 50% of fuel oil. The noncondensable gases form the gaseous fraction that can be used as a source of energy for the pyrolysis reactor. Fast pyrolysis results in 60%–75% liquid bio-oil, 15%–25% solid char, and 10%–20% noncondensable gases, by weight. This makes fast pyrolysis the preferred method for producing bio-oils. The bio-oil fraction is formed by the rapid and simultaneous depolymerization of cellulose, hemi-cellulose, and lignin. Some oligomeric species never vaporize but are blown apart into aerosols. Rapid quenching of the volatiles "freezes in" the intermediate products of fast degradation. Fast pyrolysis results in higher yields of liquids and gases, including valuable chemicals and chemical intermediates. The products from different pyrolysis methods are shown in Table 10.8.

Bio-oils are dark brown and free-flowing organic liquids and are comprised of oxygenated compounds. They are known as pyrolysis oils, wood liquids, liquid smoke, wood distillates, and liquid wood. The major compounds involved are hydroxyaldehydes, hydoxyketones, sugars, dehydrosugars, carboxylic acids, and phenolic compounds. Bio-oils also have oligomers from lignin and cellulose with molecular weights from 100 to 5000 g per g-mol. Bio-oil can be considered a micro-emulsion with a continuous aqueous phase of holocellulose decomposition products and small molecules from lignin. The discontinuous phase is the pyrolytic lignin macromolecules. The ultimate analysis for typical bio-oils is $CH_{1.9}O_{0.7}$. Typical properties of bio-oils are shown in Table 10.9.

Bio-oil has characteristics that may require it to be upgraded or treated before it can be used for many applications. Some of these characteristics and their effects on operations are shown below (A.V. Bridgwater, 1999):

TABLE 10.8

Pyrolysis Methods and Their Products

Pyrolysis Method	Time	Heating Rate	Temperature, °C	Products
Carbonization	Days	Very low	400	Charcoal
Conventional	5–30 min	Low	600	Oil, gas, char
Fast	0.5–5 s	Very high	650	Bio-oil, gas
Flash liquid	<1 s	High	<650	Bio-oil, gas
Hydro-pyrolysis	<10 s	High	<500	Bio-oil

Source: Mohan, D. et al., *Energ. Fuel.*, 20, 848, 2006.

TABLE 10.9

Typical Properties of Bio-Oils

	Reference	
Property	Mohan et al. (2006)	Oasmaa and Czernik (1999)
pH	2.5	2.0–3.7
Water content (wt %)	15–30	15–30
Solids content (wt %)	0.2–1	0.01–1
Density (kg/m³)	1.2	1.1–1.3
Viscosity (cSt at 50°C)	40–100	13–80
High heating value (MJ/kg)	16–19	—
Low heating value (MJ/kg)	—	13–80
Elemental analysis (wt %)		
C	54–58	32–49
H	5.5–7.0	6.9–8.6
N	0–0.2	0–0.2
O	35–40	44–60
S	—	0–0.0005
Ash	0–0.2	0.004–0.3

- Suspended char: erosion and equipment plugging
- Alkali metals: damage to turbine blades and deposits in boilers
- Low pH: corrosion
- High viscosity: difficult to transport in pipes
- Water content: low homogeneity and complex viscosity
- High oxygen content: low stability and low heat value
- Thermal stability: decomposition into less useful products

As a result, bio-oils may need to be upgraded to decrease the impact of the above characteristics for most applications. Upgrading may constitute filtration to remove particulates and alkali metals. This would be followed by hydrogenation or catalytic cracking to lower the oxygen content.

Bio-oils can be used as sources of fuel in a combustion boiler, the main advantage being ease of transportation and handling as compared to a solid or gaseous fuel. Bio-oils can also be fed to an existing petroleum refinery, where they can be converted into various transportation fuels. The ability to feed in a petroleum refinery is a significant advantage that bio-oils have over most biofuels, because no additional infrastructure is necessary for the use of this bio-based fuel oil. Small pyrolysis plants can be built to supply large operational facilities, which minimizes the impact of transporting large quantities of low-density biomass.

Three types of reactors have been suggested for the fast pyrolysis of biomass. In ablative pyrolysis, the wood is pressed and moved against a hot

surface as shown in Figure 10.30. The wood melts
at the heating surface to form a film of liquid oil,
which then evaporates to form the bio-oil. This
process can use large wood chips but requires the
availability of large heated surfaces. Bubbling and
circulating fluidized-bed reactors are also used
for fast pyrolysis reactions. Typical fluidized-bed
reactors are shown in Figures 10.31 and 10.32.
Fluidized beds can transfer heat to the biomass
very quickly. However, the particles need to be less
than 3mm in diameter to achieve the high yields
of bio-oil. Auger reactors have also been used
for rapid pyrolysis. The auger reactor consists of
a reactor pipe and an auger. The auger is used to
move the biomass through the pyrolysis zone in
about 30–50s at 450°C. Heat is usually provided
from an outside source.

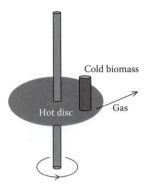

FIGURE 10.30
Ablative process for the production pyrolysis oil.

10.8 Conclusions

Biomass can be converted to transportation fuels or power using both the
biochemical and thermochemical pathways. The choice will depend on
the end product desired and the biomass source. However, thermochemical conversion processes are not very sensitive to the biomass composition
when compared to the biochemical pathway for the production of fuels. The
three thermochemical methods of converting biomass to power and fuels
discussed in this chapter include combustion, gasification, and pyrolysis.
All of these processes include multiple reactions, where heat is initially
transferred to the biomass to evaporate water, and then the heat increases

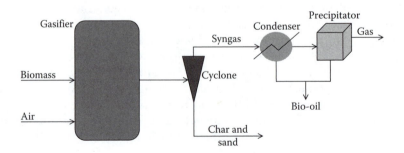

FIGURE 10.31
Bubbling fluidized-bed system for the production of pyrolysis oil.

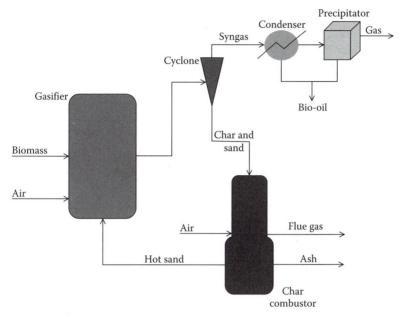

FIGURE 10.32
Circulating fluidized-bed reactor for the production of pyrolysis oil.

the temperature of the biomass to cause pyrolysis to occur and to produce volatiles. In combustion, exothermic reactions occur to convert the volatiles to H_2O and CO_2 in the presence of excess air, with heat being the primary product. In gasification, the volatiles undergo both gas–solid and gas–gas reactions to produce the final components of the low to medium heating value gas. The syngas generated can be used to produce power, transportation fuels, and chemicals. In pyrolysis, thermal decomposition occurs in the absence of oxygen to produce bio-oil, char, and gas. The bio-oil can be used as a feed to a petroleum refinery for the production of various transportation fuels or burned in a boiler to produce heat. The thermochemical conversion of biomass to various energy products may play an important role in dealing with growing energy demands.

10.9 Problems

1. Discuss the importance of moisture in thermochemical conversion especially in combustion and gasification.
2. What is the difference between low and high heating value, and how does it impact the operation of a boiler?

3. Discuss the steps and the reactions in combustion.

4. What are the steps in the Rankine cycle, and how are these related to the operation of a boiler and a turbine?

5. How are the following air emissions decreased in a boiler?

    a. Particulates

    b. CO

    c. NO_x

6. Discuss the steps and reactions in gasification.

7. Discuss how a updraft gasifier works (include the various zones and if possible the main reactions in the different zones).

8. Discuss the differences between an updraft and a downdraft gasifier.

9. What are the pollutants from gasification, and how are they removed from the syngas?

10. Discuss why an IGCC plan has a higher overall efficiency than a boiler producing steam.

11. Discuss how Fischer–Tropsch fuels can be made from syngas.

12. Describe the process for fast pyrolysis of biomass for the production of bio-oil.

References

Atchison, J. E. and J. R. Hettenhaus. 2004. Innovative methods for corn stover collecting, handling, storing and transporting. Subcontractor report NREL/SR-510-33893. National Renewable Energy Laboratory, Golden, CO.

Bridgwater, A.V. 1995. The technical and economic feasibility of biomass gasification for power generation. *Fuel*, 74 (5), 631–653.

Bridgwater, A. V., D. Meier, and D. Radlein, 1999. An overview of fast pyrolysis of biomass. *Organic Geochemistry*, 30(12), 1479–1493.

Brown, R. 2003. *Renewable Resources: Engineering New Products form Agriculture*. Blackwell Publishing, Ames, IA.

Channniwala, S. A. 1992. *Biomass Energy Foundation: Proximate/Ultimate Analysis*. M.S. Thesis, Indian Institute of Technology, Bombay, India.

Ciferno, J. and J. Marano. 2002. *Benchmarking Biomass Gasification Technologies for Fuels, Chemicals and Hydrogen Production*. National Energy Technology Laboratory, U.S. Department of Energy, Pittsburgh, PA.

Guar, S. and T. Reed. Proximate and ultimate analysis. Available from http://www.woodgas.com/proximat.htm.

Guar, S. and T. Reed. 1998. *Thermal Data for Natural and Synthetic Fuels*. Marcel Decker, New York.

Hiler, E. A. and B. A. Stout. 1985. *Biomass Energy: A Monograph*. 1st edn. Published for the Texas Engineering Experiment Station, Texas A&M University System by Texas A&M University Press, College Station, TX.

Klass, D. 1998. *Biomass for Renewable Energy, Fuels and Chemicals*. Academic Press, San Diego, CA.

McKendry, P. 2002a. Energy production from biomass: Overview of biomass. *Bioresource Technology*, 83, 37–46.

McKendry, P. 2002b. Energy production from biomass: Conversion technologies. *Bioresource Technology*, 83, 47–54.

McKendry, P. 2002c. Energy production from biomass: Gasification technologies. *Bioresource Technology*, 83, 55–63.

Mohan, D., C. Pittman, and P. Steele. 2006. Pyrolysis of wood/biomass for bio-oil: A critical review. *Energy & Fuels*, 20, 848–889.

Oasmaa, A. and S. Czernik. 1999. Fuel oil quality of biomass pyrolysis oils—State of the art for the end user. *Energy & Fuels*, 13, 914–921.

ECN. Phyllis, Composition of biomass. Available from http://www.ecn.nl/phyllis/.

Quaak, P., H. Knoef, and H. Stassen. 1999. *Energy from Biomass: A Review of Combustion and Gasification Technologies*, World Bank Technical Paper No. 422, The World Bank, Washington, DC.

Rauch, R. 2001. *Biomass Gasification to Produce Synthetic Gas for Fuel Cells, Liquid Fuels and Chemicals*. A report to IEA (International Energy Agency) Bioenergy Commission.

Reed, T. B. 2002. *Encyclopedia of Biomass Thermal Conversion: The Principles and Technology of Pyrolysis, Gasification & Combustion*. 3rd edn. Biomass Energy Foundation Press, Golden, CO.

Rezaiyan, J. and N. Cheremisinoff. 2005. *Gasification Technologies: A Primer for Engineers and Scientists*. CRC Press Taylor & Francis, Boca Raton, FL.

Scurlock, J. 2008. Bioenergy feedstock characteristics, Oak Ridge National Laboratory, Bioenergy Feedstock Development Programs. Available from http://bioenergy.ornl.gov/papers/misc/biochar_factsheet.html.

Shafizadeh, F. 1982. Introduction to pyrolysis of biomass. *Journal of Analytical and Applied Pyrolysis*, 3, 283–305.

Walkil, E. 1985. *Powerplant Technology*. McGraw Hill, New York.

Index

A

Ablative pyrolysis, 485–486
Acclimation, 168–169, 171–172
Acetone–butanol–ethanol (ABE)
     fermentation; *see also* Butanol
     production
  feedstocks
    molasses, 280
    starch, 280
  history, 274–275
Acid-catalyzed transesterification,
     368–369
ADS, *see* Anaerobically (mesophilic)
     digested sludge
Alcoholysis
  esterase-catalyzed alcoholysis, 387
  lipase-mediated alcoholysis, 375
  methanol, 354
  purpose, 355
  in situ alcoholysis, 392
Alkanes, 8–9
Alkenes, 9–10
Alkynes, 10–11
Aluminum, 173
American Society of Biological and
     Agricultural Engineers
     (ASABE), 79
Ammonia fiber explosion (AFEX),
     225, 227
Anaerobically (mesophilic) digested
     sludge (ADS), 162
Anaerobic digestion, biogas production
  ammonia inhibition
    inhibited steady state, 168
    total ammonia nitrogen (TAN),
     168
  biogas purification
    CO_2 scrubbing, 199–200
    H_2S scrubbing, 198–199
    metal corrosion, 196–197
  composition, 152
  hydrogen effect, 163
  hydrogen production, 195–196
  kinetics
    acetate concentration, 179
    anaerobic methanogenesis, 177
    methanogens and acetogens,
     178
    microbial growth, 176
    substrate utilization, 176
  metal inhibition
    acclimation, 171–172
    aluminum, 173
    antagonism, 173
    calcium, 172
    heavy metals, 173–175
    magnesium, 173
    potassium, 171–172
    sodium, 171
  microbial metabolism, 153–154
  modeling
    ambient-temperature, 182
    cellular automata (CA), 180
    mechanistic model, 181
    monod-type kinetics, 179
  organic compounds
    degradation, 154–155
    lignins and derivatives, 176
    long-chain fatty acids (LCFAs),
     175
  other anaerobic digesters, 190–194
  pH effect
    alkalinity, 167
    chemical equilibrium constants,
     165–166
    Henry's law, 164
    methane production, 163
  stoichiometry, 155–157
  sulfate/sulfide inhibition
    solubility product, 170
    sulfate-reducing bacteria (SRB),
     169
  temperature effect
    mesophilic, 158–161
    psychrophilic, 157–158
    thermophilic, 161–163

Anaerobic packed-bed (APB), 153
Anaerobic sequencing batch reactors
    (ASBR), 157, 192–193
Antagonism, 173
Aromatic compound, 11–12
Attached-growth
  expended-bed, 189–190
  packed-bed, 188–189
Auger reactors, 486

B

Bacillus subtilis, 219
Balers
  advantages and disadvantages, 98–99
  bales
    rectangular, 98
    round, 98, 99
    square, 99
  baling mechanism, 98
  productivity, 100
  rates, 100
  rectangular balers, 100
  round balers
    fixed chamber balers, 99
    variable-chamber balers, 100
  size, 98
Batch fermentation
  butanol production, 295–297
  ethanol production, 246–247
Biochemical oxygen demand (BOD),
    156–157
Biodiesel
  acyl receptors, 354–356
  B20 blend, 341–342
  biolipid feedstocks
    animal fats, 346, 349
    microalgal oils, 351–353
    plant-derived oils, 346, 347–348,
      349
    waste oils and fats, 349–351
  diesel engine, 356
  direct vegetable oil use and blending
    advantages and disadvantages,
      357–358
    gum formation and
      polymerization, 357
    performance tests, 357
    straight vegetable oil (SVO), 357

energy demand, 338–339
esterification
  diazomethane (CH_2N_2), 366
  free fatty acid (FFA) treatment
    method, 365
  mechanism, 365–366
  sulfuric acid (H_2SO_4), 365
fatty acid alkyl monoesters, 340
microemulsification
  ionic and nonionic
    microemulsion, 358
  Shipp nonionic fuel, 358–359
  ternary phase equilibrium
    diagram, 359
oils/fats properties, 353–354
vs. petrodiesel
  CO_2 emission, 340–341
  combustion, 341
  cost, 339
  structural similarity, 341–342
pyrolysis
  catalytic vegetable oils cracking,
    362–363
  vegetable oils and fats, 359–361,
    363–365
  vegetable oil soaps, 361–362
transesterification, 339–340
  acid-catalyzed transesterification,
    368–369
  alcohol-triglyceride molar ratio
    and alcohol type, 393–394
  alkali-catalyzed
    transesterification, 369–370, 371
  catalyst type and concentration,
    394–395
  dimethyl carbonate, 355, 356
  general aspects, 367–368
  kinetics, 370–372
  lipase-catalyzed
    transesterification, 374–391
  methanol, 354–355
  methyl acetate, 355, 356
  mixing intensity, 397
  moisture and free fatty acid
    content, 395–396
  organic cosolvents, 397–398
  reactant purity, 396
  reaction temperature and time,
    396–397

in situ transesterification, 391–393
supercritical alcohol, 372–374
unit operation processes
alcohol recovery, 412–413
BIOX processing system, 416
economic analysis, 419–421
ester–glycerol separation, 409–411
ester washing and treatment,
411–412
feedstock preparation, 406
flowchart, 398, 399
glycerol refining, 413–414
high free fatty acid feedstocks
pretreatment, 404–406
high free fatty acid grease feed
process, 416–419
oils/fat feedstocks pretreatment,
398–404
plant equipment, 414–415
transesterification, 406–409
wastewater treatment, 414
use, 341
vegetable oils, 339
Biogas purification
CO_2 scrubbing, 199–200
H_2S scrubbing, 198–199
metal corrosion, 196–197
Biological process
butanol production (*see* Butanol
production)
elementary reaction
biochemical reaction, 135–136
reaction rate, 136
enzymatic reaction kinetics
competitive inhibition model, 141
Haldane model, 140–141
Mickaelis-Menten equation,
137–140
product inhibition model, 141
substrate inhibition model,
140–141
ethanol production (*see* Ethanol
production)
microbial growth
acceleration phase, 144
deacceleration phase, 147
death growth phase, 148
exponential growth phase, 144–147
lag growth phase, 143–144

microbial cell growth cycle, 143,
144
Monod model, 147
stationary growth phase, 147
microorganisms
chemotrophs and phototrophs,
143
dry matter composition, 142
elemental chemical composition,
143
fermentation, 142
foods, 142–143
heterotrophs and autotrophs, 142
organic and inorganic nutrients,
143
Boiler, *see* Combustion
Brayton cycle
efficiency, 477
elements, 475–476
turbine entry temperature, 475
Bubbling fluidized-bed reactor, 486
Butanol production; *see also* Resonant
Sonic Enhanced Mixer and
Coalescer (RSEMC) technology
biobutanol recovery
distillation, 304
gas stripping, 304, 305
integration process, 319–320
liquid–liquid extraction, 304–306
novel product-recovery
techniques, 303
perstraction, 318–319
pervaporation, 316–317
reverse osmosis, 318
commercial technology
bibutanol plant, 323–324
Resodyn Corporation's process,
324
commodity chemical, 275, 278–279
economics, wheat straw *vs.* corn
starch
economical analysis, 325–327
preliminary profit-and-loss
statement, 325, 328
processing, 324–325
total manufacturing cost, 325
feedstocks
lignocellulose, 280–281
molasses, 280

starch, 280
   upstream processes, 281–282
fermentation technology, 294–295
   batch process, 295–297
   continuous culture, 299–302
   fed-batch culture, 297–299
   immobilized cell bioreactors, 302
as fuel
   advantages, 274
   properties, 273
future development
   cyclopropane fatty acid synthase
     over-expression, 322
   downstream processing, 322–323
   metabolic engineering, 321–322
history
   acetone–butanol–ethanol (ABE)
     fermentation, 274–275
   aldol condensation, 275
   amyl alcohol, 275
   industrial processes, 279
   patent application, 276–278
   solvent-forming bacteria, 273–274
microbial cultures
   3-hydroxybutyryl-CoA
     dehydrogenase, 287
   acetate/butyrate CoA-transferase,
     288–289
   acetoacetate decarboxylase, 289
   alcohol dehydrogenase, 290–291
   aldehyde dehydrogenase, 289–290
   butyryl-CoA dehydrogenase, 288
   Clostridia fermentation, 282–283
   crotonase, 287–288
   genomes, 293–294
   metabolism and pathway,
     283–286
   operons, 291–293
   thiolase, 286–287

C

Calcium, 172
Carbohydrate chemistry
   disaccharides and glycosidic bonds,
     17–18
   monosaccharides
     classification, 15–16
     Fisher projections, 15–16
     structure, 16–17, 18
   polysaccharides, 18–19
Carbonyl group, 12–13
Cellulose
   application, 24
   hemicellulose
     applications, 26
     arabinoglucuronoxylan, 29
     galactoglucomannans, 27–29
     xylans, 26–27
   hydrogen bonding, 25–26
   plant matter, 23–24
   resources, 24
   structure, 24–25
Chemical oxygen demand (COD), 157
Circulating fluidized-bed reactor, 486,
    487
Combustion
   boiler
     Dutch oven boiler, 453
     efficiency, 451–452
     emissions, 459–461
     fixed and fluidized bed system, 453
     fluidization, 456–457
     schematics, 454–455
     stationary grate boilers, 454
     suspension burners, 454
     traveling grate system, 453
     vibrating grate system, 454
   uses
     conversion efficiencies, 459
     power conversion, 457–458
     Rankine cycle, 458–459
Continuous fermentation, 246–247,
    299–302
Continuously stirred tank reactors
    (CSTR), 183
Corn
   component, 45–46
   composition, 46, 47
   corn starch *vs.* wheat straw
     economical analysis, 325–327
     preliminary profit-and-loss
      statement, 325, 328
     processing, 324–325
     total manufacturing cost, 325
   growth factors, 45
   producer, 45
   stovers, 49–50

Covered anaerobic lagoon, 193–194
Crotonase, 287–288
CSTR, *see* Continuously stirred tank
    reactors
Cut-to-length (CTL) systems
    advantage and disadvantages, 107
    forwarder, 107
    harvester, 106–107
    machinery, 106
    woody biomass, 105
Cyclopropane fatty acid synthase
        over-expression, 322

D

Densification, 129–130
Downdraft gasifier, 468–489
Dried distillers grains and solubles
        (DDGS), 215–216
Dry milling processes, 214–215
Dutch oven boiler, 453

E

Embden-Meyerhof-Parnas (EMP)
        pathway
    biochemical, 236
    catalysis, 237
    energy-consuming, 236
    pyruvate catalysis, 239–240
    rate-limiting, 236
    reversible, 238
    tricarboxylic acid (TCA)
            cycle, 240
Esterase-catalyzed alcoholysis, 387
Esterification; *see also*
        Transesterification
    diazomethane (CH$_2$N$_2$), 366
    free fatty acid (FFA) treatment
            method, 365
    mechanism, 365–366
    sulfuric acid (H$_2$SO$_4$), 365
Ethanol production
    bioethanol production process
        cellulose-platform, 213
        ethanol purification, 213–214
        starch-platform, 213
        sugar-platform, 212

by-products, 263
dehydration
    dry corn powder adsoprtion, 262
    molecular sieve adsorption, 262
    supercritical CO$_2$ extraction, 262
fermentation process
    acetic acid, 241
    butyric acid, 241
    EMP pathway, 235–240
    enzymatic reactions, 235
    fermenters, 246–247
    glycerol, 240
    yeast microbiology, 241–246
fractionation
    azeotropic mixture, 250–251
    boiling point and dew point,
        248–250
    column, 259–261
    design, 256–257
    distillation process, 252
    feeding line, 259
    fractionating column, 252
    graphical solution, 258
    mass balance, 253
    mass balance boundary, 254
    maximum and minimum reflux
        ratio, 259
history, 210–211
purification, 247–262
saccharification and hydrolysis,
    214–234 (*see also*
    Saccharification)
Ethers, 13–14

F

Fats and waxes, 37; *see also* Vegetable oils
Fed-batch fermentation, 297–299
Feedstocks
    animal fats, 346, 349
    availability, 342
    fatty acid distributions, 342, 343–344
    lignocellulose, 280–281
    microalgal oils
        advantages, 353
        algae lines, 351
        Chlorella protothecoides, 353
        oil content, 352
        renewable biofuels, 351

molasses, 280
plant-derived oils, 346
starch, 280
upstream processes, 281–282
waste oils and fats
　grease, 350
　low-cost and profitable biodiesel,
　　349–350
　processing, 350
　soapstock (SS), 350–351
Fermentation process
　acetic acid, 241
　batch and continuous-flow
　　fermenters, 246–247
　butyric acid, 241
　enzymatic reactions
　　biochemical, 236
　　catalysis, 237
　　energy-consuming, 236
　　pyruvate catalysis, 239–240
　　rate-limiting, 236
　　reversible, 238
　　tricarboxylic acid (TCA)
　　　cycle, 240
　glycerol, 240
　other alcohols, 241
　technology, 294–295
　　batch process, 295–297
　　continuous culture, 299–302
　　fed-batch culture, 297–299
　　immobilized cell bioreactors, 302
　yeast cell concentration, 235
　yeast microbiology
　　cell structure, 242–244
　　culture preparation, 245
　　morphology, 242
　　propagation, 244
　　Saccharomyces cerevisiae, 241–242
Field capacity
　actual or effective capacity, 77
　factors, 75–76
　field operation parameters, 77–78
　formula, 76
　theoretical field efficiency, 76
　units, 75
　utilization, 76–77
Fischer–Tropsch (F–T) process, 478–479
Fuels cell

hydrogen production
　efficiency, 482
　fuel conversion, 481–482
　schematics, 482
　methanol production, 482–483

G

Gasification
　vs. coal, 461
　fundamentals
　　biomass conversion, 462–463
　　gasification models, 464
　　gas-phase reactions, 464
　　gas–solid reaction, 463–464
　　thermodynamic and equilibrium
　　　characteristics, 464
　syngas (*see also* Syngas)
　　cleanup system, 473
　　contaminants, 472
　　dust and particulates, 472–473
　　fuel transportation, 477–483
　　heat and power applications,
　　　473–477
　　tars, 472
　synthetic gas, 465, 466
　tars, 472
　technologies
　　char residue, 467–468
　　fixed-bed gasifier, 468, 469
　　fluidized-bed design, 469–471
　　gasification agents, 464–465
　　gasifier types, 465–466
　　synthetic gas, 465, 466
　　updraft gasifier, 466–468
Geographic information system (GIS)
　　analysis
　geospatial features, 123
　transportation cost calculation,
　　123–124
　transport network, 122–123
　use, 123
Global warming potential (GWP), 5
Glucokinase, 236
Green-house gas (GHG) emissions
　global climate change, 5
　renewable energy *vs.* fossil fuel
　　energy, 4

H

Harvesting, logistics
  agricultural residue recovery
    corn stover collection, 103–104
    fractionization, 102–103
    grain combine operation, 102,
      103, 105
    harvest index, 103
    one-pass and multi-pass
      collection system, 104
    processing and equipment
      options, 102, 104
  biomass logistics supply chain, 93
  forage, grass, and hay harvest
    balers, 98–100
    chopping mechanism, 101–102
    clipping, 94–95
    conditioning, 96
    cutter bar mower, 94–95
    equipment and processing
      options, 93, 94
    forage chopper, 101–102
    hay tools, 97–98
    lodged and tangled crops, 93
    mower cutting mechanisms, 97
    rakes, 97–98
    roller design, 96
    rotary mowers, 93–94, 95–96
    row crop head, 101
    shear force cutting mechanism,
      93–94
    silage chopper, 101
    swathers, 96
    tedders, 98
    weather conditions, 93
    windrower, 96
  forest biomass equipment
    biomass sources, 105
    bundling machines, 108
    chippers, 107–108
    cut-to-length (CTL) systems, 105,
      106–107
    skidder, 105–106
    tree-length systems, 105–106
  mechanized systems, 93
Heat recovery steam generator (HRSG),
  475

Henry's law, 164
Herbaceous biomass
  coastal Bermuda grass
    contents, 57
    dry matter yields, 57
    forage and hay production,
      55, 56
    hybrid variety, 55–56
    nutrient requirement, 56–57
    soil types, 56
  energy feedstock, 51
  Miscanthus
    composition, 55
    contents, 55
    nutrient requirement, 54–55
    rhizome cutting and tissue
      culture, 54
    yields, 55
  switchgrass
    biomass yields, 53
    composition, 54
    contents, 53
    cultivars, 52–53
    ecotypes, 52
    environmental benefit, 51–52
    harvest management
      strategy, 53
Hydraulic retention time (HRT)
  anaerobic reactor performance, 159,
    181
  complete-mix digesters
    disadvantage, 186
  dilution rate, 159
  immobilization benefits, 183
  suspended solids hydrolysis, 158
  VFA production, 160
Hydrocracking process,
  362–363
Hydrogen production, anaerobic
  digestion, 195–196
3-Hydroxybutyryl-CoA dehydrogenase,
  287

I

Immobilized cell bioreactors, 302
International water association, 179

L

Life cycle assessment (LCA), 4–5
Lignin
    biological role, 29
    linkages, 30–31
    monolignols, 29–30
    paper industry, 29
    structure, 29, 30
Lignocellulose, 280–281
    biological pretreatment, 228
    chemical pretreatment
        acid hydrolysis, 226–227
        alkaline hydrolysis, 227
        oxidative delignification, 227–228
        ozonolysis, 226
    enzymatic
        cellulose and hemicelluloses, 229
        endoglucanase, 230–232
        exoglucanase, 232
        other enzymes, 232–233
        product inhibition, 233–234
        β-glucosidase, 232
        substrates, 233
    physical pretreatment
        ammonia fiber explosion (AFEX),
            225
        carbon dioxide explosion,
            225–226
        mechanical comminution, 223
        pyrolysis, 226
        steam explosion (autohydrolysis),
            223–225
    pretreatment, 222
Lipase-catalyzed transesterification
    advantages, 374
    cosolvent mixture system, 383–384
    future prospects, 390–391
    hydrophilic organic solvent system,
        382–383
    hydrophobic organic solvent system,
        382
    immobilized extracellular
        transesterification, 375–377
    intracellular transesterification,
        377–379, 380–381
    ionic liquid system, 385–386
    kinetic mechanism, 386–390
    lipase, 374–375

    solvent-free system, 379, 382
    water-containing system, 384–385
Liquefaction, 218–220
Logistics; *see also* Harvesting, logistics
    cost, 72–73
    densification, 129–130
    energy conversion process, 71–72
    flowchart, 72–73
    harvesting
        agricultural residue recovery,
            102–105
        forage, grass, and hay harvest,
            93–102
        forest biomass equipment, 105–108
    inflation, 74
    machine system performance
            analysis
        criteria, 75
        equipment selection criteria,
            88–92
        machine capacity, 75–78
        machine rate analysis, 78–88
    rule, 73
    storage
        anaerobic microorganism, 127–128
        bale wrapper, 127
        biomass bales, 125–126
        cellulosic materials, 128–129
        costs, 126
        design considerations, 124
        ensiling process, 127–128
        facilities, 124
        limiting factor, 129
        method, 125
        Ritter process, 128
        solid woody biomass, 125
        sugar and starch crops, 129
    transportation, 108–110
        GIS analysis, 122–124
        maritime shipping, 121–122
        rail, 118–121
        trucking, 111–118
Long-chain fatty acids (LCFAs), 175

M

Machine system performance analysis
    accumulated hourly rate, 86, 87
    annual hourly cost, 86–88

criteria, 75
definition, 82
equipment selection criteria
  equipment size, 90
  field and machine capacity, 89
  optimization, 90–91
  task completion time, 88–89
  tractor-implement pairing, 92
  unit price function, 89
fuel costs, 84
labor costs, 85
lubrication costs, 84
machine capacity
  balance, 75
  definition, 75
  field capacity, 75–76
  field operation parameters, 77–78
  utilization, 76–77
machine rate analysis
  average annual costs, 81
  depreciation, 79, 80
  economic life, 79–80
  fixed costs, 79–82
  housing cost, 81
  insurance, 81
  interest value, 80–81
  mower conditioner, 85, 87
  ownership and operating cost
    equation, 79
  production cost, 78
  purchase price, 80–81
  salvage value, 80
  standards, 79
  taxes, 81
  TIH, 81
  tractor, 85–86
  variable costs, 82–88
mower conditioner, 85, 87
repair and maintenance costs
  accumulated cost, 82, 83
  calculation, 82
  machine aging, 82
  round baler and grain combine,
    82–83
tractor, 85–86
Magnesium, 173
Mesophilic anaerobic process
  aceto and methanogenesis, 160–161
  codigestion concept, 160

leachate recycling, 159
zero-order kinetic model, 159–160
Metabolic engineering, 321–322
Methanol production, 482–483
Methyl tertiary butyl ether
    (MTBE), 211
Mickaelis-Menten equation
  enzymatic reaction, 137–138
  kinetic rate expression, 138–139
  linearization, 139–140
Microbial cultures, butanol
  3-hydroxybutyryl-CoA
      dehydrogenase, 287
  acetate/butyrate CoA-transferase,
      288–289
  acetoacetate decarboxylase, 289
  alcohol dehydrogenase, 290–291
  aldehyde dehydrogenase, 289–290
  butyryl-CoA dehydrogenase, 288
  Clostridia fermentation, 282–283
  crotonase, 287–288
  genomes, 293–294
  metabolism and pathway, 283–286
  operons, 291–293
  thiolase, 286–287
Microemulsification
  ionic and nonionic microemulsion,
      358
  Shipp nonionic fuel, 358–359
  ternary phase equilibrium
      diagram, 359
Molasses, 280

O

Oilseeds
  oil palm, 63
  rapeseed (Canola)
    animal feed and vegetable
        oil, 61
    composition, 60
    constituents, 62
    fatty acid profiles, 61
    producer, 61
  soybean
    advantagenous characteristics, 60
    fatty acid profiles, 61
    producer, 60
    structural component, 60–61

sunflower
    content, 63
    producer, 62–63
    waste edible oil, 63
Operating costs, *see* Machine system
    performance analysis
Operons, 291–293
Organic chemistry
    alcohol, 12
    alkanes, 8–9
    alkenes, 9–10
    alkynes, 10–11
    aromatic compound, 11–12
    carbonyl group, 12–13
    ethers, 13–14
Ownership costs, *see* Machine system
    performance analysis
Oxidation–reduction potential (ORP),
    169

P

Pectin
    biological function, 31
    homogalacturonan, 31–32
Perstraction, 318–319
Pervaporation, 316–317
Petrodiesel *vs.* biodiesel
    CO_2 emission, 340–341
    combustion, 341
    cost, 339
    structural similarity, 341–342
Plug-flow reactors (PFR), 183
Potassium, 171–172
Potato, 47–48
Pyrolysis
    ablative pyrolysis, 485–486
    Auger reactors, 486
    bio-oils
        advantages, 485
        characteristics and effects,
            484–485
        compounds, 484
        properties, 485
    experimental setup, 359, 360
    fluidized-bed reactor
        bubbling, 486
        circulating, 486, 487
    vs. transesterification, 359, 360

vegetable oils and fats
    biodiesel, 359
    catalyst acidity and pore size, 365
    catalytic cracking, 362–363
    catalytic decarboxylation, 363–364
    components, 359–360
    parameter control, 360–361
    product distribution, 364
vegetable oil soaps
    apparatus, 362
    saponification, 361–362
    tung oil calcium soap, 362
    yields, 362

R

Ratooning, 442
Renewable energy
    vs. fossil fuel energy
        air quality, 3–4
        biomass, 2–3
        crude oil–consuming countries, 2
        energy independence, 3
        green-house gas (GHG)
            emissions, 4
        water quality, 4
    life cycle assessment (LCA), 4–5
    undergraduates and graduates, 1–2
Resodyn Corporation
    biodiesel pilot plant, 418
    process, 324, 416–417
    starting material and products, 418
Resonant Sonic Enhanced Mixer and
    Coalescer (RSEMC)
    technology
    butanol partition
        concentration distribution, 313
        mass balance, 312
        organic solvent selection, 313
        partition coeffcient and organic
            solvent, 314
    energy consumption calculation,
        314–315
    extraction kinetics, 315–316
    solvent toxicity
        fresh cultural media, 309–310
        microbial screen, 308–309
        solvent saturated media, 311–312

Resources
  agricultural residues
    corn stover, 49–50
    rice straw, 50
    wheat straw, 50
  herbaceous biomass
    coastal Bermuda grass, 55–57
    Miscanthus, 53–55
    switchgrass, 51–53
  oilseeds
    oil palm, 63
    rapeseed (Canola), 61–62
    soybean, 60–61
    sunflower, 62–63
    waste edible oil, 63
  starch crop
    corn, 45–46
    potato, 47–48
    sweet potato, 48
    wheat, 46–47
  sugar crop
    sugar beet, 43–44
    sugarcane, 42–43
    sweet sorghum, 44
  woody biomass, 57–59
Reverse osmosis, 318
Rice straw, 50
Ritter process, 128

S

Saccharification
  lignocellulose hydrolysis
    enzymatic hydrolysis, 229–234
    pretreatment, 223–228
  starch
    amylopectin structure, 217
    corn kernels, composition, 214
    dry milling processes,
      214–215
    liquefaction, 218–220
    oligosaccharides and dextrins,
      hydrolysis, 217
    pH effect, glucoamylase activity,
      221
    temperature effect, glucoamylase
      activity, 221
    viscosity change, 218
    wet milling processes, 216

Saccharomyces cerevisiae, 241–244
Sodium, 171
Solid residence time (SRT), 153
Solventogenesis; *see also* Butanol
      production
  acetoacetyl-CoA, 284–285
  bdhA and *bdhB* genes, 292
  butyraldehyde/butanol
      dehydrogenase, 285
  intracellular signaling, 293
  operon, biobutanol production,
      291–292
  thiolase, 292
Specific methanogenic activity
      (SMA), 161
Starch, 280
  amylopectin
    content, 21
    polymer chains, 23
    structure, 22–23
  amylose
    content, 21
    molecular weight and chain
      length, 20
    structure, 21–22
  application, 20
  granules, 20
  plant food reserve, 20
Stationary grate boilers, 454
Steam explosion (autohydrolysis)
      method, 223–225
Storage
  anaerobic microorganism, 127–128
  bale wrapper, 127
  biomass bales, 125–126
  cellulosic materials, 128–129
  costs, 126
  ensiling process, 127–128
  facilities, 124
  limiting factor, 129
  method, 125
  Ritter process, 128
  solid woody biomass, 125
  structure design considerations, 124
  sugar and starch crops, 129
Sugar beet
  producer, 44
  sucrose and lignocellulose content,
      43–44

Sugarcane
chemical composition, 443
cropping procedure, 442
growth factors, 442–443
producer, 442
Saccharum genus, 442
sucrose and lignocellulose
content, 43
temperature and climate, 442–443
Sulfate-reducing bacteria (SRB), 169
Suspended-growth, anaerobic digesters
baffled, 187–188
complete-mix digester, 186–187
continuously stirred tank reactors
(CSTR), 183
design equation, 184–186
dilution factor, 184
plug-flow reactors (PFR), 183
suspended *vs.* attached growth, 183
Suspension burners, 454
Sweet potato, 48
Sweet sorghum, 44
Syngas
cleanup system, 473
contaminants, 472
dust and particulates, 472–473
ethanol production, 479–481
fuel transportation
chemical production, 477–478
ethanol production, 479–481
Fischer–Tropsch (F–T) process,
478–479
hydrogen production, 481–482
methanol production, 482–483
heat and power applications
Brayton cycle, 475–477
efficiency, 473
gas engine, 474
gas turbine, 475
integrated gasifcation and
combined cycle, 475, 476
options, 474
tars, 472

T

Tars, 472
Terpenes and terpenoids, 36–37
Thermal cracking, *see* Pyrolysis

Thermochemical conversion,
biomass
alkali metal content, 445–446
bulk density, 444–445
combustion, 451 (*see also*
Combustion)
char combustion, 450
flame combustion, 450
heating and drying, 448–449
pyrolysis, 449–450
conversion technologies, 447–448
gasification (*see also* Gasification)
biomass conversion, 462–463
char residue, 467–468
cleanup system, 473
vs. coal, 461
contaminants, 472
dust and particulates, 472–473
fixed-bed gasifier, 468, 469
fluidized-bed design, 469–471
fuel transportation, 477–483
gasification agents, 464–465
gasification models, 464
gasifier types, 465–466
gas-phase reactions, 464
gas–solid reaction, 463–464
heat and power applications,
473–477
syngas (*see* Syngas)
synthetic gas, 465, 466
tars, 472
thermodynamic and equilibrium
characteristics, 464
updraft gasifier, 466–468
heating value, 442–444
moisture content
conversion process, 439
fuel sources, 440
intrinsic and extrinsic moisture,
440
proximate analysis
ash content, 441
volatile matter, 441–442
pyrolysis
ablative pyrolysis, 485–486
Auger reactors, 486
bio-oils, 484–485
fluidized-bed reactor,
486–487

storage, 446–447
ultimate analysis, 442
Thermophilic anaerobic process
  seed culture, 162
  specific methanogenic activity
    (SMA), 161
  suspended-growth characteristics,
    162–163
Thiolase, 286–287
Total ammonia nitrogen (TAN), 168
Transesterification, 339–340
  acid-catalyzed transesterification
    acid catalyst, 368
    mechanism, 368–369
  alcohol-triglyceride molar ratio and
    alcohol type, 393–394
  alkali-catalyzed transesterification
    vs. acid-catalyzed
      transesterification, 369
    alkaline metal alkoxide catalyst,
      369
    alkaline metal hydroxide catalyst,
      369–370
    homogeneous and heterogeneous
      catalyst, 370
    mechanism, 370, 371
    potassium carbonate reaction, 370
  catalyst type and concentration,
    394–395
  dimethyl carbonate, 355, 356
  general aspects, 367–368
  kinetics
    butanol and soybean oil, 371
    methanolysis, 372
    pseudo-second-order kinetics,
      372
    reaction variables, 370–371
    tetrahydrofuran (THF), 372
  lipase-catalyzed transesterification
    advantages, 374
    cosolvent mixture system,
      383–384
    future prospects, 390–391
    hydrophilic organic solvent
      system, 382–383
    hydrophobic organic solvent
      system, 382
    immobilized extracellular
      transesterification, 375–377

  intracellular transesterification,
    377–379, 380–381
  ionic liquid system, 385–386
  kinetic mechanism, 386–390
  lipase, 374–375
  solvent-free system, 379, 382
  water-containing system,
    384–385
  methanol, 354–355
  methyl acetate, 355, 356
  mixing intensity, 397
  moisture and free fatty acid content,
    395–396
  organic cosolvents, 397–398
  reactant purity, 396
  reaction temperature and time,
    396–397
  in situ transesterification
    alcohol reagent, 391
    enzyme catalyst, 392–393
    ethanolic esterification, 392
    methanolic esterification, 391–392
    in situ alcoholysis, 392
    sulfuric acid catalyst, 392
    supercritical fluid extraction, 393
  supercritical alcohol
    transesterification
    autoclave, 373
    vs. catalytic methanol method, 373
    catalytic method, 374
    noncatalytic method, 372–373
    reaction temperature, 373–374
Transportation, feedstock, 108
  vs. biorefining operation, 109
  costs, 108–109
  economically favorable travel
    distances, 109–110
  geographic information system (GIS)
    analysis
    geospatial features, 123
    transportation cost calculation,
      123–124
    transport network, 122–123
    use, 123
  haul distance, 109
  maritime shipping
    advantage and disadvantages, 121
    bulk ships, 121
    container ship, 122

costs, 122
transoceanic shipping and barge
transport, 121
rail
advantage, 118–119
disadvantages, 119
reliant mode, 120
satellite facilities, 120–121
vs. truck cost, 120
unit trains, 119–120
transportation modes, 109–110
trucking
capacity, 112–113, 114
cost estimation, 116–117, 118
densification, 113–115
dimensional parameters, 112
economic analysis, 116
feedstock producer, 111
intermodal transport system, 111
Traveling grate system, 453
Tricarboxylic acid (TCA) cycle, 240

U

United States Department of
Agriculture Agricultural
Marketing Service (USDA
AMS), 122
Unit operation processes, biodiesel
alcohol recovery, 412–413
BIOX processing system, 416
economic analysis
blended biodiesel, 420
feedstock cost, 419, 420
generic process model, 419–420
Kyoto protocol, 421
ester–glycerol separation
density difference, 409–410
equipment, 410–411
mixing and esterification, 410
ester treatment, 412
ester washing, 411–412
feedstock preparation, 406
filtering, 412
flowchart, 398, 399
glycerol refining
chemical refining, 413
glycerol purification, 414
physical refining, 413–414

high free fatty acid feedstocks
treatment
soaps, 405–406
techniques, 405–406
high free fatty acid grease feed
process
facility, 419
feedstock and biodiesel, 416, 418
procedure, 417
oils/fat feedstocks
bleaching, 402
deodorization, 402–403
mechanical extraction, 399–400
oil-insoluble and soluble
impurities, 400–401
refining, 401–402
rendering, 403–404
solvent extraction, 399–400
vegetable oil source, 399
plant equipment
distillation column, 415
pumps, 415
reactors, 414–415
settling tank, 415
primary alcohol, 406
selective absorption, 412
transesterification
batch processing, 406–408
continuous processing, 408
noncatalyzed transesterification
system, 409
wastewater treatment, 414
Updraft gasifier, 466–468
disadvantages and advantages, 468
vs. downdraft gasifier, 468, 469
zones, 467
Up-flow anaerobic filter (UAF), 158
Up-flow anaerobic sludge blanket
(UASB), 153, 190–192

V

van't Hoff–Arrhenius equation,
177–178
Vegetable oils
applications, 32
fatty acids
cis-trans isomers, 33, 35
classification, 32–33

oilseeds, 34
structure, 33, 34
free fatty acids, 35
major and minor non-triglycerides, 33, 35
phospholipids, 35, 36
triglycerides, 32, 33
Vibrating grate system, 454
Volatile fatty acids (VFAs), 154

W

Waste activated sludge (WAS), 162
Wet milling processes, 216
Wheat
composition, 47
producer, 46–47
straw, 50
wheat straw *vs.* corn starch

economical analysis, 325–327
preliminary profit-and-loss statement, 325, 328
processing, 324–325
total manufacturing cost, 325
Woody biomass, 58
contents, 59
pine, 57
poplar, 57, 59
softwoods and hardwoods, 57
soil types, 59

Y

Yeast extract supplemented with peptone glucose medium (YEPG), 244

Z

Zero-order kinetic model, 159–160